STATISTICS
Methods and Analyses

STATISTICS
Methods and Analyses

second edition

Lincoln L. Chao
Professor of Quantitative Systems
California State University, Long Beach

INTERNATIONAL STUDENT EDITION

McGRAW-HILL KOGAKUSHA, LTD.

Tokyo Auckland Düsseldorf Johannesburg London Mexico
New Delhi Panama São Paulo Singapore Sydney

STATISTICS: Methods and Analyses

INTERNATIONAL STUDENT EDITION

Exclusive rights by McGraw-Hill Kogakusha, Ltd. for manufacture and export. This book cannot be re-exported from the country to which it is consigned by McGraw-Hill.
I

This book was set in Times New Roman.
The editors were Thomas H. Kothman and Claudia A. Hepburn;
the designer was Ben Kann;
and the production supervisor was Sam Ratkewitch.
The drawings were done by John Cordes, J & R Technical Services, Inc.

Library of Congress Cataloging in Publication Data

Chao, Lincoln L
Statistics: methods and analysis.

1. Statistics. 2. Mathematical statistics.
I. Title.
HA29.C5425 1974 519.5 73-13896
ISBN 0-07-010525-1

TOSHO PRINTING CO., LTD. TOKYO, JAPAN

In memory of
MY PARENTS

Contents

Appendixes

Preface

Relatively few students in business and other social sciences have mathematical training beyond secondary school algebra. Many, and perhaps most, students in these fields lack the background to understand the more sophisticated techniques of statistics. On the other hand, in business, government, and other institutions there has been increasing emphasis on quantitative approaches to management decision making. Students are expected to know the techniques of manipulating numerical data and extracting useful conclusions. To that end, an understanding of the logic and theory of inferential statistics is essential. Under the circumstances, a realistic textbook in elementary statistics should be one that is comprehensible to the average undergraduate student in business and other social sciences, and at the same time provides enough theoretical material to challenge the more serious reader. Such a book would bridge the gap between the predominantly descriptive texts and the largely mathematical ones in statistics.

Although mastering the theory of statistics requires a considerable degree of mathematical preparation, a substantial amount of statistical theory can be presented meaningfully to the average college student if theoretical treatment is undertaken with clear and detailed exposition of the concepts.

This book has been written on the basis of the preceding considerations. The very favorable reception of the first edition attests to the validity of this approach.

The primary objective of the second edition remains the same as that of the first. Its emphasis is still on inferential statistics, and the approach continues to be nonmathematical. Simple algebraic manipulations, which would assist the student in better appreciating the rationale of the various methods and theories, are provided in a number of places in the text. Many nonessential proofs and derivations which were included in the first edition, however, have been deleted or presented in the supplements. Thus this edition is more streamlined with the intended result of enhanced teachability.

Another improvement is the addition of Chap. O, Orientation, which is designed to set the tone for the chapters to follow. It introduces the students to the basic concepts of statistics, with its uses and misuses, and prepares them for the thought processes which are required in most of the discussions ahead. Chapter 1 deals with the basic materials about *sets* and *functions*, which were previously presented in two separate chapters, while Chaps. 2 and 3 are devoted to the discussion of basic probability theory, which is more adequately treated than in the first edition. Numerous real-world examples and problems are added in this second edition, which should make the subject more interesting and provide the needed motivation for the student.

For a one-semester first course, a suggested program includes the first 14 chapters with the exception of Secs. 7-5, 7-6, 8-4, 9-2, 9-4, 10-4, 13-3-1 to 13-3-4, 13-4-1 to 13-4-3, 14-2, and 14-3. There will be no loss of continuity by excluding these sections. The entire book can be covered in two quarters, or possibly two semesters if the supplements are included, of three units each. The book may be used for a second course; the coverage of such a course will, of course, depend on the materials included in the first.

This edition would not have taken its present form without the very sound and generous assistance of many teachers throughout the United States and my colleagues and students at California State University, Long Beach. I am especially indebted to the following professors from other universities for their comments and suggestions: David P. Baron, Northwestern University; Michael E. Coughlin, Ohio State University; H. C. Fryer, Kansas State University; Paul A. Hatt, California State Polytechnic University, Pomona; Thomas J. Higgins, University of Wisconsin; Edward L. Marrinan, Jr., University of Dayton; Lester C. Sartorious, Georgia State University; and Andrew C. Stedry, University of Texas.

It is hardly necessary to mention that I owe my wife, Beatrice, and our sons, Moses and Martin, infinite thanks for their sacrifices during the process of revising this text.

Lincoln L. Chao

STATISTICS
Methods and Analyses

Chapter 0
Orientation

In this first chapter, we shall attempt to answer a few basic questions briefly. Among these are: What is meant by statistics and how was it originated? What are some of the underlying concepts of statistics? What are some of its most important applications, and what are some of its possible abuses?

0-1
Statistics and Its History

If you were to ask the "man in the street" what statistics means to him, he would probably say that statistics means "numbers" or "manipulation of numbers." Yes, in common parlance the word statistics is used almost synonymously with *numbers* or *figures*. For instance, the statistics as reported by the sports announcer of a football or basketball game are little more than numbers or quantitative facts. The report about deaths and births appearing in the newspapers every day is shown as "vital statistics." Likewise, some people may say that they have read about the statistics of automobile accidents in the Los Angeles basin. The term used in this sense means little more than numerical information. It ought not to be confused with the same word used in this text, which has a few quite different meanings.

We shall consider the meaning of statistics as a course of study. Before we discuss what statistics is all about, let us first look into the origin from which modern statistics is evolved.

0-1-1 The history

The study of statistics has undergone substantial changes since its inception. Two sources of developments deserve special mention. First, the origin of statistics may be traced to the need for numerical data by the new governments of western Europe emerging from the medieval society. With the transformation of the medieval society into the political state, the new government in order to be successful needed information on the resources of the country it governed. Thus it was imperative for the new rulers to obtain numerical description, such as the number of citizens of the various political units—cities, counties, and provinces—under their jurisdiction. The term statistics being derived from the Latin root *status*, which signifies a state in the political sense, was then used to designate the collection and description of such data of the state. The need for collecting and analyzing numerical data prompted the development of methods to facilitate the work, which, until the more modern era, constituted the bulk of the subject of statistics.

It is not necessary to list all those who contributed to the development of statistical methods. Mention should be made, however, of a Belgian, Adolph Quetelet (1796–1874), who was the first to apply modern methods to collected facts. Quetelet is often referred to as the "father of modern statistics" because of his continual emphasis on the importance of applying statistical methods. His outstanding contributions to statistical practice and methodology included many areas of government statistics such as census taking, the development of uniformity and comparability in statistics among nations, and the organization of the first international statistical conference. The Central Commission for Statistics which Quetelet established became a model for similar institutions in other countries.

Another source of statistics is found in the interest in gambling in the seventeenth century. Owing to the indulgence in and the prestige associated with various forms of gambling for recreation by the nobility of England and France during this period, an intense interest was generated in the *games of chance*, which inadvertently led to the development of the *theory of probability*. In the meantime, there existed a technology which was advanced enough to provide gambling devices such as dice and cards with sophistication sufficient to challenge the keen imagination of the mathematician. The gambler is concerned about and governed by the chance of error associated with a particular course of action. The actual outcome of any single trial is unknown, but the theory of probability will show the expected number of occurrences of a particular event in a very large number of trials. The modern statistician has a tremendous interest in such a theory.

The studies in probability, meanwhile, involved mathematical treatment of errors in measurement. As a result, theories and methods were developed to describe the distribution patterns of such errors. It was observed as early as the eighteenth century that repeated measurements of a certain object or phenomenon would yield a pattern of distribution of errors like a bell-shaped curve. The curve is symmetrical with the true value, which means zero error, in the center. It is symmetrical because the negative and positive deviations from the center are equal in magnitude and frequency. The larger the deviations. the smaller the frequencies will be.

A most important discovery in statistics was made in connection with the work of evaluating errors of observation in astronomy. The resulting bell-shaped symmetrical distribution of errors is referred to as the *normal curve of error*. It is also known as the *gaussian distribution of error*, named after its discoverer Karl Friedrich Gauss (1777–1855).

The normal distribution can be used as an approximation to a number of other distributions subsequently developed. These distributions have made it possible for the development of inferential statistics, a concept to be discussed shortly.

Among the contemporaries of Quetelet and Gauss who helped to advance statistics as a science were Florence Nightingale (1820–1910) and Francis Galton (1822–1911). Nightingale was a strong believer in statistical methods. She claimed that the executive must be guided by statistical knowledge if he wanted to be successful and that politicians and legislators frequently failed because of their insufficient statistical knowledge. Galton, like his cousin Charles Darwin, was deeply interested in the study of heredity, to which he applied statistical methods. Among his chief contributions was the development of such fundamental methods as regression and correlation.

Galton's work stimulated a series of investigations by Karl Pearson (1857–1936). He started the publication of the journal *Biometrika*, which has profoundly influenced the development of statistics. Many of the methods to be discussed in this text were discovered by Pearson, and the most important of these is the *chi-square distribution* which he derived in 1900.

In the twentieth century the most prominent contributors to the study of statistics have been William S. Gosset (1876–1937) and Sir Ronald Fisher (1890–1962). Gosset, writing under the name of "Student," derived the t distribution, and his special contribution was in the field of small-sample theory. Fisher developed the well-known F distribution and made continuous contributions up to 1962, many of which have had great influence on the modern procedures of statistics. Although his work was particularly in the fields of biology, genetics, and agriculture, its impact has permeated all applications of statistics.

0-1-2 Statistics defined

Let us now proceed to find out what statistics is all about. Although it is rather difficult to obtain a consensus on the definition of statistics, virtually all statisticians agree on the classification of the subject into two types, namely, *descriptive statistics* and *statistical inference*, which are, in fact, two functions of statistical analysis.

Descriptive statistics In the early days of its development, statistics involved little more than the collection and presentation of data. Even today, statistics still contains data treatment and analysis designed to summarize and describe the facts that have been collected. It usually takes the forms of tables, graphs, charts, and indexes. Inasmuch as the primary objective of this kind of data treatment is to *describe* the principal features of the assembled data, it is generally referred to as descriptive statistics. The analysis is limited to the collected data themselves and does not involve inferences or generalizations about the totality of all potential observations of which the collected information forms only a small part.

Let us imagine, for instance, that a personnel director has administered an aptitude test to a group of newly hired college graduates. Among the things which he may do with the resulting test scores by means of descriptive statistics are the following. He may rearrange the scores or classify them in such a way that at a glance at the data we shall be able to get an overall picture of them. He may compute the average of the scores and learn something about the typical aptitude of the employees. He may construct tables, graphs, and charts for visualization of the data or convert the raw scores into ranks or percentiles for comparison. He may use the average as a central point and describe the spread, or dispersion, of the data. Furthermore, if later on some measurements of job performances of these employees are obtained, he may try to describe the relation between the test scores and such measurements. As long as such a relationship is established, he may predict an employee's performance on the job on the basis of his aptitude test score.

Inferential statistics Although a description of the collected facts is sometimes an end itself, in most statistical analyses the researcher is actually nearer the beginning than the end of his work when he has finished his descriptive tasks. This is the case because the ultimate purpose of statistical work is, on the basis of the collected information, to draw useful conclusions relative to the totality of all possible observations under consideration. In other words, descriptive statistics constitutes only the preliminary work for *inferential statistics*, which is also known as statistical inference. Understanding the term statistical inference requires a knowledge of two basic concepts, the concept of a population and that of a sample.

The term *population*, which is used interchangeably with the word *universe*, refers to the totality of all potential observations or measurements under consideration in a given problem situation. In a survey of voters' preference in a presidential election, the population consists of all the responses of the registered voters. The term, however, does not necessarily refer to a collection of human beings or living organisms. Thus, if a research is to be made of yearly sales of supermarkets, then the yearly sales of all supermarkets will constitute the population. A different problem situation usually has a different population. The characteristics of a population are generally referred to as its *parameters;* a population parameter is considered the *true value*. For instance, the *arithmetic average* of the annual earnings of all the 85 million workers in America is a characteristic, or a parameter, of the population of the annual earnings of the American workers. Likewise, the proportion of all customers who express a preference for a particular brand of a certain product is a parameter of the population of all the customers; it is the *true* proportion, or *population proportion*.

Obviously, it is quite impracticable to calculate a true value of any population which has a large number of potential observations. Statistical investigation usually involves the collection of partial or incomplete information called the *sample*, on the basis of which inference is made about the true parameter value. A sample is a collection of observations taken from a given population, and any characteristic of a sample is generally referred to as a *statistic*. If 1,000 of the 85 million American workers are selected and their annual earnings are obtained, the 1,000 observations will constitute a sample and the arithmetic average of these 1,000 annual earnings will be a statistic. Similarly, if 100 of all the potential customers are selected and interviewed to see whether they are in favor of a particular brand of a product, these 100 customers will constitute a sample. If 70 of the 100 customers are in favor of the brand, then the *sample proportion* will be 0.70. A sample proportion is, of course, a *sample statistic*.

Statistical inference is a technique by which conclusions or generalizations are drawn about the parameter or parameters of a population on the basis of the statistic or statistics of a sample taken from the population. It is recognized, however, that there is a certain probability that the generalization may be wrong. No statistical inference with respect to the parameter of a population will be necessary if all the potential observations of the population have been taken. Because of the constraints of time and resources, however, it is incumbent upon us to make decisions relative to the parameter value of a population on the basis of the statistic value of a sample. This indicates the important role that inferential statistics has been playing in modern scientific endeavors.

Having considered the two functions of statistics, let us now summarize the meaning of statistics as follows:

> *Statistics is a body of theories and methods which have been developed for handling the collection, analysis, and description of sample data for drawing useful conclusions. Its primary function is to assist the investigator in making decisions relative to the parameter of the population from which the sample data are taken.*

In other words, modern statistics embraces both descriptive statistics and statistical inference. Of the two, the emphasis on the latter has become more pronounced in the past few decades.

Exercise 0-1

1 In what context is the word "statistics" usually used by the "man in the street"?

2 In what context is the word "statistics" usually used in sports reports and by the news media?

3 What are the sources of the study of statistics?

4 State briefly the contributions made by each of the following statisticians:

a Adolph Quetelet.
b Karl Friedrich Gauss.
c Florence Nightingale.
d Francis Galton.
e Karl Pearson.
f William S. Gosset.
g Ronald A. Fisher.

5 Explain the difference between the concepts of a population and a sample.

6 Explain the difference between the methods of descriptive statistics and statistical inference.

7 A survey has been conducted to determine whether the housewives in the United States prefer a particular brand of detergent. Among the 50 housewives interviewed 30 have indicated that they prefer this brand.

a What constitutes the sample?
b What constitutes the population?
c What is the sample statistic, and what is the population parameter?
d Identify the method that has to be used to arrive at each of the following statements:

1 The sample proportion is 0.60.
2 The population proportion is between 0.50 and 0.70 with a certain probability that it may lie beyond these limits.

8 A researcher reported that five 100-watt light bulbs of brand *A* burned out after 700, 855, 879, 905, and 951 hours of continuous use, while five 100-watt bulbs of brand *B* burned out after 901, 919, 980, 1,000, and 1,050 hours of continuous use. Which of the two statistical functions is required for reaching each of the following statements:

a The five bulbs of brand *A* averaged 858 hours of service life, and the five bulbs of brand *B* averaged 970 hours.
b The average length of service lives of brand *A* bulbs is shorter than that of brand *B* bulbs.
c The five bulbs of brand *A* have greater dispersion in service lives than those of brand *B*.
d The service lives of brand *A* bulbs are more variable than those of brand *B* bulbs.
e If the price of brand *A* bulbs is the same as that of brand *B* bulbs, you will prefer to purchase bulbs of brand *B*.

9 Explain the reason that statistical inference has been playing such an important role in modern scientific investigations.

10 Use your own words to explain the meaning of statistics as a discipline of study.

0-2 Uses and Abuses of Statistics

Statistical theories and methods have found applications in a variety of intellectual fields. They have been used as tools of research and analysis not only in the various social sciences but in engineering and the natural and physical sciences as well. There are, however, many abuses of statistics, either through ignorance or by design. In this section we shall briefly examine the typical applications of statistics, especially inferential statistics, and cite a few of the often committed abuses.

0-2-1 Applications of statistics

Statistics has become a crucial subject for students in business, economics, sociology, anthropology, and other behavioral sciences. It is nowadays impossible to earn a degree in these fields without at least one course in statistics. This is the case because statistical methods have proved to be of great help in a wide range of endeavors involving the behavior of individuals, groups, and institutions.

The growing complexity of the modern economy has created an increasing degree of uncertainty about future operations of the business enterprise. Business firms have to make decisions in spite of such uncertainties. Sound and rational decision making requires careful analysis and interpretation of factual information, in which statistical techniques have proved to be especially useful. In business today, statistics is no longer considered merely the analysis of sales and accounting records, but rather has become an integral part of managerial decision making. In marketing research and forecasting of economic trends, for instance, the use of regression analysis and other statistical methods is virtually indispensable.

A few other examples will be sufficient to illustrate the importance of statistical inference in business. Suppose that a manufacturing company wishes to decide whether a shipment of purchased goods should be accepted or rejected. A portion of the shipment is then inspected in order to determine its quality. If 3 of the 30 units inspected are found to be defective, does this result show sufficient evidence to indicate that the shipment is of poor quality and should be rejected? Inferential statistics is especially designed to answer this kind of question.

A flour manufacturing company packs flour into paper bags, each of which is supposed to hold 25 pounds. If the packing process is under control, the average weight of the bags will be 25 pounds. Suppose that a sample of bags is periodically taken to check the packing process. If a sample of 50 bags yields a sample mean of 24 pounds and 12 ounces, we may apply the method of statistical inference to determine whether the process is in control.

As another example, let us consider a new cold vaccine developed by a pharmaceutical company. The company claims that the new vaccine is 95 percent effective, or 95 out of 100 people who have used it would survive the winter without a cold. If, in a sample of 30 people who have received the vaccine, 27 persons survive the winter without a cold, is there sufficient evidence to support the company's claim?

Similarly, statistics plays an important role in education and psychology. An educator may want to know whether there is a significant relation between the scores of a scholastic aptitude test and the grade point averages for a group of students. If such a relationship exists, he will be able to make predictions about the grade point average for a student on the basis of his aptitude test score. Also, two different teaching methods may be compared to determine their relative effectiveness. Two groups of students with comparable ability may be selected and taught by the two methods in a given subject. A standard test is administered to the two groups at the end of the instructional period. On the basis of the average test score of each group, we can draw a conclusion whether the two methods are equally effective or one method is, in a statistical sense, significantly more effective than the other.

In the past several decades, the number of studies and surveys on training and education has increased at an accelerated pace. The need to analyze and interpret the information that has been collected makes it necessary for students in education and psychology to have at least some basic understanding of statistical methods.

Statistical techniques have long been used in the biological sciences. In agronomy, for instance, they can assist in determining the effects of seed strain, insecticide, and fertilizer on yields. They have been used to produce improved livestock through special arrangements of feeding and breeding. In medicine, statistical methods can be used to identify the possible side effects or effectiveness of drugs and to improve the methods of controlling the spread of social diseases. They can also be applied, with fruitful outcomes, to genetics, forestry, and the general area of ecology. It is not surprising that statistics is a highly recommended, if not yet required, course for students in these fields.

Likewise, statistics has found increasing application in physics and chemistry, where it has been used to test hypotheses on the basis of laboratory data. The research work of the physical scientist has expanded the arena of experimental design, an important statistical technique. In engineering, the use of statistical tools to control production quality has been a rewarding experience for several decades.

The rapid growth in the application of statistical techniques in recent years can be, in part at least, attributable to the increasing ease of handling large quantities of numerical data. Electronic computation has made it possible to analyze, in a tremendously reduced amount of time, gigantic quantities of data which would have been practically unmanageable otherwise. Although this text does not deal with the use of computer programming in solving statistical problems, the student should avail himself of the opportunity of using the computer whenever it is available and practical.

0-2-2 Abuses of statistics

The statistical worker should be always on the alert to avoid any misuse of his collected data. Erroneous use of data or careless and illogical reasoning will mislead you and destroy the value of a study. A few brief examples of improper procedures or fallacious interpretations may help to clarify this point.

It often happens that conclusions may be based on sample data that are not representative. In fact, one of the most publicized abuses of statistical method was the application of a fallacious sampling procedure. The error was made by the *Literary Digest* in its forecast of the 1936 presidential election. The *Digest* sent straw ballots to a sample of voters from 10 million persons originally selected from telephone directories and its own subscription list. The returns indicated that 161 electoral votes would be cast for Franklin Roosevelt and 370 for Alfred Landon. The final election results were 523 electoral votes for Roosevelt and only 8 for Landon. Roosevelt carried 46 of the 48 states and many of them by a landslide. The fallacy lay in the fact that in the depression year of 1936 those who could afford telephones or magazine subscriptions did not consistute a representative sample of the American voters. The voters in the sample were in the upper economic brackets and were predominantly for Landon, and the majority of the voters were for Roosevelt. At least partially owing to this error, the *Digest* soon ceased to exist. In subsequent elections, pollsters have been more careful and have developed sampling procedures to ensure representativeness.

Sometimes erroneous conclusions are reached because the collected data are numerically insufficient. Suppose that an insurance salesman makes house-to-house calls to sell a certain insurance policy. He claims that one-fourth of his calls resulted in sales; his claim is based on the fact that he made one sale on a Saturday afternoon when he made a total of four calls. It gives the impression that 25 percent of *all* his calls are successful, which is apparently not the case. The claim would gain more credence if the salesman gave the total number of calls over a longer period of time. We should be inclined to believe him if he said that out of 100 calls 25 sales were made, or even better, if he said that out of 1,000 calls 250 sales were made. The greater the total number included, the more reliable the proportion will become. Thus, whenever a fraction or percentage is used, it is advisable to make known the total number of cases or observations to ensure reliability.

Because we have not yet presented the various methods of statistics, our discussion of the abuse of statistics is limited. Suffice it to say that virtually any statistical procedure might be carried out fallaciously or its results misinterpreted. However, it is not difficult to avoid such a fallacy or misinterpretation. We refer you to Darrell Huff's little book entitled *How to Lie with Statistics* (published by W. W. Norton & Company, Inc., in 1954), if you desire to look further into the misuses of statistics.

Exercise 0-2

1 What are some of the most important reasons that students of business are required to take at least one course in statistics?

2 How can statistics play an important role in education and psychology?

3 Explain the possible contributions that statistics can make to the biological sciences.

4 How can electronic computers contribute to the rapid growth in the application of statistical techniques?

5 Give an example to illustrate the application of statistical inference to each of the following business functions:

a Personnel management.
b Production management.
c Marketing.
d Real estate management.
e Any other function.

6 A local newspaper reports that one-third of the female faculty of a college are married to their students. Is there anything wrong with such a statement? Explain.

7 Suppose that a study reports a greater suicide rate for a certain ethnic group than for other groups. Upon further investigation the study reveals that members of this ethnic group are predominantly city dwellers. Is there anything incorrect in the original report of the study? Explain.

8 An attorney from Galveston, Texas, wrote, "Dear Abby: For the past ten years I have kept a record of the number of divorces between couples of mixed religions, and my records show that 6 out of 10 marriages between Baptists and Catholics end in divorce. I do not know the reason, but thought you would be interested in the statistics." Abby replied, "Dear Attorney: Although your minisurvey on Baptist-Catholic divorces is hardly representative of the nation, if 4 out of 10 marriages make it in Galveston, you are doing better than the state of California, where I am told they had more divorces last year than marriages!" (See page 9 of the View section of *Los Angeles Times*, September 11, 1972.) Comment on Dear Abby's reply.

Chapter I

Sets and functions

This chapter deals briefly with the topics of sets, relations, and functions. It may appear strange that a book on statistics starts off with a discussion of these subjects. It is true that a course in elementary statistics does not necessarily include a treatment of sets and functions, and books about descriptive statistics seldom do. However, set theory provides a basis for the discussion of probability theory, which forms a major part of this book. Furthermore, the notion of a set can be used to develop a precise definition of the term *function*, which pervades the discussions throughout this text. Although students have been exposed to the notion of a function ever since their high school days, they are nevertheless somewhat vague about the precise meaning of this concept unless they have a very good mathematical background. One of the objectives of this chapter is to provide a precise and specific definition of the word function. As we shall see shortly, a function is a special kind of relation, and the notion of a relation grows out of the concept of a set.

In the next section, we shall define the notion of a set and describe how a set is specified and presented. Also, the concepts of universal sets and subsets will be discussed, and they will be compared with the terms population and sample dealt with in the previous chapter. The section will conclude with a presentation and illustration of set operations. Just as the operation of numbers results in new numbers, the operation of sets leads to new sets.

The final section of this chapter discusses relations and functions. A knowledge of the notion of a relation will help students to understand better the meaning of the word function, which is one of the most pervasive concepts in mathematics as well as in statistics. The meaning of a function, its domain and range, and the commonly used notations for a function will be presented.

1-1 Sets

In this section we shall briefly examine the concept of a set, its notations and specifications, some of the fundamental notions about sets, Venn diagrams, and the operations with sets.

1-1-1 Notation and specification

The notion of a set, important as it is, is not a complex one. A *set* is simply a *well-defined collection of distinct objects.* The chairs in this classroom form a set; so do the desks and the various books you possess. The English alphabet is a set of 26 letters; the *natural* numbers between 1 and 10 inclusive is a set of 10 intergers.[1]

The individual objects of a set are *elements*, or *members*. The set is the *collection* of its elements. If a certain object is in a set, it is said to be an element, or member, of the set.

The elements of a set must be *distinct;* that is, each must appear only once. Once an object is recorded in a set, it is not repeated. For instance, the set of letters in the word "statistics" is a set with the five distinct letters, a, c, i, s, and t as its members. The set of price quotations of IBM common stock during a given day, 650, 645, 651, 650, 647, and 651, does not contain all the six numbers, but the four distinct numbers 645, 647, 650, and 651.

A set must be *well defined.* This means that it must be *specified* so that you can determine whether a certain object is a member of the set. For any given object, there must be an unequivocal "yes" or "no" answer to the question: Is this object an element of the set? Of these two answers, one and only one is correct.

In set notation we follow the convention of using a capital letter, such as I, A, R, S, T, or U, to designate the set, and a small letter, such as a, b, x, y, or z, to symbolize a particular element of the set. The symbol "$\in$" is used to denote "is an element of," or "belongs to." Thus $x \in S$ is read, "x is an element of the set S," or "x belongs to S."

A slash through $\in$ signifies negation. When we want to show, "y is not a member of S," we write "$y \notin S$."

There are two different ways in which sets may be specified. One of these is the *tabular*, or *enumeration*, method that simply lists all the elements of the set. The names of individual elements are separated by commas and are enclosed in *braces* { }. For example,

$$A = \{a, b, c, d, e\}$$

is a complete itemization of the set A, which indicates that A consists of the

[1] Natural numbers are positive integers, such as 1, 2, 3, and 4.

letters *a*, *b*, *c*, *d*, and *e*. Another set specified similarly is $B = \{$a book, a chair, a desk, a pen, a cup of tea, a college professor$\}$. Here again, the respective elements of the set are simply listed. It is important to note that the members of a set are not necessarily of the same kind. Also, the order in which the elements of a set are listed is immaterial; the set $\{x, y, z\}$ is equal to the set $\{z, x, y\}$.

In many cases, however, it is impossible to list all the members of a set, for example, the set of points on a line or a line segment, or the set of *real* numbers.[2] Under such circumstances, another method of specifying a set is used. That is, a *rule* or a *description* is provided, by which you can decide what objects are or are not members. Thus the set S may be specified by

$$S = \{x \mid x \text{ is a real number}\}$$

The symbol "|" is read "such that," and the above expression indicates that S is the set of all elements x such that x is a real number. Similarly, the expression

$$M = \{i \mid i \text{ is an integer and } 0 \le i \le 100\}$$

is read "M is the set of all elements i such that i is an integer between 0 and 100 inclusive."

In general, when the set S is specified by a rule or a descriptive sentence, we write "$S = \{x \mid (\text{some condition about } x)\}$." It is read "$S$ is the set of all x such that the condition about x is stated." There may be more than one condition about x, for example,

$$S = \{x \mid x^2 = 2, x > 0\}$$

is the set of a single number, namely, $\{\sqrt{2}\}$.

A slight modification of the above notation is frequently used. If R stands for the set of all real numbers, then

$$A = \{x \in R \mid x^2 = 4\}$$

is read "the set of those real numbers x such that $x^2 = 4$." This is the same as $A = \{-2, 2\}$.

Specifying a set by a rule or description is most frequently encountered, and it is obviously far more convenient than listing all the elements entailed by the tabular method.

1-1-2 Basic concepts

A number of concepts frequently used in the discussion of set theory will now be considered. They include the universal set, subsets, empty set, and finite and finite sets.

[2] Real numbers are all the numbers between $-\infty$ and ∞.

Universal set In discussing sets, the notion of a universal set is important and indispensable. A *universal set* is the set of all elements that are considered in a particular discussion. It is the frame of reference for the discussion. Sometimes it is a small set consisting of only a few elements; for example, in the toss of two dice to see the sum of spots uppermost, the universal set U is

$U = \{x \mid x$ is the total number of spots uppermost on the two dice$\}$

or

$U = \{2, 3, 4, 5, 6, 7, 8, 9, 10, 11, 12\}$

Some universal sets contain still fewer elements than this. On the other hand, a universal set may be infinitely large. For instance, in our study of mathematics, the universal set of numbers with which we began was the set of natural numbers. As our knowledge increased, the set of numbers expanded to include integers, rational numbers, real numbers, and complex numbers. When we discuss a problem, we should have clearly in mind the universal set with which we are concerned. (For the system of numbers, see Fig. 1-1.)

Subsets If every element of the set S is also an element of the set T, then S is a *subset* of T, or S is contained in T, written "$S \subseteq T$." That is,

The set S is a subset of the set T if for each $x \in S$, then $x \in T$.

For example, if T is the set of all integers and S is the set of positive integers then S is a subset of T.

If there is at least one element of T that is not in S, then S is a *proper subset* of T, written "$S \subset T$." In the above example, S is also a proper subset of T since some integers (negative) are not positive integers.

The symbols "$\subseteq$" for "is a subset of" and "$\subset$" for "is a proper subset of" are similar to "$\leq$" for "is less than or equal to" and "$<$" for "is less than." A set can be a subset of another set even though the two are identical, while it can be a proper subset of another only if the former does not contain some of the objects that are members of the latter.

The definition of a subset implies that a set can be a subset of itself, that is, $S \subseteq S$. Accordingly, two sets S and T are said to be equal, or $S = T$, if $S \subseteq T$ and $T \subseteq S$.

Empty set Any set that has no element in it is called an *empty*, or *null*, set, designated by $\emptyset$, a zero with a slash through it. (It is a Scandinavian letter and not the Greek letter phi.) The theory of sets requires the concept of an empty set just as ordinary mathematics needs the number zero. The null set is a subset of every set.

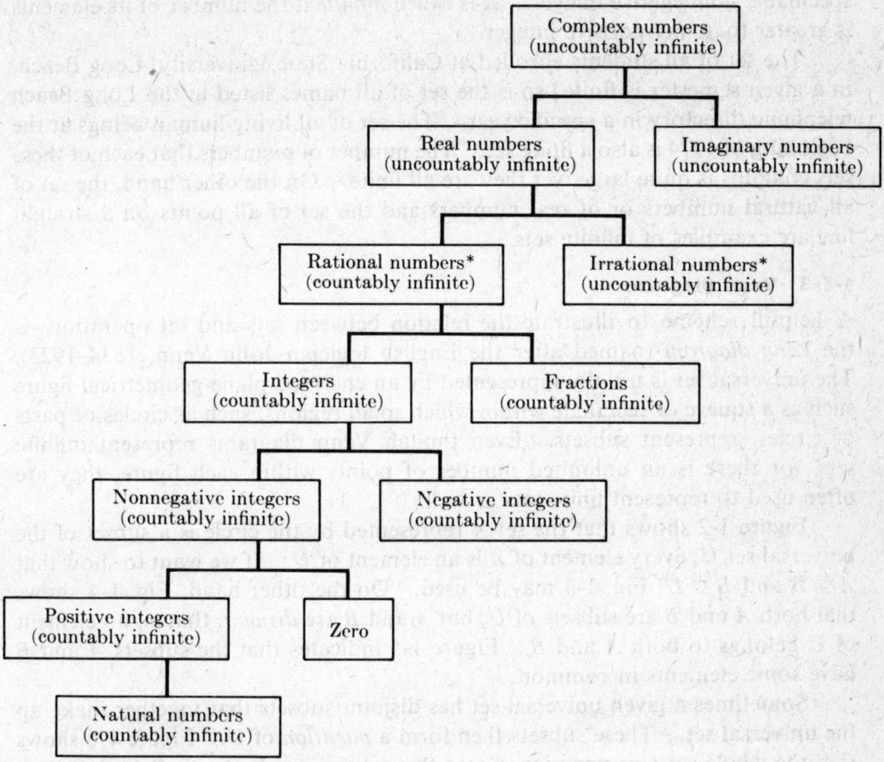

Figure 1-1

The number system. (*Rational numbers can be written as finite decimal expansions such as $\frac{1}{4} = 0.25$ or as infinite decimal expansions such as $\frac{1}{6} = 0.1666666\ldots$ and $\frac{2}{7} = 0.28571428571428\ldots$. In infinite decimal expansions of rational numbers, the digits after a given position repeat themselves as "6" and "285714" do above. The decimal expansions of irrational numbers are also infinite, as, for example, $e = 2.7182818284\ldots$. However, they do not possess the property of periodic repetition of decimals as do rational numbers.)

When we write $A = \emptyset$, we mean "A is an empty set." When we write $A = \{0\}$, we mean "A is the set whose only member is zero." The latter is not an empty set. The set of people who have traveled from the earth to Mars is an empty set. So is the set of women presidents of the United States.

Finite and infinite sets A set may contain a finite or an infinite number of objects. A set is called *finite* if its elements are equal in number to some specifiable nonnegative integer. It is called *infinite* if the number of its elements is greater than any positive integer.

The set of all students enrolled at California State University, Long Beach, in a given semester is finite; so is the set of all names listed in the Long Beach telephone directory in a specific year. The set of all living human beings at the beginning of 1974 is also a finite set. The number of members that each of these sets contains is quite large, yet they are all finite. On the other hand, the set of all natural numbers or of real numbers and the set of all points on a straight line are examples of infinite sets.

1-1-3 Venn diagram

A helpful scheme to illustrate the relation between sets and set operations is the *Venn diagram* (named after the English logician John Venn, 1834–1923). The universal set is usually represented by an enclosed plane-geometrical figure such as a square or rectangle within which small regions, such as circles or parts of circles, represent subsets. Even though Venn diagrams represent infinite sets, for there is an unlimited number of points within each figure, they are often used to represent finite sets as well.

Figure 1-2 shows that the set A represented by the circle is a subset of the universal set U; every element of A is an element of U. If we want to show that $A \subseteq B$ and $B \subseteq U$, Fig. 1-3 may be used. On the other hand, Fig. 1-4 shows that both A and B are subsets of U, but A and B are *disjoint*, that is, no element of U belongs to both A and B. Figure 1-5 indicates that the subsets A and B have some elements in common.

Sometimes a given universal set has disjoint subsets that together make up the universal set. These subsets then form a *partition* of U. Figure 1-6 shows that the whole set U is partitioned into three disjoint subsets, A, B, and C.

1-1-4 Operations with sets

Just as operations on numbers result in new numbers, operations on sets lead to new sets. Suppose that the sets A, B, C, and so on are subsets of the universal set U; then these subsets can be operated on to form new sets, which are also subsets of U.

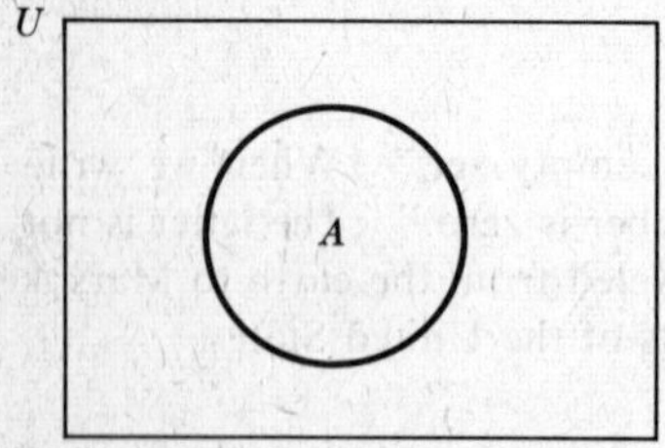

Figure 1-2
A subset.

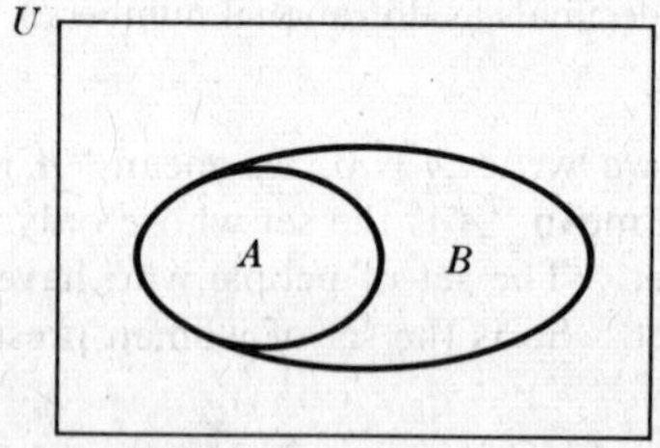

Figure 1-3
Subset of a subset.

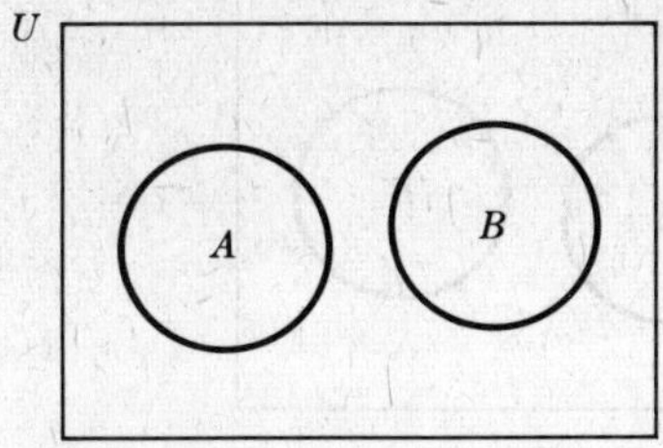

Figure 1-4
Disjoint subsets.

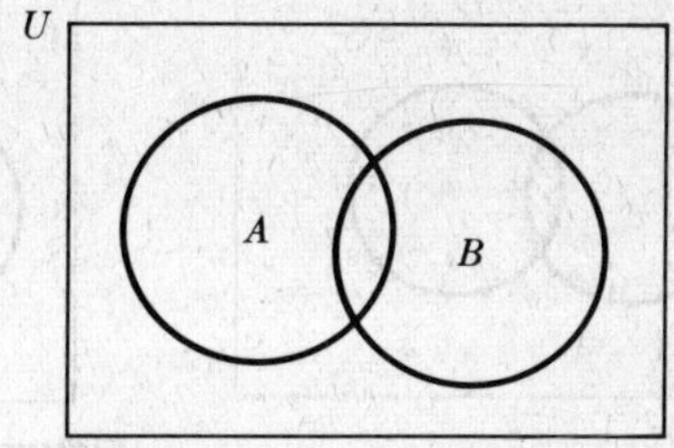

Figure 1-5
Overlapping subsets.

Intersection One of the operations is the intersection of two or more sets. Given the sets A and B, which are subsets of U, the *intersection* of A and B, written "$A \cap B$," is the set of all elements that are members of both sets. ("$A \cap B$" is read "A cap B," "A intersect B," or "A intersection B.") Formally,

$$A \cap B = \{x \mid x \in A \textit{ and } x \in B\} \qquad (1\text{-}1)$$

For example, if

$U = \{x \mid x \text{ is a positive integer}\}$

$A = \{x \mid x \text{ is a positive even integer}\}$

$B = \{x \mid x \text{ is a positive integer smaller than } 10\}$

then

$A \cap B = \{x \mid x \text{ is a positive even integer smaller than } 10\}$

or

$A \cap B = \{2, 4, 6, 8\}$

since these are the elements that are members of *both A and B*.

The shaded area in Fig. 1-7 represents the intersection of A and B. If A and B have no elements in common, their intersection is an empty set, or $A \cap B = \emptyset$, which is shown in Fig. 1-8. Then A and B are said to be disjoint, or *mutually exclusive*.

Figure 1-6
A partition.

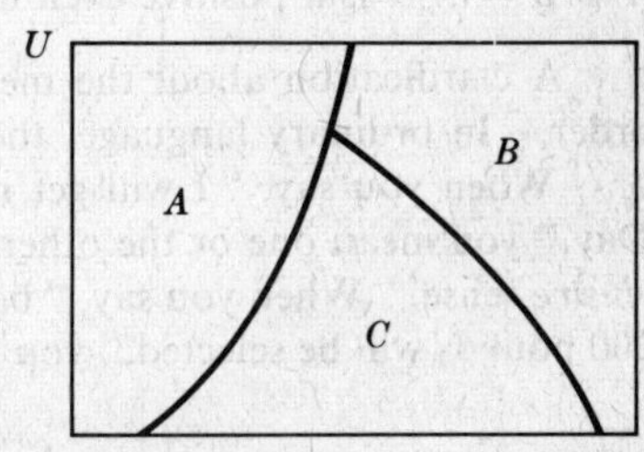

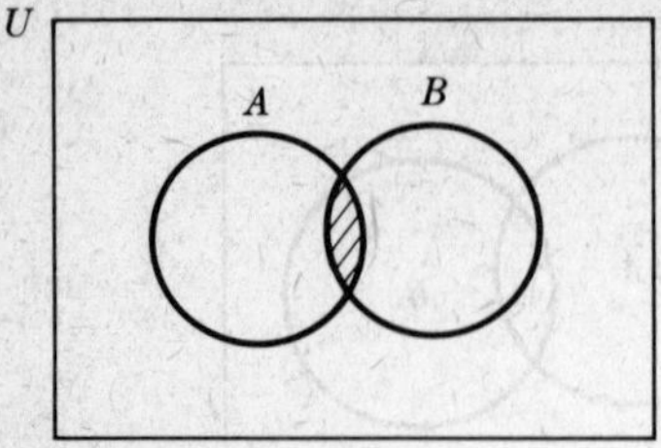

Figure 1-7
Intersection of two sets.

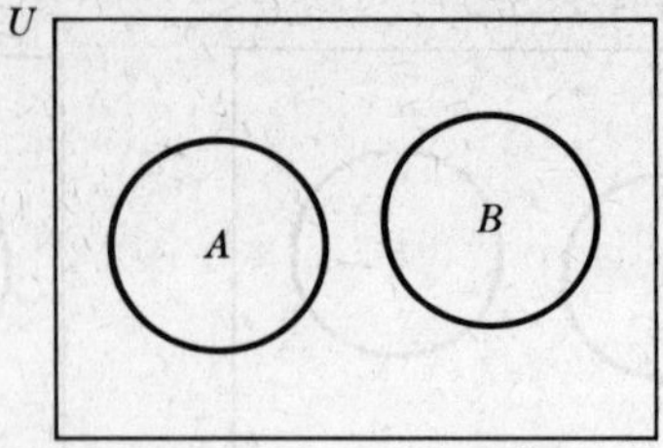

Figure 1-8
Intersection of two sets as an empty set.

The intersection may involve more than two sets. For example, the intersection of A, B, and C is

$$A \cap B \cap C = \{x \mid x \in A \text{ and } x \in B \text{ and } x \in C\}$$

Thus if A were all female students, B were all those 19 years of age, and C were those enrolled in a statistics class, then $A \cap B \cap C$ would be the set of all girls 19 years of age enrolled in a statistics class.

Union Another operation is the union of two or more sets. The *union* of two sets A and B is the set of all elements belonging to A or B or both. The union of A and B is denoted by $A \cup B$, read "A cup B" or "A union B." Formally,

$$A \cup B = \{x \mid x \in A \text{ or } x \in B\} \tag{1-2}$$

For example, if

$U = \{x \mid x \text{ is a positive integer}\}$

$A = \{x \mid x \text{ is a positive even integer}\}$

$B = \{x \mid x \text{ is a positive even integer smaller than } 10\}$

then

$A \cup B = \{x \mid x \text{ is a positive even integer}\}$

A clarification about the meaning of the word *or* in set theory is now in order. In ordinary language, there are two ways to interpret the meaning of or. When you say, "I will get married on Christmas Day or on New Year's Day," you mean one or the other date *but not both.* Here *or* is used in the *exclusive* sense. When you say, "boys who are 6 feet tall or who weigh more than 200 pounds will be selected," you mean *one or the other or both.* Here the word

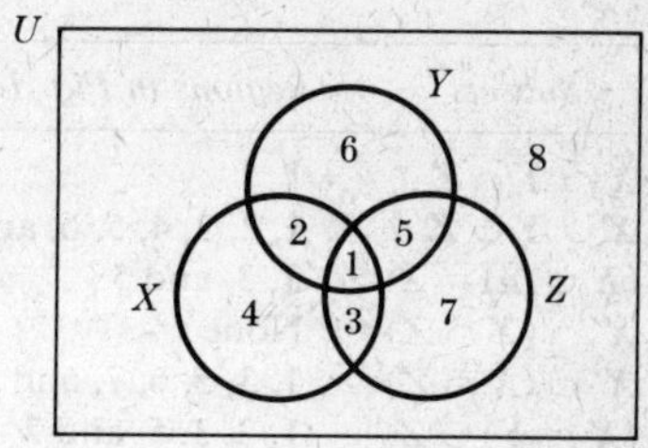

Figure 1-9
Partition of U into eight regions.

is used in the *inclusive* sense, in which it is synonymous with "and/or." In our discussion of sets, *or* used in the definition of union is to be taken in the inclusive sense. It is the key word to be remembered just as *and* is the key word in the definition of intersection.

Complementation Another operation is complementation. If the set A is a subset of a universal set, we can speak of the set of elements that are not in A. This set is called the *complement* of A, symbolized by A'. (Some authors use $\bar{A}$, $\tilde{A}$, or $\sim A$.) Formally,

$$A' = \{x \mid x \notin A\} \tag{1-3}$$

For example, if we are discussing integers and I is used to designate the universal set of all integers, N the set of negative integers, and P the set of positive integers plus zero, then $N' = P$ and $N = P'$,

$$N + N' = N + P = N' + P' = P + P' = I$$

New sets We have shown that new sets are obtained through each of the three basic operations on sets. We shall next show, by using the Venn diagram, that a combination of two or all three of these operations will also lead to new sets.

Let X, Y, and Z be three subsets of some universal set U. In Fig. 1-9, the eight nonoverlapping regions of the rectangle are labeled 1, 2, ..., 8, corresponding to the following eight possibilities for any element $x \in U$:

$x \in X \cap Y \cap Z$	region 1	$x \in X' \cap Y \cap Z$	region 5
$x \in X \cap Y \cap Z'$	region 2	$x \in X' \cap Y \cap Z'$	region 6
$x \in X \cap Y' \cap Z$	region 3	$x \in X' \cap Y' \cap Z$	region 7
$x \in X \cap Y' \cap Z'$	region 4	$x \in X' \cap Y' \cap Z'$	region 8

Observe that many new subsets of U can be obtained by forming complements, unions, and intersections of the three original subsets. The following subsets are a few examples:

Subsets	*Regions in Fig. 1-9*
$X \cap Y \cap Z$	1
$X \cup Y \cup Z$	1, 2, 3, 4, 5, 6, and 7
$(X \cup Y) \cap Z$	1, 3, and 5
$X' \cap (X \cap Z)$	None
$X' \cup (X \cap Z)$	1, 3, 5, 6, 7, and 8
$(X \cap Y) \cup Z$	1, 2, 3, 5, and 7
$(X \cup Y) \cap Z'$	2, 4, and 6

Exercise 1-1

1 Specify each of the following sets by the tabular method:

a Letters used in the English word "statistics."
b Digits used in the number 780,000,000,000.
c Letters used in the English term "statistical inference."
d The most recent *five* presidents of the United States.

2 By the use of a rule, specify each of the four sets in Prob. 1.

3 For each of the following groups of sets, determine which are the same and which are different:

a $X = \{t \mid t + 2 = 0\}$, $Y = \{t \mid t^2 + 2t = 0\}$, $Z = \{t \mid t^2 + t - 2 = 0\}$.
b $A = \{8, 7, 6\}$, $B = \{8, 9, 6\}$, $C = \{6, 7, 8\}$.
c $F = \{1, 2, 3, 4, 5\}$, $G = \{x \mid x \text{ is an integer larger than 1 and smaller than 5}\}$, $H = \{4, 3, 2\}$

4 Suppose that the set U is a sample of 1,000 security buyers interviewed about their buying habits. The distribution of the buyers by sex and type of securities bought is indicated below.

Type of securities bought	*Men*	*Women*	*Total*
Speculative	300	500	800
Investment	100	100	200
Total	400	600	1,000

Express each of the *eight* classes of buyers in set notation.

5 If I is the set of all nonnegative integers, which of the following subsets are infinite?

a The set of integers from 10 to 60.
b The set of all nonnegative even integers.
c The set of all integers above 1,000.
d The set of all integers that are perfect squares.
e The set of all nonnegative integers below 1,000.

6 Indicate whether each of the following sets is finite or infinite, and express it in set notation:

a The set of all corporations in the United States.
b The set of baby hogs that have been born or will be born on earth.
c The set of letters in the English alphabet.
d The set of odd integers.

7 Given that the set $U = \{p, q, r, 7, 9\}$, list four sets that are subsets of U.

8 Identify those of the following statements that involve a subset relation. Express each in set notation by using the capital letters for set symbols.

a All colleges (C) are educational institutions (E).
b All manufacturers (M) are business firms (B).
c All stockholders (S) are creditors (C).
d All bondholders (B) are creditors (C).

9 Let the universal set $I = \{a, b, c\}$. List the subsets of I in set notation.

10 If $C = \{1, 2, 3, 4\}$, what are the possible subsets of C?

11 If S contains n elements, how many subsets of S are there?

12 In a survey of 100 athletes, the numbers of those who specialized in various sports were found to be baseball, 35; basketball, 50; football, 40; baseball, and basketball 15; baseball and football, 10; basketball and football, 15; and all three sports, 5.

a How many athletes specialized in no sports?
b How many athletes specialized in football only?

(*Hint:* Draw a Venn diagram with three circles representing the three sports.)

13 Another survey of 100 athletes stated that the numbers of those who specialized in the various sports were, in all three sports, 5; baseball and basketball, 10; football and basketball, 8; baseball and football, 20; basketball, 30; baseball, 23; and football, 50. The surveyor who turned in this report was fired. Why?

14 In a survey of the mathematics background of 50 students enrolled in statistics, it was found that the numbers of those having studied the various mathematical subjects were as follows: matrix algebra, 23; analytical geometry, 18; finite mathematics, 13; matrix algebra and analytical geometry, 3; matrix algebra and finite mathematics, 6; analytical geometry and finite mathematics, 3; and all three subjects, 1.

a How many students have never taken any of these three subjects?
b How many students have taken only matrix algebra? Only analytical geometry? Only finite mathematics?
c How many students have taken only matrix algebra and analytical geometry? Only matrix algebra and finite mathematics? Only analytical geometry and finite mathematics?

15 Let the universal set $I = \{1, 2, 3, 4, 5, 6\}$, $A = \{1, 2, 3\}$, $B = \{2, 3, 4\}$, and $C = \{4, 5, 6\}$. Find:

a The overlapping subsets.
b The disjoint subsets.
c The subsets that partition I.

16 Partition $I = \{1, 2, 3, 4, 5, 6\}$ into:

a Two subsets. *c* Four subsets.
b Three subsets. *d* Five subsets.

17 Let $A = \{0, 1, 2, 3, 4, 5, 6, 7, 8, 9\}$, $B = \{0, 2, 4, 6, 8\}$, $C = \{3, 5, 7\}$, and $D = \{3, 7\}$. Find:

a $A \cup B$. *g* $A \cup (C \cap D)$.
b $C \cup D$. *h* $A \cap (C \cup D)$.
c $B \cup D$. *i* $(B \cup C) \cap B$.
d $A \cap C$. *j* $(B \cup D) \cup A$.
e $B \cap A$. *k* $(B \cup C) \cap D$.
f $C \cap D$. *l* $(\emptyset \cup A) \cap \emptyset$.

18 $I = \{1, 2, 3, 4, 5, 6, 7, 8, 9, 10\}$, $X = \{3, 5, 7, 9\}$, and $Y = \{2, 4, 6\}$, where I is the universal set. Find:

a $(X \cap Y)'$.
b $X' \cup Y'$.
c $(X \cap Y) \cup (X' \cap Y) \cup (X \cap Y')$.

19 With the sets A, B, and C shown in the Venn diagram below, represent the following sets by drawing the Venn diagram and shading the areas:

a $A \cap B'$.
b $A \cup B'$.
c $(A' \cap B)' \cup A$.
d $(A \cup B)' \cap C'$.
e $(A \cup C)' \cup C$.
f $(A \cap B)' \cup C$.

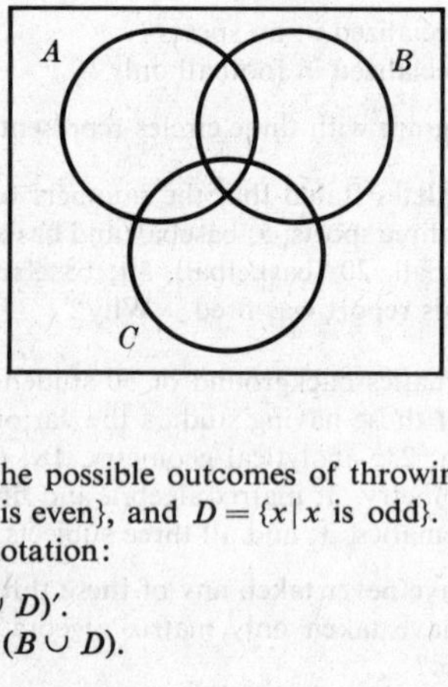

20 Let the universal set be the possible outcomes of throwing one die and $A = \{x \mid x < 4\}$, $B = \{x \mid x \leq 4\}$, $C = \{x \mid x \text{ is even}\}$, and $D = \{x \mid x \text{ is odd}\}$. List the elements of each of the following sets in set notation:

a $A \cap C$. *d* $(A \cup D)'$.
b $B \cap C$. *e* $A \cap (B \cup D)$.
c $A \cap B$.

21 Let the universal set $I = \{a, b, c, d\}$, and the sets $R = \{a, b, c\}$, and $S = \{b, c, d\}$. List the elements of each of the following sets:

a R'. *g* $(R \cup S)'$.
b S'. *h* $(R \cap S)'$.
c $R \cup S$. *i* $(R \cup S')'$.
d $R \cap S$. *j* $R' \cup S'$.
e $R \cup S'$. *k* $R' \cap S'$.
f $R' \cup S$. *l* $(R' \cup S')'$.

1-2
Relations and Functions

The theory of sets provides a basis for the study of another concept, namely, the notion of a *relation*, which is merely an extension of the set theory discussed in the previous section. In addition to developing a precise definition of relation, a special kind of relation generally known as *function* will be considered in this section. The concept of a function is one of the most important notions in statistics. We shall examine the meaning of a function, its domain and range, and its commonly used symbols and notations.

1-2-1 Relations

Relation, in one of its meanings, signifies a family tie between two persons, such as "is the father of," "is the sister of," "is the wife of," etc. In everyday conversation and in mathematics it is common to use the expressions "is parallel to," "is greater than," "is equal to," and many others like these. "Mark is taller than Mary," "$X > 3$," and "$y = x$" are examples of relations. Since these are relations between two numbers or objects, they are called *binary relations*. In this section we are concerned only with binary relations, that is, with relations between two elements.

Before we define the concept of a mathematical relation, it is necessary to understand a particular kind of set, the *cartesian product set*. Just as a collection of objects constitutes a set, a collection of *pairs of objects* constitutes a special kind of set, the cartesian product set. Let us use (x, y) to symbolize a pair of objects. If the first object is a member of some set S and the second object is a member of some set T, that is, $x \in S$ and $y \in T$, then (x, y) is called an *ordered pair*. It is ordered because the order in which the two objects are listed is essential. This ordered pair (x, y) is different from the set $\{x, y\}$. We know that $\{x, y\}$ represents the same set as $\{y, x\}$, since order is immaterial in listing members of a set. On the other hand, an ordered pair (x, y) is not necessarily the same as the pair (y, x). Two ordered pairs are equal if their first objects are the same; so are their second objects.

Let S and T represent two distinct finite sets. It is then possible to identify all the possible pairs (x, y); each pair relates an element of S to an element of T. These possible pairs constitute a set called the *cartesian product set* of S and T (named after René Descartes, 1596–1650, one of the mathematicians who invented analytic geometry.) The cartesian product set of S and T, denoted by $S \times T$ (read "S cross T"), is formally defined as follows:

$$S \times T = \{(x, y) \mid x \in S \text{ and } y \in T\} \tag{1-4}$$

It is really not a "product" as you know the meaning of this word in arithmetic. Here product merely refers to the fact that the total number of possible pairs is always equal to the number of elements in S multiplied by the number of elements in T.

Suppose that we have a set $S = \{1, 3, 5\}$ and a set $T = \{2, 4, 6\}$. Then we can form all the possible ordered pairs (x, y) by taking as the first object an element from S and the second from T. Thus the cartesian product set is the collection of the nine (3×3) ordered pairs as follows:

$$\begin{Bmatrix} (1, 2) & (1, 4) & (1, 6) \\ (3, 2) & (3, 4) & (3, 6) \\ (5, 2) & (5, 4) & (5, 6) \end{Bmatrix}$$

The set of ordered pairs can also be shown graphically as in Fig. 1-10. In this graph, there are several pairs of perpendicular lines. One pair of the perpendicular lines is called the axes. The line represented by x is the *horizontal axis*, and the line represented by y is the *vertical axis*. The point where x and y intersect is called the *origin*. Each point in the figure is represented by a pair of numbers called the *coordinates*. The first coordinate is called the *abscissa*, which represents the horizontal distance of the point from the vertical axis. The second coordinate is called the *ordinate*, which represents the vertical distance of the point from the horizontal axis. The cartesian product set of nine ordered pairs or elements, or $S \times T$, is represented by the nine points on the graph.

The mathematical concept of a relation is defined in terms of the cartesian product set as follows:

If S and T are any two sets, then a relation in $S \times T$ is any subset of $S \times T$.

Suppose that $S = \{\text{John, Mark, David}\}$ and $T = \{\text{Mary, Diana, Elsie}\}$. The cartesian product set $S \times T$ then is

$S \times T = \{$(John, Mary), (John, Diana), (John Elsie), (Mark, Mary), (Mark, Diana), (Mark, Elsie), (David, Mary), (David, Diana), (David, Elsie)$\}$

The set $\{$(John, Mary)$\}$ is a relation in $S \times T$. So is the set $\{$(John, Diana), (Mark, Diana), (David, Diana)$\}$. Both sets are subsets of the cartesian product set $S \times T$.

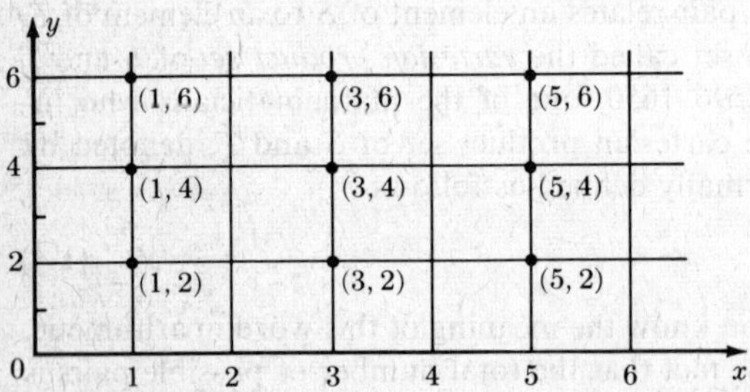

Figure 1-10
A cartesian product set $S \times T$, where $S = \{1,3,5\}$ and $T = \{2,4,6\}$.

Just as a set may be specified, so also can a mathematical relation be specified either by complete listing or by providing a rule or description. Let $S = \{1, 2, 3\}$ and $T = \{2, 3, 4\}$. Then, by the listing method, the relation that the two numbers are equal is the set L, where

$$L = \{(2, 2), (3, 3)\}$$

Moreover, if the relation that the first number is odd and the second even is M, we have

$$M = \{(1, 2), (1, 4), (3, 2), (3, 4)\}$$

Or, instead of listing all the relevant pairs, we may use a statement to specify a relation. Thus, if we let x stand for any element of S, and y any of T, and

$$S \times T = \{(x, y) \mid x \in S, y \in T\}$$

then the relations mentioned in the preceding paragraph can be specified as follows:

$$L = \{(x, y) \in S \times T \mid x = y\}$$

and

$$M = \{(x, y) \in S \times T \mid x \text{ is odd and } y \text{ is even}\}$$

where $S = \{1, 2, 3\}$ and $T = \{2, 3, 4\}$. Still another relation in $S \times T$ is $N = \{(x, y) \in S \times T \mid x < y\}$, which is equivalent to

$$N = \{(1, 2), (1, 3), (1, 4), (2, 3), (2, 4), (3, 4)\}$$

Just as ordinary sets and cartesian product sets, relations can be expressed in graphic forms. The nine dots in Fig. 1-11 constitute the cartesian product set $S \times T$, where $S = \{1, 2, 3\}$ and $T = \{2, 3, 4\}$. The circled dots represent the relation that x is odd and y is even.

If the ordered pairs are too numerous to plot, a line or curve in a graph may be used to display a relation. For example, let R be the set of real numbers; then the cartesian product set $R \times R$ is made up of all possible pairs of real numbers or, graphically, of all points in a two-dimensional plane. Any point, a line, a curve, or a region on this plane is a subset of the entire plane, and thus is a relation.

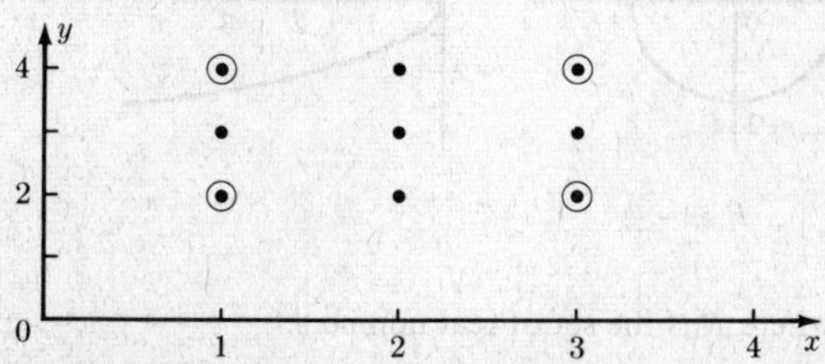

Figure 1-11
A cartesian product set $S \times T$, where $S = \{1,2,3\}$ and $T = \{2,3,4\}$.

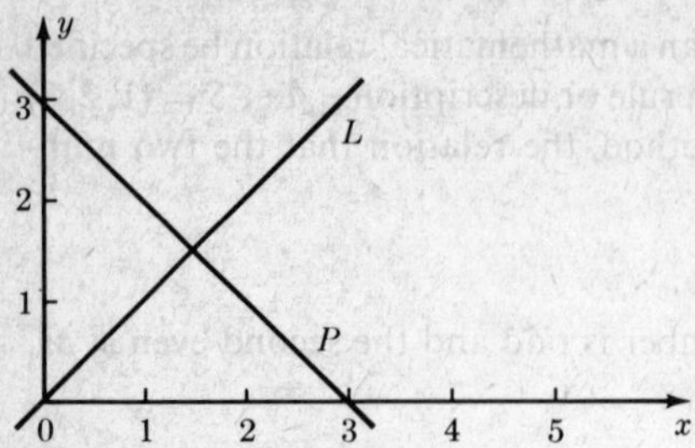

Figure 1-12
Two relations.

For instance, as shown in Fig. 1-12, the relation

$$L = \{(x, y) \in R \times R \mid x = y\}$$

is represented by the straight line L, which goes through the origin and bisects the first and the third quadrants. The line P represents another relation, which is specified as $\{(x, y) \in R \times R \mid x + y = 3\}$.

Similar mathematical expressions and graphs can be used to specify countless relations of real numbers. As additional examples, we have

$$A = \{(x, y) \in R \times R \mid 2 \leq x \leq 4, 1 \leq y \leq 3\}$$
$$B = \{(x, y) \in R \times R \mid x^2 + y^2 = 1\}$$
$$C = \{(x, y) \in R \times R \mid y^2 = 2x\}$$

The corresponding graphs of these relations are shown in Fig. 1-13.

1-2-2 Functions

A function is a special kind of relation. If each element of a set S is paired with *exactly one* element of another set T, than this relation is called a *function*. Some authors refer to a function as a *single-valued function*, and a relation as a *multiple-valued function*. Most authors, however, call only the single-valued function a function.

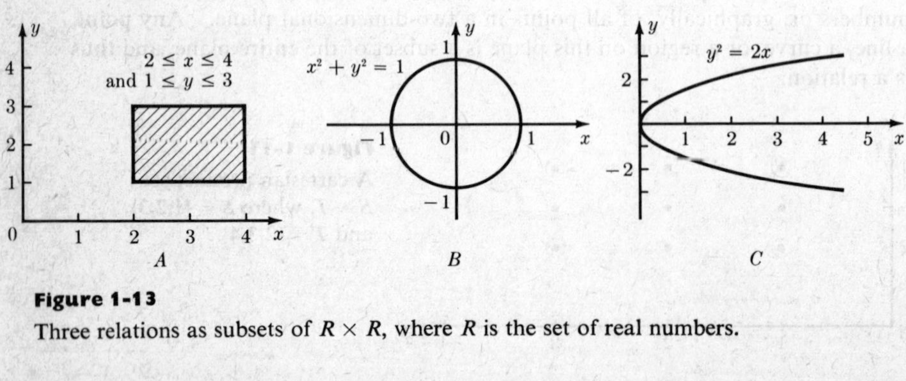

Figure 1-13
Three relations as subsets of $R \times R$, where R is the set of real numbers.

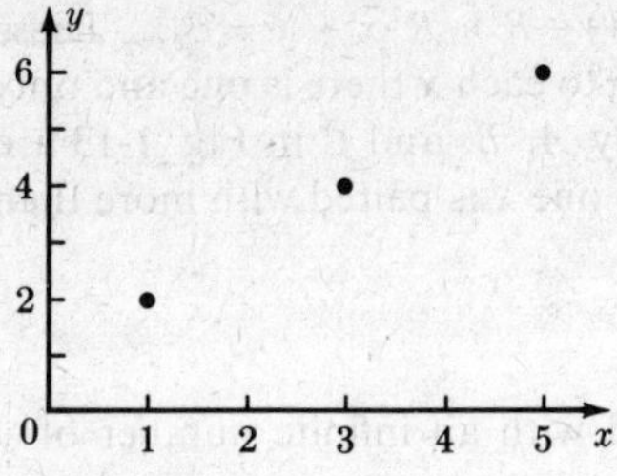

Figure 1-14

A function

$H = \{(1,2),(3,4),(5,6)\}$.

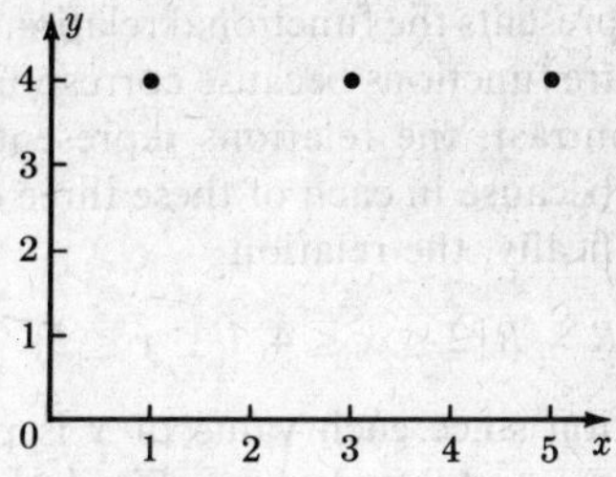

Figure 1-15

A function

$M = \{(1,4),(3,4),(5,4)\}$.

A few examples will make the notion of a function clearer. Let S be the set of all the states of the United States and T the set of their governors. Since every state has one and only one governor, the relation that each state is paired with a person as its governor is a functional relation, or a function. Also, with each state in S can be associated its population, or its territory in square miles. In each of these cases, some object or number (only one) in T is paired or associated with each element of S.

As another example, if we let $S = \{1, 3, 5\}$ and $T = \{2, 4, 6\}$, we may have a relation $H = \{(1, 2), (3, 4), (5, 6)\}$. Then H represents an association that is also a function since each of the elements of S is paired with one and only one element of T. Figure 1-14 depicts this functional relation. Similarly, Fig. 1-15 shows the function $M = \{(1, 4), (3, 4), (5, 4)\}$. On the other hand, the relation $N = \{(1, 2), (1, 4), (1, 6)\}$ is not a function. Here more than one element from the range is paired with the same element of the domain. Graphically, if any vertical line crosses the graph of a relation in two or more points, the relation is not a function. This is shown in Fig. 1-16.

As an additional example, the relation

$$L = \{(x, y) \in R \times R \mid x = y\}$$

is a function represented by the line L in Fig. 1-12. Likewise, the line P in the

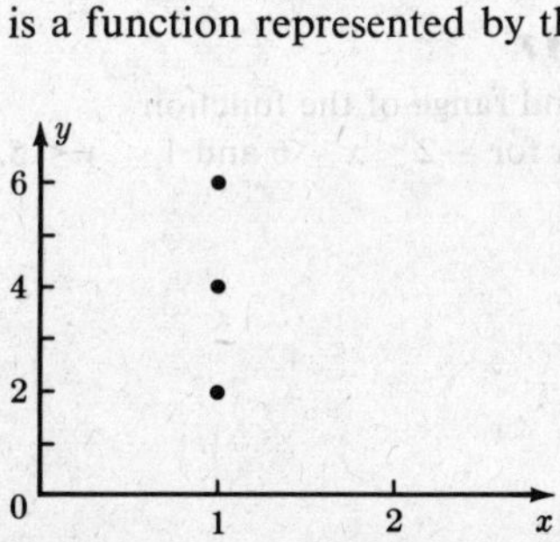

Figure 1-16

A relation $N = \{(1,2),(1,4),(1,6)\}$.

same figure represents the functional relation $\{(x, y) \in R \times R \mid x + y = 3\}$. These two relations are functions because corresponding to each x there is one and only one y. In contrast, the relations represented by A, B, and C in Fig. 1-13 are not functions because in each of these three cases one x is paired with more than one y. Specifically, the relation

$$A = \{(x, y) \in R \times R \mid 2 \leq x \leq 4, 1 \leq y \leq 3\}$$

is not a function since each value of x is paired with an infinite number of y values between 1 and 3 inclusive. Similarly, the relations

$$B = \{(x, y) \in R \times R \mid x^2 + y^2 = 1\}$$

and

$$C = \{(x, y) \in R \times R \mid y^2 = 2x\}$$

are not functions since in each of these two cases two values of y are associated with each and every value of x.

Domain and range The domain of a function is the set of which the first object or number of an ordered pair, or x, is a member. The range of a function is the set of which the second object or number of an ordered pair, or y, is a member. Thus the function

$$F = \{(1, 2), (3, 4), (5, 6)\}$$

has its domain D and range R as follows:

$$D = \{1, 3, 5\} \quad \text{and} \quad R = \{2, 4, 6\}$$

The line segment W in Fig. 1-17 represents the function

$$W = \{(x, y) \in R \times R \mid y = 2 + \tfrac{1}{2}x \text{ for } -2 \leq x \leq 6 \text{ and } 1 \leq y \leq 5\}$$

The domain of this function is the set of all real numbers starting with -2 and

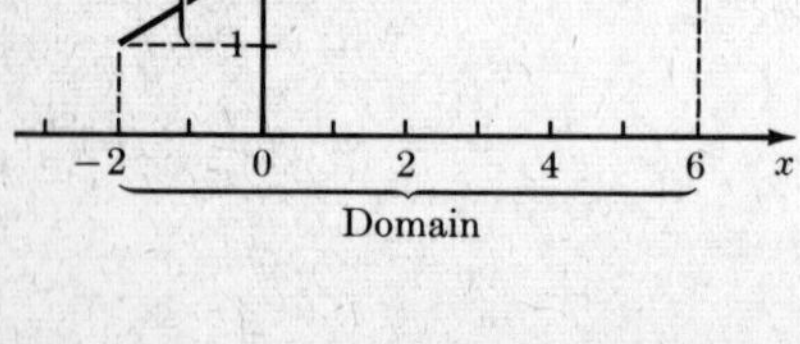

Figure 1-17

Domain and range of the function $y = 2 + \frac{1}{2}x$ for $-2 \leq x \leq 6$ and $1 \leq y \leq 5$.

ending at 6; the range of the function is the set of all real numbers between 1 and 5 inclusive.

The domain of a function may be a finite set or an infinite set; the same can be said about the range of a function.

Functional notations Single letters such as f, g, F, P, or, in some cases, double or even triple letters such as FR, PF, and RCF, are frequently used to designate functions. If we use f to designate a function, and x is an element of the domain of f, then $f(x)$ is usually used to denote the second element of the ordered pair. The notation $f(x)$ is called a *functional notation*. (It is read "f of x" or "the value of f at x" and must not be mistaken for "f multiplied by x.") Thus instead of (x, y), we may write $(x, f(x))$ for the ordered pair, for $f(x)$ stands for the value of y, or $y = f(x)$.

The expression $y = f(x)$ signifies that there is a rule that associates each possible value of x with one and only one value of y. The letter f preceding (x) simply indicates that there exists a *function rule* specifying a subset of all (x, y) pairs. Given the value of x and the function rule, the value of y is determined.

Suppose that we define a function f by the rule

$$f(x) = x^2 + 2x + 4$$

If we want to find the real number that pairs with x, substitute the value of x in the formula and perform the required algebraic operations on x. Thus

$$y = f(x) = 2^2 + 2(2) + 4 = 12 \qquad \text{for } x = 2$$

For $x = 3$, then $y = 19$; and for $x = 0$, then $y = 4$.

The functional notation $f(x)$ can be used together with the summation notation $\sum$. If a function f is defined on a set of nonnegative integers with $n + 1$ elements, namely, 0, 1, 2, ..., n, we may have the expression

$$F(x) = \sum_{x=0}^{n} f(x) = f(0) + f(1) + \cdots + f(n) \tag{1-5}$$

where $\sum_{x=0}^{n}$, which may also be written $\sum_{x=0}^{n}$, symbolizes the summation from 0 to n of x.[1] The capital letter F is customarily used to refer to a cumulative function, and $F(x)$ is the accumulated value of y paired with each value of x. For instance, if $F(x) = 2x + 3$ for $x = 1, 2, 3, 4$, and 5, then

$$F(2) = [2(1) + 3] + [2(2) + 3] = 5 + 7 = 12$$

and

$$\begin{aligned} F(5) &= [2(1) + 3] + [2(2) + 3] + [2(3) + 3] + [2(4) + 3] + [2(5) + 3] \\ &= 5 + 7 + 9 + 11 + 13 = 45 \end{aligned}$$

[1] For further clarification of the summation notation, see Sec. 4-2-2.

Dependent and independent variables In the discussion of relations and functions we frequently used the two letters x and y. As used, x and y are called *variables*. Before leaving this chapter, let us examine the concept of a variable a bit further. First, we define this concept as follows:

A variable is a symbol used to stand for (in the place of) any one of the elements of some specified set.

It must be noted that a variable can be replaced by any element of a certain set. Thus, in the equation $y = 2x^2 - x$, x stands for an unspecified real number. The value of y is determined when a number is substituted for x.

The symbol that stands for a value in the domain of a function, the first element of the pair (x, y), is usually referred to as the *independent variable*. It is called independent because it can assume any value of the domain of the function. Once the function rule is given and an x value is fixed, the value of y is determined. This is the reason why y is customarily called the *dependent variable*.

Our discussion so far has been on binary functions, with one independent and one dependent variable. However, the notion can be extended to two or more independent variables. Given n independent variables $x_1, x_2, \ldots, x_n$, each standing for a real number, the statement

$$y = f(x_1, x_2, \ldots, x_n)$$

indicates that once the function rule and the values $x_1, x_2, \ldots, x_n$ are known, the value of y is completely determined. For example, if $x_1 = 2$, $x_2 = 1$, and $x_3 = 4$, the statement

$$y = 2x_1{}^2 + x_2 + 3x_3$$

yields a value of y, namely, 21.

Exercise 1-2

1 Use your own words to define the following concepts as fully as you can:

a Cartesian product set.
b Relation.
c Domain of a function.
d Range of a function.

2 Let $P = \{1, 2, x\}$, $Q = \{a, x, y\}$, $R = \{x, y, 2\}$; find the following:

a $P \times Q$.
b $P \times R$.
c $Q \times P$.
d $Q \times R$.
e $R \times P$.

3 In each of the following pairs of sets, is the first set equal to the second?

a $\{a, b\}$ and $\{b, a\}$.
b $\{a, b\}$ and $\{(b, a)\}$.
c $\{(a, b)\}$ and $\{a, b\}$.
d $\{(b, a)\}$ and $\{(a, b)\}$.
e $\{b, a\}$ and $\{(a, b)\}$.
f $\{(a, b), (b, a)\}$ and $\{(b, a), (a, b)\}$.

4 If P has three elements, Q four, and R two, how many elements does the cartesian product set $P \times Q \times R$ have? Give any two subsets of $P \times Q \times R$.

5 Identify the elements of set P, if set $Q = \{1, 2, 3\}$ and $P \times Q = \{(4, 1), (4, 2), (4, 3), (5, 1), (5, 2), (5, 3), (6, 1), (6, 2), (6, 3)\}$.

6 Draw and scale x and y axes, and then plot the following relations:

a $A = \{(x, y) \mid x^2 + y^2 = 4\}$.
b $B = \{(x, y) \mid 1 \leq x \leq 5 \text{ and } 1 \leq y \leq 5\}$.
c $C = \{(x, y) \mid 1 \leq x \leq 5 \text{ or } 1 \leq y \leq 5\}$.

7 Given the following equations, find x and y:

a $(x + 1, y - 2) = (2, 1)$.
b $(x + y, 1) = (5, x - y)$.
c $(x + y, 1) = (3, x - y)$.
d $(2x - 2, +5) = (x, y + 1)$.

8 Given the following equation, find x, y, and z:

$(5 - y, z + y, 6 - z) = (z, x + z, x + y)$.

9 Given that the digits 0, 1, 2, 3, 4, 5, 6, 7, 8, 9 are contained in each of three sets, how many ordered triples of digits are there?

10 Given that $f(x) = x^2$, what is:

a $f(0)$?
b $f(r)$?
c $f(2 + y)$?
d $f(4)$?
e $f(-2)$?
f $f(r + 2)$?
g $f(-6)$?
h $f(2y)$?

11 Given that $X \neq Y$, $X \neq Z$, and $Y \neq Z$, determine whether each of the following sets of ordered pairs is a function. If not, explain why not.

a $\{(X, 0), (Y, 1), (Z, 2)\}$.
b $\{(X, 4), (Y, 4), (Z, 4)\}$.
c $\{(2, X), (3, Y), (2, Z)\}$.
d $\{(X, 1), (Z, 2), (Y, 7)\}$.
e $\{(1, X), (1, Y), (1, Z)\}$.
f $\{(Z, 1), (Z, 2), (Z, 4)\}$.
g $\{(1, X), (2, Y), (3, Z)\}$.
h $\{(2, Y), (3, Z), (2, X)\}$.

12 Determine whether each of the following sets of ordered pairs is a function. If not, explain why not.

a $\{(a, b) \mid a \text{ and } b \text{ are numbers and } a = b^2\}$.
b $\{(a, b) \mid a \text{ and } b \text{ are numbers and } a^2 + b^2 = 1\}$.
c $\{(c, d) \mid c \text{ and } d \text{ are numbers and } c^2 - 4d^2 = 4\}$.
d $\{(k, l) \mid k \text{ and } l \text{ are numbers and } l = k^2 + 4k - 6\}$.

13 If S is the set of all students in your school, and if $y = G(x)$ means that y is a grandparent of x for x in S, is G a function?

14 If S is the set of all students in your school, and if $W(x)$ stands for the weight of student x for x in S, is W a function? Estimate the range of W.

15 Let S be the set of all students in your school, and $A(x)$ be the age of student x. Is A a function? Estimate the range of A.

16 Identify the domain and the range in each of the following functions:

a $A = \{(x, 1)\}$.
b $B = \{(2, 3), (3, 4), (5, 6)\}$.
c $C = \{(x, 1), (2, y), (3, z)\}$.
d $D = \{(1, 1), (2, 2), (3, 3)\}$.
e $E = \{(0, 2), (1, x), (2, y), (3, 2)\}$.
f $F = \{(x, y), (w, z), (u, v), (p, q)\}$.

17 Determine the range of $y = 2x - 6$ if the domain is $0 \leq x \leq 10$.

18 Assume that f is a function defined by $f(x) = x^2 + x - 2$; what is:

a $f(-2)$? *c* $f(0)$? *e* $f(2)$?
b $f(-1)$? *d* $f(1)$? *f* $f(a-3)$?

19 Given that f is a function defined by $f(x) = 3x + 2$, what is:

a $f(x+1)$? *c* $f(4y+3)$?
b $f\left(\frac{y+2}{3}\right)$? *d* $f(3z-2)$?

20 Let $f(x) = x + 1$ and $g(x) = x^2$; find the following values:

a $f(-2)$. *e* $f(g(1))$.
b $g(-2)$. *f* $g(f(-1))$.
c $f(g(0))$. *g* $g(f(0))$.
d $f(g(-1))$. *h* $g(f(1))$.

21 Name the independent and dependent variables in the following functions:

a $y = f(x) = x$. *c* $y = f(x) = 2x^2 - 3x + 5$.
b $y = f(x) = x^2 + 2$. *d* $z = f(x, y) = 2x^2 + 3xy + y^2$.

Chapter 2

Counting rules and probabilities

The primary objective of this chapter is to examine the concept of probability. We shall start with the notions of an event, random event, sample point, and sample space. The counting rules with which most students are familiar through their study of algebra will be reviewed. This is done because, as indicated later in the chapter, the computation of probabilities is basically a process of counting events and possibilities.

2-1 Events and Possibilities

A fundamental concept of the theory of probability is that of an *event*, or *random event*. An event, according to Webster's Dictionary, is a happening or occurrence. In our discussion it is used synonymously with the words *result*, *outcome*, and *consequence*. Furthermore, what we are most interested in is the notion of a random event. In this section we shall first examine the concept of a random event and then the idea of the totality of all possible events, known as the *sample space*, of a given problem situation will be considered.

2-1-1 Random events

Random events are closely related to what is known as *random phenomena*. In fact, one of the reasons why probability theory is important and has diverse applications is the random nature of many social and physical phenomena. A phenomenon is referred to as *random* if only chance factors determine its occurrence or nonoccurrence; the outcome occurs haphazardly without plan or choice. Among such phenomena are the toss of a coin to see whether heads or tails occurs, the roll of a cubic die to see which side shows up, the number of defective products in a production run, the density of telephone traffic in a certain hour of the day, the number of automobile accidents on some freeway during a certain period, and countless others. Each of these phenomena possesses the common property that its observed outcome may not always be the same even under the same set of circumstances. In other words, there is no *deterministic regularity* in its outcome. For instance, the number of defectives produced by the same set of workers, machines, and other factors may vary from run to run though conditions are identical. The number of fatal accidents on the San Diego freeway in 1 month may be quite different from those in another month though circumstances are similar.

However, the observed outcomes may differ in such a way that there is a *probabilistic regularity*. This means that in a sufficiently large number of observations there exist numbers between 0 and 1 representing the proportion of the time in which the different outcomes may be observed. A phenomenon that possesses probabilistic, though not deterministic, regularity is often referred to as a *random phenomenon*.

Knowing the meaning of a random phenomenon, we are in a position to define the concept of a random event. Briefly speaking.

> *A random event is a possible distinct outcome, or a combination of such outcomes, of some experiment of a random phenomenon.*

In statistics the word *experiment* is used in quite a different context from that in the laboratory sciences. By experiment we mean some *act*, *trial*, or *operation* that results in one of a set of possible outcomes. The roll of two dice to note the sum of spots uppermost, the toss of two coins to see the number of heads that turn up, and the inspection of a sample of products to observe the number of products that are defective are all examples of an experiment. In such an experiment, the outcome of a single trial is often called an *elementary event*. Thus the roll of two dice resulting in a sum of, say, 12 spots uppermost is an elementary event and so is the toss of two coins resulting in two heads.

Frequently, however, what we are interested in is not any particular elementary event, but rather the category of the outcome an observation represents. In other words, what we are concerned with is the sets of elementary events. Any such set is simply called an *event*, or a *compound event*. For instance, the six possible outcomes in the rolling of a cubic die can be classified into sets in various ways. The odd numbers constitute an event; so do the even numbers. An event may be the set of all points above 3 or below 4. Each of these events is a compound event; it consists of more than one elementary event.

In short, a random event is a *subset* of the universal set of all the elementary events of an experiment. The observation of any element of an event E leads us to conclude that event E has occurred. For instance, if event E is the odd number of spots of the die and the face of the die shows three spots after it is rolled, which is an element of the set (or subset) E, we say that event E has occurred.

2-1-2 Sample space

The universal set of any experiment referred to above is known as the *sample space* of the experiment. The concept of a sample space is of paramount importance in discussing the theory of probability, because we can predict *what is probable* only if we know *what is possible*, and the sample space is the totality of all possibilities in a given problem situation. In predicting the outcome of a basketball tournament, for instance, we need to know which teams are going to participate. Similarly, in predicting the winner in a beauty pageant, we need to know who are going to enter the contest. Likewise, we need to know which horses are going to run if we want to predict the winner of the Kentucky Derby. In other words, if we want to know what the probable outcome will be, we need to know what the possible outcomes are.

We may now define the concept of a sample space as follows:

A sample space of an experiment is the universal set of all possible distinct outcomes for that experiment.

Each element of the sample space is called a *sample point*, which is an elementary event. It is one possible outcome of a single act or trial of the experiment. For instance, the experiment of rolling one cubic die to see the number of spots uppermost has the sample space $S = \{1, 2, 3, 4, 5, 6\}$. The tossing of two coins to see the number of heads (H) or tails (T) come up has the sample space $U = \{(H, H), (H, T), (T, H), (T, T)\}$. Each number in S is a sample point, or an elementary event; so is each of the pairs of letters in U.

In order to clarify the relation between the idea of an event and that of a subset, let us consider again the experiment of rolling a cubic die and its sample space $S = \{1, 2, 3, 4, 5, 6\}$. Furthermore, let $E = \{1, 3, 5\}$ and $F = \{2, 4, 6\}$. It is obvious that both E and F are subsets of S; they are also random events. Moreover, the sets S and $\emptyset$ are subsets of S, and so they are random events. The event S is sure to happen and the event $\emptyset$ is sure not to happen in any trial of the experiment, because all the possible events are in S and none in $\emptyset$.

The events E and F as defined in the last paragraph have no points in common. That is, the intersection of E and F is a null set, or $E \cap F = \emptyset$; it is impossible to observe an event that is in *both E and F*. This means that the events E and F cannot both happen; such two events are said to be *mutually exclusive*. On the other hand, if the union of two sets is identical with the sample space, then the two sets are said to be *collectively exhaustive*. Thus the events E and F are collectively exhaustive, since $E \cup F = S$. If the intersection of two events is a null set, and at the same time their union is identical with the sample space, we say that the sample space is *partitioned* into the two events.

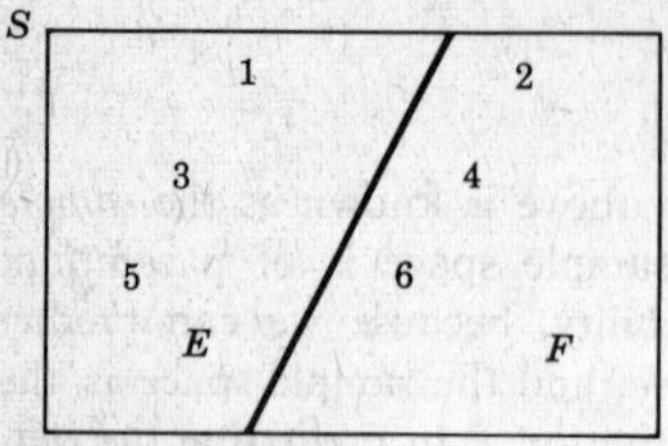

Figure 2-1
Partition of the sample space $S = \{1,2,3,4,5,6\}$ into the event $E = \{1,3,5\}$ and the event $F = \{2,4,6\}$.

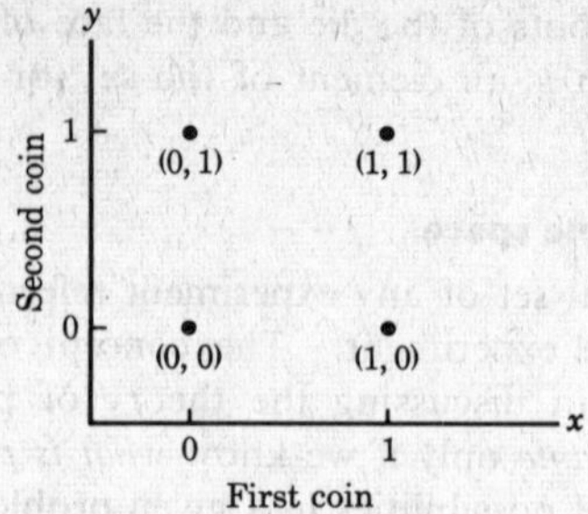

Figure 2-2
The sample space of the experiment of tossing two coins to see the number of tails (0) or heads (1) show up.

Thus the sample space S is partitioned into E and F, because $E \cap F = \emptyset$ *and* $E \cup F = S$ as shown in the Venn diagram in Fig. 2-1.

We have used the term *sample points* to refer to the individual outcomes of an experiment. The use of points makes it possible to represent the sample space in a graphic form. In the toss of two coins, for example, we can visualize all the possible outcomes by the use of the two-dimensional plane. Let the points 0 and 1 of the x coordinate respectively designate tails (T) and heads (H) of the first coin, and the points 0 and 1 of the y coordinate designate tails and heads of the second coin. Then, as shown in Fig. 2-2, we have:

The point (0, 0) represents the outcome that both coins turn up tails, or (T, T).

The point (0, 1) represents the outcome that the first coin turns up tails and the second coin turns up heads, or (T, H).

The point (1, 0) represents the outcome that the first coin turns up heads and the second coin turns up tails, or (H, T).

The point (1, 1) represents the outcome that both coins turn up heads, or (H, H).

Observe that the first number in each of the parentheses is the x coordinate, and the second number is the y coordinate.

In the roll of two cubic dice to see the sum of spots uppermost, the sample space consists of 36 possible outcomes, which are shown in Fig. 2-3. Similarly, this sample space can be presented by the use of the two-dimensional plane. Let the x coordinate represent the number of spots of the first die, and y coordinate the number of spots of the second die. The 36 sample points are shown

First die

+	1	2	3	4	5	6
1	2	3	4	5	6	7
2	3	4	5	6	7	8
3	4	5	6	7	8	9
4	5	6	7	8	9	10
5	6	7	8	9	10	11
6	7	8	9	10	11	12

Second die

Figure 2-3
Sum of spots on the faces of two dice after being rolled.

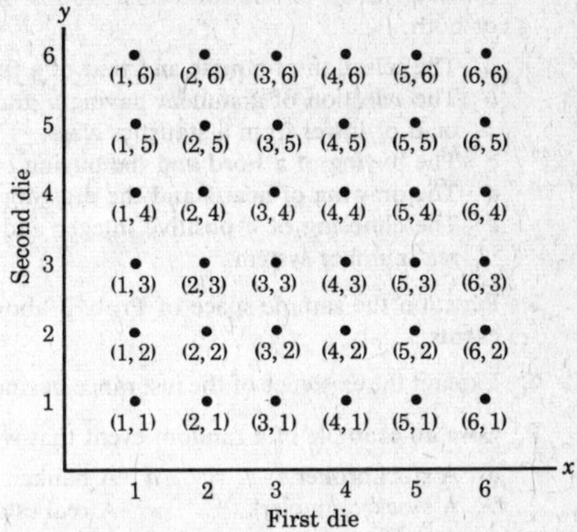

Figure 2-4
The sample space of the experiment of rolling two cubic dice.

in Fig. 2-4. The point (1, 2) represents the outcome that the first die shows one spot and the second die shows two spots, making a sum of three spots; the point (3, 2) represents the outcome that the first die shows three spots and the second die shows two spots, making a sum of five spots; and so on.

If there are more than two coins, or more than two dice, used in the experiment, it would be very difficult or even impossible to represent the sample space in a graphic form. This is so because each sample point is then represented by three or more numbers requiring a three- (or more) dimensional system.

Exercise 2-1

1 Explain the meaning of each of the following terms:

a Random phenomenon.
b Random event.
c Experiment.
d Sample space.

2 Distinguish between the two concepts in each of the following pairs:

a Elementary event and compound event.
b Sample point and sample space.
c Mutually exclusive events and collectively exhaustive events.

3 Three coins are tossed simultaneously to see how many heads are obtained. Let H designate heads and T designate tails. List all the elementary events of the sample space. Give an example of a compound event in this experiment.

4 Indicate which of the following groups of events are mutually exclusive or exhaustive or both:

a The selection of a male and that of a female student from your college.
b The selection of a student having a grade of B or higher and of one having a grade of B or lower from a statistics class.
c The buying of a Ford and the buying of a Chevrolet from a used-car lot.
d The drawing of hearts and the drawing of clubs from an ordinary deck of cards.
e The choosing of a positive integer and the choosing of a negative integer from the real-number system.

5 Partition the sample space of Prob. 3 above into two mutually exclusive and exhaustive events.

6 Explain the existence of the insurance business in the light of the discussion in this section.

7 Give an example of a random event that would be studied by:

a A stockbroker.
b A stockroom clerk.
c A traffic manager.
d A banker.
e A real estate broker.
f A marketing researcher.

8 Suppose that a housewife and her husband are interviewed by a marketing researcher to determine each person's "like" or "dislike" of a certain product. How many possible responses are there from this couple? Represent the sample space graphically by the use of the two-dimensional plane.

9 John has five cards marked 1, 2, 3, 4, and 5, and Mary has five cards also marked 1, 2, 3, 4, and 5. If John's cards are to be paired with Mary's, how many possible pairings are there? Represent the sample space of this experiment graphically by the use of the two-dimensional plane.

10 Partition the sample space in Prob. 9 above into:

a Two mutually exclusive but *not* exhaustive events.
b Two exhaustive but *not* mutually exclusive events.
c Two events that are not mutually exclusive or exhaustive.

11 Let G be the event that an employee is a college graduate, M be the event that an employee is married, and C be the event that an employee continues his education while working. With reference to the Venn diagram below, identify the regions which represent the following events:

a The event that an employee is a college graduate and married but does not continue his education.

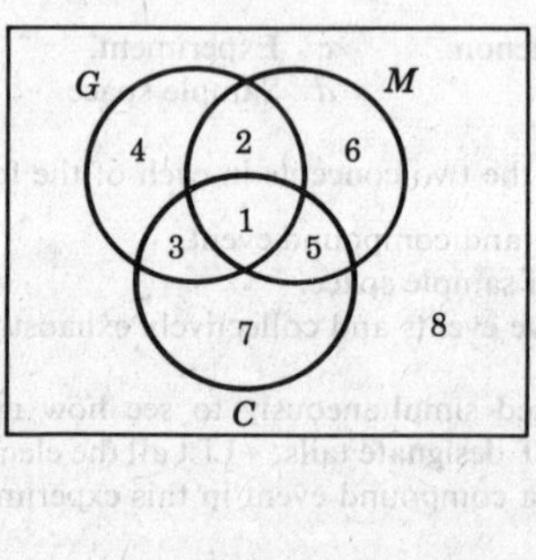

b The event that an employee is a college graduate but neither is married nor continues his education.

c The event that an employee does continue his education but remains unmarried and not yet graduated from college.

d The event that an employee is not a college graduate.

12 Referring to the Venn diagram in Prob. 11 above, use your own words to explain what events are represented by the following regions:

a Region 1. *c* Region 8.
b Region 6. *d* Regions 2, 4, 6, and 8.

2-2
Counting Rules

In the previous discussion we have tried to enumerate all the possible events in the sample space of a given experiment. In most practical situations, however, the number of all possible events is so large that it is troublesome, or even impossible, to list or enumerate them. Fortunately, most problems do not repuire a complete enumeration, and it is quite sufficient to know only the total number of possible events in the sample space. We shall now attempt to examine the various rules of counting the number of possible events without enumerating them one by one. These rules include the law of multiplication and the rules of permutation and combination.

2-2-1 The law of multiplication

If $A_1, A_2, \ldots,$ and A_n are n different acts that can be performed in $k_1, k_2, \ldots, k_n$ ways, respectively, then the total number of ways in which $A_1, A_2, \ldots, A_n$ can be performed in succession is given by

$$k_1 \times k_2 \times \cdots \times k_n \tag{2-1}$$

Suppose that you were to take a trip from San Diego to Seattle via Los Angeles and San Francisco and that you had four ways to choose between San Diego and Los Angeles, three between Los Angeles and San Francisco, and two between San Francisco and Seattle, as follows:

San Diego to Los Angeles A_1		*Los Angeles to San Fransicso* A_2		*San Francisco to Seattle* A_3	
Bicycle	(B)	Car	(C)	Air	(A)
Greyhound bus	(G)	Horse	(H)	Steamer	(S)
Motorcycle	(M)	Train	(T)		
Walk	(W)				

The total number of ways in which you could travel from San Diego to Seattle would be $4 \times 3 \times 2 = 24$, as indicated below:

A_1	A_2	A_3	A_1	A_2	A_3	A_1	A_2	A_3	A_1	A_2	A_3
B	C	A	G	C	A	M	C	A	W	C	A
B	C	S	G	C	S	M	C	S	W	C	S
B	H	A	G	H	A	M	H	A	W	H	A
B	H	S	G	H	S	M	H	S	W	H	S
B	T	A	G	T	A	M	T	A	W	T	A
B	T	S	G	T	S	M	T	S	W	T	S

Observe that the 24 ways of traveling from San Diego to Seattle are in fact 24 *sequences*, each of which is treated as an elementary event. The 24 sequences constitute the sample space of the experiment.

Figure 2-5 presents a tree diagram showing the sample space of these 24 elementary events. The diagram shows that there are, first of all, four branches

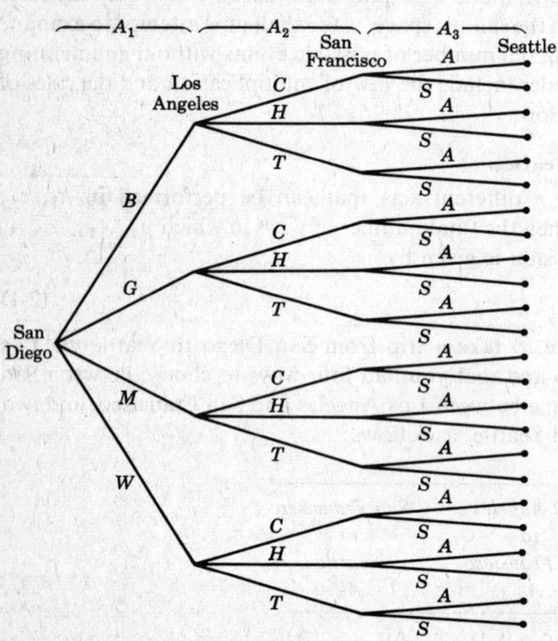

Figure 2-5
A tree diagram showing the sample space of all possible sequences for traveling from San Diego to Seattle.

representing four ways, namely, B, G, M, and W, in which to travel from San Diego to Los Angeles. From Los Angeles to San Francisco there are three ways, C, H, and T, available; they are represented by the three branches growing out of each of the four branches indicated above. As a result, there are $(4)(3) = 12$ branches representing 12 ways in which to travel from San Diego to San Francisco. Furthermore, there are two ways, A and S, in which to travel from San Francisco to Seattle, which are represented by two branches growing out of each of the 12 branches. Thus there are altogether 24 branches, each of which consists of three segments, representing 24 paths from left to right. That is, there are 24 different possible outcomes in the sample space.

Furthermore, if the number of ways in which each of the n acts $A_1, A_2, \ldots, A_n$ can be performed remains the same, say k, the total number of ways will be

$$k \times k \times \cdots \times k = k^n \tag{2-2}$$

In the above example, if the number of ways in which you wish to travel remain four each for A_1, A_2, and A_3, the total number of ways will be

$$4 \times 4 \times 4 = 4^3 = 64$$

In any experiment, if there are n trials, each of which can have k mutually exclusive and exhaustive results, then the total number of distinct outcomes or sequences of observations will be k^n. In the roll of a cubic die, for example, the number of possible results is 6, or $k = 6$. If you roll twice ($n = 2$), there will be 6^2 possible outcomes; if you roll three times, there will be 6^3 possible outcomes; etc.

2-2-2 Permutations

A *permutation* is an ordered arrangement; the permutation of a set of objects is the ordered arrangement of these objects. If there are n objects in the set, the number of permutations will depend on r, the number of objects you want to select and arrange.

First, let us assume that r is less than n. Suppose that there are five candidates ($n = 5$) for three positions ($r = 3$), namely, a president (P), a vice president (V), and a secretary (S). In how many ways can these three positions be filled? The answer to this question is shown as follows:

Positions	$P \quad V \quad S$
Ways to fill positions	$5 \times 4 \times 3 = 60$

There are evidently five ways to select a president ($n = 5$). Then there are four candidates ($n - 1$) remaining, so there are four ways to select a vice president. Afterward, three ways are left to select a secretary. Thus the total number of ways to fill the three positions is 60. In general, out of n objects, the number of permutations of r objects, written ${}_nP_r$, is given by

$$
\begin{aligned}
{}_nP_r &= n(n-1)(n-2)\cdots[n-(r-1)] \\
&= n(n-1)(n-2)\cdots(n-r+1) \qquad (2\text{-}3)
\end{aligned}
$$

We may express ${}_nP_r$ in terms of the symbol $n!$ (n factorial), which is

$n(n-1)(n-2)(n-3)\cdots(3)(2)(1).$

Since the last factor in the above equation is $(n-r+1)$, it is necessary to multiply the right side of the equation by $(n-r)!$ to make it $n!$ In order to keep the quantity unchanged, we need to divide $n!$ by $(n-r)!$ Thus we have[1]

$$
\begin{aligned}
{}_nP_r &= \frac{n(n-1)(n-2)\cdots(n-r+1)(n-r)(n-r-1)\cdots(2)(1)}{(n-r)(n-r-1)\cdots(2)(1)} \\
&= \frac{n!}{(n-r)!} \qquad (2\text{-}4)
\end{aligned}
$$

Next, let us assume that r is equal to n. If the total number of n objects is equal to the number of objects to be selected and arranged, the number of permutations is simply $n!$ If the above three positions are to be filled from a committee of three persons A, B, and C, the total number of different ways to fill these three positions will be 3!, or 6. The possible arrangements are:

P	*V*	*S*
A	*B*	*C*
A	*C*	*B*
B	*A*	*C*
B	*C*	*A*
C	*A*	*B*
C	*B*	*A*

Each permutation is an arrangement. The sample space in such a problem is the set of all possible arrangements with order taken into consideration. If three persons A, B, and C are to fill three positions, president, vice president, and secretary, then the above-mentioned six arrangements form the sample space. Similarly, if there are five candidates for the three positions, the sample space will consist of all the $5(4)(3) = 60$ arrangements.

2-2-3 Combinations

If we are interested only in the number of ways that r objects can be selected out of a total of n objects, and are not interested in the order or arrangement of them, our problem will be one of *combination*. In the last section we learned that

[1] Recall that $n! = n(n-1)(n-2)\cdots(n-r)! = n(n-1)(n-2)\cdots 3(2)(1)$. For instance, $5! = 5(4)(3!) = (5)(4)(3)(2)(1) = 120$. By definition $0! = 1$.

the number of ways of selecting r objects from n objects and ordering them is $n!/(n-r)!$. Now let $\binom{n}{r}$ be the number of combinations of n things taken r at a time irrespective of order. We have noted that for each set of r things there are $r!$ permutations. Since a combination of r things is a set with r elements, $\binom{n}{r}r!$ must be equal to the number of permutations of n things taken r at a time. Thus we have

$$\binom{n}{r}r! = \frac{n!}{(n-r)!}$$

or

$$\binom{n}{r} = \frac{n!}{(n-r)!\,r!} \tag{2-5}$$

Suppose that there are 20 people and a committee of 3 is to be selected from them. In how many ways can this be done? It is evident that the order of the selected 3 people is of no significance. Therefore this is a problem of combination. The number of ways to select this committee is

$$\binom{20}{3} = \frac{20!}{(20-3)!\,3!} = \frac{20!}{17!\,3!} = \frac{20 \times 19 \times 18}{3 \times 2 \times 1} = 1{,}140$$

That is, each combination of 3 people constitutes an elementary event, and there are altogether 1,140 elementary events in the sample space.

It is important to remember the distinction that a permutation is an *ordered arrangement* of objects, and a combination is a *set* of objects irrespective of the order in which the objects appear. In a problem of a combination of n objects taken r at a time, any combination of r objects is *a subset of the universal set of n objects.* The sample space in such a problem is the set of all subsets of r objects each, or the set of all combinations of r objects taken from a total of n objects.

Suppose that 10 candidates are running for the board of directors of a certain corporation, which consists of 7 members. In how many ways can 7 directors be chosen from a total of 10 candidates? To answer this question all we need to know is the number of subsets of 7 people each that can be formed from a set of 10 people. It is the number of combinations of 10 people taken 7 at a time. Thus we have

$$\binom{10}{7} = \frac{10!}{(7!)(3!)} = \frac{10(9)(8)}{3(2)(1)} = 120$$

That is, there are 120 different ways of selecting 7 directors for the board from a total of 10 candidates.

In determining the number of combinations, sometimes *additions* are required and sometimes *multiplications*, depending on the particular problem. For clarifying these ramifications, let us consider another example.

Suppose that the work force of a certain department consists of 10 men and 5 women. A task force of four persons is to be selected to carry out a particular project. Find (*1*) the number of ways to select two men *and* two women for the task force, and (*2*) the number of ways to select *either* four men *or* four women.

1. Each way of selection here is a combination since the order in which a person is selected is immaterial. The number of combinations of 10 men taken 2 at a time is

$$\binom{10}{2} = \frac{10(9)}{2(1)} = 45$$

and the number of combinations of 5 women taken 2 at a time is

$$\binom{5}{2} = \frac{5(4)}{2(1)} = 10$$

Since there are 45 ways to perform one act and 10 ways to perform another act, by the law of multiplication the total number of ways to carry out both acts is

$$45(10) = 450$$

That is, the number of ways to select two men *and* two women for the task force is 450.

2. There are $\binom{10}{4}$ ways to select four men and $\binom{5}{4}$ ways to select four women. Since the question includes the expression "either . . . or . . . ," its answer requires the union of the two procedures of selection. Thus the number of ways to select four men *or* four women for the task force is

$$\binom{10}{4} + \binom{5}{4} = \frac{10(9)(8)(7)}{4(3)(2)(1)} + \frac{5}{1} = 210 + 5 = 215$$

Exercise 2-2

1 Use your own words to explain the difference between permutations and combinations.

2 Five dice are thrown. In how many possible ways can they fall?

3 A coin is tossed four times. In how many possible ways can it fall in the four tosses? Use a tree diagram to show all the possible outcomes.

4 California auto license plates have three letters followed by three digits or three digits followed by three letters. How many distinct plates can be made in California?

5 The policyholders of a life insurance company are classified by age—under thirty years (A_1), between thirty and forty-five years (A_2), and over forty-five years (A_3)—, by marital status—single (S), married (W), and unmarried (U)—, and by sex— male (M) and female (F). In how many ways can the policyholders be classified? What constitutes the sample space of this experiment? Construct a tree diagram to represent the sample space.

6 A restaurant has a menu with a choice of soup (S) or salad (A) for the appetizer, a choice of beef (B), pork (P), or other meat or seafood (O) for the main dish; and a choice of cake (C), pie (I), or fruit (F) for dessert. A complete dinner includes one choice in each of the three courses. What constitutes the sample space of this experiment? Construct a tree diagram to represent the sample space of all possible complete dinners.

7 Given that a set X has five elements, how many permutations are there if the elements of X are taken two at a time!?

8 If 10 books are being considered for a literary award, in how many ways can the first 3 different prizes be awarded?

9 In the Miss America Beauty Pageant, 15 semifinalists are usually chosen first, and then 5 finalists are selected. In how many ways can the 5 positions, Miss America, a first runner-up, a second runner-up, a third runner-up, and a fourth runner-up, be filled from the 15 semifinalists?

10 If each letter can be used only once, how many permutations of two letters can be taken from the English alphabet? How many three-letter permutations? Answer these two questions under the assumption that each letter can be used more than once.

11 The Far West Football League consists of six teams. How many different outcomes can there possibly be for the season? (Assume that no team finishes the season in a tie with another team.)

12 From a deck of ten cards numbered 0, 1, 2, . . . , 9, five cards are drawn. In how many different orders can the five cards be drawn with replacement? Without replacement?

13 The board of directors of the ABC Company consists of 15 members. In how many ways can the 3 officers, chairman, vice chairman, and secretary, be selected?

14 From a group of seven candidates, a five-man committee is to be selected. In how many ways can this be done? If the seven candidates are to fill five different positions, in how many ways can the positions be filled?

15 How many 5-man basketball teams can be chosen from a squad of 12 men if the positions played by the members of the squad are not considered?

16 A club has 15 members. In how many ways can a committee of 3 members be selected?

17 Give an expression to show the number of different bridge hands that can be dealt.

18 A statistics class contains 30 students, 24 men and 6 women. In how many different ways can a committee of 4 students be selected? In how many ways if this committee must include 2 women? One woman?

19 In the Senate of a certain state there are 20 Democrats and 10 Republicans. A committee of 5 senators is to be selected. In how many ways can this be done? In how many ways if there are to be 3 Democrats and 2 Republicans? In how many ways if all are from the same party?

2-3
Computation of Probabilities

In this section we attempt to see how the counting rules discussed in the last section can be applied to the computation of probabilities. First, let us review the meaning of the word probability.

Earlier in this chapter, we discussed the concept of a random event. An event is referred to as "random" because of the uncertainty of its occurrence. A widely used method of measuring the uncertainties of random events is to assign them *probabilities*. We are not certain, for instance, that a football team X will beat another team Y, but we may say that the probability of X's beating Y is 0.80. The probability assigned reflects the *strength of our belief* regarding the outcome that X will beat Y, and thus represents our *subjective evaluation*. We cannot be sure that X will win in any particular game, but it would be fair to bet $8 against $2, or at *odds* of 8 to 2, that X will beat Y in any game. In fact, the concepts of probabilities and odds are closely related. If you consider it equitable to bet m dollars against n dollars that a certain event will happen, then you are actually assigning the event a probability $m/(m+n)$. Thus if you consider it equitable to bet $8 against $2 that X will beat Y in the next game, you are in fact assigning to the event that X will win the game a probability of $8/(8+2) = \frac{8}{10} = 0.8$.

We can also view the probability of an event as the *relative frequency*, which means the *proportion of the times*, that the event will occur in the long run. For instance, if we say the probability that a product turned out by a given production process will be defective is 0.05, we mean that in the long run 5 percent of all the products turned out by the production process will be defective. If we say that the probability of drawing a spade from an ordinary deck of playing cards is 0.25, we mean that in the long run one-fourth of the cards drawn will be spades when the card is replaced and the deck is shuffled before each of the next draws. We shall now examine the procedure of determining the probability of an event when the concept of a probability is the proportion of the times that the event will happen in the long run.

2-3-1 Computation procedure

In many problems, the computation of probabilities is merely a matter of counting. This is especially true when each of the elementary events has the same probability of occurring. Although there is no generalization which can apply to every problem, familiarity with the following procedure will prove helpful.

Identify the sample space In any problem it is important to know the sample space or the total number of elementary events in the sample space. In other words, you must know the total number of distinct outcomes that may possibly occur relative to the problem under consideration.

As previously shown, an elementary event may be a sequence of observations. Although such a sequence or series of observations apparently has more than a single outcome, the nature of the problem may require that the entire sequence be treated as an elementary event. For instance, the toss of one coin three times would have the following possible sequences, or sample space:

Sequence	*First toss*	*Second toss*	*Third toss*
1	H	H	H
2	H	H	T
3	H	T	H
4	T	H	H
5	H	T	T
6	T	H	T
7	T	T	H
8	T	T	T

It is convenient to treat each of these eight sequences as an elementary event.

In general, if a trial of some experiment leads to one of n mutually exclusive and exhaustive consequences, designated $C_1, C_2, \ldots, C_n$, and if the trial is repeated t times, then the outcome of any series of t trials will be $(C_i^1, C_i^2, \ldots, C_i^t)$ for $i = 1, 2, \ldots, n$. Each superscript represents a trial in sequence, or a position in which one of the n possibilities may appear. For instance, one possible outcome, or sequence, in the above example is (H, T, H), which can be considered (C_2^1, C_1^2, C_2^3), where the subscript 1 denotes T and 2 denotes H.

Define the event The next step is to determine the particular event or events for which you want to compute the probabilities. You need to know the number of elementary events that make up the particular event. In the roll of two ideal dice, for instance, if the event is $X = 7$, the number of elementary events will be six, namely (1, 6), (2, 5), (3, 4), (4, 3), (5, 2), and (6,1). Note that each elementary event in this case is a sequence of two observations ($t = 2$), and that each of the two positions may be filled with one of six numbers ($n = 6$) representing the six faces of the die.

Calculate the probability of the event If the elementary events are equally probable, that is, if they have the same probability of occurring, the probability of an event A is simply the ratio of the number of elementary events in A to the total number of elementary events possible. Let $P(A)$ designate the probability that the event A will occur and S designate the sample space; then we have

$$P(A) = \frac{\text{number of elementary events in } A}{\text{total number of elementary events in } S} \tag{2-6}$$

In the roll of two dice, for example, since the total number of elementary events in the sample space is 36, and the sum 7 can be obtained in 6 of the 36 possible outcomes, the probability of obtaining a sum 7 is therefore $\frac{6}{36} = \frac{1}{6}$.

2-3-2 Some applications

The above-mentioned rules and procedures are often applied in solving practical probability problems. We now present three examples to illustrate their application.

Suppose that the ABC Company has developed a new product and wishes to conduct a nationwide contest for naming it. Anyone who wants to enter the contest must submit a five-letter word—the second and the fourth letters are to be consonants, and the other three vowels. Letters may be repeated. A person may submit more than one word; however, his entry will be void if the same word has previously been submitted. One word will be selected at random to determine the winner. The primary objective of the contest is to advertise the new product, and so the company may not use the word thus selected. Assume that a certain person has submitted 10 valid words. What is his chance of winning the contest?

In this problem each possible word is a sequence, that is, an elementary event. The sequence has five positions, in the first of which are 5 (number of vowels) possible observations, in the second 21 (number of consonants) observations, and so on. Thus we have

$$5 \times 21 \times 5 \times 21 \times 5 = 55{,}125$$

possible entries for the contest, each of which has the same chance of being selected. On the assumption that this was exactly the total number of valid entries the company received, the probability of one person's winning the contest would be

$$10 \times \frac{1}{55{,}125} = \frac{2}{11{,}025}$$

As another application, assume that from a well-shuffled ordinary deck of 52 cards a person draws a hand of 13 cards containing four aces, four kings, four queens, and one jack. What is the probability of drawing such a hand?

First of all, we must know the total number of different hands that can be possibly drawn. Remember that the order in which the cards appear in the hand is immaterial. By the rule of combination, the total number of combinations of 13 cards from a deck of 52 cards appears to be

$$\binom{52}{13} = \frac{52!}{13!\,39!}$$

In other words, there are 52!/(13! 39!) different possible hands each of which is equally probable. That is, the ratio

$$\frac{1}{52!/(13!\,39!)}$$

is the probability of drawing any given hand.

Since there are four cards of the same kind, by Eqs. (2-1) and (2-5) the total number of ways of combining four aces, four kings, four queens, and one jack is

$$\binom{4}{4}\binom{4}{4}\binom{4}{4}\binom{4}{1} = \frac{4!}{4!\,0!} \times \frac{4!}{4!\,0!} \times \frac{4!}{4!\,0!} \times \frac{4!}{1!\,3!}$$
$$= 1 \times 1 \times 1 \times 4 = 4$$

Thus the probability of drawing the above-specified hand is

$$\frac{4}{52!/(13!\,39!)} = \frac{13!\,39!\,4}{52!}$$

which can be worked out by writing out the factorials, canceling in denominator and numerator, and then dividing.

Consider another example. Suppose that a TV retail shop has received a shipment of 12 new sets, 8 of model A and 4 of model B. If four sets are sold, what is the probability that of the four sets sold two are of model A and two are of model B? What is the probability that all four sets sold are of the same model?

The total number of ways to select 4 objects from a total of 12 objects is

$$\binom{12}{4} = \frac{12(11)(10)(9)}{4(3)(2)(1)} = 495$$

The number of ways to select two objects from eight objects is $8!/(2!\,6!) = 28$, and the number of ways to select two objects from four objects is $4!/(2!\,2!) = 6$. Thus the number of ways to select two sets from the eight sets of model A and two sets from the four sets of model B is $28 \times 6 = 168$, and the probability of selling two sets of each model is

$$\frac{168}{495} = \frac{56}{165}$$

The number of ways to select four objects from eight objects is $8!/(4!\,4!) = 70$, and the number of ways to select four objects from four objects is $4!/(4!\,0!) = 1$. Consequently, the probability that all four sets sold are of the same model is

$$\frac{70+1}{495} = \frac{71}{495}$$

Exercise 2-3

1 A stockbroker believes that 3 to 1 are fair odds that the price of a certain common stock will rise by at least 10 percent during the next month. What is his subjective probability that this will be the case?

2 If it is widely believed that the odds are 7 to 3 that the president of a company will get to a meeting late, what is the probability that he will get there on time?

3 The manager of a retail store has observed that out of the 3,000 people who entered the store during the previous month, 450 made some purchases. What is the probability that the next customer who enters the store will buy something?

4 A club consists of 15 members, 3 of whom are women, named X, Y, and Z. If a committee of 3 members is selected at random, what is the probability that all 3 women will be included on the committee? What is the probability that X, Y, and Z will be selected, at random from the total membership, to the offices of president, vice president, and secretary-treasurer, respectively?

5 Three balanced coins are tossed. Find the probability that:

a Fewer than two heads will occur.
b More than two tails will occur.
c Two heads or two tails will occur.

6 Two ideal dice are rolled. Find the probability that:

a The sum of spots uppermost is less than 4.
b The sum of spots uppermost is greater than 7.
c The sum of spots uppermost is greater than 10 or less than 5.

7 An automobile dealer has 10 new cars, 3 of model *A*, 3 of model *B*, and 4 of model *C*. What is the probability of selling two cars of the same model if sales of different models are made at random?

8 A dealer of television sets will accept a shipment of 15 sets if a sample of 4 sets contains no defectives. What is the probability that he will accept the shipment if it contains 3 defective sets?

9 If you draw 13 cards at random from a well-shuffled ordinary deck of cards, find the probability of drawing a perfect bridge hand, that is, a hand containing 13 cards of the same suit.

10 The board of directors of a corporation contains nine members; five are known to be in favor of the acquisition of another company and four are against it. When a formal board meeting is convened to consider the matter, only five members are present. A decision to purchase the company requires a simple majority of those present. What is the probability that the acquisition will be made?

11 In a poker hand of 5 cards drawn from an ordinary deck of 52 cards, give an expression for the probability of drawing a flush, that is, all 5 cards of the same suit.

Chapter 3

Probability rules and distributions

The discussion of the preceding chapter dealt primarily with the counting of random events. It showed how the *rules of counting* can be applied to the computation of probabilities. The principal purpose of this chapter is to review the widely used *rules of probability* and the procedures by which these rules are applied. Furthermore, the basic probability distributions will be considered.

Among the probability rules to be examined are, first of all, the *addition rules*, which include the *special* and *general* addition rules. The *multiplication rules*, which are also often classified as special and general, will be considered next. The discussion of the *rule of Bayes'*, which is an extension of the general multiplication rule, will then follow. The two basic probability distributions, *discrete* and *continuous*, will be briefly illustrated in the final section of the chapter.

3-1 Addition Rules

As indicated in the preceding chapter, the word *probability* is used in connection with a random event. Thus the rules of probability must be stated in terms of events. Let S be the sample space of an experiment that is partitioned into mutually exclusive and exhaustive events, which may be elementary or compound. The probability of any event E in S then is a real number associated with E such that the following theorems are satisfied:

Theorem 1 The probability of any event must be 0 or a positive real number not greater than 1. Symbolically,

$$0 \leq P(E) \leq 1$$

Theorem 2 The probability of any event can be equal to 1. If an event is identical with the sample space, or $E = S$, then the probability of such an event is 1, which implies that the event is certain to occur. That is,

$$P(S) = 1$$

Theorem 3 The probability of an "impossible" event, or an empty set, is 0. A probability of 0 implies that the event is certain not to occur. This is to say that

$$P(\emptyset) = 0$$

Theorem 4 If E_1 and E_2 are two events in S and if the occurrence of E_1 implies that E_2 occurs, that is, E_1 is a subset of E_2, then the probability of E_1 is less than or equal to the probability of E_2; or:

$$\text{If } E_1 \subseteq E_2, \text{ then } P(E_1) \leq P(E_2)$$

The above theorems are the most fundamental rules of probability. We accept them as truth without proof, even though their proofs can be carried out. On the basis of these theorems, other probability rules have been derived. Among the latter are the addition rules, about which this section is primarily concerned.

3-1-1 Special addition rule

A very important rule which can apply to many probability problems is known as the *special addition rule.* It is referred to as *special* because it applies *only to events that are mutually exclusive.* The rule may be briefly stated as follows:

If two events are mutually exclusive, then the probability of the union of the two events is equal to the sum of their respective probabilities. Symbolically, if the two events are designated E_1 and E_2, we have

$$P(E_1 \cup E_2) = P(E_1) + P(E_2) \tag{3-1}$$

Let us again use the example of rolling a cubic die to illustrate this rule of probability. The sample space of this experiment is $S = \{1, 2, 3, 4, 5, 6\}$. Assume that the die is an ideal one, that is, each of the six sides has the same probability of showing up. If E_1 denotes the event of two or fewer spots' showing up and E_2 denotes the event of four or more spots' showing up, then E_1 and E_2 are mutually exclusive. As shown in Fig. 3-1, the probability of their union is

$$\begin{aligned} P(E_1 \cup E_2) &= P(E_1) + P(E_2) \\ &= \frac{2}{6} + \frac{3}{6} = \frac{5}{6} \end{aligned}$$

As another example, if the probabilities that an automobile dealer will sell two and three cars next Saturday are respectively 0.35 and 0.25, then the prob-

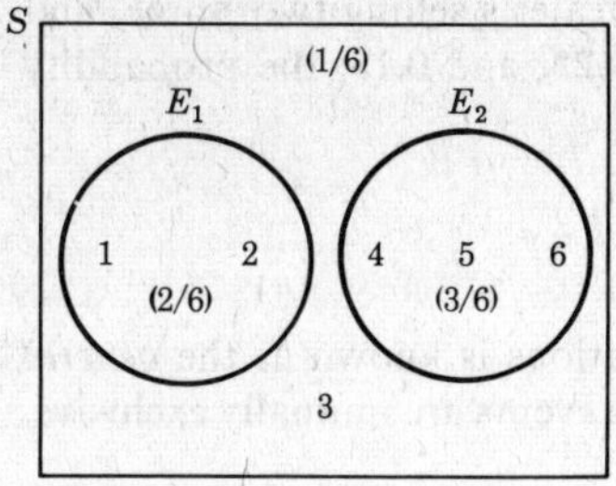

Figure 3-1
Probability of the union of two mutually exclusive events.

ability that he will sell two *or* three cars next Saturday is $0.35 + 0.25 = 0.60$. Equation (3-1) is applied here, because the event of selling two cars and that of selling three cars cannot both happen next Saturday; they are mutually exclusive.

Another important rule closely related to Eq. (3-1) is one for *complementary events*. Let E be an event in the sample space S, and E' be the complement of E. Then the sum of the probabilities of E and E' is always equal to 1. That is,

$$P(E) + P(E') = 1 \tag{3-2}$$

For instance, in the experiment of rolling an ideal cubic die if E denotes the event that an odd number of spots occurs and E' denotes the event that an even number of spots occurs, then the sum of the probabilities of these two complementary events is

$$P(E) + P(E') = \frac{1}{2} + \frac{1}{2} = 1$$

Similarly, if E designates the event that two or fewer cars will be sold and E' the event that three or more cars will be sold next Saturday by an automobile dealer, it is obvious that the sum of the probabilities of E and E' must be equal to 1, since these two events comprise any possible number of cars to be sold.

Furthermore, Eq. (3-1) can be expanded to include more than two events. If there are n events that are mutually exclusive, then the probability of the union of these events is equal to the sum of their respective probabilities. Symbolically, if $E_1, E_2, \ldots, E_n$ designate these events, we have

$$P(E_1 \cup E_2 \cup \cdots \cup E_n) = P(E_1) + P(E_2) + \cdots + P(E_n) \tag{3-3}$$

In the experiment of rolling an ideal die, for instance, if E_1, E_2, and E_3 designate the events that one or two spots, three or four spots, and five spots occur, respectively, then the probability of the union of these three events, that is, the probability of any one of the five sides' showing up, is the sum of the probabilities of the three events. Accordingly, we have

$$\begin{aligned} P(E_1 \cup E_2 \cup E_3) &= P(E_1) + P(E_2) + P(E_3) \\ &= \frac{2}{6} + \frac{2}{6} + \frac{1}{6} = \frac{5}{6} \end{aligned}$$

Likewise, if the probabilities of the automobile dealer's selling two, three, and four cars next Saturday are, respectively, 0.35, 0.25, and 0.15, the probability that two, three, *or* four cars will be sold is

$$0.35 + 0.25 + 0.15 = 0.75$$

3-1-2 General addition rule

Another addition rule which has broader applications is known as the *general addition rule.* It applies regardless of whether the events are mutually exclusive. Briefly, the rule may be stated as follows:

> *The probability of the union of two events is equal to the sum of their individual probabilities* minus *the probability of their intersection regardless of whether they are mutually exclusive. Symbolically, if the two events are designated E_1 and E_2, we have*
>
> $$P(E_1 \cup E_2) = P(E_1) + P(E_2) - P(E_1 \cap E_2) \tag{3-4}$$

Referring again to the experiment of rolling an ideal die, let E_1 be the event of three or fewer spots' showing up and E_2 be the event of three or four spots' showing up. As shown in Fig. 3-2, the event of three spots belongs to *both* E_1 and E_2; it is the intersection of E_1 and E_2 and is counted twice in computing the individual probabilities of the two events. The probability of this intersection must be subtracted from the sum of the individual probabilities in order to obtain the correct probabillity of the union of the two events. Thus, by Eq. (3-4),

$$P(E_1 \cup E_2) = \frac{3}{6} + \frac{2}{6} - \frac{1}{6} = \frac{5}{6} - \frac{1}{6} = \frac{4}{6} = \frac{2}{3}$$

The general addition rule also applies to events that are mutually exclusive. Refer to Fig. 3-1 above. Since the two events are mutually exclusive, their intersection is a null set. The probability of a null set is 0 (theorem 3). Consequently, the probability of the union of E_1 and E_2 in Fig. 3-1 is

$$\frac{2}{6} + \frac{3}{6} - 0 = \frac{5}{6}$$

which is the same as obtained under the special addition rule.

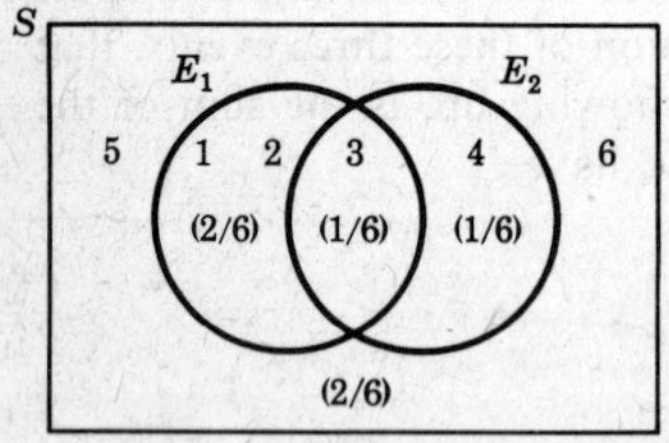

Figure 3-2
Probability of the union of two non-mutually exclusive events.

Consider another example. Suppose that 80 percent of all Americans who are vacationing in Europe will visit Paris, 70 percent of them will visit Berlin, and 60 percent of them will visit both Paris and Berlin. What is the probability that an American vacationing in Europe will visit Berlin or Paris or both? What is the probability that he will visit neither city?

The probability that an American vacationing in Europe will visit either Berlin or Paris or both is obtained by applying Eq. (3-4) as follows:

$$0.80 + 0.70 - 0.60 = 1.50 - 0.60 = 0.90$$

Consequently, the probability that he will visit neither Berlin nor Paris is

$$1.00 - 0.90 = 0.10$$

Exercise 3-1

1 Explain why there is a mistake in each of the following statements:

a The probabilities that a certain truck driver would have no, one, and two or more accidents during the year are 0.90, 0.02, and 0.09.

b A stockbroker claims that the probability that a stock's price will go up is 0.38, remain unchanged 0.52, and go down 0.12.

c The probabilities that there would be no, one, two or three rainy days next week are 0.54, 0.28, 0.14, and 0.04, respectively.

2 The probabilities that during the next week an automobile dealer will sell 0, 1, 2, 3, or 4 more cars are 0.14, 0.21, 0.48, 0.12, and 0.05, respectively. If X is the number of cars he will sell next week, find the probability that:

a $X = 0$, 1, or 3. *d* $X < 3$.
b $X = 2$ or 3. *e* $X > 2$.
c $X < 4$. *f* $X \geq 3$.

3 Suppose that each of the 52 cards of an ordinary deck has the same probability of being drawn. What is the probability of drawing:

a A black king? *d* Either a 10 or a face card?
b Either a spade or the king of diamonds? *e* A red jack or a black king?
c A queen, king, or ace? *f* Either a spade or an ace?

4 A die is biased so that the probability of each side of the die turning up is proportional to the number of dots on it; that is, a 4 is twice as likely as a 2, a 3 is three times as likely as a 1, a 6 is twice as likely as a 3, etc. Find the probability of obtaining:

a A number less than 3.
b A number equal to or less than 4.
c An odd number.
d A number greater than or equal to 4.

5 Among the 200 employees of a department 150 are college graduates, 60 of the whole group devote at least part of their time to statistical work, and 40 of the 150 college graduates devote at least part of their time to statistical work. If one of these employees is chosen at random, what is the probability that the one chosen is not a college graduate and does not do any statistical work? [*Note:* Draw a Venn diagram, and represent the two groups of employees by two circles G (college graduates) and S (doing statistical work).]

6 According to her grandfather's will, Mary is to receive $10,000 from his estate upon his death. Her grandfather recently passed away. Mary is considering to use the money for one of the following alternatives: (1) college education, (2) investment, or (3) an extended around-the-world tour. The probabilities of taking these alternatives are, respectively, 0.5, 0.3, and 0.1.

a What is the probability that she will do one of these things?

b What is the probability that she will forget these three possibilities and spend the money for something else?

7 An integer is randomly selected from the first 30 positive integers. What is the probability that the integer drawn is divisible by 4 or 6?

8 A college graduate applies for employment to company X and another company Y. It is estimated that the probability of his being hired by X is 0.7, the probability of his being hired by Y is 0.5, and the probability of at least one of his applications' *being rejected* is 0.60. What is the probability that he will then be employed by at least one of the companies?

9 Equation (3-4) can be expanded for computing the probability of occurrence of *at least one* of three events G, M, and C by using the following formula:

$$P(G \cup M \cup C) = P(G) + P(M) + P(C) - P(G \cap M) - P(G \cap C) - P(M \cap C) + P(G \cap M \cap C) \qquad (3\text{-}5)$$

Use the Venn diagram in Prob. 11 of Exercise 2-1 to verify that each region within the three circles is counted only once in the expression on the right of Eq. (3-5).

10 A card is drawn at random from a deck of ordinary playing cards. What is the probability that the card drawn is a diamond (D), a face card (F), or a 9 (N)? [*Note:* Use Eq. (3-5) in Prob. 9 above.]

3-2 Multiplication Rules

In this section we shall discuss the *multiplication rules* and the problems to which these rules are applied. They are *rules of probability*, not to be confused with the law of multiplication discussed in Sec. 2-2-1, which is a *rule of counting* events. Like the addition rules, the multiplication rules of probability can be classified into two types. The *special multiplication rule* applies only to those events that are *independent* in nature, and the *general multiplication rule* applies to *both independent and dependent events*.

3-2-1 Special multiplication rule

Inasmuch as the special rule of multiplication applies only to events that are independent, we shall in this subsection focus our attention on problems with only *independent events*. Two or more events are said to be *independent* if they have no relation to one another. More specifically, if A and B represent two events and if the occurrence of A does not affect, and is not affected by, the occurrence of B, then A and B are said to be independent. If A and B are independent events, the probability that *both A and B* will occur is equal to the product of their respective probabilities. That is,

$$P(A \cap B) = P(A) \times P(B) \tag{3-6}$$

Equation (3-6) is referred to as the *special multiplication rule* because its application is limited to independent events only.

Suppose, for instance, that two ideal coins are tossed. Let A be the event "heads on the first coin" and B be the event "heads on the second coin"; then the probability that both coins will turn up heads is $\frac{1}{2} \times \frac{1}{2} = \frac{1}{4}$. Equation (3-6) is applied here because the occurrence of A in no way affects the probability that B will occur, and vice versa; in other words, A and B are independent. Similarly, in the roll of two ideal dice if A designates the event that an odd number of spots will show up on the first die and B designates the event that an even number of spots will show up on the second die, then because of the independence between A and B, the probability that both A and B will occur is $\frac{1}{2} \times \frac{1}{2} = \frac{1}{4}$.

Let us use another example to illustrate this concept still further. Suppose that a company hires both male and female employees for virtually the same kind of work. After a period of employment, some of each sex group are promoted, and some are not. Let A and A' stand for male and female employees, and B and B' for promoted and not promoted. Table 3-1 shows the proportions of the company's employees among these classes. Thus the probability of selecting a male employee from the company, or $P(A)$, is 0.60, and that of selecting a female, or $P(A')$, is 0.40. The probabilities of finding a promoted and a nonpromoted person, or $P(B)$ and $P(B')$, are 0.30 and 0.70, respectively. Since these probabilities appear in the margins of Table 3-1, they are called *marginal probabilities.* Observe that the marginal probability of A is the probability of the event A; it is the sum of the probabilities in which A is involved.

Two or more events form a *joint event* if all of them occur. The probability that two or more events all occur is called the *joint probability* of these events. In the above example, the events of *A and B* form a joint event, and the probability that both will occur is a joint probability, or the probability of their intersection. Thus,

$$P(A \cap B) = P(A) \times P(B) = 0.60 \times 0.30 = 0.18$$

Table 3-1

Proportions of the employees among various classes—independent events

Promotional status	*Sex* A	A'	*Total*
B	0.18	0.12	0.30
B'	0.42	0.28	0.70
Total	0.60	0.40	1.00

Table 3-1 is deliberately arranged so that sex and promotional status of employees are independent of each other. Thus the probability of "male" *and* "promoted" is the product of their separate or marginal probabilities. Similarly, the probability of each of the other joint events is the product of their corresponding marginal probabilities.

Sex and promotional status are independent because knowing one does not affect the probability that the other will occur. Given a male employee, the probability of his being promoted is 0.18/0.60 or 0.30. The latter is exactly the probability of anyone being promoted without specifying sex. This leads us to another definition of independence: If events A and B are independent, then

$$P(A) = P(A|B)$$

or

$$P(B) = P(B|A)$$

where $P(A|B)$ is the probability of A when it is known that event B has occurred, and $P(B|A)$ is the probability of B, after event A has occurred. $P(A|B)$ [or $P(B|A)$] is called *conditional* probability, for it is the probability of A (or B) on the condition that B (or A) has occurred.

3-2-2 General multiplication rule

It has been pointed out that if events A and B are *independent*, then the joint probability of A and B is equal to the product of their respective probabilities. That is, $P(A \cap B) = P(A) \times P(B)$. Evidently, if A and B are not independent, or if A and B are *dependent*, then the joint probability of A and B is no longer equal to the product of their respective probabilities. That is, for dependent events, $P(A \cap B) \neq P(A) \times P(B)$. This implies that the occurrence of A will affect the probability that B will occur. Accordingly,

$$P(A) \neq P(A|B) \quad \text{and} \quad P(B) \neq P(B|A)$$

Suppose that a reexamination of the personnel records of the above-mentioned company shows that the marginal probabilities in Table 3-1 remain as they are, but the joint probabilities are different, as indicated in Table 3-2. As shown in this table, employees' sex and promotional status are not indepen-

Table 3-2

Proportions of the employees among various classes—dependent events

Promotional status	*Sex*		
	A	*A′*	*Total*
B	0.12	0.18	0.30
B′	0.48	0.22	0.70
Total	0.60	0.40	1.00

dent. The joint probability of male and promoted, or 0.12, is *not* equal to the product of their respective marginal probabilities, namely, 0.18. Given the event A, the conditional probability of B, or $P(B|A)$, is not equal to $P(B)$. Likewise, after B is known, the probability of A, or $P(A|B)$, is different from $P(A)$. That is,

$$P(B|A) = \frac{0.12}{0.60} = 0.20 \neq P(B)$$

p(B) = 0.30

and

$$P(A|B) = \frac{0.12}{0.30} = 0.40 \neq P(A)$$

p(A) = 0.60

For dependent events A and B, the joint probability of both events is given by

$$P(A \cap B) = P(A) \times P(B|A) \tag{3-7}$$

or

$$P(A \cap B) = P(B) \times P(A|B) \tag{3-8}$$

Equation (3-7) or (3-8) is referred to as the *general multiplication rule*, for it applies to both independent and dependent events. When A and B are independent, $P(B|A) = P(B)$ and $P(A|B) = P(A)$, and Eqs. (3-7) and (3-8) will be identical with Eq. (3-6).

From Eq. (3-7) we obtain the conditional probability of B, given A, as follows:

$$P(B|A) = \frac{P(A \cap B)}{P(A)} \tag{3-9}$$

Likewise, the conditional probability of A, given B, is

$$P(A|B) = \frac{P(A \cap B)}{P(B)} \tag{3-10}$$

We may use these formulas to calculate conditional probabilities for the above example. The probability of "promoted" after a male employee has been identified is

$$P(B|A) = \frac{P(A \cap B)}{P(A)} = \frac{0.12}{0.60} = 0.20$$

Similarly, if a promoted employee has been chosen, the probability that the person is a male is calculated as follows:

$$P(A|B) = \frac{P(A \cap B)}{P(B)} = \frac{0.12}{0.30} = 0.40$$

The joint probability of both events, a promoted male employee, can be calculated by either Eq. (3-7) or (3-8):

$$P(A \cap B) = P(A) \times P(B|A)$$
$$= 0.60 \times 0.20 = 0.12$$

or

$$P(A \cap B) = P(B) \times P(A|B)$$
$$= 0.30 \times 0.40 = 0.12$$

For visualization, the tree diagram may be used to illustrate the procedure of computing joint probabilities for dependent events. As shown in Fig. 3-3, the marginal probabilities of male (A) and female (A') are presented as two main branches of the tree. On the condition that a male or female employee has been selected, the conditional probabilities are presented as branches growing out of the main branch. Each joint probability is obtained by multiplying the two probabilities along the sequence of a main branch and one of its subsequent branches.

Consider another example. Suppose that two groups of machines A and B are designed to produce identical nails. Group A consists of three machines, and group B consists of two machines. It is known that 3 percent of the nails produced by machines of group A are defective (D), and that 98 percent of the nails produced by machines of group B meet specifications (D'). If a machine is selected at random and one of the nails produced by the machine thus selected is inspected, what is the joint probability of selecting a machine from group A and a defective nail? What are the other possible joint probabilities?

Since $P(A) = \frac{3}{5} = 0.60$ and $P(B) = \frac{2}{5} = 0.40$ and the conditional probabilities are

$$P(D|A) = 0.03 \qquad P(D'|A) = 0.97$$
$$P(D|B) = 0.02 \qquad P(D'|B) = 0.98$$

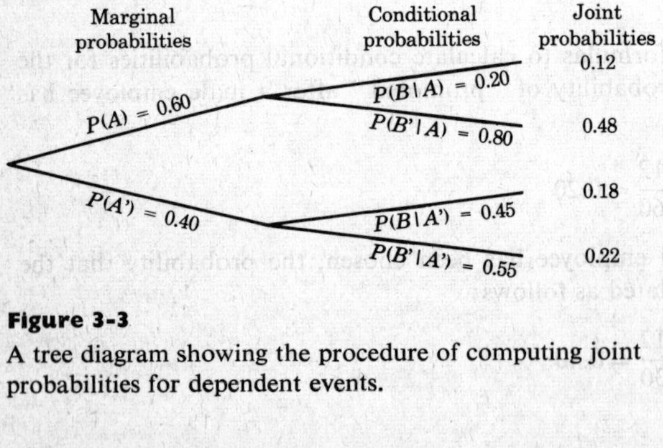

Figure 3-3
A tree diagram showing the procedure of computing joint probabilities for dependent events.

the joint probability of selecting a machine from group A and a defective nail is

$$P(A \cap D) = P(A) \times P(D|A) = 0.60 \times 0.03 = 0.0180$$

Similarly, the other possible joint probabilities are

$$P(A \cap D') = P(A) \times P(D'|A) = 0.60 \times 0.97 = 0.5820$$

$$P(B \cap D) = P(B) \times P(D|B) = 0.40 \times 0.02 = 0.0080$$

and

$$P(B \cap D') = P(B) \times P(D'|B) = 0.40 \times 0.98 = 0.3920$$

When sampling is taken from a finite set of objects, the event of drawing a certain object *with replacement* and the event of a subsequent draw are *independent*, since the former does not affect the probability that the latter will occur. On the other hand, when sampling is taken *without* replacement, the event of drawing a certain object will affect the probability of an event specified in a subsequent draw; such events are therefore dependent in nature. The following example illustrates the probability computation for such dependent events.

Suppose that two cards are drawn in succession from a deck of 52 cards; the first card shall not be replaced when the second card is drawn. What is the probability that the first draw is a king and the second is a queen?

Since there are four kings in the deck of 52 cards, the probability of drawing a king is $\frac{4}{52} = \frac{1}{13}$. After the first draw there are 51 cards left in the deck, of which 4 are queens. Thus the probability of drawing a queen is $\frac{4}{51}$. Consequently, the probability of drawing a king and then a queen is

$$\frac{1}{13} \times \frac{4}{51} = \frac{4}{663}$$

Exercise 3-2

1 Explain the difference between the special and general multiplication rules.

2 Distinguish between each of the following pairs of terms:

a Independent event and dependent event.
b Joint probability and conditional probability.

3 Two ideal coins and an ideal die are tossed together, and X is the sum of the number of heads and the number of spots on the die. Find the probability for each of the following:

a $X = 8$. *c* $X = 4$. *e* $X > 6$.
b $X = 1$. *d* $4 \leq X \leq 6$.

4 A salesman X is chosen from one group of salesmen. The probability that X will meet the sales quota is 0.5. Another salesman Y is chosen from a different group of salesmen, independent of the other, and the probability of his meeting the quota is 0.3. Find the probability of each of the following:

a X will not meet the quota, but Y will. 0.15
b Both X and Y will meet the quota. 0.15
c At least one will meet the quota. 0.65
d Neither will meet the quota. 0.35
e Y will not meet the quota, but X will. 0.35
f At most one will meet the quota. 0.85

5 From an ordinary deck of cards we draw card X, put it back, shuffle, and draw a second card Y. Find the following probabilities:

a $P(X$ is a heart and Y is a king$)$.
b $P(X$ is a 4 and Y is a 5$)$.
c $P($neither X nor Y is a spade$)$.
d $P(X$ is a jack and Y is a jack$)$.
e $P(X$ is a face card and Y is a 10$)$.

6 Suppose that the digits 0, 1, 2, 3, 4, 5, 6, 7, 8, and 9 are each written on a card. The 10 cards are put into a hat and thoroughly mixed. Draw a card, and note its number X. Replace the card, mix them again, draw another card, and note its number Y. Graph the sample space on a two-dimensional plane, and then find the probability of:

a $P(X=3 \text{ and } Y=4)$.
b $P(X \neq 3 \text{ and } Y=3)$.
c $P(X \leq 3 \text{ and } Y>1)$.
d $P(X>3 \text{ and } Y>3)$.
e $(PX+Y<6)$.
f $P(X>7 \text{ and } Y>7)$.

7 Suppose that a box contains five red and three green balls. A ball X is drawn, and its color is noted. The ball is replaced before another ball Y is drawn. Find the following probabilities:

a $P(X$ is red and Y is green$)$.
b $P(X$ is red or green$)$.
c $P($either X or Y is green$)$.
d $P($neither X nor Y is green$)$.
e $P($both X and Y are red$)$.
f $P($at least one of the balls is green$)$.
g $P(X$ is red or Y is green$)$.
h $P(X$ is not red and Y is not green$)$.

8 A large company hires people either with or without college training for doing the same kind of work. The company records show that 40 percent of the employees have college training (C), or $P(C)=0.40$, and 60 percent of them have no college training (C'), or $P(C')=0.60$. The employees are rated by their supervisors 6 months after they were hired. The records indicate that the proportions of those rated excellent (R_1), average (R_2), and poor (R_3) are 0.30, 0.50, and 0.20, respectively; that is, $P(R_1)=0.30$, $P(R_2)=0.50$, and $P(R_3)=0.20$. If it is assumed that college training or the absence of it and performance rating are independent, what are the possible joint probabilities? What is the probability that an employee is rated excellent on the condition that he is college trained?

9 In Prob. 8 above it is assumed that the six possible joint probabilities are as follows:

$P(C \cap R_1)=0.08$ $\quad P(C' \cap R_1)=0.22$
$P(C \cap R_2)=0.24$ $\quad P(C' \cap R_2)=0.26$
$P(C \cap R_3)=0.08$ $\quad P(C' \cap R_3)=0.12$

a Is performance rating independent of college training or the absence of it?
b Find the conditional probabilities $P(R_1 | C)$, $P(R_2 | C')$, and $P(R_3 | C)$.

10 Suppose that we have two boxes: box I containing three black and five green balls; and box II containing two black, one green, and two red balls. A box is selected at random, and then one ball is drawn from the box thus selected. What is the probability that the ball selected will be a black ball? Use a tree diagram to show how you arrive at your answer.

11 Suppose that a special deck of cards contains six cards numbered 1, 2, 3, 4, 5, and 6. Let X be the number of the card drawn at random after the deck is shuffled. The card is *not* replaced; consequently, only five cards are left in the deck. Let Y be the number of the card at the second draw. Find the probability of:

a $X > 3$ *and* $Y > 3$.
b $X < 3$ *and* $Y < 3$.
c X is odd *and* Y is even.
d X is odd *and* Y is odd.

12 Consider each of the following poker hands of five cards drawn from a normal deck of cards:

a The hand contains four clubs and one spade. If the spade is discarded and another card drawn, what is the probability that the next card will be a club?
b The hand contains a 5, 6, 8, 9, and king. If the king is discarded and another card drawn, what is the probability that the next card will be a 7?
c The hand contains two aces, two kings, and a queen. If the queen is discarded and another card drawn, what is the probability that the next card will be an ace or a king?
d The hand contains a 5, 6, 7, 8, and 10. If the 10 is discarded, what is the probability that the next card will be either a 4 or a 9?

13 If the digits 0, 1, 2, 3, 4, 5, 6, 7, 8, and 9 are written on a card as in Prob. 6 above but the first card is *not* replaced when the second card is drawn, what is the sample space of this experiment? Find the following:

a $P(X$ and Y are both even).
b $P(X$ is even and Y is odd).
c $P(X \geq 7$ and $Y \geq 7)$.
d $P(X = 4$ and $Y = 5)$.
e $P(X \geq 7$ and Y is even).

3-3
The Rule of Bayes

Bayes' rule has gained wide use and popularity in recent years, and the concept is not a difficult one. Bayes' rule is merely a formula for computing conditional probabilities. When a joint-probability table such as Table 3-2 is given, it is a simple matter to compute conditional probabilities. The process is only the dividing of a joint probability by the marginal probability, which is the total of all the joint probabilities in a particular row or column, depending on the event that is assumed to be given. If a joint-probability table is not given, then it is necessary to set up such a table to determine the conditional probability in question. Let us use a brief example to illustrate the procedure.

Suppose that three bags are placed in a dark room. In bag 1, or B_1, there are six red balls (R) and four white balls (W); in bag 2, or B_2, there are seven red balls and three white balls; and in bag 3, or B_3, there are five red and five white balls. If someone went to the room and selected a bag at random from which he obtained a red ball, what is the probability that the red ball he has drawn is from B_1? From B_2? From B_3?

On the assumption that the bags are identical, the probability that any one of the three will be selected is $\frac{1}{3}$, which is the marginal probability. The conditional probabilities for red and white balls given each bag are shown in Table 3-3.

Table 3-3
Conditional probability

Bag	Red ball (R)	White ball (W)	Total
B_1	$P(R\mid B_1) = 0.60$	$P(W\mid B_1) = 0.40$	1.00
B_2	$P(R\mid B_2) = 0.70$	$P(W\mid B_2) = 0.30$	1.00
B_3	$P(R\mid B_3) = 0.50$	$P(W\mid B_3) = 0.50$	1.00

Our problem requires the computation of three conditional probabilities, namely, $P(B_1\mid R)$, $P(B_2\mid R)$, and $P(B_3\mid R)$. For this purpose we need to set up a joint-probability table, which, by Eq. (3-7) or (3-8), is obtained by multiplying the conditional probabilities by their corresponding marginal probabilities. That is,

$$P(B_1 \cap R) = P(B_1) \times P(R\mid B_1) \qquad P(B_1 \cap W) = P(B_1) \times P(W\mid B_1)$$
$$P(B_2 \cap R) = P(B_2) \times P(R\mid B_2) \qquad P(B_2 \cap W) = P(B_2) \times P(W\mid B_2)$$
$$P(B_3 \cap R) = P(B_3) \times P(R\mid B_3) \qquad P(B_3 \cap W) = P(B_3) \times P(W\mid B_3)$$

Since $P(B_1) = P(B_2) \times P(B_3) = \frac{1}{3}$, the computations require only the multiplication of $\frac{1}{3}$ by each of the cell entries in Table 3-3, which is shown in Table 3-4. Thus, the probability of drawing a red ball, or $P(R)$, is $\frac{18}{30}$, and that of drawing a white ball, or $P(W)$, is $\frac{12}{30}$. Our answers to this problem are obtained by dividing the probability of R into each of the joint probabilities in the R column as follows:

$$P(B_1\mid R) = \frac{\frac{6}{30}}{\frac{18}{30}} = \frac{6}{18}$$

$$P(B_2\mid R) = \frac{\frac{7}{30}}{\frac{18}{30}} = \frac{7}{18}$$

$$P(B_3\mid R) = \frac{\frac{5}{30}}{\frac{18}{30}} = \frac{5}{18}$$

Table 3-4
Joint probability

Bag	Red ball R	White ball W
B_1	$\frac{1}{3}(0.6) = \frac{6}{30}$	$\frac{1}{3}(0.4) = \frac{4}{30}$
B_2	$\frac{1}{3}(0.7) = \frac{7}{30}$	$\frac{1}{3}(0.3) = \frac{3}{30}$
B_3	$\frac{1}{3}(0.5) = \frac{5}{30}$	$\frac{1}{3}(0.5) = \frac{5}{30}$
Total	$\frac{18}{30}$	$\frac{12}{30}$

In general, if there are n B's and we wish to compute the conditional probability, or $P(B_i \mid R)$, for $i = 1, 2, \ldots, n$, Bayes' formula runs as follows:

$$P(B_i \mid R) = \frac{P(B_i \cap R)}{P(R)} = \frac{P(B_i)P(R \mid B_i)}{P(B_1)P(R \mid B_1) + P(B_2)P(R \mid B_2) + \cdots + P(B_n)P(R \mid B_n)} \tag{3-11}$$

where the B's are events of some type, and R is an event of another type, or *observation*. Thus, Bayes' formula is designed to compute the conditional probability *in reverse*. That is, if we know $P(R \mid B_i)$, we compute the probability $P(B_i \mid R)$. Since R is a sort of observation and $P(B_i \mid R)$ is the probability of B_i after an observation has been made, such a probability is often referred to as the *posterior*, or *a posteriori*, *probability*, while the probability of any B, or $P(B_i)$, is called the *prior*, or *a priori*, *probability*, suggesting that it is the probability *before* any observation is made.

Since the Bayes rule is very widely applied, and beginning students may find it not too easy to grasp the concept, let us use another example to illustrate its application.

For applying Bayes' formula, let us assume that a manufacturer has three machines, designated X, Y, and Z, which turn out identical parts. It is known that 5 percent of the parts produced by machine X are defective (D), and the percentages of defective parts turned out by Y and Z are 10 and 15 percent, respectively. The products of all three machines are mixed up; there is no way to identify each with the machine that produced it. It is known, however, that the three machines have equal capacity and are operated at the same rate. If a part is selected at random and found to be defective, what is the probability that this part was produced by machine Y? That is, what is $P(Y \mid D)$?

For this problem, we know

$$P(X) = \frac{1}{3} \qquad P(Y) = \frac{1}{3} \qquad P(Z) = \frac{1}{3}$$

$$P(D \mid X) = 0.05 \qquad P(D \mid Y) = 0.10 \qquad P(D \mid Z) = 0.15$$

Thus

$$P(Y \mid D) = \frac{P(Y) \times P(D \mid Y)}{P(X) \times P(D \mid X) + P(Y) \times P(D \mid Y) + P(Z) \times P(D \mid Z)}$$

$$= \frac{\frac{1}{3}(0.10)}{\frac{1}{3}(0.05) + \frac{1}{3}(0.10) + \frac{1}{3}(0.15)}$$

$$= \frac{1}{3}$$

Exercise 3-3

1 Use your own words to explain the difference between prior probability and posterior probability.

2 In Prob. 10 of Exercise 3-2, suppose that a person told you that he has a black ball drawn from one of the two boxes. Refer to the tree diagram showing all the possible joint probabilities, and then compute the conditional probability that the black ball he has drawn is from box I, or $P(\text{box I} \mid \text{black ball})$.

3 The detection of hepatitis is usually done through a kind of blood test. The reliability of the blood test is assumed to be as follows: Of people with hepatitis, 80 percent of the blood tests detect the disease, but 20 percent fail to detect it. Of people without hepatitis, 5 percent are erroneously diagnosed as having the disease, and 95 percent of the tests give the correct diagnoses. Suppose that one person is taken at random from a large group of people of whom only 1 percent have hepatitis, and a blood test shows that he has hepatitis. What is the probability that he actually does have the disease?

4 Suppose that there are three identical boxes, A, B, and C. Box A contains two pennies, B contains one penny and two nickels, and C contains one dime, two nickels, and two pennies. Suppose further that you choose a box at random and then take out a coin from the box that you have chosen. If the coin is a penny, what is the probability that the coin was taken from box A? From box B? From box C?

5 The sales manager of an insurance company told his newly hired salesman that about 30 percent of the people who inquire about life insurance policies actually buy such a policy, and 70 percent do not. According to the records of the company, 40 percent of the people who inquired and bought life insurance had an annual income between \$10,000 and \$30,000 while only 20 percent of those who inquired but did not buy life insurance had the same level of income. One person who inquired about insurance had an income of \$15,000. What is the probability that this man will buy an insurance policy? What is the probability that an inquirer with an income of \$50,000 will buy an insurance policy?

6 Three different machines, designated I, II, and III, are used to produce identical parts. According to the records, 10 percent of the parts turned out by machine I, 15 percent by machine II, and 20 percent by machine III are defective. All the parts produced by the three machines are mixed up and can no longer be identified with the machine that produced them. A part is selected at random from the mixed-up stock and found to be defective. What is the probability that this part was produced by machine I? By machine II? By machine III? (Assume that the capacities of the machines are the same.)

7 Suppose that you become very ill in the middle of the night and ask your sleepy wife to get some drug for you from the medicine cabinet. Two kinds of drugs are available, A and B. There are only four bottles in the cabinet; one bottle contains A, and three bottles contain B. If A is taken, there is a 90 percent chance of dizziness (D); but if B is taken, the probability of dizziness is only 10 percent. Infuriated by being awakened from a sound sleep, your wife grabs some pills and hands them to you without knowing whether they are A or B. A few minutes later, you become very dizzy. What is the probability that the drug you took was A? Or B?

8 The faculty of a certain college are classified as follows: 50 percent of them hold the doctoral degree (D), 30 percent hold the master's degree (M), and 20 percent hold only the bachelor's degree (B). Of those who hold the doctoral degree, 60 percent are male (F') and

40 percent are female (F); of those who hold the master's degree, 40 percent are male and 60 percent are female; and of those who hold the bachelor's degree, 70 percent are male and 30 percent are female. If a faculty member is selected at random, what is the probability that the person is female? Given that the person selected is female, what is the probability that she holds a master's degree?

9 A cosmetics saleslady conducts her business by making house-to-house calls. Through the years she has accumulated the following data: Of all the calls she has made, 20 percent resulted in what she considered large sales (L), 30 percent in small sales (S), and 50 percent in no sales (N) at all. Furthermore, of those who made large purchases, 60 percent live in two-story single-family homes (T); of those who made a small amount of purchases, 50 percent live in such homes; and of those who made no purchases at all, 20 percent live in such homes. If the next family she is going to call on lives in a two-story single-family home, what is the probability that the next call will result in large sales? Small sales? No sales?

3-4 Basic Probability Distributions

In the last chapter and this one so far, we have discussed the concept of a random event and the rules and procedures of computing probabilities of random events. In the remainder of this chapter we shall consider another closely related concept, the notion of a random variable, and the basic probability distributions of random variables. Inasmuch as random variables are usually classified as *discrete* and *continuous*, probability distributions may be likewise described as discrete and continuous.

3-4-1 Random variable

As already indicated at the end of Chap. 1, a *variable* is a symbol such as X used to stand for any one of a set of values. A variable is described as *random* if it is related to a random event. The value that a random variable will take on in any single trial of an experiment is determined only by chance factors; it occurs abruptly and haphazardly without plan or choice. Thus the concept may be briefly defined as follows:

> *If the values which a symbol takes on are associated with the elementary random events in the sample space of a given experiment and hence depend on chance factors for their occurrences, then the symbol is referred to as a* random variable.

Let X designate the random variable of a particular experiment. It is not difficult to see that a value of X may be assigned to each possible elementary event in the sample space of the experiment. For instance, in a survey of consumers' attitude toward a certain product or service, a "no" answer to a question may be assigned the value 0 and a "yes" answer may be assigned the value 1. The sales volume attributed to any particular salesman during the next

Table 3-5
The random variable X as the number of heads in the toss of two coins

Sample space	Value of the random variable X
TT	0
TH	1
HT	1
HH	2

year may be $150,000, the defectives found in a batch of 200 parts may number 11, the rainfall in a given area during the next rainy season may be 20 inches, and so on. Each of the values mentioned here is not predetermined but rather depends on chance factors for its occurrence. It is a value of X, the random variable of a given experiment.

Consider another familiar example for further clarification of this concept. Let X designate the number of heads obtained in the toss of two coins; then X has a sample space with four elementary random events as follows: {(TT), (TH), (HT), (HH)}, where T designates tails and H heads. Each of these four possible elementary events can be paired with a value of the random variable X, as shown in Table 3-5. There are only three distinct values of X, although there are four possible elementary events in the sample space.

It is important to remember that any value of X is also a random event since there must be some elementary event or compound event with which the value is associated. Not only any single value of X but also two or more values of X are considered an event. Thus in the example of tossing two coins, each of the following expressions for X constitutes an event: $X = 0$, $X = 1$, $X = 2$, $X \leq 1$, $X \geq 1$, and $0 \leq X \leq 2$.

3-4-2 Probability distributions

On the basis of the notion of a random variable, we can now proceed to discuss the concept of probability as a function. The probability of an event can be considered a function because corresponding to each distinct possible value of a random variable X, which is in fact an event, there is one and only one real number between 0 and 1 inclusive called its probability.

Let us begin by considering the following situation. Suppose that there is a certain experiment that has a finite number of possible outcomes, or elementary events. If the symbol X is used as the random variable to designate the outcome of any single trial of this experiment, and each possible outcome is associated with one of m distinct values for X designated by the lower-case letters x_1, x_2, $\ldots$, x_m, then $S = \{x_1, x_2, \ldots, x_m\}$ becomes the domain of the function of X. If P stands for a function rule and $P(x)$ represents the probability paired with a point x in S, we say that $P(x)$ is the probability that $X = x$, or $P(X = x)$.

Suppose, for example, that our experiment concerns the toss of two *ideal* coins to see the number of heads obtained. Let X be the number of heads observed; the domain of the function of X is $S = \{0, 1, 2\}$, which is the set of all possible values that X can take on. Since the coins are ideal and there are four equiprobable elementary events in the sample space, the symbol $P(0)$ represents the probability of $X = 0$ which is $\frac{1}{4}$, or $P(X = 0) = \frac{1}{4}$.

As pointed out above, there is a probability that can be assigned to each possible value of X, which is either an elementary or compound event. If the values of X are mutually exclusive and collectively exhaustive, then the probability function for each and every value of X can be together referred to as a *probability distribution*. That is:

Given an experiment, a probability distribution is an expression of a probability function which has as its domain a set of mutually exclusive and exhaustive values generated by the experiment.

The nature of any particular probability distribution is dependent on the random variable under consideration. The probability distribution for a discrete random variable together with its cumulative distribution will be discussed first, followed by an examination of the distribution of a continuous random variable.

Discrete probability distribution A random variable, the values of which differ from each other by finite amounts, is called a *discrete random variable*. Usually such a random variable has a finite number of points in its sample space. The probability distribution of a discrete random variable can be illustrated by the example of tossing two ideal coins mentioned above. Since X, the number of heads obtained, has the sample space {(T, T), (T, H), (H, T), (H, H)}, each element of which has the probability of $\frac{1}{4}$ to occur, and since the function of X has the domain $\{0, 1, 2\}$, the probability distribution of X can be expressed as in Table 3-6. Its graphic representation is shown in Fig. 3-4.

It is important to note that the probability distribution can be concisely summarized by the *cumulative probability distribution* (cpd), which gives the probabilities for $X \leq x$, where x represents each and every value in the domain of the function of X. The cpd of X for the coin example is indicated in Table 3-7.

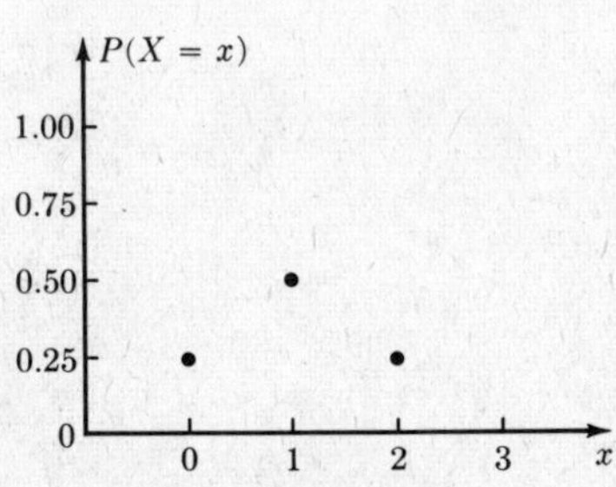

Figure 3-4
Probability distribution of X, the number of heads in a toss of two ideal coins.

Table 3-6

Probability distribution of X, the number of heads in a toss of two ideal coins

x	*Probability* $P(x)$
0	0.25
1	0.50
2	0.25
Total	1.00

Table 3-7

Cumulative probability distribution of X, the number of heads in a toss of two ideal coins

x	*cpd*
0	0.25
1	0.75
2	1.00

As another example, let X be the sum of the spots on the faces of two ideal dice. The set of all possible outcomes of the experiment was shown in Fig. 2-3 in Sec. 2-1-2. There are altogether 36 possible elementary events, which can be grouped into 11 events, or compound events. The corresponding values of X for these 11 events form the domain of the function of X, or $S = \{2, 3, 4, 5, 6, 7, 8, 9, 10, 11, 12\}$. Of the 36 possible outcomes, X takes on the value of 2 once, 3 twice, and so on, as indicated in Table 3-8. The probability distribution and the cumulative probability distribution for X are also shown in Table 3-8. Figure 3-5 shows the probability distribution.

Table 3-8

Probability distribution and cpd of X, the sum of the points on the faces of two ideal dice

x	*Number of outcomes*	*Probability distribution*	*cpd*
2	1	$\frac{1}{36}$	$\frac{1}{36}$
3	2	$\frac{2}{36}$	$\frac{3}{36}$
4	3	$\frac{3}{36}$	$\frac{6}{36}$
5	4	$\frac{4}{36}$	$\frac{10}{36}$
6	5	$\frac{5}{36}$	$\frac{15}{36}$
7	6	$\frac{6}{36}$	$\frac{21}{36}$
8	5	$\frac{5}{36}$	$\frac{26}{36}$
9	4	$\frac{4}{36}$	$\frac{30}{36}$
10	3	$\frac{3}{36}$	$\frac{33}{36}$
11	2	$\frac{2}{36}$	$\frac{35}{36}$
12	1	$\frac{1}{36}$	1.00
Total	36	1.00	

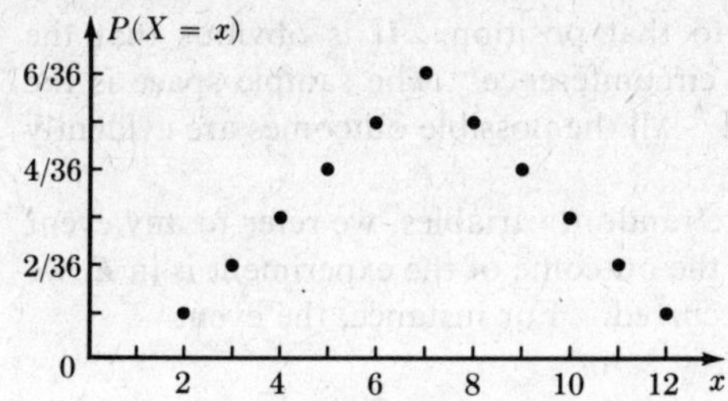

Figure 3-5
Probability distribution of X, the sum of the points on the faces of two ideal dice.

Continuous probability distribution Up to now our attention has been focused on random variables that have a finite number of points in the sample space. There are many experiments, however, that have an infinite number of possible outcomes. The random variable of such an experiment takes on values that differ from each other by an infinitesimal amount. For instance, if we are concerned with the weights of human beings, select a person at random, and learn his or her weight X, it is most convenient to use as the sample space a set of real numbers labeled as pounds which differ from each other by 1 pound. This is, of course, not strictly realistic because we can record only a limited number of weights as the possible outcomes of this experiment. It is easy to realize that this same problem still remains with us if X stands for ounces or even smaller subdivisions.

To be more realistic, we use an interval of real numbers as the sample space of the random variable. Such a random variable is *continuous* in the sense that there is an infinite number of points in any interval of real numbers, and the values within the interval differ from each other by an infinitesimal amount.

For illustration, let us consider an experiment with a dial as indicated in Fig. 3-6. The circumference of the circle is marked with selected numbers between 0 and 1, namely, 0.10, 0.20, . . . , 0.90. The scale between the numbers is even, so that 0.75 is three-quarters around the dial from 0. When the pointer (assumed to be perfectly balanced) at the center is spun, it circles around the dial and eventually stops at some position. Let X be the random variable

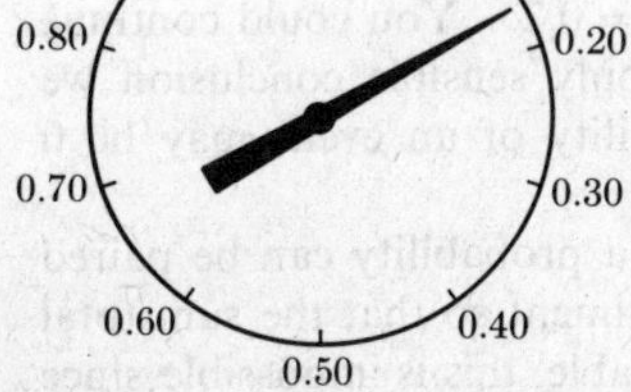

Figure 3-6
A dial.

equal to the number corresponding to that position. It is obvious that the pointer can stop at any point on the circumference. The sample space is the set of all possible readings on the dial. All the possible outcomes are evidently in this set.

Just as we have done with discrete random variables, we refer to any event E as a subset of the sample space. If the outcome of the experiment is in E, we shall conclude that the event E has occurred. For instance, the event

$$E = \{x \mid 0.75 < x \text{ or } x < 0.25\}$$

is said to have occurred if the pointer comes to rest on the upper half of the circle. Likewise, the event $F = \{x \mid 0 < x < 0.50\}$ occurs if the pointer ends up on the right-hand half of the circle. The events E *and* F both occur if X is in the set of the intersections of E and F, or $E \cap F = \{x \mid 0 < x < 0.25\}$. The events E *or* F occur if X is in the set of the union of E and F, or $E \cup F = \{x \mid 0.75 < x \text{ or } 0 < x < 0.50\}$.

Since E and F are both events, we should be able to find the probabilities of these events. Assume that the event E is still $\{x \mid 0.75 < x \text{ or } x < 0.25\}$, then the probability of E, or $P(E)$, will be $0.25 + 0.25 = 0.50$. Similarly, if the event F remains $\{x \mid 0 < x < 0.5\}$, then $P(F) = 0.50$. Also, $P(E \cap F) = 0.25$, and

$$P(E \cup F) = P(E) + P(F) - P(E \cap F) = 0.50 + 0.50 - 0.25 = 0.75$$

In general we can say that the probability of $X < a = a$, and the probability of $a < X < b = b - a$.

By now you should be aware of the essential difference between the discrete and continuous random variables. There is another important difference that deserves careful attention. In the case of rolling a cubic die, it is entirely correct to say that $P(X = 2) = \frac{1}{6}$; in the case of tossing two coins, $P(X = 2) = \frac{1}{4}$. However, if X is a continuous random variable, it is erroneous to say $P(X = a) = a$ because in this last case $P(X = a) = 0$. As indicated in the dial example, we have

$$P(0.1999 < X < 0.2001) = 0.0002$$

$$P(0.1999999 < X < 0.2000001) = 0.0000002$$

and

$$P(0.1999999999 < X < 0.2000000001) = 0.0000000002$$

All these intervals may be regarded as the point $X = 0.2$. You could continue to make the interval infinitely small. Thus the only sensible conclusion we have is $P(X = 0.2) = 0$. Surprisingly, the probability of an event may be 0 even though, in a practical sense, it can occur.

As indicated, for a discrete random variable, a probability can be paired with each and every possible outcome of the experiment so that the sum total will be equal to 1. For a continuous random variable, this is impossible since the probability for each point in the domain is 0. It is necessary then to develop a new approach to treat the probability of a continuous random variable.

The probability distribution for a continuous random variable X is a *probability density function* for X, if, for all a and b ($a \neq b$), $P(a < X < b)$ is represented by the area under the curve or line $y = f(x)$ and over the abscissa from a to b. The total area under $f(x)$ and over the domain is equal to 1. The set of all real numbers is assumed to be the domain of the probability density function. However, if the domain is not the set of all real numbers, then the probability density function is defined as 0 for all real numbers outside the domain.

Let us again use the dial problem to illustrate this point. The density shown in Fig. 3-7 is equal to 1 for the abscissa from 0 to 1, and is equal to 0 for other abscissas. The probability of X's falling within any interval from 0 to 1 is the length of the interval. For example,

$$P(0.20 < X < 0.40) = 0.20$$

since $a = 0.20$, $b = 0.40$, and $b - a = 0.20$. The shaded area in Fig. 3-7 gives the probability of $(0.20 < X < 0.40)$. Note that $P(0.20 < X < 0.40)$ is equal to $P(0.20 \leq X \leq 0.40)$ since $P(X = 0.20) = P(X = 0.40) = 0$.

The probability density function $f(x)$ generated by the dial in Fig. 3-6 and represented in Fig. 3-7 can be generalized as follows: Since the height of the function, or $f(x)$, multiplied by the width from a to b, which represents the domain, must equal 1, or $(b - a)(f(x)) = 1$, we have

$$f(x) = \begin{cases} \dfrac{1}{(b-a)} & \text{for } a \leq x \leq b \\ 0 & \text{elsewhere} \end{cases} \tag{3-12}$$

The probability distribution for a continuous random variable can be also summarized by the cumulative probability distribution (cpd or F), which is given by

$$F(x) = P(X \leq x) \tag{3-13}$$

for all values of x.

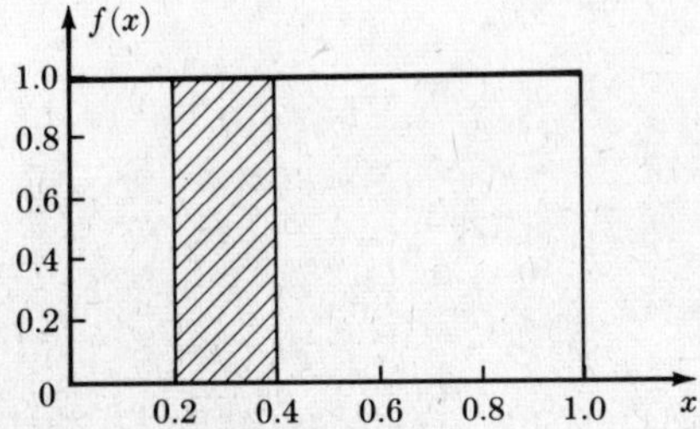

Figure 3-7
Probability density function for X, the outcome of spinning an ideal dial.

Exercise 3-4

1 Use your own words to distinguish between:

a Discrete and continuous random variables.
b Discrete and continuous probability distributions.

2 Distinguish between the concepts of probability function and probability density function.

3 Find the probability distribution and the cumulative probability distribution for X, the number of heads obtained from the toss of three well-balanced coins. Graph the probability distribution of X.

4 Find the probability distribution and the cumulative probability distribution for Y, where $Y = X^2$ and X is the number of heads obtained in the toss of three well-balanced coins.

5 The probability density function shown in Eq. (3-12) is an expression of the so-called *rectangular distribution*. If X has a rectangular distribution over the interval from a to b, what is the value of $f(x)$ as well as that of $F(x)$:

a For $a = 2$ and $b = 4$? *b* For $a = -10$ and $b = 10$?

6 Graph $f(x)$ and $F(x)$ in Prob. 5 above.

Chapter 4

Measures of central tendency

The concepts of a set, sample space, and probability covered in the preceding chapters are useful notions in the discussion of statistical methods and analyses. The theory of probability is basic to the principles of statistical inference; and the relation between a subset and a universal set is similar to that between the sample and the population, the two fundamental concepts in the subject of statistics.

Statistical methods and analyses are primarily concerned with the collection, organization, and description of numerical data which form the sample, or subset, and, on the basis of the characteristics of the sample, with making inferences about the characteristics of the population, or universal set. In this chapter and the one immediately following, we shall examine the most frequently encountered characteristics of the sample and the population.

The first section of this chapter is designed to show how to organize the sample data into a frequency distribution, which is one of the most useful techniques in statistical analysis. After that, we shall see how the data can be summarized in a single value. Such a value representing an entire set of data must be a number at which the data have a tendency to concentrate. In other words, it is a central value about which all the data in the set are distributed. It is for this reason that such summary values are often called *measures of central tendency*. They include among others the arithmetic mean, the median, and the mode. They are usually referred to as *averages*.

4-1 Organization of Data

A business analyst who tries to identify the sales patterns of a company would learn very little by merely looking through the invoices issued during a certain period. A list of outputs in units produced by the workers in a factory would convey only a very vague idea of the productivity of the group. Likewise, an enumeration of the test scores of the students enrolled in a certain class would provide very little insight into the essential nature of the mass of data. Each of these examples indicates that unorganized raw data are virtually meaningless. In order to use them better, data organization becomes necessary.

Suppose that a researcher wishes to determine the pattern of the heights of female workers in America. For this purpose a sample of 50 women is selected from the work force in the United States, and their heights in inches are obtained. Let the random variable X designate the height of any female worker, and the 50 observations for X are shown in Table 4-1.

The simplest device used in processing data is to form an *array*, an arrangement of items in the order of their magnitude. It can be formed either in descending or ascending order. Table 4-2 presents an array, in ascending order, of the heights of the 50 female workers as rearranged from Table 4-1.

Such an array has distinct advantages over the data in the unorganized form. First, it enables us to learn at once that the range from the shortest, 53 inches, to the tallest, 72 inches, is 19 inches. Second, it gives us some idea about the general characteristics of the distribution. The array, however, still offers only a cumbersome form of data organization, especially when the number of items is large. It is therefore more desirable to summarize the data into classes. The arrangement of data to show the frequency of occurrence of values or objects

Table 4-1

Heights (in inches) of 50 female workers in the United States

Worker	*Height X*	*Worker*	*Height X*	*Worker*	*Height X*	*Worker*	*Height X*	*Worker*	*Height X*
1	65	11	64	21	64	31	63	41	64
2	63	12	65	22	65	32	65	42	64
3	65	13	64	23	64	33	63	43	63
4	63	14	72	24	71	34	70	44	69
5	69	15	68	25	68	35	67	45	67
6	67	16	66	26	66	36	66	46	66
7	53	17	55	27	56	37	57	47	58
8	58	18	57	28	59	38	59	48	60
9	60	19	60	29	61	39	61	49	61
10	61	20	62	30	62	40	62	50	62

Table 4-2
Array of heights of 50 female workers (in inches)

53	60	63	64	67
55	60	63	65	67
56	61	63	65	67
57	61	63	65	68
57	61	63	65	68
58	61	64	65	69
58	62	64	66	69
59	62	64	66	70
59	62	64	66	71
60	62	64	66	72

Source: Table 4-1.

in each class is called a *frequency distribution.* Even though the frequency distribution does not show all the details of an array, much is gained by such a condensation.

4-1-1 Construction of a frequency distribution

The construction of a frequency distribution depends on the data at hand. In any event, the work is simple and will include the following two principal steps.

Defining class interval The selection of class intervals is an important step because it may affect the usefulness of the frequency distribution. An improper selection may obscure the nature of the distribution pattern and lead to incorrect or even distorted conclusions. In order to define class intervals appropriately, we should observe the following principles:

First, *the number of classes should be such that the true nature of the distribution is made manifest.* When the number of classes is excessively large, and consequently the class intervals too small, it would not reveal the true pattern of the distribution because each class would contain too few items, or none at all. On the other hand, if the number of classes is excessively small and class intervals too wide, it may conceal the fact that relatively large numbers of items are concentrated in a few classes with narrower intervals. This explains why distributions are seldom found to contain fewer than 6 or more than 20 classes. Very often a choice of about 10 classes proves quite satisfactory.

Second, *the intervals should be of equal width.* If the length of intervals is not uniform, comparison between classes is almost meaningless. Occasionally, items of extreme magnitude are grouped in *open-end* classes labeled "less than" and "or more." In such cases, however, they are used for presentation only, and not for computation purposes.

In the above example, the range is 19 inches. If we group the data in eight classes, the length of each interval should be close to, and larger than, $\frac{19}{8} =$ 2.375 inches.

Third, *the interval length should be an odd number with the midpoints each having exactly the same number of digits as the rounded data.* Since the midpoint is halfway between the lower and upper limits of each interval, this forces the class limits to have an extra decimal digit, 5. Doing this has certain advantages. For one thing, it ensures that each item falls within an interval and not on any of the boundaries. For another, any rounded figure will be put in the same class as it would have been if it had not been rounded.

For the above data, an interval of 3 inches proves appropriate; however, there are many possible sets of intervals with length 3 inches. In selecting just which particular set to use, it should be remembered that the midpoints ought to be numbers easy to work with and that every observation should fall into one of the intervals. With these considerations, we select 52, 55, . . . , 73, to be the midpoints, with 50.5 and 53.5, 53.5 and 56.5, . . . , 71.5 and 74.5 as the lower and upper boundaries.

Tabulation of data The next step is to distribute the data within the classes already selected. We may use a tally sheet, or if the number of items is large, punched cards. In any event, the number of observations falling into each class is counted, and a frequency distribution is established. Table 4-3 shows the result of the tallying procedure. It is constructed by reading down the array in Table 4-2 and, on reading 53, entering a diagonal stroke in the class 50.5 to 53.5, then for a number 55, entering another in the class 53.5 to 56.5, etc. After the tally is completed, the diagonal strokes in each class are counted and the number is entered in the last column as the frequency of the class.

Table 4-3
Tally sheet for the data on heights of 50 female workers

Heights, inches	*Midpoints*	*Tally*	*Number of workers*
50.5–53.5	52	/	1
53.5–56.5	55	//	2
56.5–59.5	58	~~////~~/	6
59.5–62.5	61	~~////~~ ~~////~~/	11
62.5–65.5	64	~~////~~ ~~////~~ ~~////~~/	16
65.5–68.5	67	~~////~~////	9
68.5–71.5	70	////	4
71.5–74.5	73	/	1
Total			50

Source: Table 4-2.

4-1-2 Frequency distributions and notations

For computation of a frequency distribution, it is usually assumed that the frequency associated with any particular class is the number of times that the value of the midpoint of the class has occurred.[1] Since the symbol X stands for the height in inches, we may say that X takes on the value of 52 once, 55 twice, etc. Moreover, if the lower-case letter x is used to designate any of the midpoints, we may use the functional notation $FR(x)$ to designate the *frequency* with which each value of x is taken on. In Table 4-4 each of the midpoints, given the heading x, is entered in the first column, while the frequency $FR(x)$ is entered in the second column. $FR(x)$ is a value of the function that has as its domain the midpoints of all the classes.

Similarly, $RF(x)$ is used to stand for the *relative frequency* of x, that is, the *proportion* of times that the value of x is taken on. In our example, this is arrived at by dividing the frequency of each class by 50, the total number of observations. In general, if n is the total number of items observed, the relative frequency is defined by

$$RF(x) = \frac{FR(x)}{n} \tag{4-1}$$

$CF(x)$ is used to designate the *cumulative frequency* at x. Each entry in column 4 is the total number of times that X takes on a value equal to or less

Table 4-4
Frequency distributions of heights of 50 female workers (in inches)

x (1)	$FR(x)$ (2)	$RF(x)$ (3)	$CF(x)$ (4)	$RCF(x)$ (5)
52	1	0.02	1	0.02
55	2	0.04	3	0.06
58	6	0.12	9	0.18
61	11	0.22	20	0.40
64	16	0.32	36	0.72
67	9	0.18	45	0.90
70	4	0.08	49	0.98
73	1	0.02	50	1.00
Total	50	1.00		

Source: Table 4-2.

[1] This assumption likely results in some discrepancies in the computation since it is rarely true that the midpoint is the arithmetic mean of the items in the class. However, the discrepancies are generally not large, for the errors tend to offset each other. As we shall see in Sec. 4-2, advantages gained by using the grouped-data method outweigh the disadvantage of slight inaccuracy.

than x. For instance, $CF(61)$ is 20 because X takes on the value of 61 or less twenty times, made up of the following: 52 once, 55 two times, 58 six times, and 61 eleven times. Note that the cumulative frequencies are obtained by adding successively the numbers in column 2.

Likewise, $RCF(x)$ designates the *relative cumulative frequency* at x. Thus $RCF(x)$ is the proportion of times that X takes on a value that does not exceed x. For instance, $RCF(61)$ is equal to 0.40 because the proportion of times that X takes on the value of 61 or less is 0.40. In other words, 40 percent of the female workers have a height of 61 inches or less. The relative cumulative frequencies in the last column can be obtained by adding successively the relative frequencies in column 3, or by using the following formula:

$$RCF(x) = \frac{CF(x)}{n} \tag{4-2}$$

Like $FR(x)$, each of these last three distributions is also a function defined over the midpoints as its domain. Given any one of the four distributions and the sample size n, the other three can be found. Since they are based on a *sample* of n observations, they are called *empirical distributions*, which are *sample distributions*, or distributions of the sample.

4-1-3 Graphic representation of distributions

Frequency distributions may be presented in graphs such as histograms, polygons, or other forms. In constructing each graph, the midpoints and class intervals are represented by the abscissa, or the x axis, and the frequencies are measured by the ordinate, or the y axis.

The histogram The histogram, sometimes called a *column diagram*, is obtained as follows: The class intervals shown in Table 4-3 are first marked off along the horizontal axis. Then, for each interval, we draw a bar as high as the frequency or relative frequency of the class. Figure 4-1 is a histogram showing the distribution of the heights of the 50 workers. The frequency is measured by the left-hand scale, and the relative frequency by the right-hand scale.

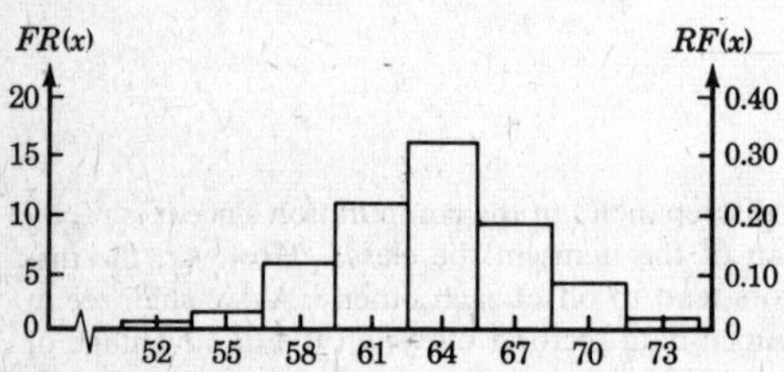

Figure 4-1
A histogram for the data in Table 4-4.

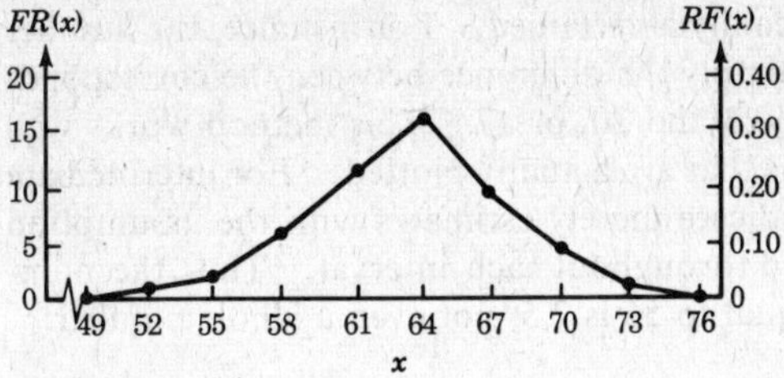

Figure 4-2
A frequency polygon for the data in Table 4-4.

The polygon Another kind of graph used to present frequency distributions is the *frequency polygon.* It is constructed by plotting a point as high as the frequency or relative frequency of a class directly above the midpoint of the class and then joining the points with straight lines. The figure also includes the midpoints of the intervals just beyond the limits of the series so that it is enclosed. Figure 4-2 is a frequency polygon for the data in Table 4-4.

The ogive The polygon is also used to present the cumulative frequency and the relative cumulative frequency, as shown in Fig. 4-3. This is done as follows: For observations at or below 50.5, the lower limit of the lowest class, a dot is put at the point showing 50.5 on the horizontal axis. This is done because no observation is of this magnitude or lower. At the point showing 53.5, a dot is plotted at the height of 1 since one observation has a value less than or equal to 53.5. Directly above the point 56.5, a dot is plotted at the height of 3 since three observations have a value less than or equal to 56.5. The plotting is continued in this fashion until a dot is put as high as 50 directly above the point 74.5 on the x axis since all the 50 observations are less than or equal to 74.5. Then the dots are joined by straight lines, and the graph thus obtained is called a *cumulative frequency polygon*, which is also known as an *ogive*.

As we shall see in Sec. 4-3, the ogive, or cumulative frequency polygon, has the advantage of providing a convenient way to estimate the median and the percentiles of a sample. In addition it has the advantage that the number of

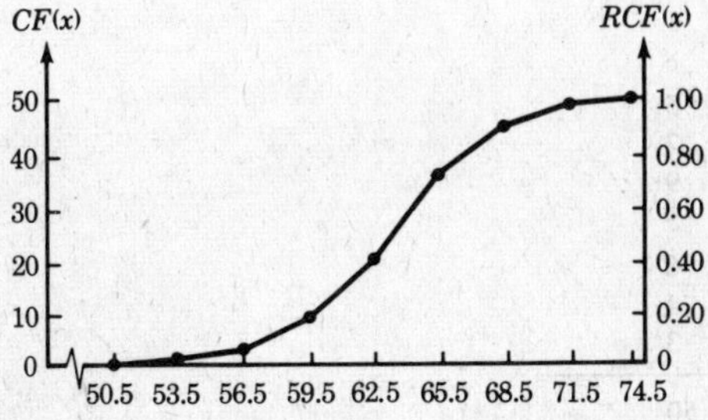

Figure 4-3
An ogive for the data in Table 4-4.

items between two values can be readily ascertained. For instance, the number of items between 56.5 and 62.5 is simply the difference between the corresponding cumulative frequencies, namely, 3 and 20, or 17. This method works well only for those values and the points that are actually plotted. For intermediate values, the points on the graph indicate merely estimates with the assumption that the items are evenly distributed throughout each interval. Thus the number of observations less than or equal to 56 is 2.5, not even a whole number.

Exercise 4-1

1 What is meant by the range of a sample of observations?

2 Explain why the collected raw data should be organized for analytic purposes.

3 For organizing the collected raw data into a frequency distribution, answer the following questions:

a Why should the number of classes be not too many nor too few?
b Why should the intervals be of equal width?
c Why should the length of each class be an odd number?

4 For the data in Table 4-2, construct a table similar to Table 4-3 with the lower limit of the lowest class equal to 51.5, the interval still having a width of 3, and the upper limit of the highest class equal to 75.5.

5 For the data as organized in Prob. 4 above, construct a table similar to Table 4-4 in the text.

6 The frequency distribution, the relative frequency distribution, and other distributions shown in Table 4-4 are referred to as *empirical* distributions. Explain why this is so.

7 Why are the organized data often represented in graphic forms?

8 Two cubic dice are rolled together for 50 times, and the sums of spots uppermost on the two dice are noted and recorded. The following frequency distribution is obtained:

Sum of spots uppermost on the two dice, x	*Frequency distribution FR(x)*
2	1
3	1
4	1
5	8
6	5
7	7
8	9
9	5
10	6
11	4
12	3
Total	50

a Find the relative, cumulative, and relative cumulative frequency distributions for the above data.
b Construct a histogram, a polygon, and an ogive.

9 The weekly earnings of 200 college students who work part time are distributed as follows:

Weekly earnings, $	*Number of students*
77.5–82.5	5
82.5–87.5	12
87.5–92.5	13
92.5–97.5	22
97.5–102.5	30
102.5–107.5	35
107.5–112.5	32
112.5–117.5	20
117.5–122.5	15
122.5–127.5	10
127.5–132.5	6
Total	200

Tabulate the data as in Table 4-4, and draw a histogram, a polygon, and an ogive.

10 Twenty-five new Fairyear tires are tested to determine their service lives. The following is the result of the test:

Service life, thousands of miles	*Number of tires*
36	2
37	1
38	3
39	4
40	5
41	4
42	2
43	3
44	1
Total	25

a Find the relative, cumulative, and relative cumulative frequency distributions for the service lives of the 25 Fairyear tires.
b Construct a histogram, a polygon, and an ogive.

11 Five coins are tossed simultaneously for 50 times, and the number of heads obtained in each toss is noted and recorded. The following results are given:

0	1	1	2	2	2	2	3	3	4
0	1	1	2	2	2	3	3	3	4
1	1	1	2	2	2	3	3	3	4
1	1	2	2	2	2	3	3	4	4
1	1	2	2	2	2	3	3	4	5

a Find the empirical distributions of the frequency, relative frequency, cumulative frequency, and relative cumulative frequency.

b Construct a histogram, a polygon, and an ogive.

12 Suppose that a standard aptitude test is administered to all the applicants at the public employment agencies of a certain state. A sample of 100 applicants is randomly selected, and their test scores are as follows:

77	44	49	33	38	33	76	55	68	39
44	59	36	55	47	61	53	32	65	51
29	41	32	45	83	58	73	47	40	26
59	43	66	44	41	25	39	72	37	55
34	47	66	53	55	58	49	45	61	41
55	92	83	77	45	62	45	36	78	48
54	50	51	66	80	73	57	61	56	50
45	82	71	48	46	69	38	72	56	64
38	45	51	44	41	68	45	92	43	12
37	16	44	57	63	71	40	64	57	51

a Make an array in ascending order.

b Assume that these data are organized into 10 classes such as 10.5 to 19.5, 19.5 to 28.5, . . . , 82.5 to 91.5, and 91.5 to 100.5. Let x stand for the midpoint of each class. Use a tally sheet to obtain the frequencies for x. Also tabulate the data as in Table 4-4.

c Using the above data, draw a histogram, a polygon, and an ogive.

4-2 The Arithmetic Mean

The *arithmetic mean*, which is sometimes simply referred to as "the mean," is the most commonly used average. It is the sum of the values observed divided by the number of observations summed. The arithmetic mean of a sample is usually designated by the symbol $\bar{X}$, which is read "X bar." In order to deal with this measure, we need to introduce two more symbols, the summation notation $\sum$ and the subscript i. The summation symbol $\sum$ is the Greek capital letter sigma and is read "the sum of." In handling numerical data it is a common practice to treat the observations as being arranged in some order. The individual observations may be identified by a serial number: 1 for the first, 2 for the second, . . . , n for the nth. The symbol i is commonly used to designate this general serial number and is written as a subscript. Thus X_1 which is read "X sub 1" or simply "X 1" is the first observation, and X_i is the ith, etc. If there are n observations in the sample, the sum of these observations may be represented either by $\sum_{i=1}^{n} X_i$, which may also be written $\sum\limits_{i=1}^{n} X_i$, read "summation from 1 to n of X sub i," or by $X_1 + X_2 + X_3 + \cdots + X_n$. In terms of these symbols, the arithmetic mean denoted by $\overline{X}$ is given by

$$\overline{X} = \frac{\sum_{i=1}^{n} X_i}{n} = \frac{X_1 + X_2 + \cdots + X_n}{n} \tag{4-3}$$

4-2-1 Computing the arithmetic mean

In the data of Table 4-1 there are 50 observations, the arithmetic mean of which is computed as follows:

$$\overline{X} = \frac{65 + 63 + \cdots + 62}{50} = \frac{3{,}160}{50} = 63.2$$

This method of computing the arithmetic mean is used when the data are not organized. If a frequency distribution is constructed, the computation will be somewhat simplified.

When the arithmetic mean is computed from a frequency distribution, the midpoint of each class is taken as the representative value of that class. The various midpoints are multiplied by their respective frequencies, the products are added, and the sum of the products is then divided by the total number of observations. Symbolically, if $x_1, x_2, \ldots, x_m$ represent the midpoints, and $FR(x_1), FR(x_2), \ldots, FR(x_m)$ represent the corresponding frequencies, where the subscript m stands for the number of classes, then

$$\overline{X} = \frac{\sum_{j=1}^{m} x_j FR(x_j)}{n} \qquad \text{for } j = 1, 2, \ldots, m \tag{4-4}$$

Table 4-5 shows the computation of the arithmetic mean from a frequency distribution. The obtained value, 63.22, is slightly different from the one computed from the ungrouped data. Such a discrepancy, however, is not unusual. In most cases we can expect a slight difference between the arithmetic mean calculated from the ungrouped data and that computed from a frequency distribution since the items of each class are often not evenly or symmetrically distributed throughout the interval.

4-2-2 Rules of summation and properties of the mean

A few simple rules of summation must be understood if we desire to learn and interpret the properties and computations of the arithmetic mean in symbolic terms. Among these rules the following are the most commonly encountered:

1. If c is a constant over n observations, then

$$\sum_{i=1}^{n} c_i = \underbrace{c + c + c + \cdots + c}_{n\text{ times}} = nc \tag{4-5}$$

Table 4-5

Computation of the arithmetic mean for data of Table 4-3

Heights, inches	*Midpoints* x_j	*Frequency* $FR(x_j)$	*Products of frequency and midpoints* $x_j FR(x_j)$
50.5–53.5	52	1	52
53.5–56.5	55	2	110
56.5–59.5	58	6	348
59.5–62.5	61	11	671
62.5–65.5	64	16	1,024
65.5–68.5	67	9	603
68.5–71.5	70	4	280
71.5–74.5	73	1	73
Total		50	3,161

$$\bar{X} = \frac{\sum_{j=1}^{m} x_j FR(x_j)}{n} = \frac{3,161}{50} = 63.22$$

2. If c is a constant multiplied by each of the n observations X_i, then the sum of the products is equal to c multiplied by the sum of the observations. Symbolically,

$$\begin{aligned}\sum_{i=1}^{n} cX_i &= cX_1 + cX_2 + \cdots + cX_n \\ &= c(X_1 + X_2 + \cdots + X_n) \\ &= c\sum_{i=1}^{n} X_i \qquad (4\text{-}6)\end{aligned}$$

3. If X and Y are two random variables, then

$$\sum_{i=1}^{n}(X_i + Y_i) = \sum_{i=1}^{n} X_i + \sum_{i=1}^{n} Y_i \qquad (4\text{-}7)$$

The proof of this runs as follows:

$$\begin{aligned}\sum_{i=1}^{n}(X_i + Y_i) &= (X_1 + Y_1) + (X_2 + Y_2) + \cdots + (X_n + Y_n) \\ &= (X_1 + X_2 + \cdots + X_n) + (Y_1 + Y_2 + \cdots + Y_n) \\ &= \sum_{i=1}^{n} X_i + \sum_{i=1}^{n} Y_i\end{aligned}$$

4. If X and Y are two random variables, then

$$\sum_{i=1}^{n}(X_i - Y_i) = \sum_{i=1}^{n} X_i - \sum_{i=1}^{n} Y_i \tag{4-8}$$

Rules 3 and 4 can be extended to cases with three or more random variables.

Knowing these rules of summation, we can proceed to examine the properties of the arithmetic mean. One of the important properties is that the sum of the deviations from the mean is equal to 0. That is,

$$\sum_{i=1}^{n}(X_i - \overline{X}) = 0 \tag{4-9}$$

which is proved as follows:

$$\begin{aligned}\sum_{i=1}^{n}(X_i - \overline{X}) &= \sum_{i=1}^{n} X_i - \sum_{i=1}^{n} \overline{X} && \text{by (4-8)}\\ &= \sum_{i=1}^{n} X_i - n\overline{X} && \text{by (4-5)}\\ &= n\overline{X} - n\overline{X} && \text{by (4-3)}\\ &= 0\end{aligned}$$

Another important property of the mean is expressed in the following relationship: If

$$Y_i = a + bX_i, \text{ then } \overline{Y} = a + b\overline{X} \tag{4-10}$$

Here X_i and Y_i represent two paired numbers in two sequences of observations, and a and b are constants. The proof of this requires only a summation of each side of the original equation and then dividing it by n. Since $Y_i = a + bX_i$, then

$$\begin{aligned}\sum_{i=1}^{n} Y_i &= \sum_{i=1}^{n}(a + bX_i)\\ &= \sum_{i=1}^{n} a + \sum_{i=1}^{n} bX_i && \text{by (4-7)}\\ &= na + b\sum_{i=1}^{n} X_i && \text{by (4-5) and (4-6)}\end{aligned}$$

therefore

$$\frac{\sum_{i=1}^{n} Y_i}{n} = \frac{na}{n} + b\frac{\sum_{i=1}^{n} X}{n}$$

and

$$\overline{Y} = a + b\overline{X}$$

Suppose, for instance, that there is the relationship $Y = 3 + 2X$ and X takes on the value of 1, 3, 5, 7, or 9. Then we have:

X	$Y = 3 + 2X$
1	5
3	9
5	13
7	17
9	21
Total 25	65
$\bar{X} = \frac{25}{5} = 5$	$\bar{Y} = \frac{65}{5} = 13$

The value of $\bar{Y}$, however, can be computed directly from the relationship $Y = 3 + 2X$. Since $Y = 3 + 2X$, then by Eq. (4-10), $\bar{Y} = 3 + 2\bar{X} = 3 + 2(5) = 3 + 10 = 13$.

This relationship is commonly interpreted as saying that the mean is preserved under a *linear transformation of scale*. With this in mind, it is easy to see that

$$\text{If } Y_i = bX_i \text{, then } \bar{Y} = b\bar{X} \tag{4-11}$$

and

$$\text{If } \bar{Y}_i = a + X_i \text{, then } \bar{Y} = a + \bar{X} \tag{4-12}$$

Finally, we are going to prove that

$$\text{If } W_i = X_i + Y_i \text{, then } \bar{W} = \bar{X} + \bar{Y} \tag{4-13}$$

Let $X_1, X_2, \ldots, X_n$ and $Y_1, Y_2, \ldots, Y_n$ be two sequences of numbers. Since

$$W_i = X_i + Y_i$$

then

$$\sum_{i=1}^{n} W_i = \sum_{i=1}^{n} (X_i + Y_i) = \sum_{i=1}^{n} X_i + \sum_{i=1}^{n} Y_i$$

thus

$$\frac{\sum_{i=1}^{n} W_i}{n} = \frac{\sum_{i=1}^{n} X_i}{n} + \frac{\sum_{i=1}^{n} Y_i}{n}$$

and

$$\bar{W} = \bar{X} + \bar{Y}$$

More generally, we can say that the mean of the sum is equal to the sum of the means.

Since an average is often referred to as a measure of central tendency, it should somehow reflect the "center" of a series of observations. An arithmetic mean is very sensitive to the influence of a few large numbers in a set of predominantly small numbers; thus it sometimes fails to indicate the center of a series of values. An example will help to make this point clear. Suppose that a survey of household income is conducted in a town with 1,000 families. Among these 1,000 families, 3 have an income of $1 million each and 997 have incomes of only $3,000. The arithmetic mean of family incomes is therefore $5,991, but 99.7 percent of all the families have incomes only slightly over half of that amount. While such sensitivity of the mean is sometimes desirable, in some instances it is not. In the latter cases the use of some other measures such as the median will be more appropriate.

Exercise 4-2

1 How is the arithmetic mean defined? Why is the mean computed from grouped data usually different slightly from the mean computed from the ungrouped for the same set of data?

2 Explain the drawbacks of the arithmetic mean as a measure of central tendency.

3 Suppose that we have two series, X and Y, which are paired as follows:

X:	1	2	3	4	5
Y:	2	4	6	8	10

Find the value for each of the following expressions:

a $\sum (X - Y)$.
b $\sum (X + Y)$.
c $\sum (2X + 3Y)$.

4 Suppose that we have two series, X and Y, which are paired as follows:

X:	1	3	5	7	9
Y:	10	8	6	4	2

Find the value for each of the following expressions:

a $(\sum X)(\sum Y)$.
b $(\sum X^2) - (\sum X)^2$.
c $(\sum XY) - (\sum X)(\sum Y)$.

5 Use the summation symbol to represent the following expressions:

a $X_1{}^2 + X_2{}^2 + X_3{}^2 + X_4{}^2$.
b $aX_1 + a2X_2 + a3X_3 + a4X_4 + a5X_5$.
c $c(X_1 + 1) + c(X_2 + 2) + c(X_3 + 3) + \cdots + c(X_n + n)$.
d $(X_1 - 1)1^2 + (X_2 - 2)2^2 + (X_3 - 3)3^2 + (X_4 - 4)4^2 + (X_5 - 5)5^2$.

6 Find the value for each of the following expressions:

a $\sum_{i=1}^{8} i.$ *b* $\sum_{j=1}^{5} (j^2 + 3).$ *c* $\sum_{k=1}^{4} (k + 2)^2.$

7 Assume that $\sum_{i=1}^{4} X_i = 14$ and $\sum_{i=1}^{4} X_i^2 = 54$; evaluate the following:

a $\sum_{i=1}^{4} (X_i - 2).$ *c* $\sum_{i=1}^{4} (X_i - 1)^2.$

b $\sum_{i=1}^{4} (X_i + 2).$ *d* $\sum_{i=1}^{4} 3(2X_i^2 - 3X_i + 1).$

8 Compute the arithmetic mean for the data in Prob. 8 of Exercise 4-1.

9 Compute the arithmetic mean for the data in Prob. 9 of Exercise 4-1.

10 Compute the arithmetic mean for the data in Prob. 10 of Exercise 4-1.

11 Find the arithmetic mean for the data in Prob. 11 of Exercise 4-1 by:

a Equation (4-3). *b* Equation (4-4).

12 Find the arithmetic mean for the data in Prob. 12 of Exercise 4-1 by:

a Equation (4-3). *b* Equation (4-4).

13 Suppose that the series of X consists of the following values 2, 4, 6, 8, 10, and 12. Verify that $\sum (X - \bar{X}) = 0$.

14 Let C° designate the temperature measured by the centigrade thermometer and F° designate the temperature measured by Fahrenheit. We know that the following relation holds:

$$F° = 32 + \frac{9}{5} C°$$

a If the temperature today is 20 degrees in centigrade, what is the temperature in Fahrenheit?

b If the temperature today is 86 degrees in Fahrenheit, what is the temperature in centigrade?

c During the month of December, the average temperature in Los Angeles is 70 degrees Fahrenheit. What is the December average temperature in Los Angeles as measured by centigrade thermometer?

15 Suppose that we have two series, X and Y, which are paired as follows:

X:	2	4	6	8	10
Y:	1	2	3	4	5

a Find the mean of W, or $\bar{W}$, if $W = X - Y$.

b Find the mean of W, or $\bar{W}$, if $W = 2X + 3Y$.

4-3 Other Averages

In this section we shall consider some other measures of central tendency and related concepts. Among these are the median and the mode.

4-3-1 The median

Since in a set of predominantly small numbers the arithmetic mean is very sensitive to the influence of a few large numbers, it is, under such circumstances, more appropriate to use the median as a measure of central tendency. Briefly, the *median* may be defined as follows:

> *The median is the* value *that divides a set of observations ordered with respect to magnitude of values so that the number of items above it is equal to the number of items below it.*

As with the arithmetic mean, the method of determining the median depends on whether the data are grouped or ungrouped.

The median for ungrouped data Suppose that we have a set of nine numbers, 1, 2, 3, 4, 5, 6, 7, 8, and 9. It is evident that the median is the value 5, for there are four numbers above it and four numbers below it. If the set includes also the number 10, then any value greater than 5 and less than 6 can be taken as the median. It is a common practice, however, that the average of the two central numbers is taken as the median when the total number of items in the list is even. In this instance, 5.5 would be considered the median. An examination of Table 4-2 reveals that 63.5, the average of the two middle values, is the median.

In general, if $X_1, X_2, \ldots, X_n$ are a list of numbers arranged in the order of increasing (or decreasing) magnitude, and n is odd, the median (abbreviated Med) is the number occurring in the middle of the list. Symbolically,

$$\text{Med} = X_{(n+1)/2} \tag{4-14}$$

If n is even, the median is given by

$$\text{Med} = \frac{X_{n/2} + X_{(n/2)+1}}{2} \tag{4-15}$$

Suppose, for instance, that we have 11 (an odd number) observations for a variable X as follows:

X_1	X_2	X_3	X_4	X_5	X_6	X_7	X_8	X_9	X_{10}	X_{11}
0	1	1	2	3	4	6	7	8	8	9

Here $n = 11$ and $(11 + 1)/2 = 6$. By Eq. (4-14), the median of these 11 observations is

$$\text{Med} = X_{(n+1)/2} = X_6 = 4$$

If another observation $X_{12} = 9$ is added, there will be 12 (an even number) observations. By Eq. (4-15), the median is computed as follows:

$$\text{Med} = \frac{X_{n/2} + X_{(n/2)+1}}{2} = \frac{X_6 + X_7}{2} = \frac{4 + 6}{2} = 5$$

Table 4-6

Some distributions of the data on heights of 50 female workers

Heights, inches	*Midpoints* x	*Frequency* $FR(x)$	*Cumulative frequency* $CF(x)$	*Relative cumulative frequency* $RCF(x)$
50.5–53.5	52	1	1	0.02
53.5–56.5	55	2	3	0.06
56.5–59.5	58	6	9	0.18
59.5–62.5	61	11	20	0.40
62.5–65.5	64	16	36	0.72
65.5–68.5	67	9	45	0.90
68.5–71.5	70	4	49	0.98
71.5–74.5	73	1	50	1.00
Total		50		

The median for grouped data To find the median for grouped data, the method of interpolation is usually used. The method requires the construction of a cumulative frequency distribution or a relative cumulative frequency distribution. It is based on the assumption that the items within each class are evenly spread throughout the interval. To illustrate this method of determining the median for the data on the heights of female workers, Tables 4-3 and 4-4 are partially reproduced in Table 4-6.

The first step in finding the median from grouped data is to identify the *median class*; it is the class in which the median is located. It is apparent that the class of 62.5 to 65.5 contains the median because the twenty-first through the thirty-sixth items, which include the twenty-fifth of the 50 numbers, fall within this class. In general, the median is located in the class with the *smallest* midpoint for which the cumulative frequency is more than $n/2$ or the relative cumulative frequency is larger than 0.5. If the distribution of the data is such that there is a class interval for which the cumulative frequency is exactly equal to $n/2$ or the relative cumulative frequency is exactly 0.5, then the upper limit of that class would be the median and no interpolation would be needed.

After the class is identified, the next step is to estimate the position of the median within the class. This is done by subtracting the cumulative frequency of the next lower class from $n/2$, and expressing this difference, divided by the frequency of that class, as a fraction of the class interval. This fraction multiplied by the length of the class interval gives the position of the median within the class.

Finally, the value thus determined is added to the lower limit of the class to provide the actual value of the median. As a formula, the method appears as follows:

$$\text{Med} = \frac{n/2 - CF(x_{k-1})}{FR(x_k)} w + L \tag{4-16}$$

where n = total number of items in distribution
$CF(x_{k-1})$ = cumulative frequency below lower limit of the median class
$FR(x_k)$ = frequency of median class
w = class interval
L = lower limit of median class

The median for the data on the heights of female workers can be calculated by using the above formula as follows:

$$\text{Med} = \frac{\frac{50}{2} - 20}{16} 3 + 62.5$$

$$= \frac{5}{16} 3 + 62.5 = \frac{15}{16} + 62.5 = 0.94 + 62.5 = 63.44$$

The discrepancy between this value and the one obtained directly from Table 4-2 is due to the assumption for the grouped data that the items of the median class are evenly distributed within the class.

By the use of the ogive of Fig. 4-3, we can also find approximately the value of the median. This is done by drawing a horizontal line through the point 0.50 on the right-hand scale, or 25 on the left-hand scale, and drawing from the intersection of this line and the ogive another line perpendicular to the x axis. The point on the x axis that this perpendicular line would run through is, approximately, 63.5, the value of the median calculated above.

Quartiles and percentiles Other measures that are allied to the median because they are also based on their position in a series of observations includes the quartiles and percentiles. There are three *quartiles*, usually designated Q_1, Q_2, and Q_3, which divide the series into four equal parts. The *second quartile* Q_2 is identical with the median. The *first quartile* Q_1 is the value at or below which one-fourth of all the items in the series fall; the *third quartile* Q_3 is the value at or below which three-fourths of the items lie.

With ungrouped data, a quartile, as does the median, assumes either the value of one of the items, or the value halfway between two items. If n is divisible by 4, the first quartile has the value halfway between the $n/4$th and the next higher number. If $n/4$ is not an integer, the first quartile has the value of the next higher integer. By substituting $3n/4$th for $n/4$th, this technique is used to determine the value of the third quartile.

For the data in Table 4-2, for example, the first quartile is found to be 61, and the third quartile, 66.

With grouped data, the method of estimating the first and third quartiles is similar to that of estimating the median. As a formula, the calculation of the kth quartile appears as follows:

$$Q_k = \frac{kn/4 - CF(x_{k-1})}{FR(x_k)} w + L \tag{4-17}$$

where n = total number of items in distribution
w = class interval
$CF(x_{k-1})$ = cumulative frequency below lower limit of class in which kth quartile is located
$FR(x_k)$ = frequency of kth quartile class
L = lower limit of kth quartile class

Thus for the value of the first quartile of the data in Table 4-6, we have

$$Q_1 = \frac{\frac{50}{4} - 9}{11}(3) + 59.5 = \frac{3.5(3)}{11} + 59.5 = 60.49$$

Similarly, the third quartile is computed

$$Q_3 = \frac{3(50)/4 - 36}{9}(3) + 65.5 = \frac{1.5(3)}{6} + 65.5 = 0.5 + 65.5 = 66$$

The three quartiles are regarded as the 25th, 50th, and 75th *percentiles*, which are commonly designated P_{25}, P_{50}, and P_{75}. There are 99 percentiles, which divide the distribution into 100 equal parts. The percentiles are determined by techniques similar to those used for finding the median and the quartiles. With ungrouped data, the percentile takes either the value halfway between two observations or the value of one of the observations, depending on whether n is divisible by 100. If a frequency distribution has been constructed a formula similar to those for determining the median and the quartiles may be used.

Equation (4-17) can be slightly modified for locating the percentile of a set of observations. For instance, the 60th percentile, denoted by P_{60}, is the value that separates the lower 60 percent from the upper 40 percent of all the observations. Refer to the data of Table 4-6. The class within which the 60th percentile falls is 62.5 to 65.5. The cumulative frequency to the 60th number is $60n/100 = 60(50)/100 = 30$, the cumulative frequency up to the lower limit of the 60th percentile class is 20, and the frequency of the 60th percentile class is 16. Thus we have the 60th percentile of the data of Table 4-6 computed as follows:

$$P_{60} = \frac{60(50)/100 - 20}{16}(3) + 62.5 = \frac{30}{16} + 62.5$$
$$= 1.875 + 62.5 = 64.375$$

4-3-2 The mode

Frequently, business data are classified into groups that are not numerical. For example, a company's employees may be classified according to their sex, marital status, occupations, etc. Under such circumstances, it does not make sense to ask for the mean or median of employees' sex, marital status, or occupations, unless the various groups of each of these variables are assigned numerical values. It does make sense, however, to ask which occupation, if any, includes the most employees. This occupation is then called the *modal* occupation. Thus, in addition to the arithmetic mean and median, we need to introduce another measure of central tendency, the mode.

The *mode* is defined as the value that occurs most frequently in a distribution. To determine the value of the mode for a frequency distribution, it is necessary to identify the *modal class* in which the mode is located. In general the modal class is one with the highest frequency of the distribution. It is evident, for example, that the modal class for the distribution in Table 4-6 is the class of 62.5 to 65.5, since it contains more items than any other class.

Once the modal class is identified, the next step is to locate the mode within the class. The midpoint of the modal class would be used as the modal value if the distribution included an infinite number of items and the class intervals were infinitely small. The rationale behind this is that when the class interval is reduced, the modal value is delineated within a smaller range of values, and thus tends to be closer to the midpoint of the class. In practice, however, this ideal situation is virtually nonexistent. The value of the mode therefore is usually estimated by the method of interpolation.

In approximating the modal value by interpolation, we must realize that the items usually gravitate toward the point of the greatest density. There is a tendency for this point of gravity to pull the modal value toward either the upper or the lower class limit, depending on whether the *postmodal* or the *premodal* class has the greater frequency. For the distribution in Table 4-6, the frequency of the premodal class is 11 and that of the postmodal is 9. This suggests that the modal value is pulled toward the lower limit of the modal class. As a formula, the approximation of the value of the mode (Mo) appears as follows:

$$\text{Mo} = \frac{d_1}{d_1 + d_2} w + L \qquad (4\text{-}18)$$

where d_1 = *absolute* difference between frequency of modal and premodal class

d_2 = absolute difference between frequency of modal and postmodal class

w = length of class interval

L = lower limit of modal class

Applying this formula for the data in Table 4-6, we find

$$\text{Mo} = \frac{5}{5+7} 3 + 62.5 = \frac{15}{12} + 62.5 = 1.25 + 62.5 = 63.75$$

This approximation of the modal value from a frequency distribution has a number of problems. First of all, it may be observed that the modal value estimated from a highly skewed distribution is too close to one end of the data, and thus fails to be a good representative of the series. Moreover, the location of the modal class, and consequently the value of the mode, depends on the ways in which the data are classified. The position of the mode varies with the selected length of the class interval. Thus, while the mode may be a useful measure of central tendency in some instances, it remains the most volatile among all the averages.

Furthermore, the mode is almost useless if a distribution is *not unimodal.*[1] (A unimodal distribution is one that has a single modal class.) However, if two adjacent classes are the modal classes, then the two may be combined as a single class in applying the above formula for finding the modal value.

4-3-3 Comparison of the averages

The arithmetic mean, median, and mode are regarded as the most important averages because of their simplicity and usefulness. However, as briefly indicated in the previous sections, the three averages are not equally applicable to any conditions. In the remainder of this section, we shall see that the relative positions of these three measures depend on the skewness of the distribution.

Figure 4-4 shows the positions of the arithmetic mean, the median, and the mode. If the distribution is symmetrical,as demonstrated in Fig. 4-4*a*, the three measures of central tendency are of identical value. If the distribution is skewed, as shown in Fig. 4-4*b* and *c*, the three average values diverge. Nevertheless, for a unimodal distribution, the relative positions of the three measures will be such that the median is always between the mode and the mean regardless of the direction of skewness.

Figure 4-4*b* shows that the distribution is positively skewed, or skewed to the right—the longer tail of the distribution is to the right. The median takes the position such that half of the distribution lies above it and half lies below it. Since the long tail is to the right, the mode, which is still at the peak of the curve, must be pulled to the left of the median, that is, toward the low values of the distribution. The arithmetic mean, being most sensitive to extreme values, is pulled to the right of the median, or toward the high values of the distribution. The reverse is true when the distribution is negatively skewed, as in Fig. 4-4*c*. With a distribution of moderate skewness, the median tends to be approximately one-third as far away from the mean as from the mode.

The reader may be wondering which average to use to characterize a given set of data. Generally, the average we should use depends on the pattern of the distribution of the data and the purpose for which it is intended.

If the distribution is symmetrical, or nearly so, it does not make much difference which measure is used. If it is skewed, positive or negative, it may be

[1] A distribution with two modes is called a *bimodal distribution.* In some instances where bimodal values occur, they may be taken as an indication that the population consists of two different subpopulations.

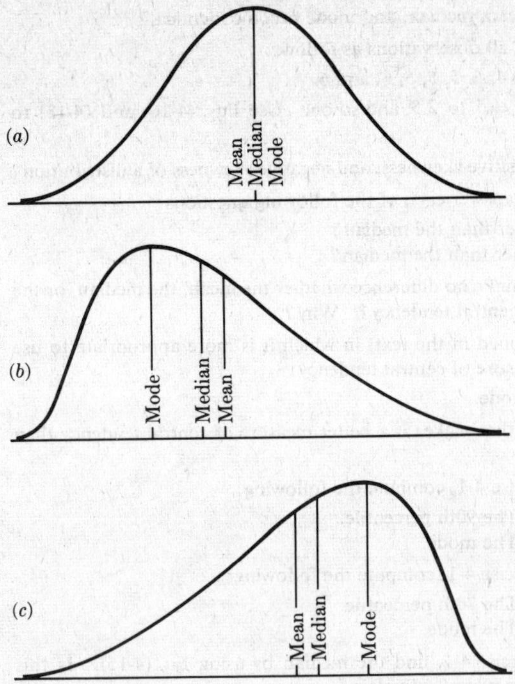

Figure 4-4
The position of the mean, the median, and the mode as shown in (*a*) symmetrical distribution, (*b*) positively skewed distribution, and (*c*) negatively skewed distribution.

more appropriate to use the mode of the median, since the mean is not a typical value under the circumstance.

If the measure is used for obtaining a total value, the mean should be used. For example, if a passenger airplane is designed to carry 20,000 pounds, it will be expected to load 100 people on the basis that the average weight including luggage per person is 200 pounds. If the desire is to determine the typical household expenditure on food, the mode should be used. On the other hand, the relative academic achievement of a student in his class may be indicated by stating whether his test score is above or below the median, or the 50th percentile.

Other measures of central tendency, less frequently used, include the mid-range, the quadratic mean, the log normal mean, the geometric mean, and the harmonic mean. The computation and application of the last two measures are illustrated in Probs. 14, 15, and 16 of Exercise 4-3.

Exercise 4-3

1 Identify the median and the mode for each of the following lists of numbers:

a 2, 4, 5, 6, 6, 6, 9, 10, 13, and 15.
b 1, 3, 5, 7, 7, 7, 9, 9, 10, 10, 11, and 12.

2 Under what condition do the mean, median, and mode become identical?

3 Suppose that a series consists of 20 observations as follows:

1, 2, 2, 3, 3, 3, 3, 3, 4, 4, 4, 4, 4, 4, 5, 5, 5, 5, 6, and 6.

Let the class limits be 0.5 to 1.5, 1.5 to 2.5, and so on. Use Eqs. (4-16) and (4-18) to find the median and the mode.

4 What are meant by skewness, positive skewness, and negative skewness of a distribution?

5 For a given set of observations, answer each of the following questions:

a When will the mean be greater than the median?
b When will the mean be smaller than the median?

6 Under what condition would it make no difference whether the mean, the median, or the mode is used as the measure of central tendency? Why?

7 Describe a situation (not mentioned in the text) in which it is more appropriate to use each of the following as the measure of central tendency:

a The median. *b* The mode.

8 Identify a property of the mean that makes it a better measure of central tendency than any of the other measures.

9 For the data in Prob. 9 of Exercise 4-1, compute the following:

a The median.
b The third quartile.
c The 90th percentile.
d The mode.

10 For the data in Prob. 10 of Exercise 4-1, compute the following:

a The median.
b The first quartile.
c The 70th percentile.
d The mode.

11 For the data in Prob. 11 of Exercise 4-1, find the median by using Eq. (4-15). Is this median larger or smaller than the mean? Explain.

12 For the data in Prob. 12 of Exercise 4-1, compute the following:

a The median by Eq. (4-16).
b The 75th percentile by Eq. (4-17).
c The mode by Eq. (4-18).

13 One of the advantages of the median and the mode over the arithmetic mean is that the median and the mode can be easily determined for an open-end distribution, while such a distribution makes it difficult to compute the mean. An *open-end distribution* is a frequency distribution that has one or both of the end classes unspecified. The following is an example of an open-end frequency distribution:

Family annual income, \$	*Number of families*
Less than 3,000	25
3,000–4,999	31
5,000–6,999	42
7,000–8,999	45
9,000–10,999	52
11,000–12,999	42
13,000–14,999	35
15,000 or over	28
Total	300

Since the lower limit of the first class and the upper limit of the last class are unstated, it is impossible to compute the mean by the usual procedure. The median and the mode can still be determined for such a distribution. Find the median and the mode for this distribution.

14 The arithmetic mean, the median, and the mode are considered the more important measures of central tendency because of their broad usefulness. Under certain circumstances, however, other measures of central tendency are more appropriate. One of these is called the *geometric mean*, commonly designated (G), which is defined as the nth root of the product of a set of n values. Symbolically,

$$G = \sqrt[n]{X_1 X_2 \cdots X_n}$$

One of the most useful applications of the geometric mean is to average *rates of change*. Suppose, for example, the sales of a growing company had increased from $2 million in 1972 to $4 million in 1973 and $6 million in 1974. The increase was 200 percent from 1972 to 1974, with an arithmetic mean of 100 percent per year. This is obviously wrong since the average rate of increase per year was much less. To obtain the average rate of growth in sales, the geometric mean should be used. Since the sales in 1973 were twice as high as 1972, and the sales in 1974 were 1.5 times as high as in 1973, the geometric mean, or G, of the two values 2.0 and 1.5 is computed as follows:

$$G = \sqrt{(2.0)(1.5)} = \sqrt{3} = 1.7325$$

The average rate of growth in sales is therefore 73.25 percent per year for the 2-year period. Find the geometric mean for each of the following sets of values:

a 1, 3, 9.
b 2, 4, 8.
c 2, 8.
d 4, 4, 4, 4.
e 1; 10; 100; 1,000; 10,000.

15 Another measure of central tendency, which is sometimes used, is the *harmonic mean*, commonly designated H. It is defined as the reciprocal of the arithmetic mean of the reciprocals of the individual values; that is,

$$H = \frac{n}{1/X_1 + 1/X_2 + \cdots + 1/X_n}$$

The applications of the harmonic mean may be illustrated as follows: Suppose that a person spent $1 in each of three stores. In the first store, he bought socks at 20 cents a pair; in the second, socks at 25 cents a pair; and in the third, socks at 50 cents a pair. What is the average price he paid per pair of socks? The data are expressed as "so many pairs of socks per dollar," while we wish to know "the amount of money per pair of socks." It will be erroneous to compute the average price by first computing the average price for each shop and then the average of all the three averages. The correct answer is obtained by applying the method of the harmonic mean as follows:

$$H = \frac{3}{1/0.2 + 1/0.25 + 1/0.5} = \frac{3}{5 + 4 + 2} = \frac{3}{11} = 27\frac{3}{11} \text{ cents}$$

In general, the harmonic mean is used if the observations are expressed *inversely* to what is expressed in the required average. More specifically, if the average cost per unit of product is required but the data are expressed as so many units of product for a certain amount of cost, the harmonic mean should be used. Find the harmonic mean for each of the sets of values in Prob. 14 above.

16 Suppose that the productivity of three workers, A, B, and C, is shown as follows:

Worker	*Hourly output*
A	30 units
B	20 units
C	40 units

If they work an 8-hour day, what is the average time required per unit of output?

4-4
The Mean of the Population

So far, our interest has been limited to characteristics of the sample, which are known as sample *statistics*, meaning statistical measures or descriptions of a set of collected data. We shall now consider the mean of the sample space, or universal set. It is commonly referred to as the *mean of the population* and designated by the symbol μ, the Greek letter read "mu." It is known as a *parameter* of the population.

If there is a finite number of potential observations in a given problem situation, that is, the population is finite, then the definition of the mean of the population is similar to Eq. (4-3) with only the sample size n replaced by the population size. When the population is infinite, however, the sum of all potential observations divided by infinity obviously cannot be used for expressing the mean of the population; some other device must be found to define the population mean μ. The answer is to consider the mean μ the *expected value*, or *expectation*, of the random variable. The notion of expectation has broad applications in statistical analysis. Before we discuss the properties of the expected value of a random variable, let us first see how the concept is formally defined.

4-4-1 Expected value defined

As indicated in Sec. 3-4, a probability distribution is the theoretical distribution of a random variable. The probability distribution for a population is analogous to the relative frequency distribution of a sample. Assume that a discrete random variable X has the possible values $x_1, x_2, \ldots, x_m$ with corresponding probabilities $P(x_1), P(x_2), \ldots, P(x_m)$. Then the expected value of the random variable X, which is usually designated by the symbol $E(X)$, is defined as follows:

$$E(X) = \sum_{j=1}^{m} x_j P(x_j) = x_1 P(x_1) + x_2 P(x_2) + \cdots + x_m P(x_m) = \mu \qquad (4\text{-}19)$$

That is, the expected value of the discrete random variable X is equal to the sum of the products of all the possible values of X multiplied by their respective probabilities. It is equal to the population mean of X, or μ.

Suppose, for example, that X is used to designate the decimal digits, 0, 1, 2, 3, 4, 5, 6, 7, 8, 9. If we use the random digit table or the "goldfish bowl" method to select a random digit,[1] then each and every one of the above 10 digits has the same probability, namely, 0.1, of being drawn. The mean or expected value of this random variable is therefore

$$E(X) = 0(0.1) + 1(0.1) + 2(0.1) + \cdots + 9(0.1) = 4.5$$

That is, the *long-run* average of the digit numbers drawn in this experiment will be 4.5.

4-4-2 Properties of expected value

If there exists a function of the random variable X defined on a domain $\{x_1, x_2, \ldots, x_m\}$ with corresponding probabilities $P(x_1), P(x_2), \ldots, P(x_m)$, we can speak of the expected value of the function $f(x)$ as follows:

$$E(f(x)) = \sum_{j=1}^{m} f(x_j)P(x_j) \tag{4-20}$$

This is the general expression of the expected value of a function. Suppose that we have a function $f(x) = x^2$, where x stands for the number of spots uppermost after an ideal die is rolled. The probability distribution of x is therefore *uniform*, since each of the six sides has the same probability of occurring, namely, $\frac{1}{6}$. Thus the expected value of $f(x)$ is

$$\begin{aligned} E(f(x)) &= E(x^2) = \sum x^2 P(x) \\ &= 1^2(\tfrac{1}{6}) + 2^2(\tfrac{1}{6}) + 3^2(\tfrac{1}{6}) + 4^2(\tfrac{1}{6}) + 5^2(\tfrac{1}{6}) + 6^2(\tfrac{1}{6}) \\ &= \tfrac{1}{6}(1 + 4 + 9 + 16 + 25 + 36) = \tfrac{1}{6}(91) = \tfrac{91}{6} = 15\tfrac{1}{6} \end{aligned}$$

Some other important properties of expectation which apply to more specific functions of a random variable are listed below:

$$E(c) = c \tag{4-21}$$

$$E(cX) = cE(X) \tag{4-22}$$

$$E(a + bX) = a + bE(X) \tag{4-23}$$

If Y is another random variable, then the following relations apply:

$$\text{If } X \geq Y, \text{ then } E(X) \geq E(Y) \tag{4-24}$$

$$E(X + Y) = E(X) + E(Y) \tag{4-25}$$

$$E(aX + bY) = aE(X) + bE(Y) \tag{4-26}$$

[1] For the random digit table and the goldfish bowl methods, see Sec. 6-2, Sampling Designs.

These relations can be expanded to cover three or more variables. Generally, if $X_1, X_2, \ldots, X_K$ are K different random variables, then

$$E(a_1X_1 + a_2X_2 + \cdots + a_KX_K) = a_1E(X_1) + a_2E(X_2) + \cdots + a_KE(X_K) \quad (4\text{-}27)$$

In short, the expected value of a sum is the sum of the expected values.

Let us use a simple example to illustrate such a relationship. Suppose that two ideal coins and one ideal die are tossed together. Let X be the number of heads obtained from the two coins and Y be the number of spots uppermost on the die. Suppose further that you will be paid in dollars equal to two times the number of heads obtained *and* three times the number of spots on the die. (Of course, your privilege of playing this game will cost you something at the outset.) What is the expected payment for each toss?

From the information given here, we know the payment is a function of X and Y. That is,

$$\text{Payment} = 2X + 3Y$$

By Eq. (4-26), we have

$$E(\text{payment}) = 2E(X) + 3E(Y)$$

The values of X are 0, 1, and 2 with probabilities $\frac{1}{4}$, $\frac{2}{4}$, and $\frac{1}{4}$, respectively. Likewise, the values of Y are 1, 2, 3, 4, 5, and 6 with the probability of $\frac{1}{6}$ each. Consequently,

$$\begin{aligned} E(\text{payment}) &= 2[0(\tfrac{1}{4}) + 1(\tfrac{2}{4}) + 2(\tfrac{1}{4})] + 3[1(\tfrac{1}{6}) + 2(\tfrac{1}{6}) + 3(\tfrac{1}{6}) + 4(\tfrac{1}{6}) + 5(\tfrac{1}{6}) + 6(\tfrac{1}{6})] \\ &= 2(1) + 3(\tfrac{21}{6}) = 2 + 3(3.5) = 2 + 10.5 = 12.5 \text{ dollars} \end{aligned}$$

Exercise 4-4

1 What is meant by an expected value?

2 A balanced die is rolled. Let X be the number of spots shown uppermost after it is rolled. Find the expected number of spots for each roll.

3 Three balanced coins are tossed. Let X be the number of heads obtained after they are tossed. Find the expected number of heads for each toss.

4 A certain company has a branch in one city with a daily sales of X dollars and another branch in a different city with a daily sales of Y dollars. It has been found, through the analysis of the company's records of long periods in the past, that the long-run averages of daily sales of X and Y are \$45,000 and \$55,000, respectively. Let T be the total daily sales of the two branches. What is the expected value of T?

5 Two brands of cigarettes are compared to determine the difference D in the nicotine content of the two brands. Let X be the nicotine content of brand I and Y be the nicotine content of brand II. It has been found that the mean of X is 25 milligrams and the mean of Y is 22 milligrams. What is the expected value of D?

6 Let X designate the number of spots showing on the face of a well-balanced die after it is rolled. Given that $Y = X^2 + 2X$, find $E(Y)$.

7 Let X be the number of heads obtained in the toss of three ideal coins. If two or more heads turn up, you collect \$2; otherwise, you pay \$1. What is your expected gain or loss per game?

8 Suppose that an ideal coin is tossed twice. If it turns up tails both times, you win nothing. If heads turns up only once, you win \$1. If both tosses result in heads, you win \$4. In other words, $Y = X^2$ for $X =$ number of heads obtained and $Y =$ number of dollars you would win. Find:

a The expected number of heads to turn up.
b The expected amount of money you would win.

9 Suppose that X is uniformly distributed on $\{1, 2, 3\}$ and Y is uniformly distributed on $\{0, 1\}$ and that X and Y are independent. If $W = 2X + 3Y$, verify that $E(W) = 2E(X) + 3E(Y)$.

10 Given the random variables X and Y in Prob. 9 and that $Z = 2X - Y$, verify that $E(Z) = 2E(X) - E(Y)$.

11 Mr. Smith has \$5 million. He has an opportunity to invest his money to end up with \$$X$ million, where X has the following probability distribution:

x	$P(x)$
3	0.1
5	0.5
7	0.2
9	0.2
Total	1.0

What will be his expected amount of money if he decides to invest his \$5 million?

12 A well-balanced die is rolled. If the number of spots showing is less than 3, you collect \$10. If it shows 3 or more, you lose \$6. What is your expected gain or loss from this game?

13 Suppose that in a lottery 1,000 tickets are sold at \$3 each, and three prizes are to be awarded. The first prize is a television set worth \$500, the second prize is a short wave radio worth \$300, and the third prize is a 10-speed bicycle worth \$200. If you plan to buy one ticket, what is your expected gain or loss from this venture?

14 Let N be the number of heads obtained in the toss of two ideal nickels, and D be the number of heads obtained in the toss of three ideal dimes. Suppose that you are given the privilege of tossing the five coins simultaneously; you will receive a sum of money in dollars that will be equal to six times the number of heads of the nickels plus three times the number of heads of the dimes after the coins are tossed. Let W be the total amount of money that you will receive for each game, that is, $W = 6N + 3D$. What is the expected value of W?

Chapter 5

Measures of dispersion

In the preceding chapter, consideration was given to certain measures of central tendency of a distribution. These measures serve to locate the "center" of the distribution, but they do not reveal how the items are spread out, or scattered, on each side of the center. This latter characteristic of a distribution is commonly referred to as the *dispersion*, also known as *scatter* or *variation*. If the dispersion is small, it indicates high uniformity of the items in the distribution. On the other hand, a large dispersion shows little uniformity. No dispersion indicates perfect uniformity, which means that all items in the distribution are identical. It is just as important to measure this property of a distribution as to locate the central values.

Various distributions may have exactly the same averages, but substantial differences in variability. For example, two departments of 10 workers each may have identical average output, nine units per hour, yet their distributions may differ as follows:

Department I		*Department II*	
Output per hour	*Number of workers*	*Output per hour*	*Number of workers*
7	1	3	1
8	2	4	1
9	4	5	1
10	2	6	1
11	1	9	2
		12	1
		13	1
		14	1
		15	1

These two distributions are certainly not identical. Their difference lies in the dispersion of their items. Workers in department I show relatively little variation in their hourly outputs, while those in department II have much greater variation. It is the primary objective of this chapter to consider the basic techniques by which this important characteristic of a distribution is measured.

5-1 Crude Measures

In this section we shall discuss briefly some of the simple and crude measures of dispersion. They include the range and the average deviation.

5-1-1 The range

The simplest and crudest measure of dispersion is the *range*, which is the difference between the lowest and the highest values in the distribution. If the smallest item of a set is 90 and the largest item is 200, the range is then 200 minus 90, or 110. Of the heights of 50 female workers in Table 4-2, the range is 72 inches minus 53 inches, or 19 inches.

For grouped data, the difference between the lower class limit of the lowest class and the upper class limit of the highest class is considered the range. Thus the range of the same data, but as presented in Table 4-3, is $74.5 - 50.5 = 24$ inches. This is obviously higher than the 19 inches stated above. Such a discrepancy is brought about by the fact that the class limits extend slightly beyond the extreme values of the data.

The range is easy to compute and is a common way to describe dispersion. It is especially useful in situations where the purpose of investigation is only to find out the extent of extreme variations. For instance, performance of stocks in the stock market is usually reported in terms of their ranges, by quoting the low and high prices during each session or over a period of time. Meteorologists usually report only the low and high temperatures rather than all the hourly readings of the day.

The range, however, has certain disadvantages. Since it depends solely on the highest and lowest values, it is extremely sensitive to unusual values in a series. Furthermore, the range does not provide measurement of the dispersion of items relative to the center value. It tends to increase as the size of the sample increases. These defects tend to limit its usefulness as a measure of variability.

5-1-2 The average deviation

The *average deviation*, sometimes called the *mean deviation*, is the arithmetic mean of the *absolute* values of the deviations from the mean or the median. The median is sometimes preferred in computing the average deviation because the sum of the absolute values of the deviations from the median is smaller than that from any other value. In practice, however, the arithmetic mean is generally used. If the distribution is symmetrical, the mean is identical with the median and the same average deviation is obtained.

If $X_1, X_2, \ldots, X_n$ form a sample of n observations, the formula for computing the average deviation, commonly designated AD, appears as follows:

$$\text{AD} = \frac{\sum_{i=1}^{n} |X_i - \bar{X}|}{n} = \frac{\sum_{i=1}^{n} |d_i|}{n} \tag{5-1}$$

where d stands for deviations from the mean, and $|\quad|$ means that the signs of the deviations are ignored.[1] Table 5-1 shows how the average deviation is computed from 10 raw scores.

It has been found that if the distribution is bell-shaped and symmetrical, generally known as *normal* (to be discussed in Chap. 8), 57.5 percent of the observations are included in the interval $\bar{X} \pm \text{AD}$. Also, if it is moderately skewed, this interval will include approximately 57.5 percent of all the observations. In other words, more than half of the items are concentrated within one unit of the average deviation around the mean.

One of the drawbacks of the average deviation is the ambiguity about the

Table 5-1

Computation of average deviation

Scores X_i	*Deviations from mean* $\lvert d_i \rvert = \lvert X_i - \bar{X} \rvert$
3	3
3	3
5	1
5	1
5	1
7	1
7	1
8	2
8	2
9	3
Total 60	18

$$\bar{X} = \frac{60}{10} = 6$$

$$\text{AD} = \frac{\sum_{i=1}^{n} |d_i|}{n} = \frac{18}{10} = 1.8$$

[1] The absolute value of a number x denoted by $|x|$ is defined as follows:

$|x| = x$ if $x \geq 0$
$|x| = -x$ if $x < 0$
$|x| < c$ if $-c < x < c$
$|x| > c$ if $x > c$ or $x < -c$

measure of central tendency that has been used for its computation. In order to avoid confusion, it is necessary to state clearly whether the mean or the median is used in computing the average deviation. The most serious defect of this measure, however, is the fact that the signs of deviations must be ignored. If the signs are not ignored, the sum of deviations from the mean is 0; it is close to 0 if the median is used as the reference point. The procedure of ignoring the signs makes the method nonalgebraic, and the measure is not amenable to mathematical manipulation.

But these drawbacks should not blind us to its practical advantages. It takes into account every item in the distribution and shows the scatter of the items around the measure of central tendency. Its simplicity in meaning and computation makes it easily comprehensible to those who are unfamiliar with statistical methods. Moreover, a knowledge of the average deviation helps the reader to understand the most important measure of dispersion, the standard deviation.

Exercise 5-1

1 Among the crude measures of dispersion is the range.

a What are the disadvantages of the range as a measure of dispersion?
b Why is it sometimes used in spite of its disadvantages?

2 Among the crude measures of dispersion is the average deviation.

a What are the drawbacks of the average deviation as a measure of dispersion?
b How useful is the concept of average deviation as a measure of dispersion?

3 Given the following sample of 10 observed values:

2 2 3 3 4 4 6 8 9 9

a Find the average deviation from the mean.
b Find the average deviation from the median.
c Draw a conclusion or generalization after comparing (*a*) and (*b*).

4 The hourly outputs of 10 workers in each of the two departments in the example at the beginning of this chapter are reproduced below:

Department I		*Department II*	
Output per hour	*Number of workers*	*Output per hour*	*Number of workers*
7	1	3	1
8	2	4	1
9	4	5	1
10	2	6	1
11	1	9	2
		12	1
		13	1
		14	1
		15	1

a Find the range of the hourly outputs for each of the two departments.
b Find the average deviation from the mean of the hourly outputs for each of the two departments.

5 Suppose that the scores received by 20 new workers in an aptitude test are shown as follows: 31, 39, 40, 43, 45, 50, 52, 58, 59, 60, 70, 71, 72, 78, 80, 85, 87, 90, 91, and 99. Find the average deviation from the mean for the 20 test scores.

6 Suppose that the test scores in Prob. 5 above are organized into a frequency distribution with classes 27.5 to 42.5, 42.5 to 57.5, 57.5 to 72.5, 72.5 to 87.5, and 87.5 to 102.5 and midpoints 35, 50, 65, 80, and 95, respectively. Find the average deviation from the mean by using the frequency distribution.

5-2
The Variance and Standard Deviation

There is another way to handle the problem of signs of deviations from the mean. Instead of ignoring them as in the computation of the average deviation, they may each be squared, and then the results are added. By the squaring operation, the deviations from the mean will sum not to 0 but to a positive number, and each deviation will contribute to the sum of squares regardless of sign. This sum of squares can be regarded as a measure of the total dispersion of the distribution. By dividing the sum by n, the number of items in the sample, we obtain the mean of squares of deviations, a measure called the *variance* of the distribution. As a formula, the variance of a sample of n observations, commonly designated s^2, is

$$s^2 = \frac{\sum_{i=1}^{n} (X_i - \bar{X})^2}{n} \tag{5-2}$$

The variance is of considerable importance in itself, and we shall discuss it in greater detail in Chaps. 12 and 13.

Because of the operation of squaring, the variance is expressed in *square units*, and not in the original unit. It is therefore necessary to extract the square root to restore the original unit. The measure of dispersion thus obtained is called the *standard deviation*. Let s stand for the standard deviation of the sample. We have

$$s = \sqrt{\frac{\sum_{i=1}^{n} (X_i - \bar{X})^2}{n}} \tag{5-3}$$

Thus, by definition, the standard deviation is the *positive* square root of the mean-square deviations of the items from the arithmetic mean. Observe that, similar to the average deviation, the standard deviation is an average variation per observation. (Appendix IV provides the square roots of numbers of 1 to 99.9 with the top row listing an extra digit for each number in the first column.)

5-2-1 Ungrouped data

With a small sample of ungrouped data, we may use Eqs. (5-2) and (5-3) to calculate the variance and the standard deviation. For the sake of simplicity, let us use the data in Table 5-1 to illustrate the computation processes. The results are shown in Table 5-2.

The process, however, is rather laborious because the arithmetic mean must be subtracted from every observation. It is especially time consuming if the mean is any number with several digits or decimal places. For this reason, a more efficient method of computation has been developed by making use of the following relationship:

$$\begin{aligned}\sum_{i=1}^{n}(X_i-\overline{X})^2 &= \sum_{i=1}^{n}(X_i^2-2\overline{X}X_i+\overline{X}^2)\\ &= \sum_{i=1}^{n}X_i^2-2\overline{X}\sum_{i=1}^{n}X_i+n\overline{X}^2\\ &= \sum_{i=1}^{n}X_i^2-2\overline{X}n\overline{X}+n\overline{X}^2 \qquad \text{by (4-3)}\\ &= \sum_{i=1}^{n}X_i^2-n\overline{X}^2\end{aligned}$$

Table 5-2

Computation of standard deviation from ungrouped data

	Scores X_i	Deviations from mean, $X_i-\overline{X}$	Squares of deviations, $(X_i-\overline{X})^2$	Squares of raw scores X_i^2
	3	−3	9	9
	3	−3	9	9
	5	−1	1	25
	5	−1	1	25
	5	−1	1	25
	7	1	1	49
	7	1	1	49
	8	2	4	64
	8	2	4	64
	9	3	9	81
Total	60		40	400

$$\overline{X}=\frac{\sum_{i=1}^{n}X_i}{n}=\frac{60}{10}=6 \qquad \text{(from Table 5-1)}$$

$$s^2=\frac{\sum_{i=1}^{n}(X_i-\overline{X})^2}{n}=\frac{40}{10}=4 \qquad \text{and} \qquad s=\sqrt{\frac{\sum_{i=1}^{n}(X_i-\overline{X})^2}{n}}=\sqrt{4}=2$$

Source: Table 5-1.

Therefore

$$s^2 = \frac{\sum_{i=1}^{n} (X_i - \overline{X})^2}{n} = \frac{\sum_{i=1}^{n} X_i^2 - n\overline{X}^2}{n} = \frac{\sum_{i=1}^{n} X_i^2}{n} - \overline{X}^2 \qquad (5\text{-}4)$$

and

$$s = \sqrt{s^2} = \sqrt{\frac{\sum_{i=1}^{n} X_i^2}{n} - \overline{X}^2} \qquad (5\text{-}5)$$

These formulas make it unnecessary to subtract the mean from each observation. The ungrouped values are simply squared, then totaled, and the sum divided by n. The result thus obtained minus the square of the mean yields the variance of the distribution. The last column of Table 5-2 provides the squared values of the raw scores. By using Eqs. (5-4) and (5-5), we obtain the variance and standard deviation for the ungrouped data as follows:

$$s^2 = \frac{1}{n} \sum_{i=1}^{n} X_i^2 - \overline{X}^2 = \frac{400}{10} - 6^2 = 40 - 36 = 4$$

and

$$s = \sqrt{\frac{1}{n} \sum_{i=1}^{n} X_i^2 - \overline{X}^2} = \sqrt{4} = 2$$

5-2-2 Grouped data

As indicated in Chap. 4, a large sample is usually grouped into a frequency distribution. For grouped data, the methods shown above cannot be used to compute the variance and the standard deviation since they are based on ungrouped data. Before we present the methods for computing the variance and the standard deviation of a frequency distribution, let us first review the procedure used in Table 5-2.

$$s^2 = \frac{\sum_{i=1}^{n} (X_i - \overline{X})^2}{n}$$

$$= \frac{(3-6)^2 + (3-6)^2 + (5-6)^2 + (5-6)^2 + (5-6)^2 + (7-6)^2 + (7-6)^2 + (8-6)^2 + (8-6)^2 + (9-6)^2}{10}$$

$$= \frac{(3-6)^2(2) + (5-6)^2(3) + (7-6)^2(2) + (8-6)^2(2) + (9-6)^2(1)}{10}$$

$$= \frac{18 + 3 + 2 + 8 + 9}{10} = \frac{40}{10} = 4$$

Observe that the numbers 2, 3, 2, 2, and 1 in the numerator of the third expression are the frequencies. Symbolically, this expression appears to be

$$\frac{(x_1 - \overline{X})^2 FR(x_1) + (x_2 - \overline{X})^2 FR(x_2) + \cdots + (x_m - \overline{X})^2 FR(x_m)}{n}$$

$$= \frac{\sum_{j=1}^{m} (x_j - \overline{X})^2 FR(x_j)}{n}$$

Thus, for the variance and the standard deviation of a frequency distribution, we have

$$s^2 = \frac{\sum_{j=1}^{m} (x_j - \overline{X})^2 FR(x_j)}{n} \tag{5-6}$$

and

$$s = \sqrt{\frac{\sum_{j=1}^{m} (x_j - \overline{X})^2 FR(x_j)}{n}} \tag{5-7}$$

Table 5-3

Computation of standard deviation from grouped data

x_j	$FR(x_j)$	$x_j - \overline{X}$	$(x_j - \overline{X})^2$	$(x_j - \overline{X})^2 FR(x_j)$	x_j^2	$x_j^2 FR(x_j)$
3	2	−3	9	18	9	18
5	3	−1	1	3	25	75
7	2	1	1	2	49	98
8	2	2	4	8	64	128
9	1	3	9	9	81	81
Total	10			40		400

$$s^2 = \frac{\sum_{j=1}^{m} (x_j - \overline{X})^2 FR(x_j)}{n} = \frac{40}{10} = 4$$

and

$$s = \sqrt{\frac{\sum_{j=1}^{m} (x_j - \overline{X})^2 FR(x_j)}{n}} = \sqrt{\frac{40}{10}} = \sqrt{4} = 2$$

Or

$$s^2 = \frac{\sum_{j=1}^{m} x_j^2 FR(x_j)}{n} - \overline{X}^2 = \frac{400}{10} - 6^2 = 40 - 36 = 4$$

and

$$s = \sqrt{\frac{\sum_{j=1}^{m} x_j^2 FR(x_j)}{n} - \overline{X}^2} = \sqrt{\frac{400}{10} - 6^2} = \sqrt{40 - 36} = \sqrt{4} = 2$$

Source: Table 5-2.

where x_j = jth value that variable X can take on for $j = 1, 2, \ldots, m$

m = number of distinct values assumed by X

$FR(x_j)$ = frequency or number of times jth value is taken on

If Eq. (5-6) or (5-7) is used, it is again necessary to subtract the arithmetic mean from each of the distinct values assumed. This operation can also be dispensed with by using the following equations:

$$s^2 = \frac{\sum_{j=1}^{m} x_j^2 FR(x_j)}{n} - \bar{X}^2 \tag{5-8}$$

and

$$s = \sqrt{\frac{\sum_{j=1}^{m} x_j^2 FR(x_j)}{n} - \bar{X}^2} \tag{5-9}$$

Table 5-3 illustrates using Eqs. (5-6) to (5-9) to compute the variance and the standard deviation for a frequency distribution.

Exercise 5-2

1 As a measure of dispersion, what are the advantages of the standard deviation over the range and the average deviation, respectively? Under what condition will the variance of a variable equal 0?

2 If the average grade of male students is 3.0 and the standard deviation of their grades is 0.25, and if the corresponding figures for female students are 3.2 and 0.25, do the grades of male students show smaller variability? Why?

3 Two salesmen selling the same product have the following records over a long period of time:

	Salesman I	*Salesman II*
Average sales volume per month	$30,000	$35,000
Standard deviation	2,500	3,600

Which salesman seems to be more consistent in the volume of sales?

4 Refer to Prob. 4 of Exercise 5-1.

a Compute the variance and standard deviation of the hourly outputs for workers in department II by Eq. (5-2), and then verify the answers by using Eq. (5-4).

b Compute the variance and standard deviation of the hourly outputs for workers in department I by Eq. (5-6), and then verify the answers by using Eq. (5-8).

5 Refer to Prob. 5 of Exercise 5-1. Compute the variance and standard deviation by Eq. (5-4).

6 Refer to Prob. 6 of Exercise 5-1. Compute the variance and standard deviation by Eq. (5-8).

7 When is the variance equal to the standard deviation? Can the variance be less than the standard deviation? Explain.

8 When the sample observations have a symmetrical and bell-shaped distribution, the range, as discussed in Sec. 5-1, can be used to compute approximately the sample standard deviation. The range divided by 6 is roughly equal to or slightly less than the standard deviation. Use this method of approximation to find the standard deviation of the hourly outputs of workers in Department II of Prob. 4, Exercise 5-1, and the standard deviation of the test scores in Prob. 5 of Exercise 5-1.

9 Find the variance and standard deviation of the data in Prob. 9 of Exercise 4-2.

10 Find the variance and standard deviation of the data in Prob. 10 of Exercise 4-2.

11 Find the variance and standard deviation of the data in Prob. 11 of Exercise 4-2.

12 Find the variance and standard deviation of the data in Prob. 12 of Exercise 4-2.

5-3
Properties of the Standard Deviation

Let us first consider a very important property of the standard deviation, which is related to the transformation of scale discussed in Sec. 4-2-2. When a variable is a function of another variable, or when one scale is transformed into another, the standard deviation of the one can be expressed in terms of that of the other. We are going to prove that this is actually the case. Let $(X_1, X_2, \ldots, X_n)$ and $(Y_1, Y_2, \ldots, Y_n)$ be two sequences of numbers related in the following fashion:

$$Y_i = a + bX_i \qquad \text{for } i = 1, 2, \ldots, n$$

Let s_X and s_Y designate the standard deviation of X and Y, respectively; then

$$s_Y = bs_X$$

The proof of this is simple. Since $Y = a + bX$, by Eq. (4-10) we obtain

$$\overline{Y} = a + b\overline{X}$$

Therefore

$$Y_i - \overline{Y} = a + bX_i - (a + b\overline{X}) = bX_i - b\overline{X} = b(X_i - \overline{X})$$

Squaring each side of the equation, we have

$$(Y_i - \overline{Y})^2 = b^2(X_i - \overline{X})^2$$

The sum of squares of deviations of Y from its mean is

$$\sum_{i=1}^{n} (Y_i - \overline{Y})^2 = \sum_{i=1}^{n} b^2(X_i - \overline{X})^2$$

which gives the mean-square deviations for Y as follows:

$$\frac{\sum_{i=1}^{n} (Y_i - \overline{Y})^2}{n} = \frac{b^2 \sum_{i=1}^{n} (X_i - \overline{X})^2}{n}$$

That is,

$$s_Y{}^2 = b^2 s_X{}^2 \qquad \text{and} \qquad s_Y = bs_X \tag{5-10}$$

Note that the constant a does not affect the standard deviation of Y. It is simply washed out in the process of the transformation. This is true because when each observation in a series is added to by a constant number, the arithmetic mean of the series is added to by the same number. Consequently, each deviation from the mean remains unchanged, and the standard deviation is unaffected. A brief example is sufficient to make this point clearer.

Suppose that there is the relation $Y = 3 + X$ and X takes on the value of 1, 3, 5, 7, and 9. Accordingly, Y takes on the value of 4, 6, 8, 10, or 12. Since the value of $\bar{X}$ is $(1 + 3 + 5 + 7 + 9)/5 = 5$, the variance of X is

$$s_X{}^2 = \frac{\sum X^2}{n} - \bar{X}^2$$

$$= \frac{1^2 + 3^2 + 5^2 + 7^2 + 9^2}{5} - 5^2 = \frac{165}{5} - 25 = 33 - 25 = 8$$

Since the value of $\bar{Y}$ is $(4 + 6 + 8 + 10 + 12)/5 = 8$, the variance of Y is

$$s_Y{}^2 = \frac{\sum Y^2}{n} - \bar{Y} = \frac{4^2 + 6^2 + 8^2 + 10^2 + 12^2}{5} - 8^2 = \frac{360}{5} - 64 = 72 - 64 = 8$$

which is the same as the variance of X in spite of the fact that a constant 3 is added to each value of X.

However, if there is the relation $Y = 3 + 2X$, then the variance of Y is equal to the variance of X multiplied by the square of the *coefficient* of X, or $2^2 = 4$. The following computation demonstrates this fact:

Computation of the variance of Y

X	$Y = 3 + 2X$	Y^2
1	5	25
3	9	81
5	13	169
7	17	289
9	21	441
Total	$\sum Y = 65$	$\sum Y^2 = 1{,}005$

$$\bar{Y} = \frac{65}{5} = 13$$

$$s_Y{}^2 = \frac{1{,}005}{5} = 13^2 = 201 - 169 = 32$$

That is, the variance of Y is equal to the variance of X, which is 8 as computed above, multiplied by 4, which is the square of 2, the coefficient of X.

Knowing the above relationship, you should have no difficulty in proving the following:

If $Y = a - bX$, then $s_Y = |b| s_X$ (5-11)

If $Y = bX$, then $s_Y = |b| s_X$ (5-12)

If $Y = a + X$, then $s_Y = s_X$ (5-13)

Equation (4-13) shows that if $W = X + Y$, then the mean of W, or $\overline{W}$, is equal to $\overline{X} + \overline{Y}$. It is natural to ask: Is it true that if $W = X + Y$, then $s_W = s_X + s_Y$? First, let us show that the answer to this question is in the negative.

$$\begin{aligned} s_W{}^2 &= \frac{1}{n}\sum_{i=1}^{n}(W_i - \overline{W})^2 \\ &= \frac{1}{n}\sum_{i=1}^{n}[(X_i + Y_i) - (\overline{X} + \overline{Y})]^2 \qquad \text{by (4-13)} \\ &= \frac{1}{n}\sum_{i=1}^{n}[(X_i - \overline{X}) + (Y_i - \overline{Y})]^2 \\ &= \frac{1}{n}\sum_{i=1}^{n}[(X_i - \overline{X})^2 + 2(X_i - \overline{X})(Y_i - \overline{Y}) + (Y_i - \overline{Y})^2] \\ &= \frac{1}{n}\sum_{i=1}^{n}(X_i - \overline{X})^2 + \frac{1}{n}\sum_{i=1}^{n}(Y_i - \overline{Y})^2 + \frac{2}{n}\sum_{i=1}^{n}(X_i - \overline{X})(Y_i - \overline{Y}) \\ &= s_X{}^2 + s_Y{}^2 + \frac{2}{n}\sum_{i=1}^{n}(X_i - \overline{X})(Y_i - \overline{Y}) \end{aligned} \tag{5-14}$$

and

$$s_W = \sqrt{s_X{}^2 + s_Y{}^2 + \frac{2}{n}\sum_{i=1}^{n}(X_i - \overline{X})(Y_i - \overline{Y})} \tag{5-15}$$

The last term of Eq. (5-14) is made of two factors, 2 and $\left[\sum_{i=1}^{n}(X_i - \overline{X})(Y_i - \overline{Y})\right]/n$. The latter factor is of great interest in statistics and is given a special name. It is called the *covariance* of X and Y, and is designated by $\text{cov}(X,Y)$. That is,

$$\text{cov}(X,Y) = \frac{\sum_{i=1}^{n}(X_i - \overline{X})(Y_i - \overline{Y})}{n} \tag{5-16}$$

Thus the variance of the sum $X + Y$ is equal to the sum of the variance of X and that of Y plus two units of the covariance.

Equation (5-14) is the general form for the variance of the sum $X + Y$. It can be shown that if X and Y are *independent* random variables, the covariance of X and Y is expected to be 0 (see Supplement III). Consequently, if the variables X and Y are independent, the standard deviation of the sum $X + Y$ is expected to equal the square root of the sum of the variance of X plus that of Y.

You may now be wondering why the variance, and hence the standard deviation, is computed in terms of deviations from the arithmetic mean rather than from any other measure of central tendency. The explanation for this lies in the fact that the variance has the smallest value if it is computed from the mean.[1] That is to say that if we use the standard deviation to measure the extent of error in guessing the value for any item in a distribution, then the magnitude of this error will be at its minimum when we guess from the arithmetic mean rather than any other value.

In contrast, if the extent of error between any item and a measure of central tendency is appraised in terms of the absolute average deviation, then the average deviation is at its minimum when the median is used.[2] Thus, when we use the median to indicate central tendency, we usually use the average deviation from the median to express dispersion. Analogously, the average deviation is to the median as the standard deviation is to the arithmetic mean.

This is one of the reasons that the standard deviation is used to measure the variability when the mean is used to indicate central tendency. In fact, the standard deviation is by far the most important and the most frequently used measure of dispersion.

Exercise 5-3

1 Find the variance and standard deviation of a sample of 11 values for X as shown below:

X_1	X_2	X_3	X_4	X_5	X_6	X_7	X_8	X_9	X_{10}	X_{11}
2	2	3	3	3	4	4	4	5	5	9

2 Let each of the values of X in Prob. 1 above be increased by 2 and each sum thus obtained be called Y as follows:

Y_1	Y_2	Y_3	Y_4	Y_5	Y_6	Y_7	Y_8	Y_9	Y_{10}	Y_{11}
4	4	5	5	5	6	6	6	7	7	11

a Find the variance and standard deviation for Y.

b What is your observation about the characteristic of the variance or standard deviation in comparing the results of this problem with the results of Prob. 1 above.

3 Suppose that each of the values of X in Prob. 1 above is multiplied by 3 and the products are designated by Z as follows:

Z_1	Z_2	Z_3	Z_4	Z_5	Z_6	Z_7	Z_8	Z_9	Z_{10}	Z_{11}
6	6	9	9	9	12	12	12	15	15	27

[1] See Supplement I for its proof.

[2] See Supplement II for its proof.

a Find the variance and standard deviation for Z.
b Compare the results obtained in (*a*) with the results in Prob. 1. Which equation in the text is verified by your comparison?
c Generalize your observations.

4 Suppose that each of the values of X in Prob. 1 above is multiplied by 3 and then has 2 subtracted. Let the new variable be called W; we have

W_1	W_2	W_3	W_4	W_5	W_6	W_7	W_8	W_9	W_{10}	W_{11}
4	4	7	7	7	10	10	10	13	13	25

a Find the standard deviation for W.
b Compare the result in (*a*) with the results in Prob. 3*a*. Which equation in the text is verified in your comparison?
c Generalize your observations.

5 Given the following data:

X_1	X_2	X_3	X_4	X_5	X_6	X_7	X_8	X_9	X_{10}	X_{11}
1	1	2	2	2	2	4	5	5	9	11

a Find the mean and the median for X.
b Find the variance of X from the mean.
c Find the variance of X from the median.
d What generalization will you make when you compare the results of (*b*) and (*c*)?

6 The expression $\sum_{i=1}^{n}(X_i - c)^2$ is at a minimum when the constant $c = \bar{X}$. Verify this by using the values for X in Prob. 1 and for $c = 3$, 4, and 5.

7 If $Z = X + Y$, then under what condition is the variance of Z equal to the sum of the variance of X and the variance of Y? Explain.

8 In the light of the discussion in this section, compare the average deviation with the standard deviation as a measure of dispersion.

9 Suppose that X has a sample mean $\bar{X} = 15$ and a variance $s_X^2 = 25$. Find the sample mean, variance, and standard deviation of Y under each of the following relations:

a $Y = 4 + 16X$. *b* $Y = 16 - 4X$. *c* $Y = \frac{1}{4} + \frac{1}{4}X$.

10 Let F° and C° designate the degrees in daily average temperature measured by Fahrenheit and centigrade thermometers, respectively. Then, the following relationship holds:

$$F^\circ = 32 + \frac{9}{5}C^\circ$$

a If the daily averages of temperature in Paris throughout a given year have a variance of 25°C, what is the variance in Fahrenheit for that year?
b If the daily averages of temperature in Los Angeles throughout a given year have a standard deviation of 8°F, what is the standard deviation in centigrade for that year?

11 The hourly output of a newly developed machine is believed to amount to four times as much as that of the old machine. If the variance of the hourly ouptut of the old machine in a period of n hours is 16, what is the standard deviation of the hourly output of the new machine in n hours?

5-4
Variance of the Population

In the preceding sections, attention was focused on the variance and standard deviation of a sample. We shall now examine how the variance of a theoretical distribution commonly known as the *variance of a population* is expressed, and investigate some of the properties of a population variance, which is usually designated σ^2. (It is a statistical convention that characteristics of the population, called *parameters*, are designated by Greek letters such as μ and σ, while those of the sample, called *statistics*, are designated by Latin letters such as $\bar{X}$ and s. The word *statistics* as used here means *statistical measure* or *statistical description*, which is quite different from its use in the title of this book.)

The population variance of a random variable X, or $\sigma_X{}^2$, is the expected value of the squared deviations of X from the mean μ. It can be expressed as the difference of the expected value of the square and the square of the expected value of X. Symbolically,

$$\begin{aligned}\sigma_X{}^2 &= E((X-\mu)^2) = E(X^2 - 2\mu X + \mu^2)\\ &= E(X^2) - E(2\mu X) + E(\mu^2) = E(X^2) - 2\mu E(X) + \mu^2\\ &= E(X^2) - 2\mu^2 + \mu^2 = E(X^2) - \mu^2\\ &= E(X^2) - [E(X)]^2 \qquad (5\text{-}17)\end{aligned}$$

For example, let X designate the decimal digits and have a uniform distribution on $\{0, 1, 2, 3, 4, 5, 6, 7, 8, 9\}$. From Sec. 4-4-1 we know that $E(X) = 4.5$. Also, by Eq. (4-20), we have

$$\begin{aligned}E(X^2) &= \sum_{j=1}^{m} x_j{}^2 P(x_j) = 0^2(\tfrac{1}{10}) + 1^2(\tfrac{1}{10}) + \cdots 9^2(\tfrac{1}{10})\\ &= \tfrac{285}{10} = 28.5\end{aligned}$$

Thus the variance of X, or $\sigma_X{}^2$, is

$$\sigma_X{}^2 = 28.5 - 4.5^2 = 28.5 - 20.25 = 8.25$$

Among the properties of the population variance and standard deviation, which are basic to later discussions, are:

$$\text{If } Y = a + bX, \text{ then } \sigma_Y{}^2 = b^2\sigma_X{}^2 \text{ and } \sigma_Y = |b|\sigma_X \qquad (5\text{-}18)$$

$$\text{If } Y = bX, \text{ then } \sigma_Y{}^2 = b^2\sigma_X{}^2 \text{ and } \sigma_Y = |b|\sigma_X \qquad (5\text{-}19)$$

$$\text{If } Y = a + X, \text{ then } \sigma_Y{}^2 = \sigma_X{}^2 \text{ and } \sigma_Y = \sigma_X \qquad (5\text{-}20)$$

The proof of each of the relations expressed in (5-18) to (5-20) is left for those who are interested.

Furthermore, by some algebraic manipulation and the use of the expectation notation, we can demonstrate another very important relation. That is,

If X and Y are two *independent* random variables, then $\sigma_{X+Y}^2 = \sigma_X{}^2 + \sigma_Y{}^2$

(5-21)

This is to say that the variance of the sum is the sum of the variances. The proof of this relation is provided in Supplement III.

Suppose, for instance, that a certain businessman owns two franchised McDonald drive-in restaurants. Let the daily sales of the two restaurants be designated by X and Y, and X and Y be assumed independent. For many years this businessman has accumulated sales data, and according to his records the variances of X and Y are \$2,400 and \$2,500, respectively. What is the standard deviation of the daily sales with the two restaurants combined?

In this problem the variance of X, or $\sigma_X{}^2$, is given at \$2,400, and the variance of Y, or $\sigma_Y{}^2$, is \$2,500. By Eq. (5-21) the variance of the sum is the sum of the variances; that is,

$$\sigma_{X+Y}^2 = \sigma_X{}^2 + \sigma_Y{}^2 = \$2{,}400 + \$2{,}500 = \$4{,}900$$

Consequently, the standard deviation of the combined daily sales is

$$\sigma_{X+Y} = \sqrt{\$4{,}900} = \$70$$

Similarly, the variance of the *difference* between two independent variables is also the sum of the variances of the variables. Symbolically,

$$\sigma_{X-Y}^2 = \sigma_X{}^2 + \sigma_Y{}^2 \qquad (5\text{-}22)$$

Consider the difference in breaking strength between the nylon cord X and the cotton cord Y, which are believed to be independent of each other. Let D be the difference in breaking strength, or $D = X - Y$. Then, the variance of D is equal to the sum of the variance of X and the variance of Y. Suppose that it is known that the variance of X is 225 pounds and the variance of Y is 216 pounds. Accordingly, we have the variance of D, or $\sigma_D{}^2$, as follows:

$$\sigma_D{}^2 = \sigma_{X-Y}^2 = \sigma_X{}^2 + \sigma_Y{}^2$$
$$= 225 + 216 = 441 \text{ pounds}$$

Furthermore, the effect of the coefficient of a random variable upon the variance of the new variable (Eq. 5-18) applies also to the variance of the sum as well as of the difference. If X and Y are two independent random variables and a and b are two constants, then we have

$$\sigma_{aX\pm bY}^2 = a^2\sigma_X{}^2 + b^2\sigma_Y{}^2 \qquad (5\text{-}23)$$

Suppose that a certain steel wire has a breaking strength (S) that equals the total of the breaking strengths of two nylon cords (X) and four cotton cords (Y)

combined, or $S = 2X + 4Y$. Given that X and Y are independent and that the variances of X and Y are 225 and 216, respectively, then the variance of S is

$$2^2(225) + 4^2(216) = 900 + 3{,}456 = 4{,}356$$

Exercise 5-4

1 Suppose that a sample of n observations is taken from a given population. Will the sample variability be the same as the population variability? Explain.

2 Explain the similarities and differences between each of the following pairs of expressions:
a $E((X-\mu)^2)$ and $[\sum(X-\bar{X})^2]/n$.
b $\sigma^2 = E(X^2) - \mu^2$ and $s^2 = (\sum X^2/n) - \bar{X}^2$.

3 Suppose that X is uniformly distributed on {1, 2, 3}, and Y is uniformly distributed on {0, 1}. If X and Y are independent and $W = 2X + 3Y$, verify that

$\sigma_W{}^2 = 4\sigma_X{}^2 + 9\sigma_Y{}^2$ (See Prob. 9 of Exercise 4-4.)

4 Given the random variables X and Y in Prob. 3 and that $Z = 2X - Y$, verify the following:

$\sigma_Z{}^2 + 4\sigma_X{}^2 + \sigma_Y{}^2$ (See Prob. 10 of Exercise 4-4.)

5 Let X designate the number of spots showing on the face of a well-balanced die after it is rolled. Given that $Y = X^2 + 2X$, find:
a $\sigma_X{}^2$. *b* $\sigma_Y{}^2$.

6 Suppose that an ideal coin is tossed twice. If it turns up tails both times, you win nothing. If heads turns up only once in the two tosses, you win \$1. If both tosses result in heads, you win \$4. In other words, $Y = X^2$ for $X =$ number of heads obtained and $Y =$ number of dollars you would win. Find:
a $\sigma_X{}^2$. *b* $\sigma_Y{}^2$. (See Prob. 8 of Exercise 4-4.)

7 Let X designate the number of heads obtained in the toss of three ideal coins, and $Y = X^2 - X + 2$. Find:
a $\sigma_X{}^2$. *b* $\sigma_Y{}^2$.

8 A certain company has a branch in one city with a daily sales of X dollars and another branch in a different city with a daily sales of Y dollars. X and Y are believed to be independent. We are told that the variances of X and Y are known to be \$3,400 and \$3,000, respectively. Let W be the total daily sales of the two branches. What is the standard deviation of W?

9 Two brands of cigarettes are compared to determine the variance of the difference D in the nicotine content between the two brands. Let X be the nicotine content of brand I, which has a variance of 5 milligrams, and Y be the nicotine content of brand II, which has a variance of 4 milligrams. That is, $D = X - Y$. It is assumed that X and Y are independent. What is the variance of D?

10 Let N be the number of heads obtained in the toss of two ideal nickels, and D be the number of heads obtained in the toss of three ideal dimes. Suppose that you are given the privilege to play the game as follows: It is agreed upon that you will reveive a sum of money in dollars that will be equal to six times the number of heads of the nickels *plus* three times the number of heads of the dimes after the coins are tossed. Let W be the total amount of money that you will receive for each game. What is the variance of W?

5-5
Uses of the Standard Deviation

The standard deviation is used in many problems in statistical inference; and it plays an essential role in the various statistical methods to be presented in the rest of this book.

As indicated earlier in this chapter, the standard deviation of a random variable is designed to measure the variability of the values of the random variable from their mean. The standard deviation of a sample is a measure of the dispersion of the *sample distribution*—the distribution of the observations within the sample—around the sample mean $\overline{X}$, while the standard deviation of a population is a measure of the dispersion of the population around the population mean μ. The larger the standard deviation, the more likely we shall obtain a value far away from the mean. On the other hand, the smaller the standard deviation, the more likely we shall obtain a value close to the mean. This important relationship is expressed in what is widely known as the *Tchebycheff theorem*, named after the Russian mathematician who first described it. Briefly, Tchebycheff's theorem is stated as follows:[1]

The proportion that the values of a random variable will lie within k standard deviations from the mean $\overline{X}$ is at least $1 - 1/k^2$, where k is equal to or greater than 1. That is,

$$RF(|X - \overline{X}| \leq ks) \geq 1 - \frac{1}{k^2} \tag{5-24}$$

where RF signifies the proportion, or relative frequency, of the sample distribution, and s is the sample standard deviation.

The Tchebycheff theorem applies not only to the sample distribution but also to the population distribution. By replacing RF, $\overline{X}$, and s by P, μ, and σ, respectively, we have

$$P(|X - \mu| \leq k\sigma) \geq 1 - \frac{1}{k^2} \tag{5-25}$$

where P designates the probability. That is, the probability that the values of a random variable will lie within $k\sigma$ from the mean μ is at least $1 - 1/k^2$.

The idea of Tchebycheff's theorem is described graphically in Fig. 5-1. Observe that by marking off a distance of ks on either side of the sample mean $\overline{X}$ an interval is constructed. The relative frequency, or proportion, of all potential observations of X's falling within this interval is at least $1 - 1/k^2$. Accordingly, the proportion of all X values to fall within $2s$ from the mean is at least

[1] See Supplement IV for the proof of Tchebycheff's theorem.

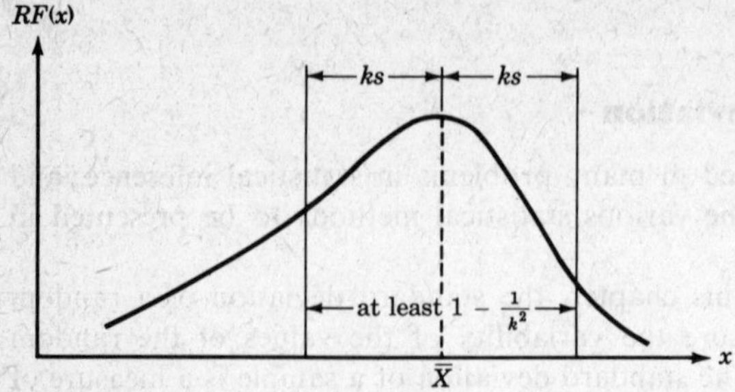

Figure 5-1
Representation of Tchebycheff's theorem.

$1 - 1/2^2 = 1 - 1/4 = 0.75$; the proportion of all X values to fall within $10s$ from the mean is at least $1 - 1/10^2 = 1 - 1/100 = 0.99$.

The Tchebycheff theorem is very conservative because it puts emphasis on the notion of "at least." When $k = 1$, then $1 - 1/k^2 = 1 - 1/1^2 = 1 - 1 = 0$. This is to say that the proportion of all X values to fall within one standard deviation from the mean is 0, which is a most uninformative result. It is, however, not erroneous. The theorem does not apply for $k < 1$. When $k = 0.5$, for instance, $1 - 1/k^2 = 1 - 1/0.5^2 = 1 - 1/0.25 = 1 - 4 = -3$, a most ludicrous statement. For $k \geq 1$, the theorem applies to all patterns of distributions, skewed or symmetrical. As we shall see later, when a distribution is symmetrical and bell-shaped, a greater proportion of all observations will fall within a given number of standard deviations from the mean than that stated under the Tchebycheff theorem.

Exercise 5-5

1 Some bell-shaped symmetrical distribution referred to in the text is known as the *normal* distribution. Is the Tchebycheff theorem wrong if it is applied to problems with a normal distribution? Explain.

2 Let the random variable X have a population mean μ and standard deviation σ. Fill in the following blanks:

a The probability that X will fall within 3σ from μ is at least ________.
b The probability that X will fall within 4σ from μ is at least ________.
c The probability that X will fall within ________ standard deviations from μ is at least 0.25.
d The probability that X will fall beyond ________ standard deviations from μ is at most 0.01.
e The probability that X will fall beyond 2σ from μ is at most ________.

3 Let the random variable X have a sample mean $\bar{X}$ and standard deviation s. Fill in the following blanks:

a At least 99 percent of the values of X will fall within ________ standard deviations from the mean.
b At least 80 percent of the values of X will fall within ________ standard deviations from the mean.
c At most 15 percent of the X values will fall beyond ________ standard deviations from $\bar{X}$.
d At most 50 percent of the X values will fall beyond ________ standard deviations from $\bar{X}$.
e At least ________ percent of the X values will satisfy the inequality $|X - \bar{X}| \leq s$.
f At most ________ percent of the X values will satisfy the inequality $|X - \bar{X}| \geq 3s$.

4 Refer to Prob. 9 of Exercise 5-2. Find the lower and upper values that bracket at least the middle 75 percent of the observations.

5 Refer to Prob. 12 of Exercise 5-2. Find the lower and upper values that bracket at least eight-ninths of the observations around the mean.

6 Let X stand for the hourly output of the production workers of a manufacturing company. Assume that before and after unionization of the company, the standard deviation of X was 40 and 30, respectively. How do you interpret the decrease in the standard deviation after unionization?

7 Use your own words to describe two different applications of the standard deviation.

Chapter 6

Random sampling and sampling distribution

Mention had been made of the concepts of a population, a sample, and statistical inference in previous discussions. Since statistical inference relative to population parameters on the basis of sample statistics is the primary concern of this text, additional examples to illustrate these and the related concepts are in order. For instance, many people have spoken out to condemn capital punishment; we sometimes wonder whether their views on this matter are shared by college students, who are presumably sensitive to important issues faced by society. We may want to know the actual percentage of students in American colleges and universities who do not approve of the death penalty. We could obtain this percentage by asking every student on each campus the question: Are you against capital punishment? and then recording the responses. The procedure, however, would be quite time consuming and probably impracticable. To overcome such difficulties, we might choose to address our question to only a portion of the student body and try to infer the attitude of all the students on the basis of the answers from this portion of students.

This is another example of *statistical inference*. The portion of students questioned is technically called a *sample*, and the entire student body, a *population*. One essential procedure of statistical inference is to take a sample *representative* of the population, and then, on the basis of information obtained from the sample, to make a useful statement about the characteristics of the population. To obtain a representative sample, it is necessary to consider the concept of a random sample.

It is obvious that numerous samples of equal size can be drawn from a given population. It is also common sense that different samples may yield different values for a sample statistic. Thus it can be expected that a given sample statistic can assume many different values. As we shall see later in this book, the probability distribution of the values of a statistic, called a *sampling distribution*, is of fundamental importance in statistical inference.

In this chapter we shall examine the notion of a random sample, the techniques of drawing such a sample, and the nature of sampling distribution of a statistic, the sample mean.

6-1 Random Sample

As indicated at the beginning of Chap. 4, a sample is simply a subset of the universal set, the population. In order to understand the meaning of a random sample, a further discussion of the notion of a population will prove helpful.

6-1-1 Population

In previous discussions the terms *sample space* and *population* have both been used to refer to the totality of all potential elementary events in a given problem. In fact, the two terms may be used interchangeably, designating the universal set of all possible outcomes of an experiment. The term population has been in use because statistics used to concern itself mostly with the numerical descriptions of human beings in the early stages of its development. As the scope of statistics expanded, the notion of a population has assumed a much broader connotation. Nowadays, it is used synonymously with the term sample space, or universal set, in most practical situations. It has become a technical term with a meaning which can be quite different from that in everyday usage.

The potential elementary events, which are also referred to as *elementary units*, under investigation have certain common *attributes*, or *characteristics*, which are of particular interest to researchers. For instance, given a large group of college students, the attribute to be investigated may be the score received by each student in a college entrance examination, the amount of money spent by each student in a given academic year, or the scholastic status of the student as passing or failing. The first two attributes are quantitative and the third is qualitative in nature. The third can be made quantitative, or numerical, by simply assigning, say, the value 0 to failing and 1 to passing.

A numerical value assigned to an elementary observation unit for a given attribute is a value of the random variable designating that attribute. The distribution of the random variable is referred to as the *population distribution*. The terms population and population distribution have become almost synonymous to most statisticians. Thus the "mean of a population" can be phrased as the "mean of a population distribution," and sampling from a population can be spoken of as sampling from a population distribution.

Let us now distinguish between populations that are finite and those that are infinite. A population is said to be *finite* if it includes a limited number of elementary events or units. Examples of a finite population are, as of any given year, individual sales of all corporations in the United States, wages received by all members of AFL-CIO, and grades received by all college students in America. Some populations may consist of only a few observations, others may consist of millions or even billions. As long as the total number of potential observations is limited, it is still a finite population.

In contrast, an *infinite* population is one that consists of an infinitely large number of observations. At least hypothetically, there is no limit to the number of units it can include. For example, the results obtained in the roll of two dice constitute an infinite population, so are the weights at birth of all human beings. Another illustration of an infinite population is the service lives of all the potential automobile tires produced in a production process that operates indefinitely under given manufacturing conditions. Still another example is the population of random digit numbers (see Sec. 6-2-1) taken *with replacement* from a set of 10 digits, namely, {0, 1, 2, 3, 4, 5, 6, 7, 8, 9}, since with replacement there is an infinitely large number of digits available for being drawn. An infinite population can always be generated from a finite set of values or units if sampling is taken with replacement.

Assuming that an infinite population is still difficult to visualize for some students, let us use another example for further clarification. Suppose that a researcher were interested in the weights of hogs at birth in the past, present, and the future. If he wished to picture the collection of all baby hogs ever born or to be born, he would be faced with "mountains" of baby hogs. If he tried to consider the weights of these hogs, he would be confronted with millions upon millions of weights. Most of these cannot even be determined, since countless hogs have yet to be born and hogs born in the past might not have been weighed. Some device has to be found to assist the researcher in looking at an infinite population like this. Fortunately, there is such a device,

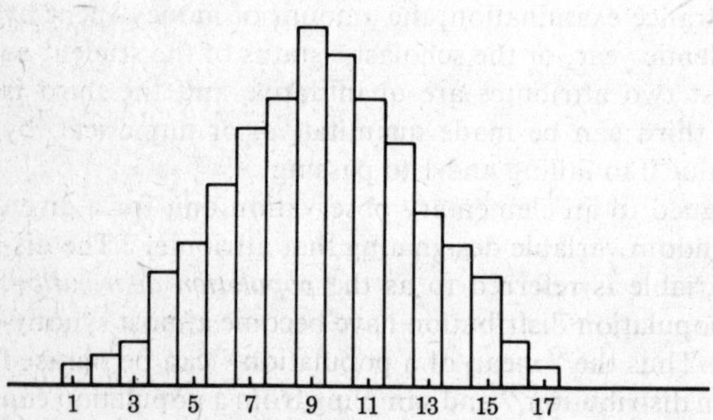

Figure 6-1
Weights at birth of hogs.

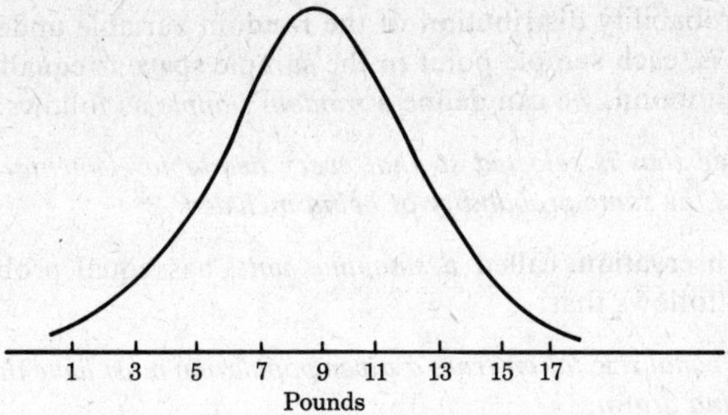

Figure 6-2
Weights at birth of hogs.

and it is to regard the innumerable weights as a relative frequency, or probability, distribution. As shown in Fig. 6-1, where the weights are rounded to the nearest pound, the distribution would tell the researcher the probability that any baby hog would weigh 2 pounds, 3 pounds, and so on. He can picture the population as a discrete probability distribution. If the weights are not rounded to the nearest pound, or any subdivision of it, he can treat the weights as values of a continuous variable and then picture the population as a continuous distribution such as the one shown in Fig. 6-2.

It is impossible to take a sample as large as an infinite population; it is possible but usually impractical to take a sample as large as a finite population. Thus, whether the population under consideration is finite or infinite, statistical investigations require generalization from sample information. To avoid invalid or illogical results, attention should be directed toward the theory of sampling.

6-1-2 Random sample defined

Any set of n elementary observation units taken from a given population can be considered a sample of size n. However, the kind of sample with which we are primarily concerned is a random sample. A random sample may be taken with or without replacement, as indicated in the preceding section. If the sample is taken *with replacement* from a population, finite or infinite, the unit drawn is returned to the population and the number of units available for future drawing is not affected. This also holds true when the sample is taken from an infinite population *without replacement*, that is, the unit selected is not returned to the population. When one item is taken each time without replacement from a finite population, the number of units remaining after each drawing will be reduced by one. Consequently, the probability of drawing any remaining unit in successive selections will be increased. This distinction must be kept in mind when an attempt is made to define the concept of a random sample, since random sampling is closely related to the theory of probability.

In this brief treatment of sampling theory, our discussion will be based on the assumption that the probability distribution of the random variable under discussion is *uniform;* that is, each sample point in the sample space is equally probable. Under this assumption, we can define a *random sample* as follows:

A random sample is one that is selected so that every available elementary unit for observation has the same probability of being included.

If each *available* unit for observation, called a *sampling unit*, has equal probability of being selected, it follows that:

Any random sample of equal size taken from a given population must have the same probability of being drawn.

Suppose that we have some specific method of selecting an elementary unit from a given population. After a unit is drawn, the value of the random variable is observed and noted. If the population is infinite, it is immaterial to specify whether the sample is selected with or without replacement because the probability of obtaining any particular unit remains unchanged. In other words, each elementary unit is sampled *independently;* the occurrence of any unit in a given trial does not affect the probability that the same or any other unit will occur on a subsequent trial. For example, in the toss of a perfect coin, the probability of obtaining heads is 0.5; given the occurrence of heads, the probability of getting heads or tails on a subsequent toss remains 0.5. A sample taken under such a condition is obviously a random sample. It is easy to see that the probability of drawing any *particular sequence* of 30 observations is $(0.5)^{30}$. Each observation has the same probability of occurring; each sample of the same size has the same probability of being drawn. Any sample drawn under such conditions is a random sample.

If the population is finite, and samples are taken with replacement, what we have just said about sampling from an infinite population still applies. The occurrence of any item does not affect the probability of occurrence of the same or another item in a future trial. Suppose, for instance, that from a college of 30,000 students, a sample of 100 students is drawn. The probability of selecting any particular student is $\frac{1}{30,000}$. After the attribute of the student is observed and noted, he is returned to the population, which remains 30,000. The probability of drawing any particular student in a subsequent selection is still $\frac{1}{30,000}$. Outcomes of all drawings are therefore independent, and the probability of selecting any *particular sequence* of 100 students is therefore $\left(\frac{1}{30,000}\right)^{100}$.

You are probably unhappy over this method because it allows the same student to be drawn more than once. To overcome this difficulty, the commonly used method of drawing a random sample is sampling without replacement. By this method, the probability of another unit's being drawn subsequently is increased, since the size of the population is decreased by *one* after each drawing. For the above example, if the first student drawn is not returned to the population, the probability of selecting any particular student among the remaining 29,999 students will be $\frac{1}{29,999}$. The probability of drawing any *particular sequence* of 100 students is then

$$\frac{1}{30,000} \times \frac{1}{29,999} \times \frac{1}{29,998} \times \cdots \times \frac{1}{29,901}$$

It is important to note that the sample mentioned immediately above refers to a *sequence* of elementary units. Recall that a sequence is an ordered arrangement; it is a *permutation*. A sample, however, is a set of elementary units, it is a *combination* without regard to the order of occurrence in the process of sampling. A set of n units without regard to the order of their arrangement constitutes only one combination, and there are $\binom{N}{n}$ possible combinations, or samples, of n units taken from a finite population of N units. Thus:

When sampling is taken from a finite population of N observation units, a sample of n units shall be called a random sample if each of the $\binom{N}{n}$ possible samples is selected with the same probability

$$\frac{1}{\binom{N}{n}} \tag{6-1}$$

Since

$$\binom{N}{n} = \frac{N!}{n!(N-n)!} = \frac{N(N-1)(N-2)\cdots(N-n+1)(N-n)!}{n!(N-n)!}$$
$$= \frac{N(N-1)(N-2)\cdots(N-n+1)}{n!}$$

we have

$$\frac{1}{\binom{N}{n}} = \frac{n!}{N(N-1)(N-2)\cdots(N-n+1)}$$

which, in effect, is

$$\frac{1}{\binom{N}{n}} = \left(\frac{n}{N}\right)\left(\frac{n-1}{N-1}\right)\left(\frac{n-2}{N-2}\right)\cdots\left(\frac{1}{N-n+1}\right) \tag{6-2}$$

Let us use another simple example to illustrate the concept of a random sample. Suppose that a certain company has on its payroll 30 salesmen ($N = 30$),

3 of whom belong to the same nationality group. They are Allen (A), Bob (B), and Carl (C). If a sample of three salesmen ($n = 3$) is to be selected at random, the probability that any combination of three salesmen will be selected is

$$\frac{1}{\binom{N}{n}} = \frac{1}{30!/3!\,27!} = \frac{3(2)(1)}{30(29)(28)} = \frac{1}{4{,}060}$$

Accordingly, the probability that salesmen, A, B, and C will be included in the sample must also be $\frac{1}{4{,}060}$. Let us proceed with the calculation more intuitively. When the first draw is made, there are 30 salesmen, and 3 of them are A, B, and C. Therefore, the probability of any one of the 3's being selected is $\frac{3}{30}$. After the first draw, there are 29 left, of whom 2 belong to that group, and the probability of any one of the 2's being selected is $\frac{2}{29}$. Finally, there are 28 left, of whom only one belongs to the group, and the probability of his being selected is $\frac{1}{28}$. As a result, the probability that salesmen A, B, and C are all included in the random sample is

$$\left(\frac{3}{30}\right)\left(\frac{2}{29}\right)\left(\frac{1}{28}\right) = \frac{1}{4{,}060}$$

which is the same as the one obtained above.

Exercise 6-1

1 What constitutes the population of X if X designates the response in a telephone interview to decide whether the respondent likes a particular TV program?

2 Give an example (not used in the text) to illustrate each of the following:

a A finite population. *b* An infinite population.

3 Define the population and the elementary events in each of the following problems:

a To find out whether a dial such as that shown in Fig. 3-6 is well balanced.
b To conduct a poll for determining voting preference in a presidential election.
c To determine whether housewives in a certain country prefer one brand of soap to another.

4 Distinguish between population and population distribution. Give an example to illustrate their difference.

5 In many situations it is impossible to secure complete information about a population. Cite at least two such situations, and indicate how you would make decisions under the circumstances.

6 Let X be a discrete random variable and Y be a continuous random variable. Is it correct to say that the population of X is finite and that of Y infinite? Explain.

7 In each of the following situations, determine whether sampling is taken from a finite or an infinite population, and then define the population:

a A sample of screws is periodically selected to determine the number of defective screws produced by an automatic machine.
b Five employees of a large company are selected as representatives of the work force to serve on a labor-management committee.
c A silver dollar is weighed 10 times, and the average weight is used as an estimate of the true weight.

8 From the 26 English letters, how many different samples of 5 letters each can be drawn?

9 From the 26 letters of the English alphabet, a random sample of 5 letters is drawn without replacement. Find the probability of drawing:

a Five vowels.
b Five *particular* consonants.
c Any other *particular* combination of 5 letters.

10 In a certain department of a company there are 20 employees, 4 of whom are women. What is the probability that a sample of 4 employees taken from this department includes all 4 women?

11 Given that each of 10 digits is written on a different ball and the 10 balls are put in a bag and thoroughly mixed, what is the probability of drawing a sample without replacement consisting of:

a The digits 1, 3, 5, 7, and 9?
b The digits 0, 2, 4, 6, and 8?
c Any other *particular* combination of 5 digits?

12 Answer Prob. 11*a*, *b*, and *c*, given that the sample is drawn with replacement.

6-2
Sampling Designs

Before we actually examine the various sampling designs, let us first consider how such designs are evaluated.

At least two criteria are generally used in evaluating a sampling design: reliability and efficiency. It can be expected that sampling involves errors. A *sampling error* is the difference between the value of a statistic obtained from a random sample and the value of the corresponding population parameter because of chance variations in the selection of elementary units. It is measured by what is commonly called reliability, or precision, of sampling, which is related to the variance of the sample statistic. The greater the variance, the smaller the *reliability* of the sample result will be.

The criterion of efficiency is related to the cost of sampling. A sampling design is considered *efficient* if the same degree of reliability is obtained at the lowest possible costs. A sampling design is considered to be more efficient than another if the former results in lower costs than the latter design, with the same degree of reliability.

As we have seen, the rules of probability are applied in the definition of a random sample. It is due to this reason that random sampling is often called *probability sampling*. By sampling design, we refer to the design of probability sampling. In this section we shall consider four most commonly used designs of probability sampling, and evaluate each by its reliability and efficiency.

6-2-1 Simple random sampling

As already pointed out, random sampling is a procedure in drawing a sample by which every unit available for sampling has an equal probability of being drawn. Alternatively, if a sample of n units is drawn at random, every possible sample of n units has the same probability of being selected. A sample drawn by this procedure is called a *simple, or unrestricted, random sample*.

The key to such a procedure is of course the technique of randomness. Achieving randomness, however, is not a simple matter. If, for example, we wish to survey the opinions of the students at a large college, we cannot select a random sample simply by interviewing students at the gate of the college commons until the desired number of interviews is obtained. Many students may never eat at the commons, and thus the probability of their being interviewed is zero. A sample thus obtained is evidently not a random sample.

One of the commonly used methods to achieve randomness in obtaining a sample is the so-called *goldfish bowl* procedure. According to this procedure, all the items in a population are numbered, and the numbers are written on cards, balls, slips of paper, or other articles that are physically homogeneous. These are put into a bowl or bag and mixed thoroughly. The size of the sample n is then specified. The articles are drawn at random one at a time until the desired number of items is obtained. If drawing is with replacement, the articles are mixed again before the next drawing and the probability of each item's being drawn remains unchanged. Without replacement, the probability of each remaining item's being drawn is increased because the total number available for selection is reduced by the number of items already drawn. However, the *remaining* items have the same chance of being drawn. In any case, the probability of obtaining a sample of size n is the same as any other combination of n items from the same population.

The goldfish bowl procedure can be simplified by the use of a table of random numbers. Appendix I is a small table of random numbers. Before we learn how to use it, let us first see how it is constructed. The table is composed of positive integers from 0 to 9. It can be obtained by the goldfish bowl method. Suppose that 10 cards or cardboard disks are numbered 0, 1, 2, 3, 4, 5, 6, 7, 8, and 9 and placed in a bowl. After they are mixed thoroughly, one is drawn. Assume that it is 4. The number 4 is entered in the table as the first digit. It is then replaced in the bowl and, after thorough stirring, another card is drawn. If the number is 7, then 7 will appear as the second digit in the table, and so on, until the entire table is filled. There are more efficient ways to construct the table, however. For instance, random digits may be generated by a dial such as the one shown in Fig. 3-6 with each of the 10 sections standing for a different digit, or by means of electronic computers.

When any two digits, say, of the same row, are combined, numbers from 00 to 99 are obtained. If three digits are combined, you get numbers from 000 to 999, and so on. The number of digits you use depends on the size of the population from which your sample is taken.

Suppose that a sample of 100 items is drawn from a population of 10,000 items. The first step is to prepare a list of the population and assign a number from 0,000 to 9,999 to each item on the list. Then a starting number with four digits is chosen in a random fashion. Once the random start is made, you can read the table vertically, horizontally, diagonally, or in any other systematic way. Record the number that has been read, but ignore the duplicated numbers. Proceed in this fashion until 100 numbers are secured. As the last step, 100 items with the corresponding recorded numbers are taken from the population, and the desired random sample is obtained.

The simple random procedure, whether the goldfish bowl or random digit method, can be used if the sampling units can be easily identified, and if the population is small and homogeneous. The method, however, is expensive and time consuming, especially when the population is large, because of the necessity of numbering all the items. As a matter of fact, many data in social sciences cannot be effectively gathered by the procedures just described. Furthermore, if the items close to each other are more homogeneous than those far apart, a simple random sample might not be representative of the population. For instance, in the survey of family incomes of a great city, simple random sampling might conceivably select a relatively large number of families with similar incomes. This would possibly result in relatively large sampling error and large variance and, consequently, in a low degree of reliability for the sample statistic. Under such circumstances, restricted random sampling designs will be more appropriate and are often employed.

6-2-2 Systematic sampling

One of the often used *restricted* random designs is *systematic sampling*. Quite frequently, a systematic procedure, rather than a randomization method, is followed if a simple random sample is desired. According to this procedure, a sample is obtained by taking every kth unit in the population after the units in the population have been numbered or arranged in some way. The letter k stands for some whole number, which is approximately the *sampling ratio* of the size of the population to the size of the sample. Thus, if the population consists of 10,000 sampling units and a sample of 500 units is desired, then

$$k = \frac{10,000}{500} = 20$$

and the sample is obtained by taking every twentieth unit in the population.

In order that each unit in the population has an equal probability of being drawn, the procedure must have a random start. With a sampling ratio of 20, we may use the goldfish bowl procedure by putting 20 balls numbered 1 to 20 into a bowl. After the balls are thoroughly mixed, one ball is drawn at random. If the ball numbered 11 is drawn, then we begin with this number and include every twentieth from it, namely, the thirty-first, the fifty-first, and so on, in the sample.

Systematic sampling can be easily employed when a list of the units in the population, such as a telephone directory, is already available. Occasionally, the samples taken for statistical quality control are systematic samples. This is the case when a sample is obtained by taking an item from the current production stream at regular intervals.

The procedure of systematic random sampling will select a more representative sample than will simple random sampling if the units in the population resemble other nearby items more than they resemble those far away. If we are to estimate the average family income of Los Angeles, for example, households at Beverly Hills and those at Watts will be in proportion and thus equally represented by using systematic sampling, while a simple random sample might select a relatively large or small number of households from each community in which the economic status of residents tends to be similar. In the latter instance, a greater sampling error will result. Since the systematic sample is not subject to this type of error, it clearly tends to be more representative.

On the other hand, the systematic sample is less representative than the simple random sample in situations where there is hidden *periodicity* in the population, a cyclical or periodic movement of the data with the length of the cycle tending to be close to the sampling ratio k. For example, if we are to estimate the average daily ticket sales for a large airline, a sample consisting of sales of every seventh day will not be representative of the daily sales of the company. The sample will include sales of, say, all Fridays or all Saturdays, and will present an erroneous picture because of the repetitive weekly pattern of ticket sales.

Furthermore, systematic sampling still has the drawback of numbering or arranging the items in a large population. This might be physically impossible if the population covers an entire country or a large geographical area.

6-2-3 Stratified sampling

Another restricted sampling design, which can be more efficient than the simple random sampling, is *stratified random sampling*. The procedure requires that we have prior knowledge about the population. The process of stratification involves dividing the population into groups or classes called *strata*. Within each such stratum are placed items more homogeneous with respect to the characteristics to be studied. From each stratum, a subsample is taken by the simple random procedure, and the overall sample is obtained by combining the subsamples for all strata.

Frequently, the sampling ratio is made the same for all strata. That is, the items in the sample are allocated among the strata in proportion to the relative number of items in each stratum in the population. A sample drawn in this way is called a *proportionate* stratified sample. Sometimes the sampling ratio in each stratum is made proportional to the standard deviation of that stratum; the more homogeneous the stratum, the smaller its standard deviation, and the smaller will be its proportion included in the sample. A sample thus obtained is called a *disproportionate* stratified sample. Often the subsamples for the various strata are obtained with methods that take both cost and variation into consideration in determining their sizes. Such a procedure of stratification is referred to as *optimum allocation*.

Stratified sampling is most effective in handling heterogeneous populations such as data on unemployment (which varies from industry to industry and from occupation to occupation), retail sales (which differ among the various geographical areas), and the attitudes of consumers toward new automobile models (influenced by such factors as sex, age, and income status). By stratification, classes are set up so that the sampling units within each class tend to be uniform and the classes tend to be different from one another. Thus the proportion of the overall sample from each stratum can be controlled and not left to chance, and the representativeness of the sample is assured.

If the variance of the observed characteristic of each stratum is smaller than that of the entire population, as is usually true because of greater uniformity within the stratum, a gain in reliability for a given sample size, or in efficiency for a given degree of reliability, will result. The greater the reduction in within-group variance, the larger will be the gain due to stratification.

The gain in reliability and efficiency can be increased by further classification of the strata into substrata, if possible. In conducting a poll for an election, for instance, we may take party affiliations as the strata. Since voting preferences are often influenced by such factors as sex, age, race, religion, and educational status, the strata may be subdivided according to these factors. This procedure of cross classification is referred to as *cross stratification*.

Regardless of how the data are classified or stratified, resulting samples are considered probability samples. Past data, intuition, expert judgment, or preliminary findings from pilot studies may be used to set up the strata. The stratified sample is therefore a combination of subsamples of the strata, which are simple random or systematic samples. As such, every available item within each stratum has the same probability of being selected. This will be the situation even if the sample is a disproportionate one, for which the probabilities of individual items of the population being selected are not equal.

6-2-4 Cluster sampling

Diametrically opposed to stratified sampling is *cluster sampling*, which consists of first selecting, at random, groups called *clusters* of individual items from the population, and then choosing all or a subsample of the items within each cluster to make up the overall sample. In order to get best results in cluster sampling design, differences among clusters are made as small as possible, while differences among individual items within each cluster are made as large as possible. Ideally, each cluster should be a miniature of the entire population, and thus a single cluster would be a satisfactory sample. In actual practice, however, these requirements are difficult to meet, but it is an important principle to remember when designing a clustered sample.

Clusters are often referred to as the *primary sampling units*. If all the items, or elementary units, of selected clusters are included in the sample, the procedure is called *single-stage sampling*. If a random subsample of items is drawn from each selected cluster, we have what is known as the *two-stage sampling*. If more than two stages are involved in obtaining the overall sample, it is called *multistage sampling*. For example, in an attitude survey of college students in this country, colleges and universities are selected as clusters or primary sampling units. This is the first stage. The second stage may be the selection of divisions within each of the chosen colleges and universities. Finally, as the third stage, students of all the selected divisions are drawn to make up the entire sample. Of course, random sampling methods are used at each stage. The objective of cluster sampling is still the study of the characteristics of the individual items or elementary units, though the primary sampling units are selected initially.

The principal advantage of cluster sampling is the great reduction in costs for a given degree of reliability. For example, the cost of taking a simple random sample of, say, 5,000 women from the population of housewives in this country would be tremendous. If a sample of 50 counties were taken first, and then a subsample taken from each of the 50 counties already selected, the sample of 5,000 could be obtained at much less cost than that required for a simple random sample of this size. This is why, when sampling from a geographically distributed population, geographical areas are taken as the primary sampling units—a procedure known as *area sampling*.

Cluster sampling is often used in statistical quality control. First, "lots" or "batches" of output are selected at random as clusters. As a single-stage sampling, all units of the selected lots may be tested. Or, a sample may be taken from each lot for testing—a two-stage sampling.

The drawback of cluster sampling lies in its relative lack of reliability for a sample of a given size. The variance of a clustered sample tends to be greater than that of a random sample taken by any other procedure discussed above. This disadvantage, however, does not lead to the conclusion that cluster sampling cannot be more efficient than other random designs. If the cost of selecting an elementary unit is much reduced, as is true with cluster sampling, the same amount of expense will enable us to select a much larger sample. As we shall see later, a larger sample will result in a smaller sampling error. It is possible to select a cluster sample large enough to achieve greater reliability with the same cost as for any other random sampling. As statisticians frequently say, the consideration of cost often leads to cluster sampling.

Exercise 6-2

1 Define the following terms in your own words as fully as you can:

a Simple random sampling.
b Systematic sampling.
c Stratified sampling.
d Cluster sampling.

2 Differentiate each of the following pairs of terms:

a Elementary sampling unit and primary sampling unit.
b Sampling reliability and sampling efficiency.
c Proportionate sampling and disproportionate sampling.

3 Define the population, and determine the most appropriate sampling design for each of the following situations:

a A production manager wishes to control the quality of a product by taking a sample for examination from time to time.
b A market researcher wishes to make a survey of housewives' attitudes toward a newly designed household appliance in a certain metropolitan area.

4 Suppose you wish to draw a sample that consists of 10 percent of the enrollment of your college. Describe the procedure you would follow in using each of the following methods:

a The random digit table.
b The goldfish bowl method.
c Systematic sampling.

5 A poll is to be conducted to determine the voting preference of the voters in a certain city. Design a sampling plan such that the sample would be representative of the population of all voters.

6 Define the following concepts in your own words as fully as you can:

a Optimum allocation.
b Area sampling.
c Cross stratification.
d Sampling ratio.
e Random sampling.

7 Compare the advantages and drawbacks of each of the following pairs of sampling designs:

a Simple random sampling and systematic sampling.
b Stratified sampling and cluster sampling.

8 Suppose that a sample of 50 workers is to be selected from the membership of a local union which consists of 8,000 workers. Describe the procedure of drawing such a random sample by using the random digit table, Appendix I.

9 Suppose that the management of a magazine wishes to conduct a survey of the attitude of its readers toward a certain feature of the magazine. There are 500,000 subscribers, and the questionnaire is to be sent to 10 percent of them. How do you draw such a sample if the systematic sampling technique is used?

10 A large company has 300,000 employees, the age distribution of whom is shown as follows:

Age, years	*Percentage*
25 or younger	15
26–35	30
36–45	25
46–55	20
56 or older	10

A sample of 2 percent of all the employees is desired. Design a sampling plan such that each age group is proportionally represented.

11 A survey of the attitude of all elementary school teachers in the United States toward a certain issue is to be conducted. Describe the procedure of drawing a three-stage sample.

6-3 Sampling Distribution

As already mentioned in Sec. 6-1, numerous different samples of equal size can be taken from the same population. The probability distribution of all the possible values of a given statistic from all the distinct possible samples of equal size is called a *sampling distribution.* In this section we shall use the theoretical distribution of a particular statistic, the sample mean, to illustrate the concept of a sampling distribution.

Let us first consider a simple hypothetical example. Suppose that random variable X takes on any of the four values 2, 4, 6, and 8. These four numbers can be used to simulate an infinite population if samples are taken *with replacement.* Suppose further that samples of size 2 are taken from this population. We may write the numbers 2, 4, 6, and 8 on four balls and use the goldfish bowl procedure to select our samples. One sample may consist of the two numbers 2 and 8, and another sample may be made up of the numbers 8 and 6, or 6 and 2, or some other combination. It is of great importance that we know all the *different* samples that could possibly result from this experiment.

There are altogether 16 different possible samples that can be selected from this population. Table 6-1 lists all the 16 samples with X_1 designating the number on the first ball drawn, and X_2 the number on the second ball. Although hypothetically an infinite number of samples could be drawn, only 16 *different* possible samples could result. Of course, if you take thousands of such samples, many of them would be identical.

Table 6-1 also shows the arithmetic mean for each of the samples. The arithmetic mean $\bar{X}$ assumes any of the values in the set of seven values, namely, {2, 3, 4, 5, 6, 7, 8}. These values are designated as $\bar{x}$; they are the possible values that $\bar{X}$ can take on. Of the 16 possible samples, one has the mean value 2, two have the value 3, three have the value 4, four have the value 5, three have the value 6, two have the value 7, and one has the value 8. The

Table 6-1
Different possible samples of size 2 for X distributed uniformly on {2, 4, 6, 8}

Sample	X_1	X_2	*Sample mean* $\bar{X}$
1	2	2	2
2	2	4	3
3	2	6	4
4	2	8	5
5	4	2	3
6	4	4	4
7	4	6	5
8	4	8	6
9	6	2	4
10	6	4	5
11	6	6	6
12	6	8	7
13	8	2	5
14	8	4	6
15	8	6	7
16	8	8	8

distribution of the sample means is shown in Table 6-2. Each of the 16 results has an equal probability $\frac{1}{4} \times \frac{1}{4} = \frac{1}{16}$ of occurring, and thus they are equally likely events. Accordingly, the probability of each of the possible values of the sample mean shown in Table 6-2 is the ratio of the number of samples associated with that value to the total number of samples, namely, 16. The last column of Table 6-2 shows the probability distribution of the sample mean $\bar{X}$.

As indicated, the probability of obtaining a sample mean 2 is $\frac{1}{16}$, that of 3 is $\frac{2}{16}$, and so forth. This does not mean, however, that out of 1,600 samples

Table 6-2
Distribution of sample mean $\bar{X}$ *with sample size* 2

Sample mean $\bar{x}$	*Number of samples*	*Probability* $P(\bar{x})$
2	1	$\frac{1}{16}$
3	2	$\frac{2}{16}$
4	3	$\frac{3}{16}$
5	4	$\frac{4}{16}$
6	3	$\frac{3}{16}$
7	2	$\frac{2}{16}$
8	1	$\frac{1}{16}$
Total	16	1.0

Source: Table 6-1.

taken, 100 samples have a mean 2, and 200 samples have a mean 3, and so on. Nor out of 16 million samples, will you get exactly 1 million with a mean 2, and 2 million with a mean 3. If 16 billion samples are drawn, however, the number of samples with a mean 2 will be close to 1 billion, or the ratio will be approximately $\frac{1}{16}$. In other words, the probability distribution is the distribution of the *theoretical* relative frequency, and can be approached only by an infinitely large number of trials. (It is hardly necessary to point out that the probability distribution shown here is an illustration of a theoretical distribution. In practice we never take so many samples, but only one particular sample, which is a part of the distribution. On the basis of the sample, inferences are made about the characteristics of the distribution.)

The sampling distribution of the sample mean as shown in Table 6-2 is graphically represented in Fig. 6-3.

To construct a distribution of the sample mean by listing all the different possible samples is, of course, an inefficient and, at times, impossible procedure. It is inefficient because there would be too many different samples to list when the random variable has a large sample space. For example, if the size of the samples is 10 and they are taken from a population of only 100 distinct items, there would be $\binom{100}{10} = 17{,}310{,}309{,}456{,}440$ different samples! It is impossible because there would be an infinite number of different samples if the random variable has an infinite sample space. It is therefore more convenient to derive the distribution of a sample statistic such as the mean in terms of the theoretical probabilities of all possible sample results on assumptions made about the population.

Accordingly, even though an empirical relative frequency distribution of a sample statistic can be constructed by drawing a large number of samples of equal size from the same population and then distributing them according to all the actual sample outcomes, the distribution of a sample statistic is more frequently stated as a functional relation between all possible outcomes and their corresponding probabilities. As already stated, the distribution of a statistic is called a sampling distribution. It is derived from the population, called *parent* population, of the underlying random variable. Specifically, if some n items are taken at random from a given parent population and some statistic is computed

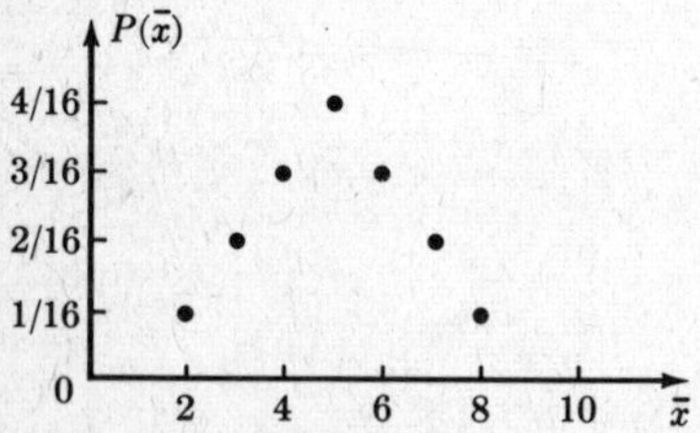

Figure 6-3

Probability distribution of the sample mean from Table 6-2.

from the sample, then the concept of a *sampling distribution* may be more formally defined as follows:

> *A sampling distribution is a theoretical distribution that expresses the functional relationship between each of the distinct values of the sample statistic and the corresponding probability for all the different possible samples of size n from the same population.*

A sample statistic is a new kind of random variable. Its distribution, the sampling distribution, has its mean and standard deviation. It is extremely important that we know the mean and the standard deviation of such a random variable, for they are the most significant attributes of a sampling distribution. We shall again use the above example to illustrate the computation of the mean and standard deviation of the sample statistic $\overline{X}$.

6-3-1 The mean of the random variable $\overline{X}$

Since the distribution of $\overline{X}$ is a probability distribution, we may apply the notion of expectation discussed in Sec. 4-4-1 to compute the mean of this new variable. By Eq. (4-19), we have

$$E(\overline{X}) = \sum_{j=1}^{m} \bar{x}_j P(\bar{x}_j) \tag{6-3}$$

Using the distribution in Table 6-2, we compute the mean of the random variable $\overline{X}$ as follows:

$$\begin{aligned}
E(\overline{X}) &= 2\left(\frac{1}{16}\right) + 3\left(\frac{2}{16}\right) + 4\left(\frac{3}{16}\right) + 5\left(\frac{4}{16}\right) + 6\left(\frac{3}{16}\right) + 7\left(\frac{2}{16}\right) + 8\left(\frac{1}{16}\right) \\
&= \frac{2 + 6 + 12 + 20 + 18 + 14 + 8}{16} = \frac{80}{16} \\
&= 5
\end{aligned}$$

We know that the mean of the parent population is $\mu = (2 + 4 + 6 + 8)/4 = 5$. Thus the mean of the sampling distribution is identical with the mean of the parent population. Or, symbolically, $E(\overline{X}) = E(X) = \mu$. Since this principle is of fundamental importance, its mathematical validity should be demonstrated. If $\overline{X}$ is derived from n observations, we have

$$\begin{aligned}
E(\overline{X}) &= E\left(\frac{\sum_{i=1}^{n} X_i}{n}\right) = \frac{1}{n} E\left(\sum_{i=1}^{n} X_i\right) && \text{by (4-22)} \\
&= \frac{1}{n} E(X_1 + X_2 + \cdots + X_n) \\
&= \frac{1}{n} [E(X_1) + E(X_2) + \cdots + E(X_n)] && \text{by (4-27)} \\
&= \frac{1}{n} [nE(X)] = E(X)
\end{aligned}$$

thus,

$$E(\bar{X}) = \mu \tag{6-4}$$

6-3-2 The standard error

The standard deviation of the random variable $\bar{X}$, as of any other random variable, measures the dispersion or variability of the sampling distribution. The standard deviation of a sampling distribution is generally referred to as the *standard error* of the distribution. The standard deviation of the mean, commonly denoted by $\sigma_{\bar{X}}$, is called the *standard error of the mean.* Before we examine the relation between the standard error of the mean and the standard deviation of the parent population, let us first compute the standard error of the mean for the hypothetical example shown in Table 6-2.

Table 6-3 shows that the variance of $\bar{X}$, designated $\sigma_{\bar{X}}{}^2$, is 2.5 and the standard error is $\sqrt{2.5}$. We know that the variance of X is

$$\sigma_X{}^2 = E(X^2) - \mu^2 = \sum x^2 P(x) - \mu^2$$

$$= \left[2^2\left(\frac{1}{4}\right) + 4^2\left(\frac{1}{4}\right) + 6^2\left(\frac{1}{4}\right) + 8^2\left(\frac{1}{4}\right)\right] - 5^2$$

$$= \frac{4 + 16 + 36 + 64}{4} - 25$$

$$= \frac{120}{4} - 25 = 30 - 25 = 5$$

and the standard deviation of X is $\sigma_X = \sqrt{5}$. Therefore the variance of the mean is equal to the variance of the population divided by the sample size; the standard error of the mean is equal to the standard deviation of the population divided by the square root of the sample size. That is,

$$\sigma_{\bar{X}}{}^2 = \frac{\sigma^2}{n} \qquad \text{and} \qquad \sigma_{\bar{X}} = \frac{\sigma_X}{\sqrt{n}}$$

This relationship can be proved mathematically. Again suppose that X is any independent random variable and $\bar{X}$ is the mean of a random sample of size n. Then we have

$$\sigma_{\bar{X}}{}^2 = \sigma^2_{(1/n)(X_1 + X_2 + \cdots + X_n)}$$

$$= \frac{1}{n^2}\sigma^2_{(X_1 + X_2 + \cdots + X_n)} \qquad \text{by (5-19)}$$

$$= \frac{1}{n^2}(\sigma^2_{X_1} + \sigma^2_{X_2} + \cdots + \sigma^2_{X_n}) \qquad \text{by (5-21)}$$

$$= \frac{1}{n^2} n\sigma_X{}^2$$

Table 6-3
Computation of standard error of mean

Sample mean $\bar{x}$	*Probability* $P(\bar{x})$	*Mean squared* $\bar{x}^2$	$\bar{x}^2 P(\bar{x})$
2	$\frac{1}{16}$	4	$\frac{4}{16}$
3	$\frac{2}{26}$	9	$\frac{18}{16}$
4	$\frac{3}{16}$	16	$\frac{48}{16}$
5	$\frac{4}{16}$	25	$\frac{100}{16}$
6	$\frac{3}{16}$	36	$\frac{108}{16}$
7	$\frac{2}{16}$	49	$\frac{98}{16}$
8	$\frac{1}{16}$	64	$\frac{64}{16}$
Total	1.0		$\frac{440}{16}$

$$\sigma_{\bar{X}}^2 = E[(\bar{X}-\mu)^2] = E(\bar{X}^2) - \mu^2$$
$$= \sum \bar{x}^2 P(\bar{x}) - \mu^2$$
$$= \frac{440}{16} - 5^2 = \frac{440-400}{16} = \frac{40}{16} = 2.5$$

and

$$\sigma_{\bar{X}} = \sqrt{2.5}$$

Source: Table 6-2.

Thus

$$\sigma_{\bar{X}}^2 = \frac{\sigma_X^2}{n} \tag{6-5}$$

and

$$\sigma_{\bar{X}} = \frac{\sigma_X}{\sqrt{n}} \tag{6-6}$$

The proof shows that the independence of the observations of the random variable plays a crucial role in the derivation of the standard error. Without it, the *additivity* of variances would not be possible.

The above result makes it possible to find the standard error of the mean without knowing the distribution for $\bar{X}$. It is necessary to know only the sample size and the standard deviation of the population. The sample size can be easily determined. The standard deviation of the population is usually unknown; but, fortunately, it can be closely estimated. Thus Eqs. (6-5) and (6-6) have great practical significance, and their application permeates the next several chapters.

It must be pointed out that the standard error of the mean *varies directly* with the standard deviation of the population, but *inversely* with the square root of the sample size. That is, given the sample size, the greater the value of σ_X, the higher the value of $\sigma_{\bar{X}}$ will be; given σ_X, the greater the value of n, the lower the value of $\sigma_{\bar{X}}$. Therefore, given σ_X, the value of $\sigma_{\bar{X}}$ can be controlled by controlling—increasing or decreasing—the size of the sample. This relationship is applied in the next section, which is concerned with one of the methods to determine the size of a random sample.

6-3-3 The law of large numbers

It is now quite obvious that if the sample size is 1, the variance of the mean is exactly identical with the population variance. If n is equal to 2, then $\sigma_{\bar{X}}^2$ is one-half as large as σ_X^2. With n equal to 100, $\sigma_{\bar{X}}^2$ will be only $\frac{1}{100}$ of σ_X^2. If n approaches infinity, $\sigma_{\bar{X}}^2$ approaches zero, which means that the sample mean is identical with the mean of the population. In general, the larger the value of n, the more probable it is that $\bar{X}$ comes arbitrarily close to the population mean μ. By choosing n large enough, we can make the probability that $\bar{X}$ is within an arbitrary value d from μ as large as we wish. This relationship is referred to as the *law of large numbers.*

This law can be illustrated by referring to Tchebycheff's theorem (see Sec. 5-5). The Tchebycheff inequality for a population is

$$P(|X - \mu| \leq k\sigma_X) \geq 1 - \frac{1}{k^2} \qquad \text{from (5-25)}$$

That is, the probability that X falls within k standard deviations from its mean is at least as large as $1 - 1/k^2$. Let us now apply this theorem to the random variable $\bar{X}$, the mean of a random sample of size n. By Eqs. (5-25) and (6-6), we have

$$P\left(|\bar{X} - \mu| \leq k\frac{\sigma_X}{\sqrt{n}}\right) \geq 1 - \frac{1}{k^2} \tag{6-7}$$

where $k(\sigma_X/\sqrt{n}) = d$, referred to in the last paragraph. Let d be any arbitrarily small positive number. If the sample is large enough, we can make the probability of $\bar{X}$'s being within d of μ as close to 1 as we wish.

For instance, given $\sigma_X = 10$, if we want the sample mean to be no more than 5 from the population mean with at least 99 percent probability, we should need a sample size of 400, which is calculated as follows. Since

$$1 - \frac{1}{k^2} = 0.99 \qquad k = 10$$

and

$$d = 5 = 10\frac{10}{\sqrt{n}}$$

$$5\sqrt{n} = 100$$

$$25n = 10{,}000$$

$$n = \frac{10{,}000}{25}$$

thus

$$n = 400$$

Exercise 6-3

1 Define the following concepts in your own words:

a Sampling distribution. *c* Law of large numbers.
b Standard error.

2 Let $\bar{X}$ be the mean of a sample $X_1, X_2, \ldots, X_n$ and $\bar{Y}$ be the mean of a sample $Y_1, Y_2, \ldots, Y_m$. Given that the two samples are independent, find:

a The mean and variance of $\bar{X}+\bar{Y}$.
b The mean and variance of $\bar{X}-\bar{Y}$.

3 If $E(X)=50$ and $\sigma_X=5$, how large a sample must you take in order that $P(|\bar{X}-50| \leq 1) \geq 0.99$?

4 Let X be an odd random digit and $\bar{X}$ be the mean of a random sample of two odd digits. Find the mean and the variance of the sampling distribution for $\bar{X}$.

5 Suppose that the random variable X has a mean μ and a variance 100. A random sample of X yields a mean $\bar{X}$. Find the required sample size n such that the probability of $\bar{X}$'s being within 0.1 from μ is at least 75 percent.

6 Suppose that a federal agency wishes to conduct a study of the average weight of airline passengers. It is assumed that the standard deviation of such passengers is 30 pounds. The agency specifies that the sample average will not deviate from the true mean by more than 5 pounds and that the probability of his achieving this degree of precision is at least $\frac{8}{9}$. How large a sample is required to meet these specifications?

7 Suppose that the U.S. Department of Labor wishes to estimate the average age of all the unemployed in the United States. The Department wants the difference between the true mean and the mean of the sample not to be greater than 2, and the chance of the sample mean's deviating from the true mean by more than 2 not to exceed $\frac{1}{16}$. It is known that the population variance of the ages of the unemployed is 36 years. Find the sample size n such that these conditions are satisfied.

8 Let X designate a random digit. From Secs. 4-4-1 and 5-4, we know that $E(X) = 4.5$ and $\sigma_X^2 = 8.25$.

a Let $\bar{X}$ be the mean of a random sample of 10 digits, $X_1, X_2, \ldots, X_{10}$. Find $E(\bar{X})$ and $\sigma_{\bar{X}}^2$.

b Let Y be the mean of 20 independent $\bar{X}$'s. Find $E(Y)$ and σ_Y^2.

c Let Z be the mean of a random sample of 200 digits. Find $E(Z)$ and σ_Z^2.

d What conclusion can you draw from the answers to *a*, *b*, and *c*?

9 Assume that simple random samples of two children are selected with replacement from a population of four children with ages 3, 4, 5, and 6. Let X be the age of any child; find the following:

a The theoretical sampling distribution of $\bar{X}$, the mean age of two children in any sample.

b The mean and the standard error of $\bar{X}$.

10 Suppose that a random variable X has the following population distribution:

x	$P(x)$
3	$\frac{1}{3}$
6	$\frac{1}{3}$
9	$\frac{1}{3}$

If a sample of three numbers is taken with replacement, derive the sampling distribution of the sample mean.

Chapter 7

Discrete probability distributions

This chapter will deal with discrete probability distributions. For better understanding of the discussion, you are advised to review Chaps. 2 and 3, especially Secs. 2-2 and 3-4, before proceeding.

Many random variables are discrete in nature. As indicated in Sec. 3-4-2, discrete random variables may assume values that differ from each other by finite amounts. In other words, the sample space of a discrete random variable is made up of a finite, or countably infinite, number of values or sample points. For instance, in the toss of four coins, we may get five possible outcomes—zero, one, two, three, or four heads—and no more. The number of heads or tails observed must be integral since fractional values are not possible in this experiment. There is a clear distinction between such a random variable and those that may theoretically assume values of fractional units, or values that differ from each other by infinitely small amounts.

The most frequently encountered discrete distributions include the binomial distribution, the distribution of the sample proportion, the hypergeometric distribution, and the Poisson exponential distribution. Since these distributions are all based on the Bernoulli random process, attention is first focused on the Bernoulli process.

7-1
Bernoulli Process

A considerable group of random phenomena known as Bernoulli processes, named after Jacques Bernoulli (1654–1705), requires the simplest probability distribution—having only two *mutually exclusive* and *exhaustive* events. For

Table 7-1
Bernoulli probability model

y	$P(y)$
0	$q = 1 - p$
1	p

example, if a housewife is selected at random from a city and is asked whether she watches a particular television program, there seem to be only two possible answers: "yes" and "no." A unit chosen from a lot turned out by an automatic machine can be either good or defective. A student selected from a college can be only male or female, married or not married, and so forth. To standardize the terminology describing these and many other similar processes, we shall call one of the two possible outcomes a *success* and the other a *failure*. These terms are used only to identify the outcomes and bear no connotation of "goodness" about the outcome. Thus we can use either term for a married student or for a defective product provided that we are consistent throughout a particular discussion.

The two events, success and failure, are, of course, qualitative in nature. As discussed in Sec. 6-1, we can convert these qualitative events into numerical ones by assigning the value "1" to a success and "0" to a failure. Let Y be a Bernoulli random variable with the possible values of 0 and 1. Also, let p stand for the probability of a success, and q, which is equal to $1 - p$, be the probability of a failure. Then the *Bernoulli probability model* will have its general form as shown in Table 7-1. Since the model has only two classes of events, it is also known as the *two-point probability model.*

Similar to other variables, the Bernoulli random variable has its mean and variance. The procedures of obtaining the mean and the variance for a Bernoulli random variable are shown in Table 7-2. By the use of the expectation notation, the mean of Y is given by

$$\begin{aligned} E(Y) &= 0(1 - p) + 1(p) \qquad \text{by (4-19)} \\ &= 0 + p \\ &= p \end{aligned} \tag{7-1}$$

Table 7-2
Process of obtaining the mean and variance of Bernoulli random variable Y

y	$P(y)$	y^2	$y^2P(y)$
0	$1 - p$	0	0
1	p	1	p
Total			p

The variance and standard deviation of Y, designated $\sigma_Y{}^2$ and σ_Y, are obtained by applying Eq. (5-17) as follows:

$$\begin{aligned}\sigma_Y{}^2 &= E(Y^2) - [E(Y)]^2\\ &= 0^2(1-p) + 1^2(p) - p^2\\ &= p - p^2\\ &= p(1-p)\\ &= pq \qquad (7\text{-}2)\end{aligned}$$

and

$$\sigma_Y = \sqrt{pq}$$

That is, the mean of a Bernoulli random variable is equal to the probability of a success, and the variance is equal to the probability of a success multiplied by that of a failure.

In the toss of an ideal coin, for example, the probability of obtaining heads, termed a success, is 0.5, and the probability of obtaining tails, termed a failure, is 0.5. The mean of such a variable is therefore 0.5, the variance is $0.5 \times 0.5 = 0.25$, and the standard deviation is $\sqrt{0.25} = 0.50$.

The mean p referred to above is, of course, a population parameter. It is the *expected ratio* of the number of successes to the total number of trials. If we take a sample of 100 trials ($n = 100$) in a particular experiment, it would be coincidental that the ratio of successes to 100 is exactly equal to p. However, if the sample size is enlarged, the *sample ratio*, or *proportion*, will come closer to p. The two tend to be identical as n approaches infinity. Or, if we take many samples of size n and for each obtain a sample ratio, the sample ratios would be centered around the population parameter p. The distribution of such a sample ratio is naturally a *sampling distribution*. But we have to delay our treatment of such a sampling distribution until we have discussed another important and related concept, the binomial distribution.

Exercise 7-1

1 Define the following concepts as fully as you can:

a Bernoulli random variable. *b* Bernoulli probability model.

2 In a clothing manufacturing company 75 percent of the employees are women and 25 percent of them are men. Let Y be the random variable that takes on the value 1 when an employee randomly selected is a woman and 0 when the employee is a man. Find the mean and variance for Y.

3 An ideal coin is tossed 30 times, and 18 heads are obtained. What is the mean of Y, the Bernoulli random variable, for these 30 tosses? Is this mean a sample mean or a population mean? Explain fully.

4 Let Y be a Bernoulli random variable. Is the distribution of Y a population distribution or a sample distribution? Explain fully.

5 Suppose that 80 percent of the families in a certain city are homeowners. Let Y be the random variable that takes on the value 1 when a randomly selected family from the city is a homeowner and 0 when it is not. Find the mean and variance for Y.

6 Describe the procedure by which you can find a sample mean for a Bernoulli random variable.

7 A cold vaccine is claimed to be 70 percent effective; that is, 70 out of 100 people who have received the vaccine will survive the winter without a cold. Let W be the Bernoulli random variable in this case. Find the mean and variance of W.

8 Given that the probability of obtaining a "failure" is 0.1. Construct a Bernoulli probability model, and then compute the mean and standard deviation for the random variable.

7-2 Binomial Distribution

Suppose that a random sample of n observations is taken by the Bernoulli process. Let us designate this sample by $(Y_1, Y_2, \ldots, Y_n)$. Since Y takes on the value 1 if a trial results in a success, and 0 if the outcome is a failure, the number of 1s in n trials is simply $Y_1 + Y_2 + \cdots + Y_n$, or $\sum_{i=1}^{n} Y_i$. Let us now introduce a new random variable X which is called the *binomial random variable*. This is used to designate the number of successes in a sample of n observations taken by the Bernoulli process, that is,

$$X = \sum_{i=1}^{n} Y_i \tag{7-3}$$

This fact can be illustrated as follows: Suppose, for instance, that six people ($n = 6$) are interviewed to determine how many of them watched a particular TV program last Saturday, that a "yes" answer is considered a success, and that two of them give "yes" answers as indicated below:

Interviewee:	Y_1	Y_2	Y_3	Y_4	Y_5	Y_6
Outcome:	1	0	0	1	0	0

Then, the number of successes is

$$X = \sum_{i=1}^{6} Y_i = Y_1 + Y_2 + Y_3 + Y_4 + Y_5 + Y_6$$
$$= 1 + 0 + 0 + 1 + 0 + 0 = 2$$

7-2-1 Probability distribution

The variable X is *discrete* because the possible values it may assume are counts of successes out of n trials, which are integers from 0 through n. That is, X

has $\{0, 1, 2, \ldots, n\}$ as the set of its possible values. We now use a simple example to illustrate the probability distribution of X.

Suppose that three coins are tossed simultaneously. It is apparent that the outcomes of this experiment are *independent*. It is assumed that the coins are all perfect, or $p = 1 - p = q = 0.5$. Let the three outcomes be Y_1, Y_2, and Y_3. The binomial random variable X designating the number of heads obtained has $\{0, 1, 2, 3\}$ as the set of its possible values. Accordingly, we have

$$\begin{aligned}
P(X = 0) &= P(Y_1 = 0 \text{ and } Y_2 = 0 \text{ and } Y_3 = 0) \\
&= P(Y_1 = 0) \times P(Y_2 = 0) \times P(Y_3 = 0) \qquad \text{by (3-6)} \\
&= q \times q \times q = q^3 = (\tfrac{1}{2})^3 = \tfrac{1}{8}
\end{aligned}$$

$$\begin{aligned}
P(X = 1) &= P(Y_1 = 1 \text{ and } Y_2 = 0 \text{ and } Y_3 = 0) \\
&\qquad \text{or } P(Y_1 = 0 \text{ and } Y_2 = 1 \text{ and } Y_3 = 0) \\
&\qquad \text{or } P(Y_1 = 0 \text{ and } Y_2 = 0 \text{ and } Y_3 = 1) \\
&= P(Y_1 = 1 \text{ and } Y_2 = 0 \text{ and } Y_3 = 0) \\
&\quad + P(Y_1 = 0 \text{ and } Y_2 = 1 \text{ and } Y_3 = 0) \\
&\quad + P(Y_1 = 0 \text{ and } Y_2 = 0 \text{ and } Y_3 = 1) \qquad \text{by (3-3)} \\
&= P(Y_1 = 1) \times P(Y_2 = 0) \times P(Y_3 = 0) \\
&\quad + P(Y_1 = 0) \times P(Y_2 = 1) \times P(Y_3 = 0) \\
&\quad + P(Y_1 = 0) \times P(Y_2 = 0) \times P(Y_3 = 1) \qquad \text{by (3-6)} \\
&= p \times q \times q + q \times p \times q + q \times q \times p \\
&= pq^2 + pq^2 + pq^2 = 3pq^2 \\
&= 3(\tfrac{1}{2})(\tfrac{1}{2})^2 = 3(\tfrac{1}{2})^3 = \tfrac{3}{8}
\end{aligned}$$

$$\begin{aligned}
P(X = 2) &= P(Y_1 = 1 \text{ and } Y_2 = 1 \text{ and } Y_3 = 0) \\
&\qquad \text{or } P(Y_1 = 1 \text{ and } Y_2 = 0 \text{ and } Y_3 = 1) \\
&\qquad \text{or } P(Y_1 = 0 \text{ and } Y_2 = 1 \text{ and } Y_3 = 1) \\
&= P(Y_1 = 1 \text{ and } Y_2 = 1 \text{ and } Y_3 = 0) \\
&\quad + P(Y_1 = 1 \text{ and } Y_2 = 0 \text{ and } Y_3 = 1) \\
&\quad + P(Y_1 = 0 \text{ and } Y_2 - 1 \text{ and } Y_3 = 1) \\
&= P(Y_1 = 1) \times P(Y_2 = 1) \times P(Y_3 = 0) \\
&\quad + P(Y_1 = 1) \times P(Y_2 = 0) \times P(Y_3 = 1) \\
&\quad + P(Y_1 = 0) \times P(Y_2 = 1) \times P(Y_3 = 1) \\
&= p \times p \times q + p \times q \times p + q \times p \times p \\
&= p^2q + p^2q + p^2q = 3p^2q \\
&= 3(\tfrac{1}{2})^2(\tfrac{1}{2}) \\
&= 3(\tfrac{1}{2})^3 = \tfrac{3}{8}
\end{aligned}$$

and

$$P(X = 3) = P(Y_1 = 1 \text{ and } Y_2 = 1 \text{ and } Y_3 = 1)$$
$$= P(Y_1 = 1) \times P(Y_2 = 1) \times P(Y_3 = 1)$$
$$= p \times p \times p = p^3 = (\tfrac{1}{2})^3 = \tfrac{1}{8}$$

The probability distribution of X is shown in Table 7-3.

Observe that $P(X = 0)$ is the probability of obtaining three tails, or q^3, and no heads, or p^0. Since $p^0 = 1$, $P(X = 0)$ may be expressed as p^0q^3. Similarly, $P(X = 3)$ may be expressed as p^3q^0. Furthermore, p^0q^3 and p^3q^0 remain as they are when multiplied by a factor 1. The resulting expressions of the probabilities of X in this experiment are shown in the last column of Table 7-3. The factors 1, 3, 3, and 1 are called *binomial coefficients*; and binomial coefficients are *combinations* of n objects taken x at a time. In this example we have

$$\binom{3}{0} = \frac{3!}{0!3!} = 1 \qquad \binom{3}{1} = \frac{3!}{1!2!} = 3 \qquad \binom{3}{2} = \frac{3!}{2!1!} = 3 \quad \text{and} \quad \binom{3}{3} = \frac{3!}{3!0!} = 1$$

The fact that each of the eight sequences in Table 7-3 is a combination rather than a permutation may not be clear to some readers, and some clarification may be in order. Take HTT as an example. Here $Y_1 = 1$, $Y_2 = 0$, and $Y_3 = 0$. If the order of occurrence of these three observations were a matter of concern, then other ordered arrangements of these three observations, such as ($Y_2 = 0$, $Y_1 = 1$, $Y_3 = 0$), ($Y_3 = 0$, $Y_2 = 0$, $Y_1 = 1$), ($Y_1 = 1$, $Y_3 = 0$, $Y_2 = 0$), ($Y_2 = 0$, $Y_3 = 0$, $Y_1 = 1$), and ($Y_3 = 0$, $Y_1 = 1$, $Y_2 = 0$), would be considered different from one another. But they are in fact identical because the order of their

Table 7-3

Binomial distribution for number of heads X in a toss of three ideal coins

Number of heads x	*Possible sequences*	*Probability of each sequence*	*Probability* $P(x)$
0	TTT	$\frac{1}{8}$	$1p^0q^3 = \frac{1}{8}$
1	HTT	$\frac{1}{8}$	
	THT	$\frac{1}{8}$	$3p^1q^2 = \frac{3}{8}$
	TTH	$\frac{1}{8}$	
2	HHT	$\frac{1}{8}$	
	HTH	$\frac{1}{8}$	$3p^2q^1 = \frac{3}{8}$
	THH	$\frac{1}{8}$	
3	HHH	$\frac{1}{8}$	$1p^3q^0 = \frac{1}{8}$
Total		1.0	1.0

arrangements is *not* a matter of concern. That is, together they constitute *one* combination. This explains why the sequence HTT, or any other sequence, is referred to as a combination.

The expressions in the last column of Table 7-3 may be generalized as follows:

$$P(X=0) = \binom{n}{0} p^0 q^{n-0}$$

$$P(X=1) = \binom{n}{1} p^1 q^{n-1}$$

$$P(X=2) = \binom{n}{2} p^2 q^{n-2}$$

....

$$P(X=n) = \binom{n}{n} p^n q^{n-n}$$

Thus, by replacing 0, 1, 2, . . . , n by x, we obtain the *binomial probability function* as follows:

$$P(X=x) = \binom{n}{x} p^x q^{n-x} \tag{7-4}$$

In tossing four ideal coins simultaneously, the probabilities of $X = x$ for $x = 0$, 1, 2, 3, and 4, the number of heads showing up, can be obtained by directly applying Eq. (7-4) as follows:

$$P(X=0) = \binom{4}{0}\left(\frac{1}{2}\right)^0\left(\frac{1}{2}\right)^{4-0} = 1\left(\frac{1}{2}\right)^4 = \frac{1}{16}$$

$$P(X=1) = \binom{4}{1}\left(\frac{1}{2}\right)^1\left(\frac{1}{2}\right)^{4-1} = 4\left(\frac{1}{2}\right)^4 = \frac{4}{16}$$

$$P(X=2) = \binom{4}{2}\left(\frac{1}{2}\right)^2\left(\frac{1}{2}\right)^{4-2} = 6\left(\frac{1}{2}\right)^4 = \frac{6}{16}$$

$$P(X=3) = \binom{4}{3}\left(\frac{1}{2}\right)^3\left(\frac{1}{2}\right)^{4-3} = 4\left(\frac{1}{2}\right)^4 = \frac{4}{16}$$

and

$$P(X=4) = \binom{4}{4}\left(\frac{1}{2}\right)^4\left(\frac{1}{2}\right)^{4-4} = 1\left(\frac{1}{2}\right)^4 = \frac{1}{16}$$

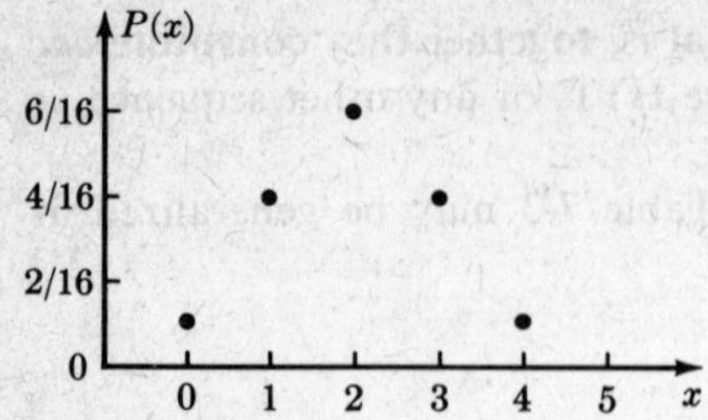

Figure 7-1
Probability distribution of the number of heads X in a toss of four ideal coins.

The probability distribution of X in this experiment is graphically shown in Fig. 7-1, and the sum of the probabilities for all distinct possible results is as follows:

$$\left(\frac{1}{2}+\frac{1}{2}\right)^4 = \sum_{x=0}^{4} \binom{4}{x}\left(\frac{1}{2}\right)^x\left(\frac{1}{2}\right)^{4-x}$$
$$= \frac{1}{16}+\frac{4}{16}+\frac{6}{16}+\frac{4}{16}+\frac{1}{16}$$
$$= \frac{16}{16} = 1$$

In summary, if X is a random variable designating the number of successes in a sample of n observations taken by the Bernoulli process with $\{0, 1, 2, \ldots, n\}$ as the set of its possible values, then X is called the *binomial random variable.* Furthermore, if the probability of a success on a single observation is p, the distribution for X, derived from Eq. (7-4) for $x = 0, 1, 2, \ldots, n$ and $q = 1 - p$, is referred to as the *binomial distribution.*

7-2-2 The use of binomial tables

The binomial distribution is a *sampling* distribution since the binomial variable X is a sample statistic that, like the sample mean, is derived from a sample of n observations.

When the value of n is large, however, the evaluation of $\binom{n}{x}p^x q^{n-x}$ becomes quite troublesome. Fortunately, large tables of binomial distributions for various values of n and p are available. A few such tables for specified values of n and p are given in Appendix II.

For illustrating the use of Appendix II, let us consider the following example: Suppose that 70 percent of the employees of a company are homeowners. What is the probability that a random sample of 20 employees will contain 15 homeowners? Here we have the probability $p = 0.70$ and the sample size $n = 20$. The probability that the sample will contain 15 homeowners is found in the intersection of row 15 and column 0.7 in the table for $n = 20$ in Appendix II. That is,

$$P(X = 15 \mid n = 20, p = 0.7) = 0.1789$$

It should be obvious by now that the binomial distribution is completely defined by n and p, the *parameters* of the distribution. The value of n and that of p specify a particular distribution among all possible binomial distributions. If X is a binomial random variable with n and p as the parameters of its distribution, then we can concisely say, "X has the distribution $B(n, p)$" or simply "X is $B(n, p)$."

In statistical inference, the binomial distribution is often put into cumulative form. Thus the probability that X falls *at or below* a certain value x is the sum of the probabilities for all the values at *and* below x or, symbolically, $P(X \leq x)$. In tossing four ideal coins, for example, the probability of getting two *or fewer* heads is

$$\begin{aligned} P(X \leq 2) &= P(X = 0 \text{ or } X = 1 \text{ or } X = 2) \\ &= P(X = 0) + P(X = 1) + P(X = 2) \qquad \text{by (3-3)} \\ &= \frac{1}{16} + \frac{4}{16} + \frac{6}{16} = \frac{11}{16} \end{aligned}$$

In the general case, the probability of getting k or fewer successes in n trials is given by

$$\begin{aligned} P(X \leq k) &= P(X = 0 \text{ or } X = 1 \text{ or } \cdots \text{ or } X = k) \\ &= P(X = 0) + P(X = 1) + \cdots + P(X = k) \\ &= \sum_{x=0}^{k} \binom{n}{x} p^x q^{n-x} \qquad \text{(7-5)} \end{aligned}$$

Since $P(X \leq x)$ is a function of x, we use the symbol $CP(x)$, wihch is called the *cumulative probability function* of x, to designate $P(X \leq x)$. A few tables of the values of $CP(x)$ are provided as Appendix III. The first column of each table lists all the possible values of x. The entry in each cell in the intersection of row x and column p for a given n is the probability that X is *equal to* or *less than* x. Subtracting this entry from 1, we get the probability that X is *larger than* x.

Consider another example. Suppose that 10 percent of the parts produced by an automatic machine are known to be defective. If a sample of 20 parts is selected at random, (1) what is the probability that the sample will contain *at most* three defectives? (2) What is the probability that the sample will contain *at least* three defectives?

1. The phrase *at most* means *equal to or less than.* Let X be the number of defective parts in a sample of 20 ($n = 20$). The probability that X is equal to or less than 3 is found in the intersection of row 3 and column 0.1 in the table for $n = 20$ in Appendix III. That is,

$$P(X \leq 3 \mid n = 20, p = 0.1) = 0.8671$$

2. The phrase *at least* means *equal to or more than.* The probability that X is equal to or more than 3 is identical with

$$1 - P(X \leq 2)$$

Consequently, we have

$$P(X \geq 3 | n = 20, p = 0.1) = 1 - P(X \leq 2 | n = 20, p = 0.1)$$
$$= 1 - 0.6770 = 0.3230$$

Exercise 7-2

1 Distinguish between the two concepts:

a Bernoulli random variable and binomial random variable.
b Bernoulli probability model and binomial probability distribution.

2 Suppose that at your school 20 percent of the students are seniors. A committee of 20 students is selected at random; find the probability that the committee consists of:

a Eleven seniors.
b More than 10 seniors.
c Five or fewer seniors.
d No seniors.

3 Let X be the number of men in a random sample of 30 students taken from your school, where 50 percent of the students are men. Find the probability that the sample consists of:

a Fifteen women.
b Twenty men.
c At most 10 women.
d At least 10 men.
e Between 12 and 24 men inclusive.

4 Out of 3,200 families with five children each, how many families do you expect to have:

a Five boys?
b Five girls?
c At least two boys?
d At most two boys?
e One boy or two girls?
f All boys or all girls?
g Two boys or three boys?

5 Suppose that 10 percent of the screws produced by an automatic machine are defective. Find the probability that out of 30 screws selected at random, there are:

a At most 3 defectives.
b At least 3 defectives.
c Between 2 and 4 defectives inclusive.

6 Let X be the number of "yes" answers obtained in 50 interviews, where a "yes" answer is as likely as a "no" answer. Find the following:

a $P(X = 25)$.
b $P(X \leq 30)$.
c $P(21 \leq X \leq 30)$.
d $P(20 < X < 30)$.
e P(at least 26 "yes" answers).
f P(at most 30 "yes" answers).
g P("yes" answers are more than "no" answers).
h P("yes" answers are at least twice as many as "no" answers).

7 A perfect die is rolled five times. Let X be the number of times that the spot "one" shows up; find the probability distribution for X.

8 According to the records accumulated over a long period of time by an insurance company, 10 percent of inquirers who responded to a mail advertisement have purchased an insurance policy. A certain salesman of the company received four such inquirers on a given day. Let X be the number of sales this salesman can possibly make on the basis of these inquiries. Find the probability distribution for X.

9 Suppose that it is known that 70 percent of all housewives who let a door-to-door cosmetics saleslady into their homes will end up buying some cosmetics. Find the probability that no more than 2 of 10 housewives who let such a saleslady into their homes will end up buying some cosmetics:

a By using Appendix III.
b By using Appendix II.

10 Suppose that 30 percent of all patients who have received a given medication will have the side effect of rising temperature. Find the probability that of 15 patients who have received this medication:

a At most 5 patients will have such side effect.
b At least 5 patients will have such side effect.
c No less than 2 and no more than 4 patients will have such side effect.

11 Let X be $B(n, p)$ and $q = 1 - p$. Write the binomial expansion of $(q + p)^n$ for

a $n = 3$. *b* $n = 5$. *c* $n = 10$.

12 For each of the three cases in the preceding problem, if $p = q = \frac{1}{2}$, find the probability distribution for X. Also, find the cumulative probability distribution of X for $n = 3$ and $n = 5$.

7-3 Properties of the Binomial Distribution

The binomial distribution, in common with most other distributions, has three significant attributes. They are the arithmetic mean, the standard deviation, and the pattern or shape of the distribution. These properties will be examined in this section.

7-3-1 The mean

The *mean* of a binomial random variable X, designated μ or $E(X)$, is the theoretical expected number of successes in n trials. In other words, the mean of a binomial variable is the average number of successes when the number of samples of size n taken by the Bernoulli process approaches infinity. Symbolically,

$$\mu = E(X) = \sum_{x=0}^{n} xP(x) \tag{7-6}$$

That is, the mean of X is the sum of the products of the values that X can assume multiplied by their respective probabilities. In the toss of four ideal coins, for example, the expected number of heads is

$$\mu = \sum_{x=0}^{4} xP(x)$$

$$= 0\left(\frac{1}{16}\right) + 1\left(\frac{4}{16}\right) + 2\left(\frac{6}{16}\right) + 3\left(\frac{4}{16}\right) + 4\left(\frac{1}{16}\right)$$

$$= \frac{0 + 4 + 12 + 12 + 4}{16} = \frac{32}{16} = 2$$

This method, however, is tedious especially when the sample size n is large. The following is a much simplified and convenient method for obtaining the mean of X:

$$\mu = np \tag{7-7}$$

which is proved as follows:[1]

$$\mu = E(X) = E\left(\sum_{i=1}^{n} Y_i\right) \qquad \text{by definition}$$

$$= E(Y_1 + Y_2 + \cdots + Y_n)$$

$$= E(Y_1) + E(Y_2) + \cdots + E(Y_n)$$

$$= p + p + \cdots + p \qquad \text{by (7-1)}$$

In the above example, $n = 4$ and $p = 0.5$, we have

$$\mu = 4(0.5) = 2$$

which checks with the value obtained in the last paragraph.

7-3-2 The standard deviation

The standard deviation of the binomial random variable measures the dispersion of the binomial distribution, which, as indicated before, is a *theoretical* probability distribution. Since the binomial random variable X has $\{0, 1, 2, \ldots, n\}$ as the set of its possible values, the variance of the binomial probability distribution is, as in Eq. (5-17), expressed as follows:

$$\sigma_X{}^2 = E(X^2) - \mu^2 = \sum_{x=0}^{n} x^2P(x) - \mu^2 \tag{7-8}$$

and the standard deviation is

$$\sigma_X = \sqrt{\sum_{z=0}^{n} x^2P(x) - \mu^2}$$

[1] For a different method of proof, see Supplement V.

Applying Eq. (7-8), we can compute the standard deviation of X, the number of heads in the toss of four ideal coins, as follows:

$$\sigma_X = \sqrt{\left[0^2\left(\frac{1}{16}\right) + 1^2\left(\frac{4}{16}\right) + 2^2\left(\frac{6}{16}\right) + 3^2\left(\frac{4}{16}\right) + 4^2\left(\frac{1}{16}\right)\right] - 2^2}$$

$$= \sqrt{\frac{0 + 4 + 24 + 36 + 16}{16} - 4}$$

$$= \sqrt{\frac{80}{16} - 4}$$

$$= \sqrt{5 - 4} = 1$$

The standard deviation of the binomial random variable X can also be computed by a much simplified procedure. We shall demonstrate that

$$\sigma_X{}^2 = npq \qquad (7\text{-}9)$$

Since $X = \sum_{i=1}^{n} Y_i$, where Y is a Bernoulli random variable as defined before, and since the variance of the sum of independent variables is equal to the sum of their respective variances, we have

$$\begin{aligned} \sigma_X{}^2 &= \sigma_{\Sigma Y}{}^2 && \text{by definition} \\ &= \sigma^2_{(Y_1+Y_2+\cdots+Y_n)} \\ &= \sigma_{Y_1}{}^2 + \sigma_{Y_2}{}^2 + \cdots + \sigma_{Y_n}{}^2 && \text{by (5-21)} \\ &= pq + pq + \cdots + pq && \text{by (7-2)} \\ &= npq \end{aligned}$$

Thus

$$\sigma_X = \sqrt{npq}$$

If X stands for the number of heads obtained in the toss of four ideal coins, then the standard deviation of X is

$$\sigma_X = \sqrt{4(0.5)(0.5)} = 1$$

which is the same as that computed from Eq. (7-8).

7-3-3 The pattern of the binomial distribution

The shape or pattern of the binomial distribution depends on the values of p and n. If $p = q = 0.5$, the distribution will be symmetrical regardless of the value of n. If $p \neq q$, the distribution will be asymmetrical. Given a particular n, the more the difference between p and q, the greater the skewness of the distribution will be. When p is smaller than q, the distribution will be skewed to the right, or positively skewed; when p is greater than q, it will be skewed to the left, or negatively skewed. However, as the value of n is increased, the distribution will become less skewed. When n approaches infinity, the distribution will approach symmetry irrespective of the difference between p and q.

The effects of increases in n on the shape of the binomial distribution are demonstrated in Figs. 7-2 and 7-3. In Fig. 7-2, the bars represent the terms of the expansion of $(p+q)^n$ on the assumption that $p = q$. The distribution is always symmetrical. As the value of n increases, the bars become narrower and more numerous. As n approaches infinity, the bars become vertical lines with no space in between, and the distribution becomes a bell-shaped smooth curve.

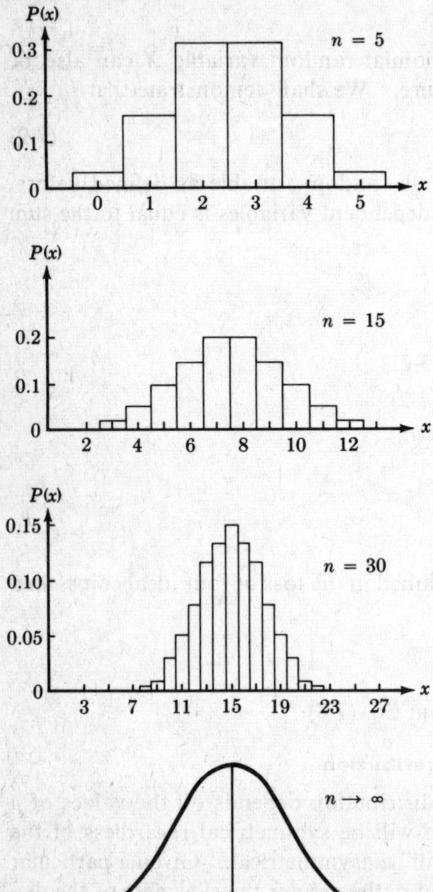

Figure 7-2
Binomial distribution, $p = 0.5$.

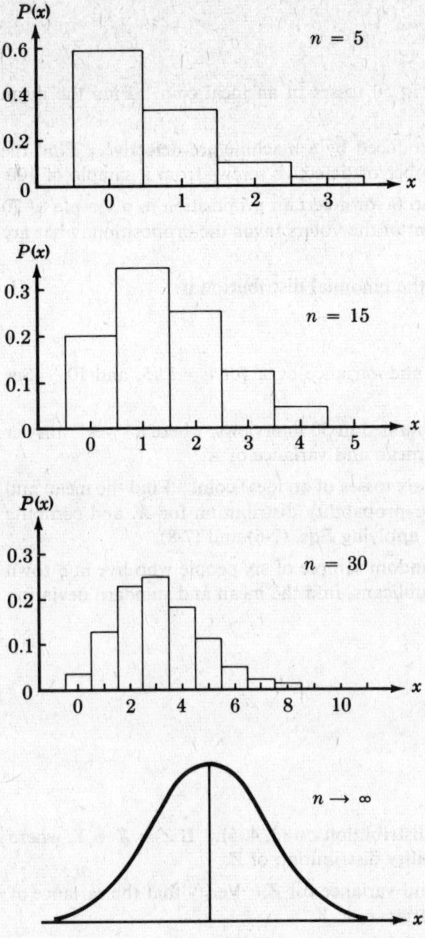

Figure 7-3
Binomial distribution, $p = 0.1$.

Figure 7-3 is drawn on the assumption that the probability of a success on a single trial is 0.1, such as the experiment of drawing a 0 from a table of random digits. It shows that as n approaches infinity, the skewness of the distribution disappears in addition to the fact that the distribution becomes continuous. Thus it is apparent that as the value of n increases, the binomial distribution becomes a continuous and symmetrical curve whether or not p and q are equal. Such a continuous bell-shaped symmetrical distribution is known as the *normal distribution*, which is the limit of the binomial when n approaches infinity.

Exercise 7-3

1 Let X be the number of heads obtained in 50 tosses of an ideal coin. Find the mean and variance for X.

2 Suppose that 10 percent of the screws produced by a machine are defective. Find the mean and variance for X if X is the number of defective screws from a sample of 100.

3 Given that X is the number of people who favor a certain proposition in a sample of 20 voters drawn from a city where 50 percent of the voters favor the proposition, what are the mean and variance of X?

4 What will the effect be on the pattern of the binomial distribution if:

a The sample size n is increased?

b The value of p increases from 0 to 1?

5 Given that X is $B(n, 0.5)$, find the mean and variance of X for $n = 3$, 5, and 10. (See Prob. 12 of Exercise 7-2.)

6 Let X be the number of "yes" answers obtained in 50 interviews, where a "yes" answer is as likely as a "no" answer. Find the mean and variance of X.

7 Let X be the number of heads obtained in six tosses of an ideal coin. Find the mean and standard deviation for X. Also, find the probability distribution for X, and compute the mean and standard deviation of X by applying Eqs. (7-6) and (7-8).

8 If X is the number of Republicans in a random sample of six people who live in a town where 40 percent of the residents are Republicans, find the mean and standard deviation for X by using Eqs. (7-6), and (7-8).

9 If X is $B(10, 0.6)$, find:

a $E(X-6)$.

b $E\left(\dfrac{X}{10}\right)$

c $E(X^2)$.

d $E\left(\dfrac{X-10}{6}\right)$.

e $E((X-6)^2)$.

10 Let X be $B(2, 0.5)$ and Y have a uniform distribution on $\{3, 4, 5\}$. If $Z = X + Y$, where X and Y are independent, find the probability distribution of Z.

11 In the preceding problem, find the mean and variance for Z. Verify that the variance of Z is equal to the variance of X plus the variance of Y.

12 The principle underlying the binomial model can be applied to situations in which the population, or sample space, is partitioned into K mutually exclusive and exhaustive classes for $K > 2$. Let the probabilities of the K classes be $p_1, p_2, \ldots, p_K$, respectively. If a random sample of n items is selected *with replacement*, then the probability of n_1 items' occurring in class 1, n_2 items' in class 2, ..., and n_K items' in class K for $n_1 + n_2 + \cdots + n_K = n$, is given by

$$\frac{n!}{n_1!\, n_2! \ldots n_K!} (p_1)^{n_1} (p_2)^{n_2} \cdots (p_k)^{n_k}$$

which is known as the *multinomial rule*. The probability distribution based on this rule is referred to as the *multinomial distribution*. Suppose that a certain department of a company has 100 employees: 50 are college trained, 30 are graduate school trained, and 20 are high school trained. A sample of 10 employees is selected at random with replacement. Find the probability that the sample contains 5 college-trained, 3 graduate school-trained, and 2 high school–trained employees.

7-4
Distribution of the Sample Proportion

In the past two sections we were concerned with the number of successes in n trials. Quite frequently, however, we are interested not in the number of successes but rather in the *proportion* of successes. If $\hat{p}$ (read as "p hat") is used to designate the proportion of successes in n trials, known as the *sample proportion*, then $\hat{p}$ is also a random variable and is defined as follows:

$$\hat{p} = \frac{X}{n} = \frac{\sum_{i=1}^{n} Y_i}{n} \tag{7-10}$$

where, as before, X is the number of successes in n trials and Y is the Bernoulli random variable. Being a proportion, the possible values of $\hat{p}$ are mostly fractional or decimal values, namely, 0, $1/n$, $2/n$, ..., 1.

Since $\hat{p}$ is a sample statistic, its distribution is a sampling distribution. The value of $\hat{p}$ depends on the values of X and n. Given the value of n, then $\hat{p}$ is completely determined if we know X. Thus the probability that a given proportion of successes occurs is identical with the probability that this number of successes occurs. That is,

$$P\left(\hat{p} = \frac{x}{n}\right) = P(X = x) = \binom{n}{x} p^x q^{n-x} \tag{7-11}$$

This is so because whenever a given number of successes occurs, so does the sample proportion of the successes. If $n = 4$, $p = q = 0.5$, the sampling distribution of the sample proportion $\hat{p}$ is given as follows:

Sample proportion $\hat{p}$	*Probability* $P(\hat{p} = x/n)$
0	$\binom{4}{0}\left(\frac{1}{2}\right)^4 = \frac{1}{16}$
$\frac{1}{4}$	$\binom{4}{1}\left(\frac{1}{2}\right)^4 = \frac{4}{16}$
$\frac{2}{4}$	$\binom{4}{2}\left(\frac{1}{2}\right)^4 = \frac{6}{16}$
$\frac{3}{4}$	$\binom{4}{3}\left(\frac{1}{2}\right)^4 = \frac{4}{16}$
1	$\binom{4}{4}\left(\frac{1}{2}\right)^4 = \frac{1}{16}$

Note that the probabilities are identical with those found in Sec. 7-2-1.

Similar to the binomial distribution, the distribution of the sample proportion has its mean and standard deviation. Using the notation for expected values, we derive the mean of the sample proportion as follows:

$$E(\hat{p}) = E\left(\frac{X}{n}\right) = \frac{1}{n}E(X) = \frac{1}{n}np = p \tag{7-12}$$

Observe that the expected value of the proportion of successes is equal to the *probability* of success for a single trial. As shown in Sec. 7-1, the lowercase letter p is the mean of the Bernoulli variable. It is referred to as the *population proportion.*

The variance of the sample proportion, designated $\sigma_{\hat{p}}^2$, is derived as follows:

$$\begin{aligned}
\sigma_{\hat{p}}^2 &= \sigma_{X/n}^2 = \sigma_{(1/n)X}^2 \\
&= \left(\frac{1}{n}\right)^2 \sigma_X^2 \qquad \text{by (5-19)} \\
&= \frac{1}{n^2} npq \qquad \text{by (7-9)} \\
&= \frac{pq}{n}
\end{aligned} \tag{7-13}$$

and the standard deviation $\sigma_{\hat{p}}$ is therefore

$$\sigma_{\hat{p}} = \sqrt{\frac{pq}{n}} = \frac{\sigma_Y}{\sqrt{n}}$$

where σ_Y is the standard deviation of the Bernoulli variable.

We may now conclude that the mean of the sample proportion is the mean of the population proportion, and its standard deviation is the standard deviation of the population divided by the square root of the sample size n. Recall that similar statements were made in Sec. 6-3 with respect to the sample mean $\bar{X}$. You should realize by now that the proportion of successes in a sample of n observations is the mean of the sample.

In the coin-tossing example, $n = 4$ and $p = q = 0.5$, the mean and the standard deviation of the sample proportion are given as follows:

$$E(\hat{p}) = p = 0.5$$

and

$$\sigma_{\hat{p}} = \sqrt{\frac{(0.5)(0.5)}{4}} = 0.25$$

If n is increased to 16, the mean of $\hat{p}$ is still 0.5, while the standard deviation is reduced to $\sigma_{\hat{p}} = 0.5/\sqrt{16} = 0.5/4 = 0.125$. Thus the larger the sample size, the smaller the standard deviation, given the value of p.

It is hardly necessary to mention that the pattern of the distribution of the sample proportion follows exactly that of the binomial distribution (see Figs. 7-2 and 7-3).

Exercise 7-4

1 Explain fully the difference and similarity between the probability distribution of a sample proportion and that of a binomial random variable.

2 Suppose that 50 percent of all the employees of a certain company are married. Let X be the number of employees who are married in a random sample of 10 employees. Find the probability distribution for the random variable $\hat{p}$, the sample proportion. Explain fully the relation between $\hat{p}$ and the Bernoulli random variable discussed in Sec. 7-1.

3 In Prob. 2 above, find the mean and variance of $\hat{p}$ by Eqs. (7-12) and (7-13). Then, compute the mean and variance by using the probability distribution of $\hat{p}$.

4 In a sample of 30 parts randomly selected from a production run, one fifth are defective. What is the probability of obtaining such a sample proportion if the true proportion of defective parts in any production run is 0.1?

5 Suppose that 20 percent of the employees in a company have had college training. In a random sample of 50 employees, it is found that 10 percent of them have had college training. What is the probability of obtaining this or a smaller proportion of college-trained employees in the sample from this company?

6 Assume that among all drivers whose cars are equipped with seat belts, only 40 percent use them whenever they drive. Find the probability that in a random sample of 50 drivers:

a At most one-fifth will use them whenever they drive.
b At least one-fifth will use them whenever they drive.
c Between 30 and 40 percent inclusive will use them whenever they drive.

7 In Prob. 6 above, find the mean and variance of $\hat{p}$, the sample proportion of the drivers who will use the seat belts in a sample of 50 drivers whose cars are equipped with seat belts.

8 Suppose that the binomial random variable X has the probability function

$$P(x) = \binom{5}{x}(0.6)^x(0.4)^{5-x}$$

List the set of possible values of $\hat{p}$, and then find the probability distribution of $\hat{p}$. Find the mean and variance of $\hat{p}$.

7-5 Hypergeometric Distribution

The binomial distribution is based on the assumptions that the population is infinite and that the random sample is taken with replacement so that the possible observations are independent of one another. The probability thus remains unchanged for each successive observation. When the population is finite and the random sample is taken without replacement, the probability will change for each additional observation. Under such circumstances, we shall have a probability distribution known as the hypergeometric distribution.

If n elements are drawn without replacement from a *set* containing N_1 elements of one kind and N_2 elements of another kind, and the two kinds are mutually exclusive and exhaustive, then by the rule of combination, the expression for the total number of ways in which n elements can be drawn from this set is

$$\binom{N_1 + N_2}{n} = \frac{(N_1 + N_2)!}{n!(N_1 + N_2 - n)!}$$

In similar manner, the expression for the number of ways in which x elements of the first kind can be drawn is $\binom{N_1}{x}$ and the number of ways in which $n - x$ elements of the second kind can be drawn is $\binom{N_2}{n-x}$. The product of these two expressions represents the total number of ways in which x elements of the first kind and $n - x$ elements of the second kind can be selected (law of multiplication).

Thus the probability that n elements randomly selected from this set of $N_1 + N_2$ elements containing x elements of the first kind and $n - x$ elements of the second kind [designated as $P(x; n, N_1, N_2)$] is given by

$$P(x; n, N_1, N_2) = \frac{\binom{N_1}{x}\binom{N_2}{n-x}}{\binom{N_1 + N_2}{n}} \tag{7-14}$$

for

$$x = \begin{cases} 0, 1, 2, \ldots, n & \text{if } n \leq N_1 \\ 0, 1, 2, \ldots, N_1 & \text{if } n > N_1 \end{cases}$$

This probability distribution is called the *hypergeometric distribution.*

Assume, for example, that a production run usually turns out 100 units, of which 90 are good and 10 are bad. What is the probability that a random sample of 20 units will contain 15 good and 5 bad units? In this example, the set contains $N_1 + N_2 = 90 + 10 = 100$ units. The total number of ways of drawing 20 from 100 units is $\binom{100}{20}$. The number of ways to draw 15 good ones is $\binom{90}{15}$, and 5 bad ones $\binom{10}{5}$. Thus the probability that this random sample of 20 will contain 15 good and 5 bad units is

$$\frac{\binom{90}{15}\binom{10}{5}}{\binom{100}{20}} = \frac{90!\,10!\,20!\,80!}{15!\,75!\,5!\,5!\,100!} = 0.0428$$

Generally speaking, given a set of N elements divided into C mutually exclusive and exhaustive classes with N_1 in class 1, N_2 in class 2, . . . , N_C in class C, if a random sample of n elements is drawn without replacement from this set, then the probability of drawing n_1 elements from class 1, n_2 from class 2, . . . , and n_C from class C is given by

$$\frac{\binom{N_1}{n_1}\binom{N_2}{n_2}\cdots\binom{N_C}{n_C}}{\binom{N}{n}} \tag{7-15}$$

where $N_1 + N_2 + \cdots + N_C = N$ and $n_1 + n_2 + \cdots + n_C = n$.

Exercise 7-5

1 Explain the circumstances under which the hypergeometric distribution rather than the binomial distribution is applied.

2 Suppose that 10 of the 15 tax returns reporting incomes of $50,000 or more are suspected of containing errors. The Internal Revenue Service decides to audit 5 of the 15 returns. What is the probability that the 5 returns audited all contain errors?

3 In Prob. 2 above, what is the probability that at least three returns contain errors?

4 Give the expression for the probability that one bridge hand taken from an ordinary deck of playing cards contains seven spades, four hearts, and two diamonds.

5 In Prob. 12 of Exercise 7-3, if the sample is drawn without replacement, what is the probability of obtaining the specified sample? (Do not evaluate.)

6 Suppose that 15 percent of the 20 parts produced by an automatic machine is defective, and that a sample of 5 parts is selected at random without replacement. Determine the following:

a The probability function for the number of defective parts in the sample.
b The probability that none of the parts in the sample is defective.
c The probability that at least one of the parts in the sample is defective.
d The probability that there is at most only one defective in the sample.

7 An urn contains four red balls and six black ones. A sample of four balls is selected from the urn without replacement. Let X be the number of red balls contained in the sample; find:

a The probability that there is at most only one red ball in the sample.
b The probability distribution for X.
c The mean for X.

8 A box contains 12 socks; 5 brown, 4 green, and 3 white. A sample of 6 socks is selected at random without replacement. Find the probability that the sample contains three different matching pairs.

7-6
Poisson Distribution

Another important family of probability distributions that are also discrete in nature is the *Poisson exponential distribution*, named after Simeon Denis Poisson (1781–1840), who first described it in 1837. The distribution has been found applicable to many processes that have an observation per unit of time or space. For example, the number of telephone calls received at a switchboard per minute, the number of customers arriving at a bank teller per 5-minute period, the number of machines breaking down during any one day, the number of requisition orders for a particular item received at the stockroom, the number of defective parts per batch shipped from the factory for assembly, and the number of pinhole defects in the insulation of electric wire per foot all follow approximately the Poisson pattern. Such processes are characterized by the expected number of *successes per unit of time or space* just as the binomial distribution is characterized by the number of successes in n trials.

The above-mentioned "per unit of time or space" is equivalent to the sample size n in a binomial distribution. We know that the binomial distribution is based on the Bernoulli process, characterized by the probability of a success in one trial. What, then, constitutes *one* trial in a Poisson process? In the case of a batch of parts, an observation of a single part to determine whether it is defective is, of course, a trial. But what about telephone calls received per minute? If we divide the minute into 60 seconds, there is still the possibility of receiving more than one call in any given second. In order that this should not happen, we need to divide the minute into much smaller subdivisions such as $\frac{1}{100}$ second. We might feel quite confident that the probability of two or more calls received in such a negligible interval of time is so small that it can be ignored for practical purposes. When a unit of time or space is divided into many subdivisions (n is large), and the probability of success on a *single trial* is small (p is small), we shall have a situation close to a Poisson distribution. Thus the Poisson distribution is one for which n is very large and p is very small, and is regarded as the limit of the binomial distribution.

Let μ be the mean or expected number of successes in a specified time or space and $\mu = np$, and the random variable X designate the number of successes in a sample of size n with the possible values $\{0, 1, 2, 3, \ldots\}$. Then the probability distribution of X is given by the formula[1]

$$P(x) = e^{-\mu}\frac{\mu^x}{x!} \tag{7-16}$$

where e is a constant and is about 2.71828, and x is any possible value that X can take on.

[1] See Supplement VI for derivation.

The Poisson distribution is rather easy to apply. Suppose, for instance, that at any time of the working day the average number of telephone calls received at a switchboard per minute is three. Let X be the number of calls received in a given minute, say from 9:00 to 9:01; the probabilities, or $P(x)$, for $x = 0, 1, 2, 3, 4, \ldots$ are given in the last column ($\mu = 3$) of the third table of Appendix V. These probabilities, together with the cumulative probabilities, or $P(X \leq x)$, are shown as follows:

Number of calls x	*Probability* $P(x)$	*Cumulative probability* $P(X \leq x)$
0	$e^{-3}\frac{3^0}{0!} = 0.04979$	0.04979
1	$e^{-3}\frac{3^1}{1!} = 0.14936$	0.19915
2	$e^{-3}\frac{3^2}{2!} = 0.22404$	0.42319
3	$e^{-3}\frac{3^3}{3!} = 0.22404$	0.64723
4	$e^{-3}\frac{3^4}{4!} = 0.16803$	0.81526

As another example, suppose that an automatic machine that turns out identical screws is known to have 2 percent of its output unusable or defective. Certain unassembled equipment requires 98 such screws for assembly and is shipped with a box of 100 screws. What is the probability that a purchaser of such equipment will not have enough good screws to complete the assembly? For $\mu = np = 100(0.02) = 2$, Appendix V gives the probabilities of 0, 1, and 2 as follows:

x	$P(x)$
0	0.13534
1	0.27067
2	0.27067
Total	$P(X \leq 2) = 0.67668$

x=3 P(x) =0.18045

P(x>3) =1-0.85715

= 0.14287

Thus

$$P(X > 2) = 1 - 0.67668 = 0.32332$$

That is, the probability that a box of 100 screws will be insufficient for the assembly is 0.32332. If a table of cumulative Poisson distribution is available, this problem can be solved more readily. The values 0.32332 can be read directly from the table as X is 3 or larger, given that the mean is 2.

Similar to the binomial distribution, the Poisson distribution is a sampling distribution and has the mean $\mu = np$. The value of μ is relatively small and always positive. The variance of the Poisson distribution is also μ. We know that the variance of the binomial distribution is equal to $np(1-p)$. As p approaches 0, the limit of the last factor $1-p$ approaches 1, and the variance approaches np, or μ. Thus the standard deviation of the Poisson distribution is simply $\sqrt{np}$, or $\sqrt{\mu}$. Observe that the Poisson distribution has only *one* parameter, namely, μ.

The Poisson distribution is positively skewed (skewed to the right). We indicated in Sec. 7-3-3 that the binomial distribution would approach a continuous distribution of symmetrical shape if n were sufficiently increased. This symmetrical distribution would be close to what is known as the normal distribution, which is the limit of the binomial. It is interesting to note that the normal distribution is also the limit of the Poisson distribution. Given the value of p,

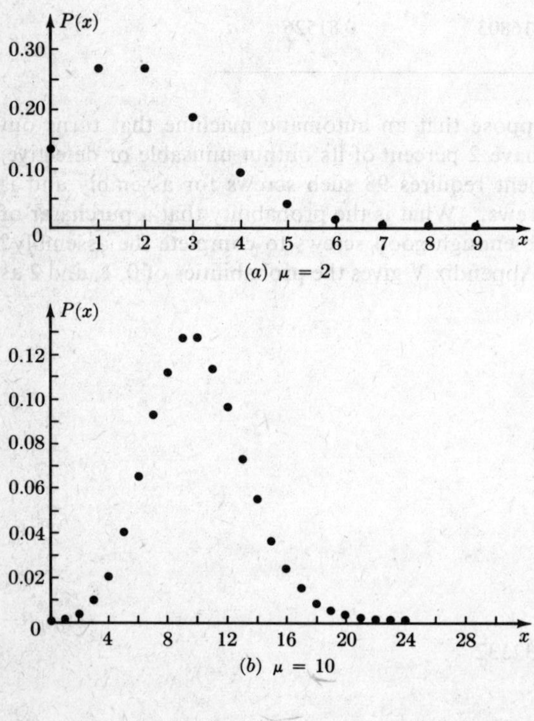

Figure 7-4
Poisson exponential distribution, $P(x) = e^{-\mu}(\mu^x/x!)$.

then np will increase as n is increased. As μ, or np, increases, the Poisson distribution will be closer and closer to a bell-shaped continuous curve.

As shown in Fig. 7-4, the distribution is skewed for $\mu = 2$. As μ is increased to 10, the distribution becomes almost symmetrical and the distance between two adjacent points becomes smaller. When n, and thus μ, is sufficiently large, the points on the graph will touch the adjacent ones and the curve will become continuous, and at the same time approach symmetry.

Exercise 7-6

1 Explain fully how the Poisson distribution is related to the binomial distribution.

2 Explain what constitutes a single trial in an experiment involving the Poisson model.

3 A Poisson distribution is given by:

$$P(x) = e^{-2}\frac{2^x}{x!}$$

Find the probabilities for $x = 0, 1, 2, 3,$ and 4.

4 Suppose that 1 percent of the bolts produced by a certain machine are defective. In a random sample of 300 bolts, find the probability that:

a All bolts are good.
b Two or fewer bolts are defective.
c Two or more bolts are defective.

5 At a certain school, exactly 10 percent of the students are girls. A sample of 50 students are selected at random. By applying the Poisson distribution, find the probability that the sample contains:

a All boys.
b Only one girl.
c Less than three girls.
d More than three girls.

Compare your answers with the results obtained under the binomial model.

6 The probability that a person who takes a certain drug would have an uncomfortable reaction is 0.002. Let X be the number of persons who suffer such a reaction in a sample of 1,000 persons who have taken the drug. Find:

a $P(X = 1)$.
b $P(X < 3)$.
c $P(X \geq 3)$.
d $P(X > 0)$.
e $P(2 \leq X \leq 5)$.

7 How is the pattern of the Poisson distribution related to that of the normal distribution?

8 Suppose that there are, on the average, four auto accidents per day on the highways in California. What is the probability that in a given day in California:

a There is no auto accident?
b There are three or fewer auto accidents?
c There are three or more auto accidents?

9 The average number of customers who appear at a window of a certain bank per minute during banking hours is *one*. Find the probability that during a given minute:

a No customer appears.
b Three or more customers appear.
c Three or fewer customers appear.

10 In a certain large school district with 2,000 teachers, the average proportion of teachers absent per school day is 0.5 percent. Find the probability that in a given day:

a All teachers are on duty.
b Two teachers are absent.
c Three or fewer teachers are absent.
d Three or more teachers are absent.

11 Compare the values of $P(X = 0)$, $P(X = 1)$, $P(X \leq 2)$, and $P(X \geq 2)$, when X is a Poisson random variable with a mean $\mu = 2$ with the values of these probabilities when X is $B(10, 0.2)$.

12 Describe at least two different situations to which the Poisson model can be applied.

Chapter 8

The normal distribution

In the previous chapter, our discussion was restricted mainly to random variables that take on a finite number of possible values, or values that differ from each other by finite amounts. Now we are going to consider another type of random variable, which may take on an infinite number of possible values, or values that may differ from each other by infinitesimal amounts. As briefly mentioned in Sec. 3-4-2, such random variables are called *continuous random variables* and their distributions are called *continuous distributions*. Many types of data are continuous in nature because the observations are obtained by measurements that may be made to any degree of accuracy desired. Measurements of time, distance, height, weight, etc., represent continuous variates since successive observations may differ only by infinitesimal amounts. The individual observation of weight, for instance, may be 165 pounds, 165.2 pounds, 165.25 pounds, 165.256 pounds, 165.2496 pounds, etc., depending on the accuracy of the measuring device used. The smaller the measurement unit, the greater the possible accuracy will be.

The distinction between discrete and continuous variables is, in most cases, a theoretical one. Clearly, even the most precise instrument available can give measurements of only a limited degree of precision. Thus in actual practice we usually can never have a distribution such that successive observations have no "gaps" in between. Consequently, in the measurement of such variables as height, weight, time, distance, and intelligence, the values obtained are really rounded numbers, or approximations. Each of such values represents not a point but an interval in which the measurement falls. Therefore a continuous distribution is made a discrete one by the limitation in measurements.

This limitation, however, should not invalidate the continuity assumption in statistical theory. The assumption of continuity of quantitative observations will enable us to apply available mathematical techniques, and thus greatly simplify our statistical work. As a matter of fact, it is often convenient to treat discrete distributions as if they were continuous. In most statistical experiments the difference between discrete and continuous variables is clear, and attention should always be paid to this distinction in using and interpreting statistical data.

One of the continuous distributions is the *normal distribution*, which occupies a central position in statistical inference. The importance of the normal distribution is derived mainly from three essential considerations.

In the first place, numerous variables seem to follow a pattern of variation that is similar to the normal distribution. Many natural traits such as the weights and heights of humans and the intelligence quotients of children are considered to have normal distribution. Numerous phenomena such as the test scores in a graduate record examination, the breaking strength of steel bars, errors of measurements, and many other variations are thought of as being normally distributed. Certain components, parts, and assemblies of machinery have been found to have normal distribution for their measures of reliability. This does not mean, however, that a relative frequency distribution actually observed follows exactly the normal rule but, on the average, the probabilities associated with the theoretical distribution apply very closely. The normal distribution is merely a theory to explain for some random variable the relation between intervals of its values and their corresponding probabilities. It can only be *approximated* by actual empirical distributions. It is remarkable that some relative frequencies do come very close to normal probabilities.

Secondly, the sampling distribution of many a sample statistic such as the mean has an approximately normal distribution regardless of the pattern of the population distribution, if the sample size is sufficiently large. Many distributions in business, economics, and other social sciences do not resemble the normal function. The distribution of the sample mean in each case, however, can be, and usually is, treated as normal as long as the sample is large. As we shall see in this chapter, the sample size that is large enough for invoking the normality of the sampling distribution of the mean is not difficult to obtain in most practical applications.

Finally, the normal function is an excellent approximation to several other sampling distributions. We shall show that this is the case in Chaps. 11 to 13. We already indicated that the Poisson and the binomial distributions approach the normal pattern as the value of n in each case is increased. It is too laborious, if not impossible, to work out the probabilities for a binomial random variable when n is large. Fortunately, probabilities for intervals of values based on the normal law can be used as approximations to binomial probabilities.

For these and other reasons, a careful examination of the normal distribution is indispensable. We shall, in this chapter, first examine briefly the normal function and the standard normal distribution. After this, a rather detailed analysis of the properties and characteristics of the normal function will be presented. In the final section, consideration will be given to the use of the normal distribution as an approximation to the binomial.

8-1 The Normal Function

The smooth and bell-shaped curve shown in Fig. 7-2 or 7-3 is referred to as the *normal curve*. It is the limit of the binomial distribution as n approaches infinity. We know that the binomial probability function is $P(x) = \binom{n}{x} p^x q^{n-x}$, where the independent variable is $B(n, p)$. When n becomes infinitely large, this function becomes the normal function with the following form:

$$f(x) = \frac{1}{\sigma\sqrt{2\pi}} e^{-(x-\mu)^2/2\sigma^2} \tag{8-1}$$

where e (about 2.71828) and π (about 3.1416) are constants. The mathematical proof of this is beyond the scope of this book. We shall, however, try to explore its ramifications and consider its properties.

The first discoverer of the normal probability function was Abraham De Moivre (1667–1754), who in 1733 derived the distribution as the limiting form of the binomial. But his work had not been discovered when the same formula was derived by Karl Friedrich Gauss (1777–1855) in connection with his work in evaluating errors of observation in astronomy. Thus the normal probability distribution is often referred to as the *gaussian distribution* or the *normal function of error*.

The symbol $f(x)$ on the left side of Eq. (8-1) stands for the *probability density* at each possible value of x; or, graphically, it is the ordinate of the curve at each possible point x on the abscissa. The entire area under the curve is equal to 1; the area cut off by an interval beneath the curve and above the abscissa is the probability of this interval.

This function rule shows that besides x, the value of the random variable, the value of the probability density $f(x)$ depends on only two parameters, μ and σ^2. The parameter μ is the mean of the normal distribution, and σ^2 is its variance. When the parameter n of the binomial distribution becomes infinitely large, its mean np will be equal to μ, and its variance npq equal to σ^2. The values of the mean μ and the standard deviation σ determine the location and shape of the normal curve. Given σ, a change in μ will shift the curve as a whole along the x axis. Figure 8-1, for instance, shows three normal curves that have identical standard deviations but different arithmetic means. On the other hand, the three normal curves in Fig. 8-2 have the same mean but different standard deviations. Figure 8-2 shows that the larger the standard deviation, the flatter

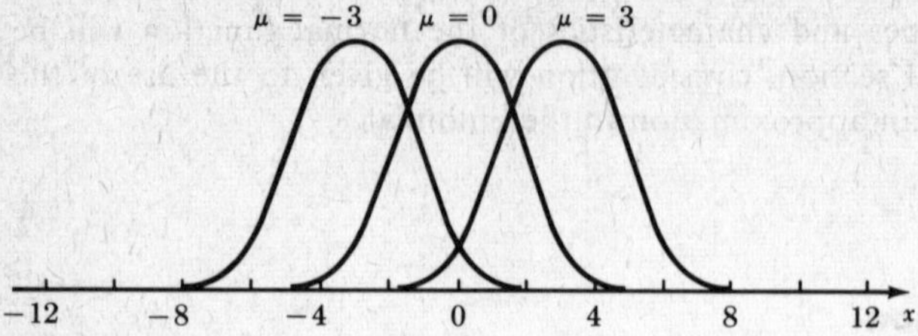

Figure 8-1
Normal distributions with different means, but with identical standard deviation $\sigma = 1.5$.

the curve will be; the smaller the standard deviation, the higher the hump. If the standard deviation were zero, there would be no distribution and all observations would have the same value, namely, the mean. It is clear that the normal distribution is completely defined by the mean μ and the standard deviation σ. The normal distribution that has the mean μ and variance σ^2 is commonly denoted by "$N(\mu, \sigma^2)$." Thus, if a random variable X has a normal distribution with a mean μ and a variance σ^2, it is simply written "X is $N(\mu, \sigma^2)$" and read as "X is normal with mean μ and variance σ^2."

Theoretically, the domain of the normal function $f(x)$ is infinite, and the normal distribution is a distribution for all values of X between $-\infty$ and $+\infty$, and each nonzero interval along the x axis has a nonzero probability. The curve is shown as approaching but never touching the x axis as x departs farther away from the mean μ. In actual practice, however, the value of $f(x)$, or the ordinate, becomes very close to zero, or almost touches the x axis, within a relatively short distance from the mean. As a matter of fact, about 99.73 percent of all observations would fall within three standard deviations from the mean. That is, $P(\mu - 3\sigma < X < \mu + 3\sigma) = 99.73$ percent. The probabilities of X's falling within one and two standard deviations from the mean are 68.27 and 95.45 percent, respectively.

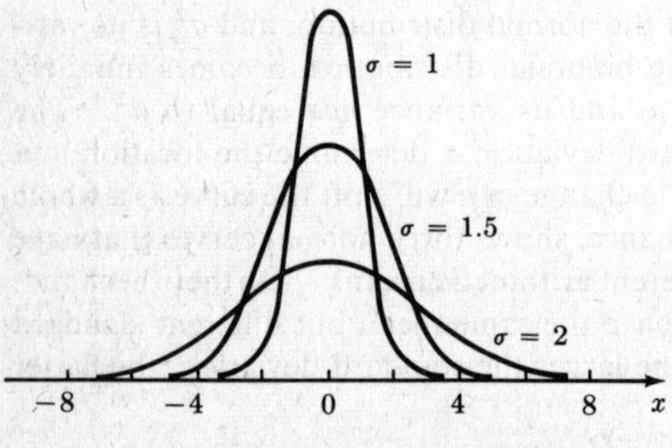

Figure 8-2
Normal distributions with different standard deviations, but with identical mean $\mu = 0$.

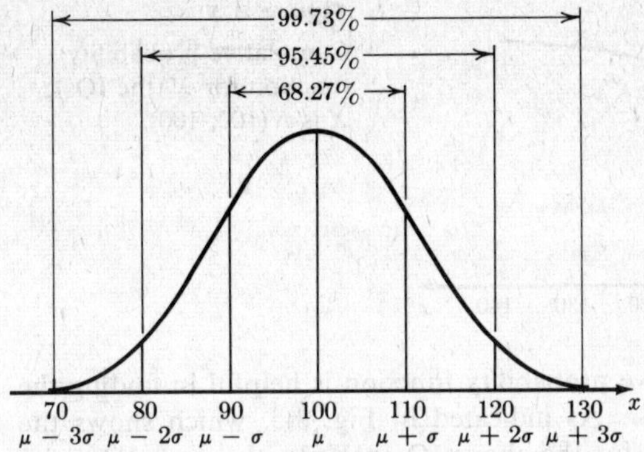

Figure 8-3
Probability density function for X (the IQs); X is $N(100, 100)$.

Intelligence quotients, for instance, seem to have an approximately normal distribution. If, in a large group of people, the mean μ of the IQs is 100 and the standard deviation σ is 10, then about 68.27 percent would have IQs between 90 and 110, about 95.45 percent between 80 and 120, and about 99.73 percent between 70 and 130. Thus, as shown in Fig. 8-3, almost all have IQs ranging from 70 to 130.

Since the probability that X assumes some exact value x is zero, any probability expression relative to a normal random variable has to be put in cumulative or interval form. The probability that X falls at or below a certain value x is the probability of the interval between $-\infty$ and x, or $P(X \leq x)$, which is also denoted by $F(x)$. It is referred to as the *cumulative probability function* of x. In practical work we often make use of the cumulative probability function in obtaining the probability that X falls within a certain interval, say, from a to b. For instance, the probability that any individual's IQ falls within the interval from 100 to 110 is given by

$$P(X \leq 110) - P(X \leq 100)$$

which is shown in Fig. 8-4 as the shaded area under the curve between 100 and 110.

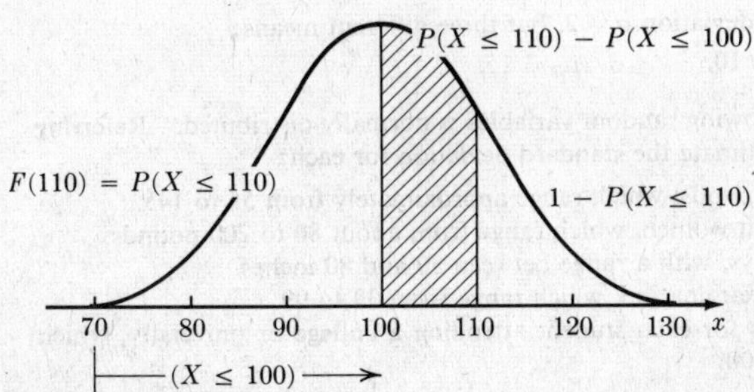

Figure 8-4
Normal distribution for the IQs; cumulative and interval probabilities are shown for $X \leq 100$ and $X \leq 110$.

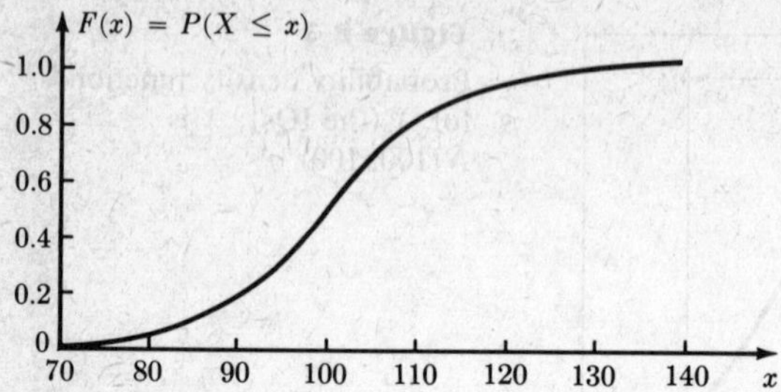

Figure 8-5
Cumulative probability function for X (the IQs); X is $N(100, 100)$.

The graph of the cumulative probability function is helpful in finding the probability of such an interval. As indicated in Fig. 8-5, which shows the cumulative probability function for the above IQ example, the probability for X at or below 100 is 50 percent and the probability for X's falling at or below 110 is about 84 percent. Thus the probability that a randomly selected individual has an IQ between 100 and 110 is approximately 34 percent.

To find the probability for a particular interval, we need such a cumulative probability distribution. However, it is rather impracticable to have a different probability distribution for each possible normal variable. In the next section we shall examine a particular normal function, to which the probability functions of all normal random variables can be reduced, and with which the evaluation of probabilities for any such variables will be greatly facilitated.

Exercise 8-1

1 Identify the parameters of the normal density function. How is the normal distribution defined by these parameters?

2 Explain as fully as you can the relationship between the binomial distribution and the normal distribution.

3 In a single graph, plot each of the two groups of normal functions for X:

a X has the same mean $\mu = 10$, but three different standard deviations.

1 $\sigma_1 = 1$. *2* $\sigma_2 = 2$. *3* $\sigma_3 = 5$.

b X has the same standard deviation $\sigma = 2$, but three different means:

1 $\mu_1 = 5$. *2* $\mu_2 = 10$. *3* $\mu_3 = 15$.

4 Suppose that each of the following random variables is normally distributed. Referring to Prob. 8 of Exercise 5-2, estimate the standard deviation for each:

a IQs of elementary school pupils, which range approximately from 55 to 145.
b Weights of adult American women, which range from about 80 to 200 pounds.
c Heights of high school boys, with a range between 50 and 80 inches.
d Scores in a midsemester examination, which range from 39 to 99.
e Total annual expenditures for each student attending a college or university, which range from \$1,500 to \$7,500.

5 Graph, approximately, each of the following normal functions:

a $N(0, 1)$. *b* $N(10, 100)$. *c* $N(100, 25)$.

6 Name at least two random variables (not used in the text), each of which has a normal distribution. Explain why you think each is distributed normally.

7 Use your own words to explain the importance of the normal distribution in statistical inference.

8 The service life of light bulbs produced by an electric company is believed to be normally distributed with a mean of 1,000 hours and a standard deviation of 50 hours. By the use of Fig. 8-3, determine the probability that:

a A randomly selected bulb will burn out in less than 900 hours.
b A randomly selected bulb will burn out in 900 to 1,050 hours.
c A randomly selected bulb will burn out in no less than 1,100 hours.

9 A production process makes bearings, the diameters of which are normally distributed with a mean of 0.5 inch and a variance of 0.0004 inch.

a What is the probability that a bearing selected at random will have a diameter of no less than 0.46 inch and no more than 0.56 inch?
b What is the probability that a bearing selected at random will have a diameter of at least 0.44 inch? (*Note:* Refer to Fig. 8-3 for answers.)

10 Refer to Fig. 8-3. If the service life of a certain fuse is believed to have a normal distribution with a mean of 2,000 hours and a variance of 10,000 hours, what is the probability that:

a A randomly selected fuse will last at most 2,200 hours?
b A randomly selected fuse will last at least 1,900 hours?

8-2 Standard Normal Distribution

The special distribution, which is abstracted for all normal random variables, is known as the *standard normal distribution.* It is the distribution of another normal variable called the *Z score*, which is defined as follows:

$$Z = \frac{X - \mu}{\sigma} \tag{8-2}$$

That is, the Z score is the difference of an observation X from the mean expressed in units of its standard deviation. For this reason, Z scores are also known as *normal deviates.*

The standard normal distribution has a mean 0 and a standard deviation 1. The expression for Z can be written as

$$Z = \frac{1}{\sigma} X - \frac{\mu}{\sigma}$$

Since $1/\sigma$ and μ/σ are constants, we have

$$E(Z) = E\left(\frac{1}{\sigma} X - \frac{\mu}{\sigma}\right)$$

$$= \frac{1}{\sigma} E(X) - \frac{\mu}{\sigma}$$

$$= \frac{\mu}{\sigma} - \frac{\mu}{\sigma} \quad \text{by (4-19)}$$

thus

$$E(Z) = 0 \tag{8-3}$$

We know that the variance of Z is equal to the variance of X multiplied by the square of X's coefficient. Let σ_Z^2 designate the variance of Z. We have

$$\sigma_Z^2 = \sigma^2_{(1/\sigma)X - \mu/\sigma}$$

$$= \left(\frac{1}{\sigma}\right)^2 \sigma^2 \quad \text{by (5-18)}$$

therefore

$$\sigma_Z^2 = 1 \tag{8-4}$$

If X in Eq. (8-2) is normal, by the linear change of scale (Sec. 8-3-2), Z will be normal, since the ratio $(X - \mu)/\sigma$ can be written $(1/\sigma)X - \mu/\sigma$, where $1/\sigma$ and μ/σ are both constants. We can thus say that Z is $N(0, 1)$.

For the standardized normal variable Z, the normal function becomes much simplified. Since $Z = (X - \mu)/\sigma$, with $\mu = 0$ and $\sigma = 1$ as noted above, substituting these quantities in Eq. (8-1) gives

$$f(z) = \frac{1}{\sqrt{2\pi}} e^{-z^2/2} \tag{8-5}$$

When $z = 0$, which is the center of the standard normal curve, $e^{-z^2/2} = 1$ and $f(z) = 1\sqrt{2\pi}$. That is, at the center the height of the curve is $1/\sqrt{2\pi}$.

The distribution of Z is usually put into cumulative form as shown in Appendix VI. The table gives the cumulative probability for various Z values. As pointed out earlier, with such a table it is unnecessary to construct a probability distribution for every normal variable. To calculate the probability of a given interval from a to b, we simply find the corresponding Z scores for a and b, and then obtain the probability by taking the difference between the cumulative probabilities of the Z scores. That is,

$$P(a \le X \le b) = P\left(\frac{a - \mu}{\sigma} \le Z \le \frac{b - \mu}{\sigma}\right) \tag{8-6}$$

To illustrate the use of the table of Appendix VI, let us suppose that we wish to determine the probability that a randomly selected person has an IQ, denoted by X, between 110 and 120. Assume that the mean of the distribution of IQs is 100 and the standard deviation is 10; we then need only to find the probability for Z's falling between positive 1 and 2. That is,

$$P(110 \le X \le 120) = P\left(\frac{110 - 100}{10} \le Z \le \frac{120 - 100}{10}\right) = P(1 \le Z \le 2)$$

$$= P(Z \le 2) - P(Z \le 1)$$

The first column of Appendix VI provides the z values with one decimal digit, and the cumulative probabilities of these values are provided in the next column headed .00. Thus, $P(Z \le 2)$ is found in the intersection of row 2.0 and column .00, namely, 0.9772, and $P(Z \le 1)$ is found in the intersection of row 1.0 and the same column, namely, 0.8413. The difference is $0.9772 - 0.8413 = 0.1359$. Consequently, $P(110 \le X \le 120) = 0.1359$.

The top row of Appendix VI provides the second decimal digit for the Z score. Thus, $P(Z \le 1.64) = 0.9495$ is found in the intersection of row 1.6 and column .04, and $P(Z \le 1.96) = 0.9750$ is found in the intersection of row 1.9 and column .06.

Observe that the table entries in Appendix VI are cumulative probabilities of Z equal to 0 or greater, as shown in the graph above the table, and thus they are 0.5000 or higher. For additional illustration of the use of Appendix VI, let us consider another example.

Suppose that a sugar manufacturing company packs refined sugar into paper bags. Let X designate the weight of each package. It is known that X is normally distributed with a mean of 80 ounces and standard deviation of 2 ounces. Find the probability that a randomly selected bag will be (*1*) at least 83.29 ounces, and (*2*) at most 76.71 ounces.

1. For computing the probability that X is equal to or more than 83.29 ounces, we first convert X into Z score, and then find the probability from Appendix VI

$$P(X \ge 83.29) = P\left(\frac{X - \mu}{\sigma} \ge \frac{83.29 - 80}{2}\right) = P\left(Z \ge \frac{3.29}{2}\right) = P(Z \ge 1.645)$$

We know that $P(Z \le 1.64) = 0.9495$ and $P(Z \le 1.65) = 0.9505$. By interpolation, the probability of Z equal to or smaller than 1.645 is of course half way between 0.9495 and 0.9505, namely, 0.9500. Thus, as shown in (*a*) of Fig. 8-6,

$$P(X \ge 83.29) = P(Z \ge 1.645) = 1 - P(Z \le 1.645) = 1 - 0.95 = 0.05$$

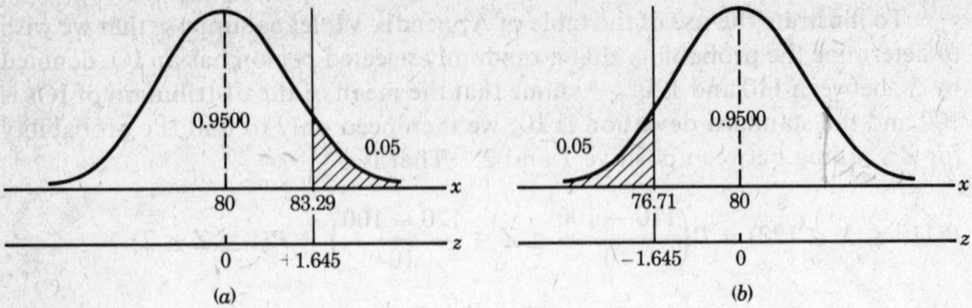

Figure 8-6
Shaded area showing (*a*) $P(Z \geq +1.645)$, and (*b*) $P(Z \leq -1.645)$.

2. Similarly, we have

$$P(X \leq 76.71) = P\left(Z \leq \frac{76.71 - 80}{2}\right) = P\left(Z \leq \frac{-3.29}{2}\right) = P(Z \leq -1.645)$$

Because of symmetry of the normal curve, the area under the curve to the left of -1.645 is equal to the area under the curve to the right of 1.645. Accordingly,

$$P(Z \leq -1.645) = P(Z \geq 1.645) = 1 - P(Z \leq 1.645) = 1 - 0.95 = 0.05$$

which is shown in (*b*) of Fig. 8-6.

For those who are still not quite sure of the circumstances under which table entries in Appendix VI should be subtracted from 1, the following rules will prove helpful:

1. The probability of Z's being equal to or *less than a positive value k* is the same as the table entry for $P(Z \leq +k)$.

2. The probability of Z's being equal to or *greater than a negative value k* is the same as the table entry for $P(Z \leq +k)$. That is,

$$P(Z \geq -k) = P(Z \leq +k)$$

In the above two cases, the table entries are the probabilities required. In the following two cases, the table entry must be subtracted from 1:

3. The probability of Z's being equal to or *greater than a positive value k* is

$$P(Z \geq +k) = 1 - P(Z \leq +k)$$

4. The probability of Z's being equal to or *less than a negative value k* is

$$P(Z \leq -k) = 1 - P(Z \geq -k) = 1 - P(Z \leq +k)$$

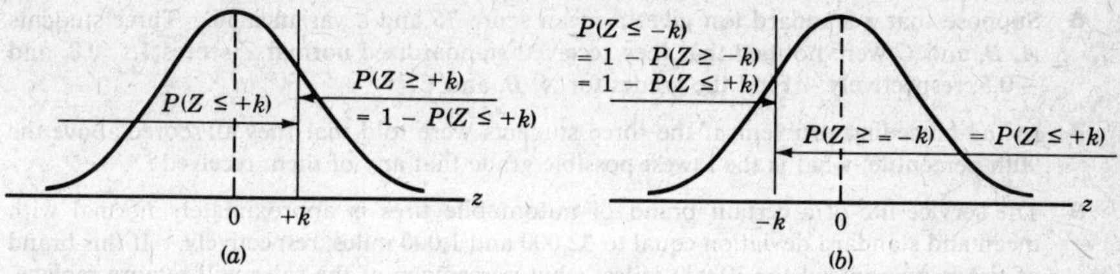

Figure 8-7

The standard normal curve showing (*a*) $P(Z \geq +k) = 1 - P(Z \leq +k)$ and (*b*) $P(Z \leq -k) = 1 - P(Z \geq -k) = 1 - P(Z \leq +k)$.

Items *1* and *3* above are shown in (*a*) of Fig. 8-7, while items *2* and *4* are shown in (*b*) of Fig. 8-7. In any case, *each expression of probability must finally include* $P(Z \leq +k)$ *in order to determine its value*, since each entry in Appendix VI is a probability accumulated from $-\infty$ to a positive value of Z.

Exercise 8-2

1 Let X be $N(100, 225)$. Find the following probabilities:

a $P(X \leq 92.5)$.
b $P(X \geq 76)$.
c $P(X \leq 107.5)$.
d $P(X \geq 124)$.
e $P(77.5 \leq X \leq 100)$.
f $P(112 \leq X \leq 128.5)$.
g $P(91 \leq X \leq 127)$.
h $P(71.35 \leq X \leq 109)$.

2 Given that Z is the standardized normal score, find the values for k:

a $P(Z \geq k) = 0.95$.
b $P(Z \geq k) = 0.99$.
c $P(Z \geq k) = 0.05$.
d $P(Z \geq k) = 0.01$.
e $P(|Z| < k) = 0.95$.
f $P(|Z| < k) = 0.99$.
g $P(|Z| > k) = 0.05$.
h $P(|Z| > k) = 0.01$.

3 Determine the values for k for the following probabilities:

a $P(Z \geq k) = 0.7438$.
b $P(Z \geq k) = 0.6045$.
c $P(Z \geq k) = 0.1200$.
d $P(Z \geq k) = 0.9984$.
e $P(Z \leq k) = 0.8000$.
f $P(Z \leq k) = 0.1250$.

4 Suppose that the distribution of weekly wages of 10,000 farm workers is normal and has a mean \$110 and a variance \$64. How many workers have wages that are:

a Equal to or less than \$100?
b Equal to or less than \$125?
c Equal to or more than \$90?
d Equal to or more than \$106?
e Between \$100 and \$120, inclusive?

5 Students of a certain high school district have a mean IQ 106, and the IQs have a variance 256. Find the proportion of students who have IQs:

a Equal to or less than 98.
b Equal to or less than 130.
c Equal to or more than 127.
d Between 94 and 118, inclusive.

6 Suppose that a standard test gives a mean score 75 and a variance 36. Three students A, B, and C were notified that they received standardized normal Z scores 1.5, 0.8, and -0.5, respectively. Find the grades for A, B, and C.

7 In the preceding problem, if the three students were told that they all scored above the 90th percentile, what is the lowest possible grade that any of them received?

8 The service life of a certain brand of automobile tires is approximately normal with mean and standard deviation equal to 32,000 and 1,000 miles, respectively. If this brand of tire is guaranteed for 30,000 miles, what percentage of the sales will require replacement?

9 The life length of a certain brand of automobile battery is believed to be approximately normal with a mean $\mu = 38$ months and a standard deviation $\sigma = 2$ months. If the company does not want to replace more than 5 percent of the batteries sold, how long should the length of guarantee be?

10 A flour manufacturing company packs flour into cloth sacks. The sack of flour is believed to be normally distributed with the mean and the standard deviation equal to 25 pounds and 0.5 pound, respectively. If a sack is randomly selected, what is the probability that the sack is:

a At most 24.2875 pounds.
b At least 26.0025 pounds.

8-3 Properties of the Normal Distribution

Because of the fact that the normal function represents an important and fundamental distribution, further and more detailed examination of its properties is in order. The following subsections deal with some of the most significant properties of the normal distribution.

8-3-1 Symmetry and bell shapedness

As pointed out repeatedly, the normal curve is *unimodal* and *symmetrical*. These attributes of the distribution can be easily seen from Eq. (8-1). Given a distribution with a known σ^2, the quantity $1/(\sigma\sqrt{2\pi})$ is a known constant. The other factor of the expression can be written as a fraction shown below:

$$\frac{1}{e^{(x-\mu)^2/2\sigma^2}}$$

where e, σ^2, and μ are all constants. The only part of the equation with which we have to be concerned is the numerator of the exponent, namely, $(x - \mu)^2$. When x is equal to μ, then $(x - \mu)^2$ is 0. Since any number raised to zero power is 1, the denominator of the above fraction, and so the fraction itself, will be equal to 1. The value of the function at the point $x = \mu$ is then simply $1/(\sigma\sqrt{2\pi})$. This quantity is therefore the height of the curve at the center point $x = \mu$.

The normal curve is unimodal or bell shaped because this quantity $x = \mu$ is the only value that has the largest ordinate. As x moves away from μ, the difference between x and μ will become greater, and so will the square of that difference. The greater the difference between x and μ, the larger will be $e^{(x-\mu)^2/2\sigma^2}$ and the smaller the fraction $1/e^{(x-\mu)^2/2\sigma^2}$ will become. Consequently, the height of the function will become lower also. The lower height, however, will not become zero unless the deviation of x from μ approaches infinity.

Observe that the difference between x and μ enters as a squared value. Thus two different values of x that give the same difference from μ will have the same probability density. This explains the symmetry of the normal curve.

Observe also that for a normal distribution, the mean, the median, and the mode all have the same identical value.

8-3-2 Linear change of scale

As pointed out previously, if X and Y are two random variables and $Y = a + X$, then the variance of Y is the same as the variance of X and the mean of Y is simply the mean of $a + X$. It can be expected that if X is normally distributed with mean μ and variance σ^2, Y must also be normally distributed with mean $a + \mu$ and variance σ^2. If the relation between X and Y is such that $Y = bX$, then the mean of Y is equal to the mean of X multiplied by b, while the variance of Y is equal to the variance of X multiplied by the square of b. Thus, if X is $N(\mu, \sigma^2)$, Y will be $N(b\mu, b^2\sigma^2)$. In general we have the following important relationship, which is accepted without mathematical proof:

If $Y = a + bX$, and X is $N(\mu, \sigma^2)$, then Y is $N(a + b\mu, b^2\sigma^2)$ (8-7)

That is, a *linear change of scale* for a random variable X that is normally distributed will result in a new variable Y that is also normally distributed.

Suppose that the length of service, designated as X, of the employees of a company is normally distributed with a mean $\mu = 15$ years and a standard deviation $\sigma = 5$ years. It has been found that the monthly earnings of the employees denoted by Y is related to X such that $Y = 1{,}000 + 50X$. Then by Eq. (8-7) Y will be normally distributed with a mean $1{,}000 + 50(15) = 1{,}750$ and a variance $(50)^2(5)^2 = 62{,}500$. That is, given the relationship $Y = 1{,}000 + 50X$, if X is $N(15, 25)$, then Y is $N(1{,}750, 62{,}500)$.

8-3-3 Linear combination of normal variables

A new variable may also result from a linear change of scale for two or more variables. Suppose that we have two distinct random variables X_1 and X_2, and a new random variable Y is the sum of X_1 and X_2, namely, $Y = X_1 + X_2$. Then the variable Y is said to be a *linear combination* of X_1 and X_2. If X_1 and X_2 are normal random variables, it can be shown that the sum Y is also normally distributed. Furthermore, if X_1 and X_2 are independent random variables, then, by Eq. (5-21), their variances will be *additive*. Consequently, the following relationship exists:

If $Y = X_1 + X_2$, X_1 is $N(\mu_1, \sigma_1^2)$, X_2 is $N(\mu_2, \sigma_2^2)$, and X_1 and X_2 are independent, then Y is $N(\mu_1 + \mu_2, \sigma_1^2 + \sigma_2^2)$ (8-8)

Similarly, by Eq. (5-22), we have

If $Y = X_1 - X_2$, X_1 is $N(\mu_1, \sigma_1^2)$, X_2 is $N(\mu_2, \sigma_2^2)$, and X_1 and X_2 are independent, then Y is $N(\mu_1 - \mu_2, \sigma_1^2 + \sigma_2^2)$ (8-9)

Suppose, for example, that X_1 is a normal random variable with a mean $\mu = 10$ and a variance $\sigma_1^2 = 3$, and that X_2 is another normal random variable with a mean $\mu_2 = 12$ and a variance $\sigma_2^2 = 4$. Suppose also that Y is a linear combination of these two variables, or $Y = X_1 + X_2$. Then, by Eq. (8-8), if X_1 and X_2 are independent, Y will be a normally distributed random variable with the mean $10 + 12 = 22$ and variance $3 + 4 = 7$, or Y is $N(22, 7)$.

Consider another example. Suppose that X_1 designates the nicotine content of one brand of cigarette and X_2 designates the nicotine content of another brand of cigarette. Suppose further that X_1 is normally distributed with a mean 25 milligrams and a variance 5 milligrams, and X_2 is normally distributed with a mean 22 milligrams and a variance 4 milligrams. If X_1 and X_2 are independent, then Y, which is equal to the difference between X_1 and X_2, or $Y = X_1 - X_2$, will also be normally distributed with a mean $25 - 22 = 3$ milligrams and a variance $5 + 4 = 9$ milligrams. That is, Y is $N(3, 9)$.

It must be pointed out that the mean of a random sample is a linear combination of random observations that make up the sample. Suppose that $\bar{X}$ is the mean of a random sample of n independent observations,

$$\bar{X} = \frac{\sum_{i=1}^{n} X_i}{n} = \frac{1}{n}(X_1 + X_2 + \cdots + X_n)$$

where $1/n$ is the coefficient of each of the X's. If X has the mean μ and the variance σ^2, then $\bar{X}$ has the mean $\sum_{i=1}^{n} (1/n)\mu_i$, or simply μ, and variance

$$\sum_{i=1}^{n} \left(\frac{1}{n}\right)^2 \sigma_i^2 = \frac{1}{n^2} n\sigma^2 = \frac{\sigma^2}{n} \qquad \text{by (4-27) and (5-23)}$$

Moreover, if the values of X are normally distributed, $\bar{X}$ is also normally distributed. That is,

$$\text{If } X \text{ is } N(\mu, \sigma^2) \text{ and } \bar{X} = \frac{\sum_{i=1}^{n} X_i}{n}, \text{ then } \bar{X} \text{ is } N\left(\mu, \frac{\sigma^2}{n}\right) \qquad (8\text{-}10)$$

In other words, if a random sample of n independent observations is taken from a normal population, the sampling distribution of the mean is also normal, *irrespective of the magnitude of n*. This is a very important principle and is applied frequently.

The sampling distribution of the arithmetic mean $\overline{X}$ has the same general shape as the density function for X. The former, however, is more concentrated about the true mean of the population than the latter is. The larger the size of the sample, the more concentrated will be the sample means about the population mean. If $n = 1$, each individual observation is, of course, the "mean" of the sample, and the sampling distribution of the mean is identical with the distribution of the population. If the sample size is larger, the standard error of the mean, which is $\sigma/\sqrt{n}$, will naturally be smaller and, consequently, the hump of the density function will be higher. Figure 8-8 shows two sampling distributions of the sample mean for $n = 9$ and $n = 80$, respectively, compared with the density function of the population ($n = 1$).

Since the mean of $\overline{X}$ is μ, or $E(\overline{X}) = E(X) = \mu$, and the standard deviation of $\overline{X}$ called standard error is the standard deviation of X divided by the square root of n, or $\sigma_{\overline{X}} = \sigma_X/\sqrt{n}$, the standard normal score Z converted from the sample mean $\overline{X}$ is expressed as follows:

$$Z = \frac{\overline{X} - \mu}{\sigma_X/\sqrt{n}} \tag{8-11}$$

Suppose that a vending machine is regulated to discharge an average of μ ounces of a soft drink. Let X be the amount of soft drink per cup, which is believed to be normally distributed with a mean $\mu = 10$ ounces and a standard deviation $\sigma = 1$ ounce. A sample of 25 cups is obtained which yields a sample mean $\overline{X}$ ounces. Is $\overline{X}$ normally distributed? What is the mean of $\overline{X}$? What

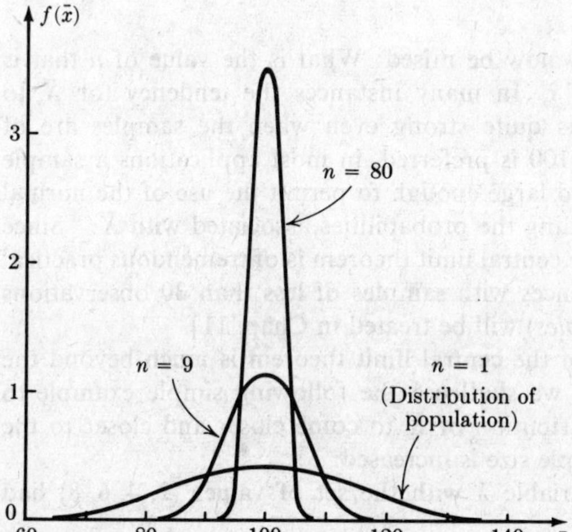

Figure 8-8

Comparison of the distribution of the population with the sampling distributions of the means of samples of sizes $n = 9$ and $n = 80$.

is the standard error of $\bar{X}$? What is the probability that $\bar{X}$ will be at least 10.3 ounces?

If X is normal, the sample mean $\bar{X}$ must be normal. The mean of $\bar{X}$ is the same as the mean of X, namely, $E(\bar{X}) = E(X) = \mu = 10$ ounces. The standard error of $\bar{X}$ is $\sigma/\sqrt{n} = 1/\sqrt{25} = \frac{1}{5} = 0.2$. Thus the probability that $\bar{X}$ will be at least 10.3 ounces is obtained as follows:

$$P(\bar{X} \geq 10.3) = P\left(Z \geq \frac{10.3 - 10}{0.2}\right) = P(Z \geq 1.5) = 1 - P(Z \leq 1.5)$$

$$= 1 - 0.9332 = 0.0668$$

8-3-4 Central limit theorem

Frequently, random samples are taken from populations that are not normal. What, then, is the shape of the density function of the sample mean when the sample is taken from a nonnormal population? A most remarkable and immensely significant fact is that when X is any random variable, discrete or continuous, with a skewed or symmetrical distribution, the mean $\bar{X}$ of the sample will have an approximately normal distribution if the sample size n is sufficiently large. This fact is known as the *central limit theorem*, which may be more formally stated as follows:

> *If X is any random variable with a mean μ and variance σ^2, the sampling distribution of the mean $\bar{X}$ of a random sample of size n is approximately normal with a mean μ and variance σ^2/n if n is sufficiently large, irrespective of the distribution of X.*

A legitimate question may now be raised: What is the value of n that is considered "sufficiently large"? In many instances the tendency for $\bar{X}$ to become normally distributed is quite strong even when the samples are of moderate size. Although $n \geq 100$ is preferred, in most applications a sample of size 30 or more is considered large enough to permit the use of the normal probability distribution for finding the probabilities associated with $\bar{X}$. Since 30 is a rather small number, the central limit theorem is of tremendous practical importance. [Statistical inferences with samples of less than 30 observations (generally known as *small samples*) will be treated in Chap. 11].

The mathematical proof of the central limit theorem is much beyond the scope of this text. However, we shall use the following simple example to show the tendency for the distribution of $\bar{X}$ to come closer and closer to the normal distribution as the sample size is increased.

In Sec. 6-3 the random variable X with the set of values {2, 4, 6, 8} had the uniform distribution:

x	$P(x)$
2	$\frac{1}{4}$
4	$\frac{1}{4}$
6	$\frac{1}{4}$
8	$\frac{1}{4}$

This is the distribution of the population, and its graphic representation is shown in Fig. 8-9. It is the distribution of the "*sample mean*" when the sample size n is equal to 1.

When a sample of size 2, that is, X_1 and X_2, is taken at random with replacement from this population, the sum of the values of the two observations for each *possible* sample is shown in one of the following cells:

+	*2*	*4*	*6*	*8*
2	4	6	8	10
4	6	8	10	12
6	8	10	12	14
8	10	12	14	16

These sums are divided by the sample size 2 to give the *possible* sample means. The probability of each sample mean is the ratio of the number of samples having that mean to the total number of possible samples of size 2, namely, $4^2 = 16$ (see Tables 6-1 and 6-2). The sampling distribution of the means of samples with size 2 is shown as follows:[1]

Sample mean $\bar{x}$	*Probability* $P(\bar{x})$
2	$\frac{1}{16}$
3	$\frac{2}{16}$
4	$\frac{3}{16}$
5	$\frac{4}{16}$
6	$\frac{3}{16}$
7	$\frac{2}{16}$
8	$\frac{1}{16}$
Total	1.0

[1] The mean and the variance of this sampling distribution are $E(\bar{X}) = E(X) = 5$ and $\sigma_{\bar{X}}^2 = \dfrac{\sigma_X{}^2}{n} = \frac{5}{2} = 2.5$ (see Secs. 6-3-1 and 6-3-2).

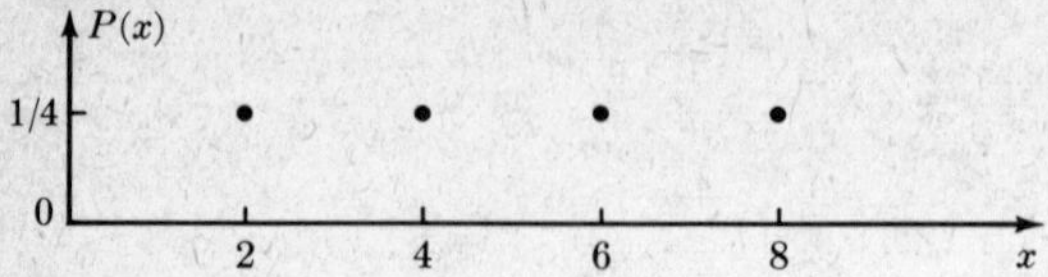

Figure 8-9

Uniform distribution of X when X has the set of possible values {2,4,6,8}.

Figure 8-10 shows the distribution of $\overline{X}$ when $n = 2$. It is clearly closer to the normal distribution than is the uniform distribution shown in Fig. 8-9.

If the sample size is increased, there will be a greater number of possible samples and sample means, and the points in the graph will be closer to their adjacent points than they are in Fig. 8-10. For instance, when $n = 3$, there will be 10 different possible samples and sample means, as shown in Table 8-1.

As enumerated, there are altogether $4^3 = 64$ *distinct* possible sequences or elementary events in this experiment. The probability of each of the sample means is, of course, the ratio of the number of samples having that mean to the total number of possible samples, namely, 64. The sampling distribution of the mean when $n = 3$ is given as follows:

Sample mean $\bar{x}$	*Probability* $P(\bar{x})$
2	$\frac{1}{64}$
$2\frac{2}{3}$	$\frac{3}{64}$
$3\frac{1}{3}$	$\frac{6}{64}$
4	$\frac{10}{64}$
$4\frac{2}{3}$	$\frac{12}{64}$
$5\frac{1}{3}$	$\frac{12}{64}$
6	$\frac{10}{64}$
$6\frac{2}{3}$	$\frac{6}{64}$
$7\frac{1}{3}$	$\frac{3}{64}$
8	$\frac{1}{64}$
Total	1.0

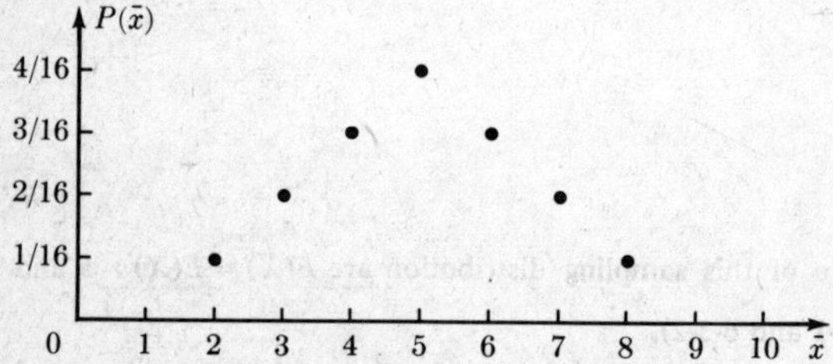

Figure 8-10

Sampling distribution of the mean $\overline{X}$ with $n = 2$, when X has the set of possible values {2,4,6,8}.

Table 8-1

Possible samples and sample means for X with n = 3, when X has the set of possible values {2, 4, 6, 8}

Possible samples X_1	X_2	X_3	Sum $\sum_{i=1}^{3} X_i$	Mean $\bar{X}$	Number of samples	Possible samples X_1	X_2	X_3	Sum $\sum_{i=1}^{3} X_i$	Mean $\bar{X}$	Number of samples
2	2	2	6	2	1	2	6	8			
						2	8	6			
2	2	4				6	2	8			
2	4	2	8	$2\frac{2}{3}$	3	6	8	2			
4	2	2				8	2	6			
						8	6	2	16	$5\frac{1}{3}$	12
2	4	4				4	6	6			
4	2	4				6	4	6			
4	4	2	10	$3\frac{1}{3}$	6	6	6	4			
2	2	6				4	4	8			
2	6	2				4	8	4			
6	2	2				8	4	4			
4	4	4				2	8	8			
2	4	6				8	2	8			
2	6	4				8	8	2			
4	2	6				4	6	8			
4	6	2	12	4	10	4	8	6	18	6	10
6	2	4				6	4	8			
6	4	2				6	8	4			
2	2	8				8	4	6			
2	8	2				8	6	4			
8	2	2				6	6	6			
2	6	6				4	8	8			
6	2	6				8	4	8			
6	6	2				8	4	4	20	$6\frac{2}{3}$	6
2	4	8				6	6	8			
2	8	4				6	8	6			
4	2	8	14	$4\frac{2}{3}$	12	8	6	6			
4	8	2									
8	2	4				6	8	8			
8	4	2				8	6	8	22	$7\frac{1}{3}$	3
4	4	6				8	8	6			
4	6	4									
6	4	4				8	8	8	24	8	1

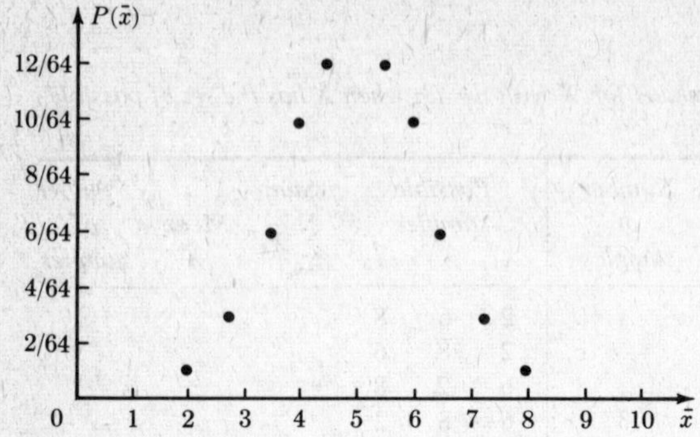

Figure 8-11
Sampling distribution of the mean $\overline{X}$ with $n = 3$, when X has the set of possible values {2,4,6,8}.

The graphic representation of this distribution is shown in Fig. 8-11. It comes closer to the normal density function with an increase of the sample size by only 1. From the above argument, we can intuitively feel that as the sample size increases, the sampling distribution of the mean becomes more and more similar to the normal distribution.

Consider another example. Suppose that the annual income per family in a certain city has a mean $\mu = \$12{,}000$ and a standard deviation $\sigma = \$3{,}000$. A sample of 36 families is randomly selected from the city, and a sample mean $\overline{X}$ is obtained. It is usually held that the annual family income, which we use X to designate, is not normally distributed. What is the probability that the sample mean $\overline{X}$ computed from the incomes of the 36 families is at most \$11,500?

Although X is not normal, by the central limit theorem $\overline{X}$ is approximately normal because the sample size $n = 36$ is large enough. Thus we have

$$P(\overline{X} \le 11{,}500) = P\left(Z \le \frac{11{,}500 - 12{,}000}{3{,}000/\sqrt{36}}\right) = P\left(Z \le \frac{-500}{500}\right)$$
$$= P(Z \le -1) = 1 - P(Z \le 1)$$
$$= 1 - 0.8413 = 0.1587$$

Exercise 8-3

1 Explain the symmetry and bell shapedness of the normal distribution.

2 If X is normally distributed with a mean 10 and variance 16, and Y is also normally distributed with a mean 10 and variance 15, and if $W = X + Y$, what will be the values of the two parameters for the distribution of W? (Assume that X and Y are independent.)

3 In Prob. 2 above, what will be the values of the two parameters for the distribution of W if $W = X - Y$?

4 Explain, in your own words, that if X is $N(\mu, \sigma^2)$, then a linear change of scale for X will lead to a new variable that is also normal.

5 Use your own words to express the central limit theorem. Why is it so important in statistical analysis?

6 Let $Y = 5X + 10$ and X be normally distributed with a mean 10 and variance 25. Find the following:

a $P(Y \geq 68)$. *d* $P(53 \leq Y \leq 59)$.
b $P(Y \leq 69.5)$. *e* $P(61 \leq Y \leq 70)$.
c $P(Y \leq 54)$. *f* $P(52 \leq Y \leq 67)$.

7 Suppose that X has a uniform distribution on $\{1, 3, 5, 7\}$, and a random sample of size 5 ($n = 5$) for X is drawn with replacement. Let $\bar{X}$ be the mean of the sample. Find the approximate values of:

a $P(\bar{X} \geq 6.5)$. *d* $P(1.5 \leq \bar{X} \leq 4.75)$.
b $P(\bar{X} \geq 5.5)$. *e* $P(1 \leq \bar{X} \leq 7)$.
c $P(\bar{X} \leq 5.55)$. *f* $P(2 \leq \bar{X} \leq 6)$.

(Assume that $\bar{X}$ is approximately normal.)

8 In a certain company the average of monthly earnings of all employees is \$800. The variance of the earnings is \$900. Let $\bar{X}$ be the mean of the earnings of 36 employees. Estimate the following:

a $P(\bar{X} \leq \$1{,}000)$. *b* $P(\bar{X} \leq \$790)$. *c* $\$790 \leq \bar{X} \leq \$810)$.

9 Suppose that X has a uniform distribution on $\{1, 2, 3, 4, 5, 6\}$. A random sample of 100 observations ($n = 100$) for X yields a mean $\bar{X}$. Determine the following:

a The pattern of the distribution of $\bar{X}$. *b* The mean and variance for $\bar{X}$.

10 Let Y be a Bernoulli random variable with the probability of a success $p = 0.5$. Given that X designates the number of successes in a random sample of 100 observations ($n = 100$) for Y, determine the following:

a The pattern of the distribution of X. *b* The mean and variance for X.

11 In the preceding problem, let the proportion X/n be denoted by $\hat{p}$. Determine the following:

a The pattern of the distribution of $\hat{p}$. *b* The mean and variance for $\hat{p}$.

12 Suppose that the nicotine content of a certain brand of cigarettes is normally distributed with a mean $\mu = 25$ milligrams and a standard deviation $\sigma = 4$ milligrams. A random sample of 25 cigarettes is selected, and it yields a sample mean $\bar{X} = 26$ milligrams. What is the probability that a sample of this size will yield a sample mean of this magnitude or greater?

13 Let X be the typing speed of the secretaries of a large company. Suppose that the variable X is normally distributed with a mean $\mu = 58$ words and a standard deviation $\sigma = 16$ words per minute. If a random sample of 16 secretaries is selected and is given a speed test, what is the probability that the sample mean $\bar{X}$ is between 50 and 70 words, inclusive?

14 A large automobile dealer suspects that his average profit margin per car sold is below the national average of \$700. It is believed that the profit margin per car sold has a standard deviation of \$80. A random sample of 64 cars sold yields an average profit margin $\bar{X}$. Find the probability that $\bar{X}$ is smaller than \$685.

15 Let X be the life length of a brand I bulb, and Y be the life length of a brand II bulb, and X and Y be independent. It is known that X is normally distributed with a mean $\mu_X = 1{,}000$ hours and a variance $\sigma_X{}^2 = 2{,}500$ hours, and Y is normally distributed with a mean $\mu_Y = 900$ hours and a variance $\sigma_Y{}^2 = 2{,}400$ hours. Let $D = X - Y$.

a How is the difference between X and Y distributed?
b Find the mean and standard deviation of D.

8-4
Normal Approximation to the Binomial

As pointed out in Sec. 7-3-3, when the sample size n increases, the binomial probability distribution approaches the smooth and bell-shaped form. This phenomenon is often used as an illustration of the central limit theorem. Recall that the Bernoulli random variable Y has $\{0, 1\}$ as the set of its values and $P(Y = 1) = p$. If a random sample of n observations, $Y_1, Y_2, \ldots, Y_n$, is taken, then the random variable X, defined as $\sum_{i=1}^{n} Y_i$, is $B(n, p)$. The sample mean $\bar{Y}$ is the sample proportion (designated $\hat{p}$ in Sec. 7-4), which, according to the central limit theorem, is approximately normally distributed with mean p and variance pq/n, provided that n is large. The binomial random variable X, or $n\bar{Y}$, is a linear function of $\bar{Y}$, and thus is also approximately normal with a mean

$$E(X) = E(n\bar{Y}) = nE(\bar{Y}) = np$$

and variance

$$\sigma_X{}^2 = \sigma_{n\bar{Y}}{}^2 = n^2\sigma_{\bar{Y}}{}^2 = n^2\frac{pq}{n} = npq$$

If X is approximately normal, its value can be transformed into a Z score by applying Eq. (8-2) as follows:

$$Z = \frac{X - E(X)}{\sigma_X} = \frac{X - np}{\sqrt{npq}} \tag{8-12}$$

where Z is $N(0, 1)$. The probability for any X can be approximated by the probability for the corresponding Z score.

It must be remembered that in the transformation of a discrete binomial variable into a continuous standardized normal deviate, it is necessary to make a *correction for continuity*. The binomial random variable X assumes positive integer values such as 0, 1, 2, 3, ..., and when its respective probabilities are desired, we merely note the following:

$$P(a < X < b) = P(a + 1 \leq X \leq b - 1) \tag{8-13}$$

where a and b are real positive numbers and $a < b$. Accordingly,

$$P(X > 2) = P(X \geq 3) = P(X > 3) + P(X = 3)$$

and

$$P(X < 3) = P(X \leq 2) = P(X < 2) + P(X = 2)$$

The probabilities for a continuous random variable, however, are stated in terms of intervals since the probability of the variable's assuming a particular number is zero (see Sec. 3-4-2). Consequently, $P(X = 3)$ has to be changed to $P(2.5 \leq X \leq 3.5)$. That is, the probability for the binomial variable to be 3 is equivalent to the probability for the corresponding continuous variable to be between 2.5 and 3.5, its cell boundaries. In general, if X is a binomial variable whose probabilities are desired, we may approximate the probabilities by applying the following correction for continuity:

$$P(a \leq X \leq b) \approx P(a - 0.5 \leq X' \leq b + 0.5) \tag{8-14}$$

where X' is the *converted* continuous normal variable. With this correction in mind, we now proceed to show that the normal approximation to the binomial becomes better as the size of the sample becomes larger.

Suppose that X is $B(10, 0.5)$ and we wish to find $P(2 \leq X \leq 4)$, or $\sum_{x=2}^{4} P(x)$. From Appendix II, for $n = 10$ and $p = 0.5$, we have

$$\begin{aligned} P(X = 2) &= 0.0439 \\ P(X = 3) &= 0.1172 \\ P(X = 4) &= \underline{0.2051} \\ P(2 \leq X \leq 4) &= 0.3662 \end{aligned}$$

These probabilities are shown as the shaded bars in Fig. 8-12, where the histogram represents the probability function for $B(10, 0.5)$.

For normal approximation to the binomial, we first need the correction for continuity. By Eq. (8-14) we have

$$P(2 \leq X \leq 4) \approx P(2 - 0.5 \leq X' \leq 4 + 0.5)$$

where X' is the converted continuous normal variable. In obtaining the probability for the right side of the equation, we need to change the converted normal variable X' into the standardized normal deviate Z by Eq. (8-12): Since

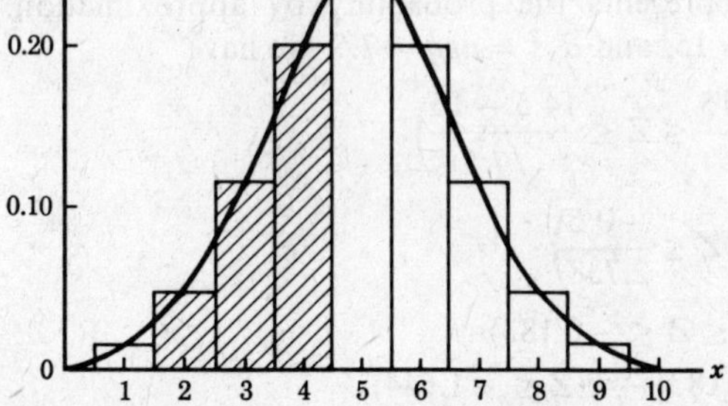

Figure 8-12 Normal approximation to binomial probability distribution, with $n = 10$, $p = 0.5$, and $\mu = np = 5$.

$n = 10$ and $p = 0.5$, the expected value is $np = 10(0.5) = 5$, and the variance is $npq = 2.5$. Thus we have

$$P(1.5 \le X' \le 4.5) = P\left(\frac{1.5 - 5}{\sqrt{2.5}} \le Z \le \frac{4.5 - 5}{\sqrt{2.5}}\right)$$
$$= P\left(\frac{-3.5}{1.58} \le Z \le \frac{-0.5}{1.58}\right)$$
$$= P(-2.215 \le Z \le -0.315)$$
$$= P(Z \le -0.315) - P(Z \le -2.215)$$

By interpolation from Appendix VI, we get

$$P(Z \le -0.315) = 1 - P(Z \le 0.315) = 1 - 0.6236 = 0.3764$$
$$P(Z \le -2.215) = 1 - P(Z \le 2.215) = 1 - 0.9866 = 0.0134$$

Thus

$$P(1.5 \le X' \le 4.5) = 0.3764 - 0.0134 = 0.3630$$

In Fig. 8-12 this probability is represented by the area under the smooth curve between 1.5 and 4.5. Compared with the exact probability 0.3662 obtained directly from the binomial distribution, the normal approximation shows an error of only $0.3662 - 0.3630 = 0.0032$, or slightly under nine-tenths of 1 percent ($\frac{0.0032}{0.3662} = 0.00874$).

Now suppose that p in this example remains unchanged but n is increased to 30. The binomial variable X is $B(30, 0.5)$. The probability of X's falling between 11 and 14, inclusive, is obtained as follows:

$$\begin{aligned} P(X = 11) &= 0.0509 \\ P(X = 12) &= 0.0805 \\ P(X = 13) &= 0.1115 \\ P(X = 14) &= \underline{0.1354} \\ P(11 \le X \le 14) &= 0.3783 \end{aligned}$$

In Fig. 8-13, the histogram is graphed as the probability function for $B(30, 0.5)$, and the shaded bars represent $P(11 \le X \le 14)$. The area under the smooth curve between 10.5 and 14.5 represents the probability by approximation. For $n = 30$, $p = 0.5$, $E(X) = np = 15$, and $\sigma_X{}^2 = npq = 7.5$, we have

$$P(10.5 \le X' \le 14.5) = P\left(\frac{10.5 - 15}{\sqrt{7.5}} \le Z \le \frac{14.5 - 15}{\sqrt{7.5}}\right)$$
$$= P\left(\frac{-4.5}{2.738} \le Z \le \frac{-0.5}{2.738}\right)$$
$$= P(-1.644 \le Z \le -0.183)$$
$$= P(Z \le -0.183) - P(Z \le -1.644).$$

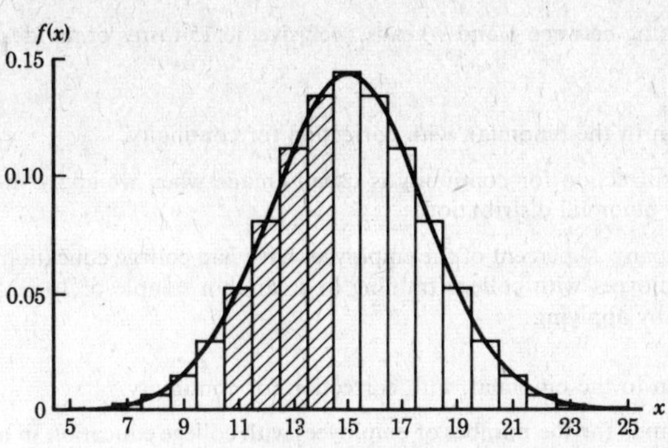

Figure 8-13

Normal approximation to binomial probability distribution, with $n = 30$, $p = 0.5$, and $\mu = np = 15$.

By interpolation, Appendix VI gives

$P(Z \leq -0.183) = 1 - P(Z \leq 0.183) = 1 - 0.5726 = 0.4274$

$P(Z \leq -1.644) = 1 - P(Z \leq 1.644) = 1 - 0.9499 = 0.0501$

Therefore

$P(10.5 \leq X' \leq 14.5) = 0.4274 - 0.0501 = 0.3773$

This differs from the exact probability by $0.3783 - 0.3773 = 0.0010$, or about $\frac{26}{100}$ of 1 percent $(\frac{0.0010}{0.3783} = 0.00264)$. It is clear that as the sample size n is increased, the binomial distribution becomes closer to the normal form.

The method of normal approximation to the binomial is discussed here because, for very large values of n, binomial tables are rarely available and, as indicated earlier, computations of probabilities for a binomial random variable are too laborious. You may wonder whether the sample size n must at least be 30 in order to apply this method. To this question we say that the value of n for satisfactory normal approximation will depend on the value of p. The generally accepted rule is:

If $p > 0.5$, then $n(1 - p) > 5$, or if $p \leq 0.5$, then $np > 5$

Accordingly, for $p = 0.5$, n must be larger than 10, and for $p = 0.9$ or 0.1, n must be larger than 50 if good approximation is desired. In general the greater the skewness of the binomial distribution, the larger the sample size n must be in order to have satisfactory results.

Exercise 8-4

1 Why is it necessary sometimes to apply the normal approximation to the binomial when you know that the exact probability is not obtained by a method of approximation?

2 What are the conditions, or rules, which have to be met when the method of normal approximation to the binomial is used?

3 Find the probability of obtaining between 6 and 9 heads, inclusive, in 15 tosses of an ideal coin by applying:

a The binomial distribution.
b The normal approximation to the binomial, with correction for continuity.

4 Explain the reason why the correction for continuity is usually made when we apply the normal approximation to the binomial distribution.

5 Suppose that in a certain company 20 percent of the employees have had college education. Let X be the number of employees with college training in a random sample of 10 employees. Find $P(3 \leq X \leq 5)$ by applying:

a The binomial distribution.
b The normal approximation to the binomial, with correction for continuity.

6 Suppose that in Prob. 5, X stands for the number of employees with college education in a random sample of 30 employees. Find $P(7 \leq X \leq 9)$ by applying:

a The binomial distribution.
b The normal approximation to the binomial, with correction for continuity.
c What conclusion can you derive after you compare the answers to this problem and the answers to Prob. 5?

7 In a certain production process, 10 percent of the units produced are defective. Let X be the number of defective units found in a sample of 100 units selected at random. Find the approximate values of :

a $P(X \leq 13)$. *c* $P(7 \leq X \leq 16)$.
b $P(X \geq 8)$. *d* $P(5 \leq X \leq 10)$.

(*Note:* Since the approximation is good for large n, the correction for continuity in this problem may be ignored.)

8 Suppose that 20 percent of all calls reaching a telephone exchange are collect calls. If 400 calls reach the exchange during a certain period, what is the probability that:

a At most 100 are collect calls?
b At least 70 are collect calls?

(*Note*: Ignore correction for continuity.)

Chapter 9

Statistical inference—estimation

In the preceding chapters we examined the basic notions of probability and sampling distribution. We are now ready to consider the methods of statistical inference, which include the procedures of estimating population parameters and testing whether a tentative statement about a population parameter is supported or disputed by evidence from the sample.

Generally speaking, there are two types of inference: deductive and inductive. A *deductive inference* is a judgment or generalization based on *axioms*, *premises*, or *assumptions*. A *conclusion* is derived by a priori reasoning or dialectical processes. For example, if two coins are assumed to be prefectly balanced, and thus the probability for each to turn up heads is 0.5 (premise), the mean, or expected, number of heads in the toss of these two coins must be 1 (conclusion). If the premises are true, the conclusions cannot be false.

An *inductive inference*, on the other hand, is a judgment or generalization derived from empirical, or experimental, observations; a conclusion is derived by a posteriori reasoning. In the toss of two coins, for example, an inductive inference is a conclusion about the mean number of heads on the basis of actual results of a sample of trials. If the outcomes of the trials are different, the conclusion will also be different. It needs no a priori assumption about the nature of the coins. Statistical inference is primarily inductive in nature. It reaches generalizations with regard to the characteristics of a population in the light of empirical observations from the sample.

A sample statistic is very likely to be different from the parameter of the population. It is only coincidental that the one is exactly the same as the other. The difference between a sample statistic and the corresponding population parameter is often referred to as the *error*. We would know the error only if the parameter of the population were known. The latter, however, is usually not known. The only way to be certain about it, in most applications, is to make all the possible observations of the entire universe. This is, of course, impossible or impracticable. As a matter of fact, the very reason for inferential statistics to exist is the lack of knowledge about the characteristics of the population. Our ignorance of such characteristics, however, does not prevent us from taking action. A moment of reflection would make us realize that scientists, businessmen, politicians, military men, and men in many other walks of life make decisions relative to the characteristics of the population, often called the "states of nature," without being at all certain of what they actually are. What most people do, though unconsciously, is to "play the odds" as they best estimate them. For example, people drive back and forth on the freeway as if they were completely sure that they would not be killed by accidents because they feel quite certain that the odds are overwhelmingly in their favor. The statistics on freeway fatalities are good indicators of these odds, or probabilities. At times, a few persons are wrong in relying on the probability that they will not become victims of traffic accidents. However, it is mainly because such accidents occur infrequently that we continue to drive on the freeway with confidence of survival.

In similar manner, statistical inferences are made in terms of odds or probabilities. From the mean of a sample, we draw inferences about the mean of the population. We do not know exactly the difference between these two means because the latter is, in most cases, unknown to us. We do know, nonetheless, that the probability for this difference to exceed, say, three or even only two standard errors is rather small. Occasionally, we are wrong in acting on this probability. That is the price we have to pay for not doing the often impossible task of investigating the entire universe in question.

The problems dealt with in inferential statistics are generally divided into two classes: the problems of estimation and those of hypothesis testing. Since, in estimating an unknown population parameter, a judgment or statement is usually made, the judgment provides only an *estimate*. It is a particular value obtained from the observations in a sample. This concept should not be confused with the term *estimator*, which refers to the *rule* or *method* of estimating a population parameter. For instance, $\bar{X}$ is called an estimator of μ because the sample mean is a method of estimating the population mean. The procedure of making judgment about a population parameter is referred to as *statistical estimation*, which is further divided into point estimation and interval estimation. In this chapter we plan to consider both types of estimation; in Chap. 10 attention will be focused on the testing of statistical hypotheses.

9-1 Point Estimation

If, from the observations in a sample, a single value is calculated as an estimate for an unknown population parameter, the procedure is referred to as *point estimation* since a single point in the set of all potential values is used as the estimate. Suppose, for example, that we wish to estimate μ, the average intelligence of all college students in the United States. Let X be the random variable designating the IQ for each student. A random sample of n students is drawn, and the mean IQ of the sample is denoted by $\bar{X}$. If we are told that a sample of 100 students is taken and $\bar{X}$ is 115, and are asked to estimate μ on the basis of this information, then such a statistic is a *point estimator* since a specific value is singled out as an estimate for μ.

In the general case, let θ stand for an unknown parameter of a distribution for a random variable X, and let X_1 $X_2, \ldots, X_n$ be a random sample of n values of X drawn from this distribution. If $\hat{\theta}$ (read as "theta hat") is an estimate of θ computed from a sample of n observations, then a value for $\hat{\theta}$ can be obtained from each random sample of n observations. Thus $\hat{\theta}$ is a sample statistic with a theoretical sampling distribution.

As pointed out earlier, an error is likely to occur because the sample is merely a small subset of the much larger set of possible observations. It is therefore hazardous to assert that the population value *is* the same as the value computed from the sample. It is true that if the number of random observations is made sufficiently large, they would yield a mean that is almost the same as the parameter. However, we are often limited in time and resources, and decisions have to be made with only meager amounts of evidence available. In order to use the available information in the best possible way, we need to identify the statistics that are "good" estimators. There are four *criteria* commonly applied in determining whether a statistic is a good estimator. They are consistency, unbiasedness, efficiency, and sufficiency, and they will be reviewed in the following subsections. The *principle of maximum likelihood*, a method often applied in choosing point estimators, will be examined in Sec. 9-2.

9-1-1 Consistency

One often-mentioned property that an estimator should possess is *consistency.* Loosely speaking, a *consistent estimator* is one that tends to have a greater probability of being close to the population parameter as the sample size becomes large. More formally,

> *If $\hat{\theta}$ is a sample statistic computed from a sample of size n and θ is the population parameter to be estimated, then $\hat{\theta}$ will be a consistent estimator of θ if, for any arbitrarily small positive number e,*
>
> $$P(|\hat{\theta} - \theta| \leq e) \to 1, \text{ as } n \to \infty \qquad (9\text{-}1)$$

That is, the probability of the sample statistic $\hat{\theta}$'s being within a distance e from the population parameter θ approaches 1 as the sample size n approaches infinity, irrespective of how small the value e is.

Some statistics such as the sample mean and variance are known to be consistent estimators since they tend to be closer to their population counterparts as the sample is enlarged. A sample statistic, however, can be an inconsistent estimator. For instance, if the value of the first observation, or the mean of the first and the last observations, of a sample is used to estimate the population mean, such an estimator is certainly not consistent since it does not tend to come closer and closer to the population value as the sample size is increased.

9-1-2 Unbiasedness

As pointed out above, $\hat{\theta}$, an estimator of θ, is a random variable. As such, it has a probability distribution with a mean and variance. Let $E(\hat{\theta})$ be the mean of $\hat{\theta}$; we can then define the *unbiased estimator* as follows:

Given that a sample statistic $\hat{\theta}$ is used to estimate the population parameter θ, $\hat{\theta}$ is said to be an unbiased estimator of θ if

$$E(\hat{\theta}) = \theta \qquad (9\text{-}2)$$

It can be expected that if we draw many samples of a given size from the same distribution, and from each obtain a value for $\hat{\theta}$, the arithmetic mean of all the values of $\hat{\theta}$ must be quite close to θ. Unbiasedness is an important property of a good estimator.

Consider, for example, the arithmetic mean of a sample as an estimator of the population mean. We have

$$\theta = \mu \qquad \text{and} \qquad \hat{\theta} = \overline{X} = \frac{\sum_{i=1}^{n} X_i}{n}$$

It is easy to demonstrate that the sample mean is an unbiased estimator of the population mean; that is,

$$E(\overline{X}) = \mu$$

The proof of this was given in Sec. 6-3-1. Similarly, the sample proportion $\hat{p}$ is an unbiased estimator of the population proportion p, since by Eq. (7-12) we have

$$E(\hat{p}) = p$$

Now we may ask: Is the sample variance an unbiased estimator of the population variance σ^2? The answer to this depends on how the sample variance is defined. If the sample variance is defined as

$$s^2 = \frac{\sum_{i=1}^{u} (X - \overline{X})^2}{n}$$

then s^2 is a biased estimator of σ^2, for as shown in Supplement VII,

$$E(s^2) \neq \sigma^2 \tag{9-3}$$

The bias, however, can be corrected by slightly changing the definition for the sample variance. Let $\hat{s}^2$ be the sample variance defined as

$$\hat{s}^2 = \frac{\sum_{i=1}^{n} (X_i - \bar{X})^2}{n-1} \tag{9-4}$$

Then, as demonstrated in Supplement VIII,

$$E(\hat{s}^2) = \sigma^2 \tag{9-5}$$

Thus the biased estimator of the population variance can always be corrected as follows:

$$\hat{s}^2 = \frac{n}{n-1} s^2 \tag{9-6}$$

This is true because

$$\frac{s^2}{n-1} = \frac{\sum_{i=1}^{n} (X_i - \bar{X})^2/n}{n-1} = \frac{\sum_{i=1}^{n} (X_i - \bar{X})^2}{n(n-1)}$$

$$= \frac{\hat{s}^2}{n}$$

therefore

$$\frac{n}{n-1} s^2 = \hat{s}^2$$

From now on we shall always place the "hat" (^) over the symbol for a sample variance if it is an unbiased estimator of the population variance. This is done to distinguish it from the biased sample variance s^2.

The computation of the unbiased estimate of the population variance by Eq. (9-4), however, is a rather laborious task inasmuch as the sample mean must be subtracted from every observed value. It is especially time consuming if the numbers in the sample and the mean have several digits or decimal places. The following formula, which is derived from Eq. (9-4), is frequently used as a more efficient substitute:

$$\hat{s}^2 = \frac{\sum X^2 - (\sum X)^2/n}{n-1} \tag{9-7}$$

Let us use an example to illustrate the procedure of computing the unbiased estimate of a population variance and its relation with the biased estimate. Suppose that the desire is to estimate the variance of X designating the number of years of formal education received by employees of a large company. A random sample of nine employees yields the following data:

X_1	X_2	X_3	X_4	X_5	X_6	X_7	X_8	X_9
11	8	7	10	9	12	8	12	13

The mean of these nine values is $\frac{90}{9} = 10$, and the sum of squared deviations is

$$(11-10)^2 + (8-10)^2 + (7-10)^2 + (10-10)^2 + (9-10)^2 + (12-10)^2 + (8-10)^2 + (12-10)^2 + (13-10)^2$$

$$1 + 4 + 9 + 0 + 1 + 4 + 4 + 4 + 9 = 36$$

Thus, by Eq. (9-4) the unbiased estimate is

$$\hat{s}^2 = \frac{36}{9-1} = \frac{36}{8} = 4.5$$

and the biased estimate is

$$s^2 = \frac{36}{9} = 4$$

Observe that the unbiased estimate is larger than the biased. When s^2 is known, $\hat{s}^2$ can be obtained by applying Eq. (9-6). In this example we have

$$\hat{s}^2 = \frac{n}{n-1}s^2 = \frac{9}{8}(4) = 4.5$$

The same answer can be obtained by applying Eq. (9-7), as indicated below:

$$\hat{s}^2 = \frac{\sum X^2 - (\sum X)^2/n}{n-1}$$

$$= \frac{11^2 - 8^2 - 7^2 - 10^2 - 9^2 - 12^2 - 8^2 - 12^2 - 13^2 - (90)^2/9}{9-1}$$

$$= \frac{936 - 900}{8} = \frac{36}{8} = 4.5$$

9-1-3 Efficiency

Another criterion for evaluating an estimator is that it should be *efficient*. The efficiency of an estimator is usually stated in relative terms as indicated below:

> *If two statistics are both used as estimators of the same parameter, the one whose sampling distribution has the smaller standard error is a more efficient estimator than the other.*

Suppose that two statistics $\hat{\theta}_1$ and $\hat{\theta}_2$ are obtained from the same sample, and both are unbiased estimators of the same parameter θ. The sampling distribution of $\hat{\theta}_1$ has a variance ${\sigma_1}^2$, and the sampling distribution of $\hat{\theta}_2$ has a variance ${\sigma_2}^2$. If ${\sigma_1}^2$ is smaller than ${\sigma_2}^2$, the values of $\hat{\theta}_1$ from samples of the same size n will be more concentrated around their expected value θ, and from any single sample $\hat{\theta}_1$ is more likely to be close to θ than $\hat{\theta}_2$ is. Thus we say that $\hat{\theta}_1$ is more efficient than $\hat{\theta}_2$ as an estimator of θ.

The efficiency of the estimator $\hat{\theta}_1$ relative to $\hat{\theta}_2$, both estimating the same parameter value, is given by the ratio ${\sigma_2}^2/{\sigma_1}^2$, which is greater than 1 since the variance of the more efficient estimator is used as the denominator. Given the variance of $\hat{\theta}_2$, the smaller the variance of $\hat{\theta}_1$, the greater the efficiency of $\hat{\theta}_1$ will be. In case there is one estimator that, among all unbiased estimators, has the smallest variance, it is then referred to as the *most efficient* estimator.

Suppose that a certain random variable X has a symmetrical and unimodal distribution, which means that the mean and the median of the distribution are identical. If a sample is drawn from this distribution, what sample statistic, the mean $\overline{X}$ or the median Med, should be used to estimate the population mean μ? The answer to this question depends on which is a more efficient estimator. Both $\overline{X}$ and Med are unbiased estimators of μ, but the standard error of the mean is known to be smaller than that of the median; that is,

$$\frac{\sigma_{\text{Med}}}{\sigma_{\overline{X}}} > 1$$

This is to say that the sample mean relative to the median is a more efficient estimator.

9-1-4 Sufficiency

The last, but not least, important attribute that a good estimator possesses is *sufficiency*. The sufficiency of an estimator can be briefly defined as follows:

> *A sufficient estimator of the parameter θ is one that exhausts all the relevant information regarding θ that is available in the sample.*

If $\hat{\theta}$ is such an estimator of θ, there is nothing we can do to improve $\hat{\theta}$ since all aspects of the sample information have already been contained in it.

When a random sample is taken by the Bernoulli process, for instance, the sample proportion $\hat{p}$ is a sufficient estimator of the population parameter p since $\hat{p}$ contains all the information about the value of p that can be obtained from the sample. Similarly, when a sample is drawn from a population that is normally distributed, the mean $\overline{X}$ is a sufficient estimator of the mean μ of the distribution. There is no information we can add to $\overline{X}$ to make it a better estimator of μ.

Suppose, for example, that a sample of 30 observations is taken for the purpose of estimating the population mean μ. The first, the last, or the fifteenth observation might be used as the estimator. Likewise, the average of the first and the thirtieth, or the average of the 10 items between 11 and 20, inclusive, can be used as an estimator of μ. They are, however, not sufficient estimators, for they do not contain all the information available from the sample. On the other hand, the sample mean $\bar{X}$ computed from the 30 observations is a sufficient estimator because every item is taken into account in obtaining the value for $\bar{X}$.

The preceding discussion has indicated that the sample mean, sample proportion, and the corrected form of the sample variance are statistics that satisfy our criteria for good estimators. We shall deal with these statistics most frequently in the rest of this book.

Exercise 9-1

1 Distinguish between the following concepts:
a Deductive and inductive inference. *b* Estimate and estimator.

2 Explain whether each of the following statistics is a sufficient estimator of the population mean:
a The median. *b* The mode.

3 Suppose that a random variable has a uniform distribution on {1, 3, 5}, and that a sample of two observations ($n = 2$) drawn with replacement from this simulated population yields a mean $\bar{X}$.
a Verify that $E(\bar{X}) = \mu$, the mean of the population.
b Verify that $E(\hat{s}^2) = \sigma_X{}^2$, the variance of the population.

4 If a consistent estimator is biased, will the bias approach zero as the sample size approaches infinity? Explain.

5 Is an estimator a random variable? Why or why not?

6 Explain why the sample mean $\bar{X}$, as well as the sample proportion $\hat{p}$, is:
a A consistent estimator. *c* An efficient estimator.
b An unbiased estimator. *d* A sufficient estimator.

7 Distinguish between the biased and unbiased estimators of the population variance. How do you compute the unbiased variance estimator if only the biased estimator is known?

8 Explain why, or why not, the biased variance estimator s^2 is a sufficient estimator for σ^2.

9 Suppose that it is desired to estimate the variance of the hourly outputs of all workers doing the same kind of work. A random sample of 10 workers yields the following results:

Worker:	X_1	X_2	X_3	X_4	X_5	X_6	X_7	X_8	X_9	X_{10}
Output:	3	4	5	6	9	9	12	13	14	15

Find the unbiased estimate of the population variance by Eq. (9-4), and then verify it by Eq. (9-7).

10 Show that the unbiased estimate in Prob. 9 above can be obtained by using Eq. (9-6) to correct the biased estimate.

9-2
The Method of Maximum Likelihood

In the preceding section we examined the characteristics of a good point estimator. The procedure by which point estimators are usually selected is known as the *method of maximum likelihood.* It involves the choosing of a statistic such that the estimated population parameter will be one for which the observed statistic has the greatest likelihood, or probability, of occurring. Such a statistic is consistent; it tends to become closer to the estimated parameter as the sample is enlarged. It is unbiased because its expected value is identical with the parameter value to be estimated. It also possesses the other properties of a good estimator. A formal treatment of these properties of the maximum likelihood estimator is quite advanced, and all we can do now is to grasp the basic concept of this method.

The procedure is to consider all the conceivable values of the population parameter, and to compute the probability that the particular sample at hand would have been obtained, given each conceivable value of the parameter. The *maximum likelihood estimate* is taken as the value, among all the values of the parameter, for which the probability of obtaining the sample statistic is the greatest.

Suppose, for example, that we should like to know the proportion p of students at a certain college who favor collective bargaining for public employees. To do this, a sample of n students is taken at random, and the question, "Do you favor collective bargaining for public employees?" is asked of each of them. If we use X to designate the number of "yes" answers in the sample, then X is $B(n, p)$. In this example, n is known, and the unknown p is to be estimated.

Assume that in this example a sample of 10 students is taken, and 4 answer "yes." That is, $n = 10$ and $X = 4$. We can then proceed to find out the probability of obtaining four "yes" answers, or a sample proportion $\frac{4}{10}$, given any specified value of p, the proportion of all students who would answer "yes" to the question. From Appendix II, we know that if $p = 0.1$, $P(X = 4) = 0.0112$, and if $p = 0.2$, $P(X = 4) = 0.0881$. Table 9-1 shows the probability of $X = 4$ for each of the specified values of p. As indicated, the greatest probability, or 0.2508, of obtaining a sample proportion $\hat{p} = 0.4$ is associated with the population proportion $p = 0.4$. It is a fact that it is more probable to obtain a sample proportion 0.4 if the population proportion is 0.4 than if it is any other value.

Of course, the population proportion p can assume other values than those specified above. For example, p may have the value of 0.55 or 0.4457. Given such a value of p, we could still find the probability of $X = 4$. Let $L(v)$ be the likelihood function, or probability density, of the particular sample observation $X = 4$; that is,

$$L(v) = P(X = 4 \mid p = v)$$

Table 9-1

Probabilities of four "yes" answers in 10 responses for specified population proportions

Population proportion p	Probability $P(X=4)$
0.0	0.0000
0.1	0.0112
0.2	0.0881
0.3	0.2001
0.4	0.2508
0.5	0.2051
0.6	0.1115
0.7	0.0368
0.8	0.0055
0.9	0.0001
1.0	0.0000

for all values of v between 0 and 1. Then, $L(v)$ is the conditional probability of $X = 4$, given that $p = v$. The likelihood function can be graphed by plotting the values appearing in Table 9-1 and then connecting the points with a smooth curve as shown in Fig. 9-1. An inspection of the curve shows that the value of p that makes the observed sample most likely is 0.4, which is exactly the same as the sample proportion.

Accordingly, we shall use 0.4 as an estimate of p. In doing so, we have made the maximum likelihood estimation for p. In the general case, a maximum likelihood estimator may be defined as follows:

> *A statistic that is selected as an estimator because its value substituted for the parameter maximizes the probablity of obtaining the observed outcome is called a maximum likelihood estimator.*

If a random variable X is $B(n, p)$, where p is to be estimated, and if a single value X is obtained in a sample of n trials, then the maximum likelihood estimator for p is given by X/n.

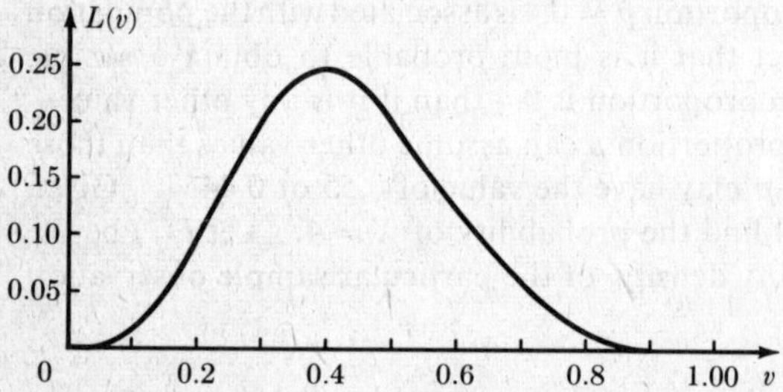

Figure 9-1

Likelihood function $L(v) = P(X = 4 | p = v)$ for $n = 10$.

It must be pointed out that any point estimator, no matter how good it is, has its drawback in not providing any measure of the *precision* of the estimate. It gives no indication of the magnitude of the *error* referred to previously. In order that some degree of confidence may be assigned to the estimate, statisticians usually use interval estimates, which we shall discuss in the next section.

Exercies 9-2

1 In a survey of church affiliation in a certain city, 30 of the 50 families included in a random sample indicated that they belong to some church. What is the maximum likelihood estimate for p, the proportion of all families in the city that are church-affiliated? Plot the likelihood function for $p = 0.1, 0.2, \ldots, 0.9$.

2 A random sample of 30 parts is taken from a box containing a large number of parts of which a proportion p is defective. If the sample contains 3 defective parts, what is the maximum likelihood estimate for p? Plot the likelihood function for p.

3 Let X be $B(50, p)$. Plot the likelihood function for p for each of the following compound events:

a $X \leq 10$.
b $X \geq 40$.
c $X \geq 20$.
d $20 \leq X \leq 30$.
e $30 \leq X \leq 35$.

4 Let p be the proportion of students in your school who are married. Suppose that in a random sample of 20 students taken from your school, 4 are married. Give the maximum likelihood estimate for p. Also plot the likelihood function for p.

5 A certain congressional district has a proportion p of the voters registered as Democrats. A random sample of 50 voters contains 25 Democrats. Give the maximum likelihood estimate, and graph the likelihood function for p.

9-3 Interval Estimation of the Mean

The procedure of determining an interval (a to b) that will include a population parameter θ with a certain probability $1 - \alpha$, is known as *interval estimation.* Here, α (the first Greek letter, read as "alpha") is *the probability that the interval does not include the true parameter value.* We make the statement $P(a < \theta < b) = 1 - \alpha$, where a and b are random variables depending on some estimator $\hat{\theta}$ of θ, and are respectively referred to as the *lower* and *upper limits* of the interval. The interval (a to b) is called the *confidence interval;* $b - a$ is a measure of the *precision* of the estimate and $1 - \alpha$ is a measure of its *reliability.*

For example, given a sample of n observations, $X_1, X_2, \ldots, X_n$, if the sample mean $\overline{X}$ is used to determine a confidence interval for the population mean μ, then we have the confidence statement for μ as follows:

$$P(\overline{X} - z_{\alpha/2}\sigma_{\overline{X}} < \mu < \overline{X} + z_{\alpha/2}\sigma_{\overline{X}}) = 1 - \alpha \tag{9-8}$$

where $\overline{X} - z_{\alpha/2}\sigma_{\overline{X}}$ = lower limit a

$\overline{X} + z_{\alpha/2}\sigma_{\overline{X}}$ = upper limit b

$z_{\alpha/2}$ = standard z score cutting off either tail by $\alpha/2$ of the distribution with $\mu = \overline{X}$

$\sigma_{\overline{X}} = \sigma_X/\sqrt{n}$ (shown in Sec. 6-3-2).

Thus the procedure of interval estimation includes the construction of a confidence interval. As indicated in Eq. (9-8), a confidence interval for μ is defined by three factors: the sample mean $\overline{X}$, the standard error of the sample mean, and the standard normal score z, which reflects the desired degree of confidence that the population mean would be included in the interval. The z score is used here because $\overline{X}$ must be normally distributed or approximately so, provided that n is large (by the central limit theorem discussed in Sec. 8-3-4).

The precision with which the population mean is estimated varies inversely with the population variance $\sigma_X{}^2$ and directly with the sample size n. Given the degree of confidence $1 - \alpha$ and the sample size n, the greater the population variance, and so the wider the confidence interval, the smaller the precision of the estimate will be. Conversely, the smaller the population variance, and so the narrower the confidence interval, the greater the precision will be. Of course, we can improve the precision by increasing the size of the sample. The greater the sample size, and so the smaller the confidence interval, the more precise will be the estimate, provided that the reliability and the population variance remain unchanged. Or, given the degree of precision and the population variance, an increase in the size of the sample will result in an improvement in the reliability of the estimate. This last statement is true because, under the specified conditions, a larger n will dictate an increase in z, which means a larger value of $1 - \alpha$, or a greater degree of reliability.

If a 95 percent confidence interval is desired, that is, $1 - \alpha = 0.95$, the value of $z_{\alpha/2}$ will be 1.96, since the cumulative probability to 1.96 from $-\infty$ is 0.9750 and to -1.96 from $+\infty$ is also 0.9750, or 95 percent of the area under a normal curve lies between $\overline{X} - 1.96\sigma_{\overline{X}}$ and $\overline{X} + 1.96\sigma_{\overline{X}}$. (You should verify this by examining the standard normal table, Appendix VI.) Thus we have

$$P\left(\overline{X} - 1.96\frac{\sigma_X}{\sqrt{n}} < \mu < \overline{X} + 1.96\frac{\sigma_X}{\sqrt{n}}\right) = 0.95$$

and the 95 percent confidence limits are

$$\overline{X} \pm 1.96\frac{\sigma_X}{\sqrt{n}}$$

Similarly, we have

$$P\left(\bar{X} - 2.575\frac{\sigma_X}{\sqrt{n}} < \mu < \bar{X} + 2.575\frac{\sigma_X}{\sqrt{n}}\right) = 0.99$$

and the 99 percent confidence limits are

$$\bar{X} \pm 2.575\frac{\sigma_X}{\sqrt{n}}$$

Here the value 2.575 is the distance above *and* below the mean in standard error units.

Suppose, for example, that a random variable X is used to designate the body weight of an airline passenger, and we are interested in knowing μ, the mean weight of all the airline passengers. Since we are limited in time and money for weighing every one, a random sample of 36 passengers is selected, and a sample mean $\bar{X} = 160$ pounds is obtained. Suppose further that the distribution of weights of airline passengers is known to have a standard deviation 30. What, then, would be the probable value of μ if the probability of our estimate's being right is 0.95? By applying Eq. (9-8), the answer to this question is

$$160 - 1.96\frac{30}{\sqrt{36}} < \mu < 160 + 1.96\frac{30}{\sqrt{36}}$$

or

$$150.2 < \mu < 169.8$$

That is, with a probability equal to 0.95, the computed interval between 150.2 and 169.8 pounds would include the mean weight of all airline passengers. However, the reader is warned against using the *erroneous* statement

$$P(150.2 < \mu < 169.8) = 0.95$$

since μ is a parameter, and no statistic, or random variable, appears inside the parentheses.

This confidence interval is graphically shown in Fig. 9-2, where the two sampling distributions I and II have the upper and lower confidence limits as their respective means μ_1 and μ_2. Distribution III has a mean $\mu_3 = \bar{X}$. The true mean lies under the curve III, but the two limits, arrived at through a random procedure, have 0.95 as the probability of including it.

Observe that the interval estimate will be different if a different mean is obtained each time when a random sample is selected. (This explains why the confidence limits a and b are random variables.) If many confidence intervals are generated by a sampling process, then it can be expected that some of these

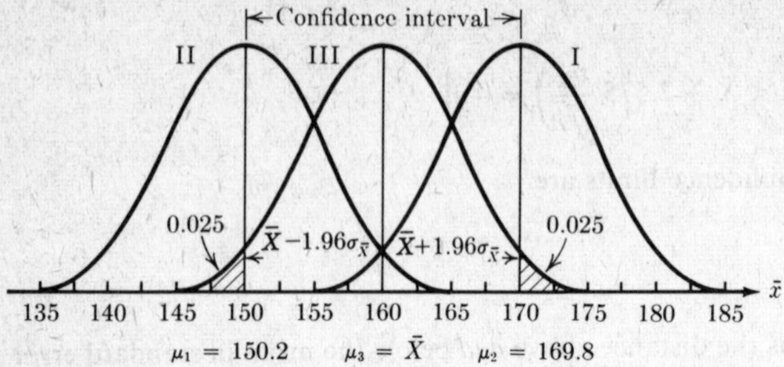

Figure 9-2
Ninety-five percent confidence inetrval of μ from sample mean $\bar{X} = 160$ with standard error $\sigma_{\bar{X}} = 5$.

intervals will actually include the true parameter value, and some will not. Generally, the level of confidence specified will show the proportion of intervals that include the true value. Thus, if a 95 percent confidence interval is specified, and if 100 intervals are constructed on the basis of 100 different samples, we should expect 95 intervals to actually include the value of the parameter.

Exercise 9-3

1 Explain the differences between a point estimate and an interval estimate. Why is an interval estimate more useful?

2 Find the 95.45 and 99.73 percent confidence intervals for the mean of X, where X is $N(\mu, \sigma^2)$ given that $\bar{X}$ is the mean of a random sample of n observations and:

a $\bar{X} = 60$, $n = 100$, $\sigma^2 = 2{,}500$.
b $\bar{X} = 30$, $n = 25$, $\sigma^2 = 144$.
c $\bar{X} = 20$, $n = 16$, $\sigma^2 = 100$.
d $\bar{X} = 10$, $n = 25$, $\sigma^2 = 64$.

3 We know that the statement

$$P(\bar{X} - 1.96\sigma_{\bar{X}} < \mu < \bar{X} + 1.96\sigma_{\bar{X}}) = 0.95$$

is correct, while the statement

$$P(140 < \mu < 160) = 0.95$$

is not. Explain fully why the latter is erroneous.

4 A standard test is administered to a large class of students. The mean score of a random sample of 100 students is 75 points. Assume that the variance of the scores for such a test is known to be 2,500. Find:

a The 98 percent confidence interval for μ, the mean score for the class.
b The upper limit of the 95 percent confidence interval for μ.
c The lower limit of the 90 percent confidence interval for μ.

5 Find the 90 and 80 percent confidence intervals for the mean of X, where X is $N(\mu, \sigma^2)$, given that $\bar{X}$ is the mean of a random sample of n observations and:

a $\bar{X} = 0.5$, $n = 25$, $\sigma^2 = 100$.
b $\bar{X} = 2.5$, $n = 16$, $\sigma^2 = 64$.
c $\bar{X} = 1.0$, $n = 16$, $\sigma^2 = 1.0$.
d $\bar{X} = 0.01$, $n = 100$, $\sigma^2 = 0.25$.

6 From all the ball bearings produced by a certain machine during a certain day, a sample of 100 is randomly selected. The mean diameter of these 100 ball bearings is found to be 0.75 inch. Given that the variance of the diameters has been 0.0025 inch, determine the 95 and 99 percent confidence intervals for μ, the mean diameter of all the ball bearings.

7 Suppose that we wish to estimate the average weight of college men in this country. For this purpose, a sample of 100 college men is randomly selected, and a sample mean $\bar{X} = 150$ pounds is obtained. Assume that the variance of the population, or $\sigma_X{}^2$, is 1,600 pounds. Find:

a The 95 percent confidence interval for μ, the mean weight of all college men in this country.

b The lower limit of the 95 percent confidence interval, with 300 pounds as the upper limit.

8 Let X be the service life in miles of a Fairyear deluxe tire, where X has a normal distribution with its mean unknown and a standard deviation equal to 4,000 miles. A random sample of 100 observations ($n = 100$) yields a mean $\bar{X} = 30{,}000$ miles, Find the 95 percent confidence interval for μ, the true mean of X. Would you advise the Fairyear Tire Company to guarantee the deluxe tire for 25,000 miles? Why or why not?

9 The confidence interval considered in this section is two-sided. Sometimes we may want only a one-sided interval. Suppose that a manufacturer using steel wires as raw materials is interested in the tensile strength of the wire. If he wants the wire to have at least a certain strength, he would be very happy if it proves to be stronger, given the price of the wire. From a shipment of the wire, a sample of 100 lengths is randomly taken and then tested. Suppose that a mean strength of 250 pounds is obtained, and that the standard deviation of the distribution of the tensile strength is 40. What, then, is the lower limit of the one-sided 95 percent confidence interval?

9-4
Interval Estimation of the Difference between Two Means

Frequently, we are interested in the *difference* between two means. The construction of a confidence interval for the true difference on the basis of an observed difference is similar to that for a single population mean.

The difference between two sample means is a random variable, and is a function of the two means. Suppose that X_1 and X_2 are two random variables with means μ_1 and μ_2 and known variances $\sigma_1{}^2$ and $\sigma_2{}^2$. Suppose further that two random samples of sizes n_1 and n_2 are drawn *independently* for the respective variables, and two sample means $\bar{X}_1$ and $\bar{X}_2$ are obtained. If $\bar{D}$ is used to designate the difference between these two means, namely,

$$\bar{D} = \bar{X}_1 - \bar{X}_2$$

then $\bar{D}$ is a linear combination of $\bar{X}_1$ and $\bar{X}_2$. If X_1 and X_2 are normally distributed, $\bar{X}_1$ and $\bar{X}_2$ will each be normally distributed, and so will $\bar{D}$ be. If X_1 and X_2 are not normally distributed, then $\bar{X}_1$ and $\bar{X}_2$ are *approximately* normal as long as n_1 and n_2 are large. Consequently, $\bar{D}$ is normally, or approximately normally, distributed regardless of the distributions of X_1 and X_2.

The statistic $\bar{D}$ has a mean $\mu_1 - \mu_2$, which we shall designate as δ (a Greek letter read as "delta"). It is now our task to estimate δ since the true difference is not known to us. In order to do this, we must know the standard error of $\bar{D}$. As indicated in Sec. 5-4, when two random variables are independent, the variance of their difference is equal to the sum of their variances. Since the statistics $\bar{X}_1$ and $\bar{X}_2$ are two random variables, we have the variance of their difference as

$$\sigma_{\bar{D}}^2 = \sigma_{\bar{X}_1 - \bar{X}_2}^2 = \sigma_{\bar{X}_1}{}^2 + \sigma_{\bar{X}_2}{}^2 = \frac{\sigma_1{}^2}{n_2} + \frac{\sigma_2{}^2}{n_2} \tag{9-9}$$

Thus the following important relationship exists:

If $\bar{X}_1$ is $N\left(\mu_1, \dfrac{\sigma_1{}^2}{n_1}\right)$ and $\bar{X}_2$ is $N\left(\mu_2, \dfrac{\sigma_2{}^2}{n_2}\right)$, and if $\bar{X}_1$ and $\bar{X}_2$ are independent,

$$\text{then } \bar{X}_1 - \bar{X}_2 \text{ is } N\left(\mu_1 - \mu_2, \frac{\sigma_1{}^2}{n_1} + \frac{\sigma_2{}^2}{n_2}\right) \tag{9-10}$$

The standard error of the difference $\bar{D}$ is therefore

$$\sigma_{\bar{D}} = \sigma_{\bar{X}_1 - \bar{X}_2} = \sqrt{\frac{\sigma_1{}^2}{n_1} + \frac{\sigma_2{}^2}{n_2}} \tag{9-11}$$

The confidence interval for the true difference δ based on the two sample means can now be constructed. Since

$$P\left(\bar{D} - z_{\alpha/2}\sqrt{\frac{\sigma_1{}^2}{n_1} + \frac{\sigma_2{}^2}{n_2}} < \delta < \bar{D} + z_{\alpha/2}\sqrt{\frac{\sigma_1{}^2}{n_1} + \frac{\sigma_2{}^2}{n_2}}\right) = 1 - \alpha \tag{9-12}$$

we have the $1 - \alpha$ confidence limits for δ as follows:

$$\bar{D} \pm z_{\alpha/2}\sqrt{\frac{\sigma_1{}^2}{n_1} + \frac{\sigma_2{}^2}{n_2}}$$

Note that it is *not* required that the two samples be of equal size. However, n_1 and n_2 must be sufficiently large; that is, each must be at least 30, and preferably larger, if X_1 and X_2 are not normal.

We shall now use an example to illustrate interval estimation of the difference between two population means. Suppose that we wish to measure the difference in sales between two types of employees in the insurance industry.

One type consists of people who are college graduates, and the other consists of high school graduates only. A random sample of 45 employees is taken from those with a college degree and the mean of monthly sales is 32 (in thousands of dollars), while the mean of a sample of 60 employees with high school diplomas only is 25. The difference between the two sample means is thus 7. It is assumed that the variance of the monthly sales is known to be 48 for college men and 56 for high school graduates. If a 95 percent confidence interval is specified, we can set up the confidence interval for the difference between the average sales of the two types of salesmen as follows:

$$7 - 1.96\sqrt{\frac{48}{45} + \frac{56}{60}} < \delta < 7 + 1.96\sqrt{\frac{48}{45} + \frac{56}{60}}$$

$$7 - 1.96(1.414) < \delta < 7 + 1.96(1.414)$$

thus

$$4.23 < \delta < 9.77$$

That is, with a probability equal to 0.95, these limits established on the basis of the randomly selected samples will include the true difference between the two means.

Exercise 9-4

1 Why is $\bar{D}$, the difference between two sample means, a random variable?

2 How is the standard error of the difference between two sample means expressed? Explain the conditions under which it is derived.

3 A manufacturing company consists of two departments producing identical products. It is suspected that the hourly outputs in the two departments are different. Two random samples of production hours are respectively selected, and the following data are obtained:

	Department 1	*Department* 2
Sample size	$n_1 = 64$	$n_2 = 49$
Sample mean	$\bar{X}_1 = 100$ units	$\bar{X}_2 = 90$ units

The variances of the hourly outputs for the two departments are known to be $\sigma_1{}^2 = 256$ and $\sigma_2{}^2 = 196$, respectively. Find the 95 percent confidence limits for δ, the true difference between the mean outputs of the two departments. What is the point estimate for δ?

4 Suppose you wish to estimate the difference between the effects of a certain sleeping pill on men and women. Let X_1 and X_2 be the number of hours of sleep for men and women, respectively. Two samples are independently taken, and the relevant data are shown below:

	Men	Women
Sample size	$n_1 = 36$	$n_2 = 64$
Sample mean	$\bar{X}_1 = 8.75$	$\bar{X}_2 = 7.25$
Variance	$\sigma_1^2 = 9.0$	$\sigma_2^2 = 4.0$

Determine the 95 and 99 percent confidence limits for δ, the true difference between the mean hours of sleep for men and women.

5 A standard test was given to all high school seniors in two states. Two independent samples of 144 students each are randomly selected, and the mean scores of the two samples are $\bar{X}_1 = 88$ and $\bar{X}_2 = 85$. Assume that the variances of the test scores are known to be 47 and 34, respectively. Find the 95 and 99 percent confidence limits for δ, the true difference between the mean scores of all high school seniors in the two states.

6 In Prob. 5, find:

a The upper limit of the 95 percent confidence interval with the lower limit extending to -100.

b The lower limit of the 99 percent confidence interval with the upper limit extending to $+100$.

7 Suppose you wish to estimate the difference between the weekly wages for machinists and carpenters. Two independent samples of 100 people each are respectively taken, and the relevant data are shown as follows:

	Machinists	Carpenters
Sample size	$n_1 = 100$	$n_2 = 100$
Sample mean	$\bar{X}_1 = 345$	$\bar{X}_2 = 340$
Variance	$\sigma_1^2 = 196$	$\sigma_2^2 = 204$

Determine the 95 and 99 percent confidence limits for δ, the true difference between the average weekly wages for machinists and carpenters.

8 In Prob. 7, determine:

a The upper limit of the 95 percent confidence interval for δ with the lower limit extending to -50.

b The lower limit of the 99 percent confidence interval for δ with the upper limit extending to $+50$.

9-5
Interval Estimation of the Proportion

Another important application of the procedure of interval estimation concerns the estimation of a proportion. Many practical problems are of the two-point model, and have the proportion p as the population parameter. Examples of such problems include many surveys in marketing research that classify the population into two types and seek to determine the proportion p of people in the large universe who favor a certain new product. In sampling, a person who favors the product may be considered a success, any other, a failure. A success is recorded as 1, and a failure as 0. If Y is a random variable with the possible values 0 and 1, and X is a binomial variable defined as $\sum_{i=1}^{n} Y_i$, which is the number of successes in a random sample of n trials, then, for each sample of size n, we obtain a sample proportion $\hat{p}$ defined as

$$\hat{p} = \bar{Y} = \frac{\sum_{i=1}^{n} Y_i}{n} = \frac{X}{n}$$

on the basis of which we can proceed to estimate the population proportion p.

As discussed in Chap. 7, the distribution of $\hat{p}$ has a mean p and a variance $p(1 - p)/n$. Since the variance, and so the standard error, of a sample proportion depends on the population proportion p, which is unknown, the construction of a confidence interval for a proportion becomes more difficult. However, if samples are *large* and by the central limit theorem the sample proportion has an approximately normal distribution, we can apply the method of normal approximation by substituting $\hat{p}(1 - \hat{p})/n$ for $p(1 - p)/n$. Thus we have

$$P\left(\frac{X}{n} - z_{\alpha/2}\sqrt{\frac{\hat{p}(1-\hat{p})}{n}} < p < \frac{X}{n} + z_{\alpha/2}\sqrt{\frac{\hat{p}(1-\hat{p})}{n}}\right) = 1 - \alpha \tag{9-13}$$

and the $1 - \alpha$ confidence limits for p are

$$\frac{X}{n} \pm z_{\alpha/2}\sqrt{\frac{\hat{p}(1-\hat{p})}{n}}$$

The confidence interval for the parameter p can be converted into an interval for μ, the mean of a binomial random variable X, by simply multiplying each term of Eq. (9-13) by n.

$$P\left(n\frac{X}{n} - nz_{\alpha/2}\sqrt{\frac{\hat{p}(1-\hat{p})}{n}} < np < n\frac{X}{n} + nz_{\alpha/2}\sqrt{\frac{\hat{p}(1-\hat{p})}{n}}\right) = 1 - \alpha$$

or

$$P(X - z_{\alpha/2}\sqrt{n\hat{p}(1-\hat{p})} < \mu < X + z_{\alpha/2}\sqrt{n\hat{p}(1-\hat{p})}) = 1 - \alpha \tag{9-14}$$

Note that the quantity $\sqrt{n\hat{p}(1 - \hat{p})}$ is the standard deviation of the binomial random variable X.

As an example, suppose that a manufacturing company wishes to know the proportion of housewives in the United States who would like a centralized vacuum cleaner, given the quality and price. A sample of 100 housewives is drawn at random, and they are interviewed by a marketing researcher from the company. Out of the 100 housewives, 20 indicate that they like the machine and 80 say that they do not. What, then, would be the true proportion p if a 95 percent confidence interval is desired?

For a 95 percent confidence interval, the value of $z_{\alpha/2}$ is again 1.96. Applying Eq. (9-13), we have the approximate confidence interval for p as follows:

$$\frac{20}{100} - 1.96\sqrt{\frac{0.20(1-0.20)}{100}} < p < \frac{20}{100} + 1.96\sqrt{\frac{0.20(1-0.20)}{100}}$$

or

$$0.1216 < p < 0.2784$$

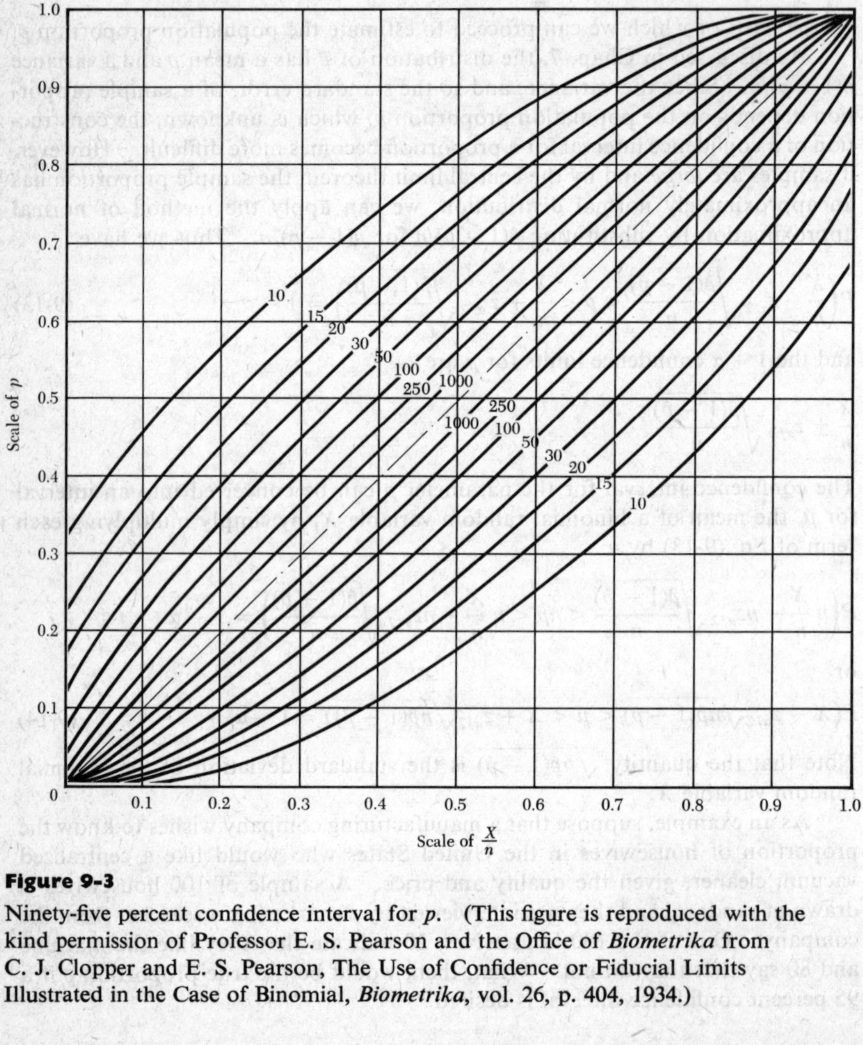

Figure 9-3

Ninety-five percent confidence interval for p. (This figure is reproduced with the kind permission of Professor E. S. Pearson and the Office of *Biometrika* from C. J. Clopper and E. S. Pearson, The Use of Confidence or Fiducial Limits Illustrated in the Case of Binomial, *Biometrika*, vol. 26, p. 404, 1934.)

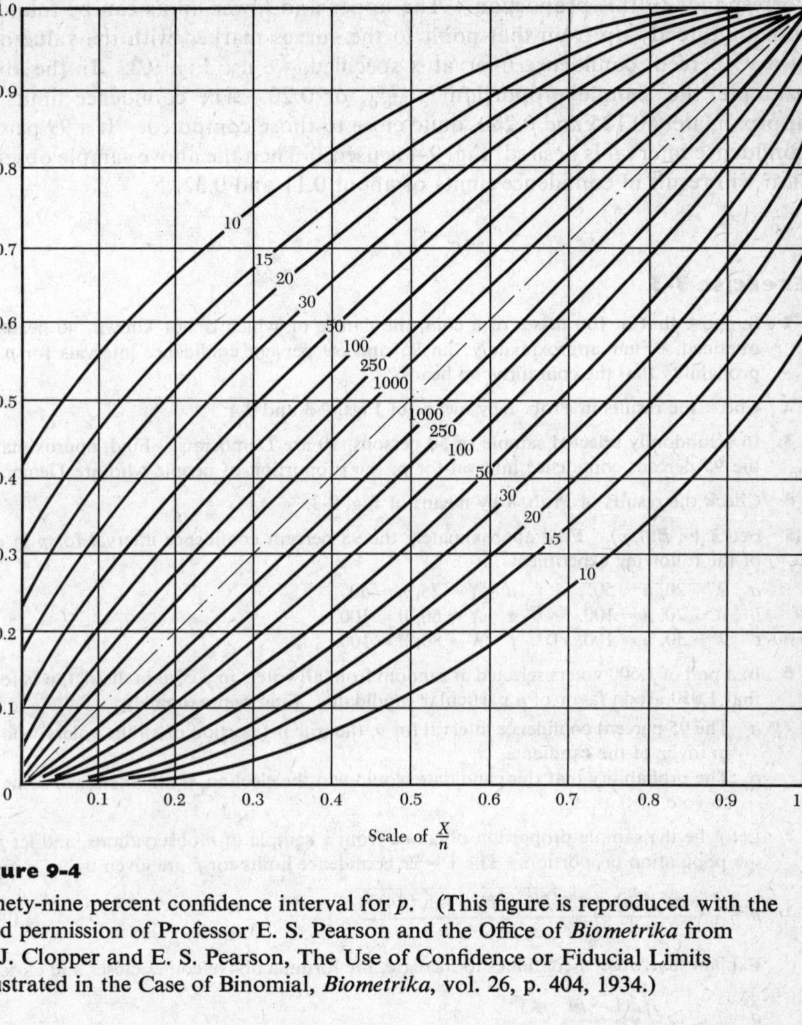

Figure 9-4

Ninety-nine percent confidence interval for p. (This figure is reproduced with the kind permission of Professor E. S. Pearson and the Office of *Biometrika* from C. J. Clopper and E. S. Pearson, The Use of Confidence or Fiducial Limits Illustrated in the Case of Binomial, *Biometrika*, vol. 26, p. 404, 1934.)

That is, with a confidence level of 95 percent, we can conclude that 12.16 to 27.84 percent of the country's housewives are interested in the cleaner.

Observe that interval estimation for p discussed here is, in fact, an approximation of an approximation. Such a confidence interval can be also estimated by the graphic method. If graphs such as those shown in Figs. 9-3 and 9-4 are available, the procedure will be an easy and simple one. First, we find the sample proportion X/n. Then we locate the point on the abscissa that corresponds to this proportion. The upper and lower limits can be found by moving directly up from that point to the curves marked with the value of n. If a 95 percent confidence interval is specified, we use Fig. 9.3. In the above example, the sample proportion is $\frac{20}{100}$, or 0.20. The confidence limits are approximately 0.125 and 0.280, quite close to those computed. If a 99 percent confidence interval is desired, Fig. 9-4 is used. Then the above sample observation will result in confidence limits of about 0.11 and 0.32.

Exercise 9-5

1 Suppose that in 100 tosses of a coin, the nature of which is not known, 40 heads are obtained. Find approximately the 95 and 99 percent confidence intervals for p, the probability that the coin turns up heads.

2 Check the results in Prob. 1 by means of Figs. 9-3 and 9-4.

3 In a randomly selected sample of 50 persons, 30 are Democrats. Find, approximately, the 95 percent confidence interval for p, the proportion of people who are Democrats.

4 Check the results of Prob. 3 by means of Fig. 9-3.

5 Let X be $B(n, p)$. Find approximately the 98 percent confidence interval for p in each of the following experiments:

a $X = 20, n = 50$.
b $X = 20, n = 100$.
c $X = 30, n = 100$.
d $X = 25, n = 50$.
e $X = 60, n = 100$.
f $X = 80, n = 100$.

6 In a poll of 1,500 voters selected at random from all voters in a certain district, it is found that 1,050 are in favor of a particular candidate. Find approximately:

a The 95 percent confidence interval for p, the true proportion of all the voters who are in favor of the candidate.

b The probability that this candidate would win the election, if all voters will come out to vote.

7 Let $\hat{p}$ be the sample proportion obtained from a sample of n observations, and let p be the population proportion. The $1 - 2c$ confidence limits for p are given by

$$p = \frac{\hat{p} + z_c^2/(2n) \pm z_c\sqrt{\hat{p}(1-\hat{p})/n + z_c^2/(4n^2)}}{1 + z_c^2/n} \tag{9-15}$$

Explain that when n continues to increase, the formula above comes closer and closer to

$$p = \hat{p} \pm z_c\sqrt{\frac{\hat{p}(1-\hat{p})}{n}}$$

8 Solve Prob. 1 by Eq. (9-15).

9 Solve Prob. 3 by Eq. (9-15).

10 In a random sample of 100 prospective customers, 50 indicate that they prefer a particular product. Estimate the true value of p of all prospective customers who prefer the product by:

a The usual approximation method.
b Eq. (9-15).

(Assume that the confidence level is set at 95 percent.)

Chapter 10

Statistical inference—hypothesis testing

In Chap. 9 we examined the procedure of estimating population parameters. In dealing with practical problems, however, it is more convenient to apply the techniques of hypothesis testing than to secure solutions within the framework of interval estimation. In this chapter we shall direct our attention to some of the basic rules and concepts of testing hypotheses.

A *statistical hypothesis* is a tentative statement about one or more parameters of a population or a group of populations. We make only *tentative* statements concerning the parameters, or the states of nature, since we are not perfectly certain about their values. Great *uncertainties* usually prevail regarding the states of nature; yet we have to make decisions relative to them. The primary function of inferential statistics is to assist us in reaching sound decisions in spite of such uncertainties. As a matter of fact, this function is of such importance that modern statistics has often been referred to as the "study of decision making in the face of uncertainty."

Generally speaking, *decision making* is the process of making a choice among alternatives. In the discussions in this and the following chapters, decision making refers to the *conclusion* reached with respect to two different hypotheses on the basis of sample observations. Given the level of confidence, we accept a hypothesis if the sample data favor it, or reject the hypothesis if sample data show little evidence to support it. This procedure of decision making through proper statistical tests between two contending hypotheses is what we refer to as *hypothesis testing*.

Before we discuss the methods of testing specific hypotheses, let us first look into the essentials of the general procedure of hypothesis testing.

10-1
Essentials of Hypothesis Testing

In testing hypotheses a number of tasks are indispensable. We shall consider these essential tasks in the various subsections below.

10-1-1 Specification of the form of population distribution

As a first step, the researcher must make an assumption about the nature of the underlying distribution of the population. Is the random variable normal or binomial? Or does it follow any other form of distribution? Many of the frequently encountered problems in statistics relate to the normal law, and the normal distribution is a good approximation for many other distributions. There are numerous problem situations, however, in which normal approximation cannot be applied. It is thus necessary to identify the theoretical probability distribution of the random variable under consideration because the decision about a hypothesis is made on the basis of probabilities of occurrence. Incidentally, a statistical procedure that requires the specification of the probability distribution of the population is called a *parametric* approach, while a *nonparametric* procedure is a distribution-free approach requiring no specification of the underlying population distribution. The nonparametric approach will be examined in Chap. 17.

10-1-2 Formulation of the null and alternative hypotheses

The hypotheses to be tested are formed in terms of selected parameters of the population. For example, we may test the hypothesis that the population mean μ is equal to μ_0 against the alternative hypothesis that μ is not equal to μ_0; that is,

$$H_0: \quad \mu = \mu_0$$
$$H_1: \quad \mu \neq \mu_0$$

where μ_0 is a particular value of the population mean. The symbol H_0 designates the *null hypothesis*, and H_1 the *alternative hypothesis*. Similarly, we may want to test the null hypothesis H_0 that the population proportion $p = p_0$ against the alternative hypothesis $H_1: p \neq p_0$, where p_0 is a particular value of the population proportion. Note that we refer to H_0 as the null hypothesis and H_1 as the alternative hypothesis. Although it is a moot question which of the two is the null hypothesis, in most cases the null hypothesis is the one that asserts the *absence* of any effect claimed for a certain action or treatment. Suppose, for example, that 70 percent of all those suffering a certain disease are cured with drug X, or $p = 0.70$. Assume that a new formula for X is used which is claimed to be more effective, or $p > 0.70$. Then it is a matter of statistical convention that we state $p = 0.70$ as H_0, the null hypothesis, and $p > 0.70$ as H_1, the alternative hypothesis. The former is the "null" hypothesis because it asserts that the new X does *not* have the effect claimed for it.

The reasoning in hypothesis testing is quite similar to the procedure used in a court trial. In trying a man for murder, the accused is presumed to be innocent, or not guilty, until proved guilty. The prosecution collects and presents evidence in an attempt to repudiate the not guilty presumption. If that is not repudiated in the light of all available evidence the accused will remain innocent and there will be no conviction. On the other hand, if his guilt is established by available evidence, there will be a conviction and the not guilty presumption will be repudiated.

The not guilty presumption is the null hypothesis, and a conviction is the alternative hypothesis. In the above example, drug X is the accused. The null hypothesis is that drug X remains 70 percent effective, which implies that the presumption "not guilty" is "X is 70 percent effective." The researcher is the prosecutor; he collects and presents all available evidence from the sample. If there is sufficient evidence to support the alternative hypothesis, which implies "proved guilty" and a conviction warranted, the decision will be to reject the null hypothesis, and the drug will be considered more than 70 percent effective. Otherwise, the null hypothesis is not rejected, and the drug will be considered still 70 percent effective, a not guilty decision.

10-1-3 Selection of the level of significance

Because of the randomness of sample observations it is possible that the sample statistic obtained might deviate so much from expectation when the null hypothesis is true that we decide to reject it. We probably want the probability of rejecting a true null hypothesis to be small. Whatever the magnitude of this probability, it will be called the *level of significance*. The term significance is used because the *difference* between the hypothetical value and the sample outcome is considered to be *significant*, or too great to be attributable to *chance*. In other words, the concept refers to the state of being "statistically significant," and is not used in a functional economic sense.

Once the level of significance is chosen, the *region of rejection*, also called the *critical region*, is decided on. A region of rejection is a set of possible values of the sample statistic that causes you to reject the null hypothesis. When the null hypothesis is, in fact, true and we reject it, we shall, of course, have made an error. This type of error is commonly referred to as *type I error*. The probability of occurrence of such an error is identical with the level of significance. We designate this probability as α.

Suppose that X is $N(\mu,\sigma^2)$, and we wish to test the hypothesis $H_0: \mu = \mu_0$ against $H_1: \mu \neq \mu_0$, and further suppose that 5 percent is chosen as the significance level, that is, $\alpha = 0.05$. We shall reject H_0 if the observed sample mean is among the extreme values in either direction from the hypothetical mean μ_0. The sample values beyond which the null hypothesis will be rejected are called the *critical values*, which are

$$\mu_0 \pm z_{\alpha/2}\,\sigma_{\bar{X}}$$

where $z_{\alpha/2}$ is the standard normal score that cuts off $\alpha/2$ in either tail of the normal curve. Since the probability of rejecting the null hypothesis when it is true is 5 percent, from Appendix VI we find that $z_{0.025} = 1.96$. Thus the probability that the sample mean will fall at or more than 1.96 standard errors *above* the mean is 2.5 percent; the probability that the sample mean will fall at or more than 1.96 standard errors *below* the mean is also 2.5 percent. The region of rejection is the union of the two intervals beyond $\pm 1.96\sigma_{\bar{X}}$ from the mean. This kind of test is known as a two-tailed test, since the region of rejection relates to the two tails of the density curve, shown in Fig. 10-1*a*.

If, instead of $\mu \neq \mu_0$, the alternative is $H_1: \mu > \mu_0$, the region of rejection will be the set of values equal to or more than 1.645 standard errors above the mean, given that the level of significance is specified at 0.05. We shall reject H_0 if the sample mean is found to be 1.645 standard errors *above* the hypothetical mean, or higher. The shaded area in Fig. 10-1*b* shows the rejection region in this case. Similarly, if the alternative hypothesis is $H_1: \mu < \mu_0$, the region of rejection will be the set of values equal to or more than 1.645 standard errors *below* the mean. This is shown in Fig. 10-1*c*. In either case the test is said to be a one-tailed test since only one of the two tails of the density curve relates to the values in the critical region.

The choosing of the level of significance, however, is an arbitrary matter. The most commonly used values for α are 0.05 and 0.01, but occasionally 0.10 or 0.001 is used. The decision on the value of α is an "executive" one and is

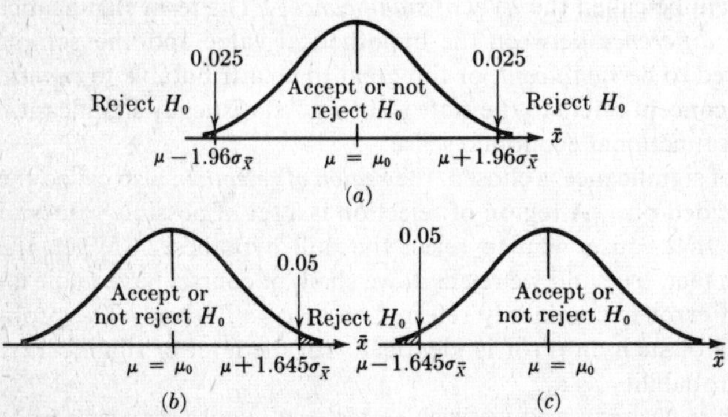

Figure 10-1
Region of rejection for testing H_0: $\mu = \mu_0$ against
(*a*) H_1: $\mu \neq \mu_0$, and (*b*) H_1: $\mu > \mu_0$, and (*c*) H_1: $\mu < \mu_0$.

not a result of statistical analysis. In any event, it must be remembered that a relatively small value should be chosen for α if the cost or risk of erroneously rejecting a true null hypothesis is too high.

The reader must remember that the meaning of α as used in hypothesis testing is quite different from its meaning in interval estimation. In hypothesis testing, it is the probability of rejecting a *true* null hypothesis, while in interval estimation it is the probability that the constructed confidence limits will *not* bracket the true parameter value.

10-1-4 Statistical decision making

With the critical value determined, the observed sample statistic is then compared with such a value. If it is a test of $H_0: \mu = \mu_0$ against $H_1: \mu \neq \mu_0$ at a significance level α, and a decision is made on the basis of a single value $\overline{X}$, the *decision rule* will be

Reject H_0 if $\overline{X} \geq \mu_0 + z_{\alpha/2}\,\sigma_{\overline{X}}$ or $\overline{X} \leq \mu_0 - z_{\alpha/2}\,\sigma_{\overline{X}}$

where $z_{\alpha/2}$ stands for the standard error units beyond which lies the critical region.

On the other hand, if the sample statistic falls within the critical value, we may accept the null hypothesis or withhold decision pending further investigation. With α as the level of significance, the probability of accepting or not rejecting a true H_0 is therefore $1 - \alpha$. If $\alpha = 0.05$, $1 - \alpha = 0.95$, which is shown as the unshaded area under each of the curves in Fig. 10-1. This is the probability of making the correct decision. It is conceivable, however, that a sample statistic may be so close to the hypothetical value that we decide to accept the null hypothesis when it is, in fact, false. If this happens, we shall have been in error. This type of error is referred to as *type II error*. The probability of making a type II error is commonly designated β (a Greek letter read as "beta"). On the other hand, if the sample statistic deviates so much from the hypothetical value that we decide to reject the null hypothesis when it is, in fact, false, we shall have made the correct decision. The probability of making such a correct decision is, of course, $1 - \beta$.

In the test of $H_0: \mu = \mu_0$ against $H_1: \mu < \mu_0$, the probability of making a type II error, or β, is shown under the lower curve of Fig. 10-2. Observe that α and β are not additive, that is, $\alpha + \beta \neq 1$. When the level of significance is increased, the critical value will become closer to the hypothetical value. Thus the shaded area represented by α will increase and the area represented by β will decrease. On the other hand, if the level of significance is reduced, the critical value will move farther away from the hypothetical value. Consequently, the shaded area represented by β will increase and the area represented by α will decrease. In short, the values of α and β will change in the opposite direction.

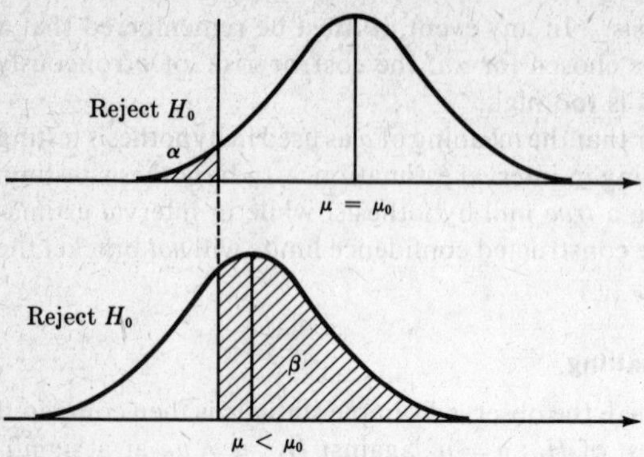

Figure 10-2

Type I and type II errors for testing H_0: $\mu = \mu_0$ against H_1: $\mu < \mu_0$.

We may now summarize the above discussions as follows: In hypothesis testing one and only one of the following four situations will occur:

1. When the null hypothesis is true, we may reject it, and the probability of rejecting a true hypothesis is α.

2. When the null hypothesis is true, we may accept or not reject it, and the probability of accepting a true hypothesis is $1 - \alpha$.

3. When the null hypothesis is false, we may accept it, and the probability of accepting a false hypothesis is β.

4. When the null hypothesis is false, we may reject it, and the probability of rejecting a false hypothesis is $1 - \beta$.

These four situations can be displayed in a tabular form as follows:

	Decision	
State of nature	*Accept H_0*	*Reject H_0*
H_0 true	Correct action ($1 - \alpha$)	Type I error (α)
H_0 false	Type II error (β)	Correction action ($1 - \beta$)

Note that when the null hypothesis is true, only α is involved and β is irrelevant. On the other hand, when the null hypothesis is false, only β is involved. Although α does not come into the picture in the latter case, it nevertheless affects the size of β. For, as pointed out above, the larger the value of α, the smaller the value of β will be, other things being equal.

Exercise 10-1

1 Define the following concepts in your own words as fully as you can:

a Statistical hypothesis. *d* Level of significance.
b Hypothesis testing. *e* Null hypothesis.
c Critical region. *f* Alternative hypothesis.

2 Under what situations do we conduct a two-tailed test? How is the alternative hypothesis stated in a two-tailed test?

3 Under what situations do we conduct a one-tailed test? How is the alternative hypothesis stated in a one-tailed test?

4 Distinguish between the following concepts:

a Type I and type II error.
b Rejection and nonrejection region.
c Statistical estimation and hypothesis testing.
d A test at α and a $1-\alpha$ confidence interval.

5 Explain why a one-tailed test is sometimes called *directional*, while a two-tailed test is referred to as *nondirectional*.

6 Are type I and type II errors *additive*, that is, does $\alpha+\beta=1$? Explain.

7 How is type I error related to type II error? Explain fully.

8 Suppose that X is $N(\mu, \sigma^2)$ and a sample of n observations yields a mean $\bar{X}$. Suppose also that we wish to test the null hypothesis $H_0:\mu=\mu_0$ against the alternatives:

a $H_1:\mu>\mu_0$. *b* $H_1:\mu<\mu_0$. *c* $H_1:\mu\neq\mu$.

What is the decision rule under each of these alternative hypotheses?

9 Suppose that the average typing speed of the secretaries of a certain company is believed to be 55 words per minute. The personnel manager of the company claims that the training program instituted some time ago has increased the typing speed of the secretaries. Formulate the null and alternative hypotheses.

10 Suppose that the tensile strength of steel wires has been known to be *at most* 1,000 pounds. The manufacturer claims that the installation of a new production process has increased the tensile strength of the wires. Formulate the null and alternative hypotheses. If the tensile strength is known to be exactly 1,000 pounds and it is challenged that the strength is different from 1,000 pounds, what are the null and alternative hypotheses?

10-2 Testing Exact Hypotheses

As stated in the beginning of this chapter, a statistical hypothesis is an assertion concerning some of the characteristics or parameters of one or more populations. We may classify hypotheses as exact and inexact. An *exact hypothesis* is one that specifies a unique value for the population parameter, such as $H: p = 0.5$ or $H: \mu = 65$. An *inexact hypothesis* is one which states that the population parameter falls within an interval of values, such as $H: p > 0.8$ or $H: \mu \neq 100$.

In this section we shall examine the problem of testing exact hypotheses with respect to parameters of both the normal and the binomial distributions.

10-2-1 Test of hypotheses about the mean

Let us first consider the procedure of testing the hypothesis about a population mean.

Suppose, for example, that the IQs of college women are suspected of averaging about 5 points higher than the mean IQ of 110 for all students in a certain college. Let X be the IQ of any female student, and a sample of 64 female students $(X_1, X_2, \ldots, X_{64})$ be drawn at random. It is assumed that the standard deviation of X is known to be 20 since this is approximately the case from past experiences, or from similar studies in other parts of the academic world. We know that human traits such as intelligence are normally distributed, and that the sample mean $\bar{X}$ has a normal distribution, or approximately so with $n = 64$, even if X should not be normal.

Let μ be the mean of X; then our problem is to decide between

$$H_0\colon \quad \mu = 110 \qquad H_1\colon \quad \mu = 115$$

where H_0 is the null hypothesis and H_1 is the alternative hypothesis. It is assumed that the two exhaust all the possibilities, and either one or the other must be true. It is obvious that both are exact hypotheses since a unique value for the population parameter is hypothesized under each of the two competing hypotheses. From our discussion in Chap. 9, we know that the sample mean $\bar{X}$ is a good estimator of the population mean. It is therefore appropriate to use $\bar{X}$ to test H_0 against H_1. Thus we can apply the standardized normal deviate

$$Z = \frac{\bar{X} - \mu}{\sigma_{\bar{X}}}$$

for evaluating the probability of occurrence of the sample mean. The denominator $\sigma_{\bar{X}}$ is the standard error and is equal to $\sigma_X/\sqrt{n} = 20/\sqrt{64} = 2.5$; and $\mu = 110$, the hypothetical mean. This is a one-tailed test since we shall not reject H_0 in favor of H_1 if $\bar{X}$ turns out to be unexpectedly small. Let us use 0.05 as the value of α. From Sec. 8-2 we know that the standardized normal score that cuts off the upper 5 percent of the normal curve is 1.645. Thus the decision rule is to reject H_0 if $Z \geq 1.645$. That is, reject H_0 if

$$\frac{\bar{X} - 110}{2.5} \geq 1.645$$

Multiplying each side of this inequality by 2.5 and then adding 110, we can express the decision rule as follows:

Reject H_0 if $\bar{X} \geq 110 + 1.645(2.5) = 114.1125$

This is the case because the probability of $\bar{X}$'s being equal to or greater than 114.1125 is 0.05 when H_0 is true; that is, $P(\bar{X} \geq 114.1125 \mid \mu = 110) = 0.05$. Therefore, if $\bar{X}$ is equal to or greater than 114.1125, we shall accept the alternative hypothesis that the female students in this college have a mean IQ of 115. If $\bar{X}$ is smaller than 114.1125, we may accept the null hypothesis or withhold judgment pending further investigation. In other words, we simply do not reject the null hypothesis if the observed sample mean $\bar{X}$ is, say, 114, lower than 114.1125.

The shaded area in Fig. 10-3*a* relates to the region of rejection, given that the null hypothesis is true. If the alternative hypothesis is, in fact, true, the probability of rejecting H_0 is greater since the sample would have been taken from a population of students with higher average IQs. To be exact, this probability is equal to

$$P(\bar{X} \geq 114.1125 \mid \mu = 115) = P\left(Z \geq \frac{114.1125 - 115}{2.5}\right) = P(Z \geq -0.355)$$

$$= P(Z \leq 0.355) = 0.6387$$

The probability of accepting the null hypothesis when it is wrong, that is, making a type II error, is $1 - 0.6387 = 0.3613$. Observe that when the alternative hypothesis is true, the distribution of $\bar{X}$ is still normal with the same standard error $\sigma_{\bar{X}} = 2.5$.

The shaded area in Fig. 10-3*b* is of importance to us for later discussions in this chapter. This probability, which is commonly designated $1 - \beta$, is referred to as the *power* of the statistical test. It is the probability of rejecting the null hypothesis when the decision rule and the true parameter under the alternative hypothesis are specified. The concept of "power" as used here is similar to that of a telescope. It reflects the capacity of a test in finding out that the true

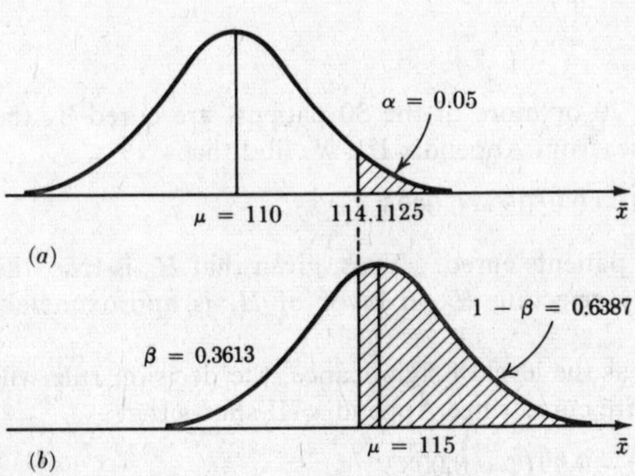

Figure 10-3
Sampling distributions of sample mean $\bar{X}$ for $\sigma_{\bar{X}} = 2.5$ showing the probability of rejecting H_0 for (*a*) $\mu = 110$ and (*b*) $\mu = 115$.

state of nature is different from the one under the null hypothesis. A high-powered telescope enables us to distinguish at a great distance, for example, a gunboat from an ordinary passenger ship on the high seas. Likewise, a high-powered statistical test can give us great assurance in concluding that the null hypothesis is wrong when it is in fact wrong. Statisticians often evaluate a statistical test in terms of the power of the test. In general, given the size of α, the higher the power, that is, the higher the value of $1 - \beta$, the better the test will be. More will be said about the power of a test in Sec. 10-4.

10-2-2 Test of hypotheses about the proportion

We now turn our attention to tests of hypotheses about the population proportion. The procedure of making decisions between two hypotheses about a proportion is similar to that of deciding between two hypotheses about a mean, especially when the sample size is large. First, let us consider a situation in which probabilities are evaluated by using the binomial distribution, and then normal approximation to the binomial will be considered.

Suppose that we have two contending theories about a certain drug. According to one, 50 percent of those who contracted a given disease are cured by applying the drug; according to the other, 80 percent are cured. These two theories can be stated as two hypotheses as follows:

$$H_0\colon\ p = 0.50 \qquad H_1\colon\ p = 0.80$$

Suppose that the level of significance is chosen at 0.0495. (Usually, the value of α consists of only two decimal digits such as 0.01 or 0.05. The probability values in Appendix III contain four decimal digits each. Accordingly, for convenience sake, we shall use values with four decimal digits for α when Appendix III is used for testing hypotheses.) If a sample of 30 patients is taken at random from all those who have used the drug, from which the sample proportion $\hat{p}$ is computed, the decision rule will be

$$\text{Reject } H_0 \text{ if } \hat{p} \geq \frac{2}{3}$$

That is, we shall reject H_0 if 20 or more of the 30 patients are cured by the drug. This is the case because, from Appendix III, we find that

$$P(X \geq 20 \mid n = 30, p = 0.5) = 1 - 0.9505 = 0.0495$$

where X is the number of the patients cured. Thus, given that H_0 is true, the probability of our erroneously rejecting H_0 in favor of H_1 is approximately 0.05.

If we have chosen 0.0082 as the level of significance, the decision rule will be to reject H_0 if 22 or more are cured since Appendix III shows that

$$P(X \geq 22 \mid n = 30, p = 0.5) = 1 - 0.9918 = 0.0082$$

Observe that in this case the lower the level of significance, the greater will be the sample proportion required to reject the null hypothesis.

The decision rule suggests that we shall accept H_0 if X is less than 20 at the 0.0495 level of significance. Thus if H_1 is, in fact, true, the probability of erroneously accepting H_0 is

$$P(X \leq 19 \mid n = 30, p = 0.8) = 0.0256$$

which is the value of β. The power of this test is $1 - 0.0256 = 0.9744$. If 0.0082 is chosen as the significance level, the probability of accepting H_0 when H_1 is true will be

$$P(X \leq 21 \mid n = 30, p = 0.8) = 0.1287$$

and the power of the test will be reduced to $1 - 0.1287 = 0.8713$.

Let us now examine the problem of testing hypotheses about a proportion by the use of a normal approximation. In the above example the random variable X is $B(30,0.5)$ if the null hypothesis is true. By Eqs. (7-7) and (7-9), the mean and variance of X are computed as follows:

$$\mu = np = 30(0.5) = 15$$

and

$$\sigma_X{}^2 = npq = 30(0.5)(0.5) = 7.5$$

X is therefore approximately normal with a mean 15 and a standard deviation $\sqrt{7.5} = 2.7386$. The hypotheses to be tested are now stated as

$$H_0\colon \quad \mu = 15 \qquad H_1\colon \quad \mu = 24$$

With $\alpha = 0.05$, the decision rule will be

$$\text{Reject } H_0 \text{ if } X \geq \mu + z_\alpha \sigma_X = 15 + 1.645(2.7386) = 19.5$$

where the condition is equivalent to "X is larger than or equal to 20," the same result as obtained above.

The method of normal approximation can also be used to test the hypothesis about a population proportion on the basis of a sample proportion. Let $\hat{p} = X/n$; then we know from Eqs. (7-12) and (7-13) that the mean and standard deviation of $\hat{p}$ are, respectively,

$$E(\hat{p}) = p \qquad \text{and} \qquad \sigma_{\hat{p}} = \sqrt{\frac{pq}{n}}$$

In the above example, the sampling distribution of $\hat{p}$ has a mean

$$E(\hat{p}) = 0.5$$

and a standard deviation

$$\sigma_{\hat{p}} = \sqrt{\frac{(0.5)(0.5)}{30}} = \frac{0.5}{\sqrt{30}} \approx 0.09$$

given that the null hypothesis is true. Again, with $\alpha = 0.05$, the decision rule becomes

Reject H_0 if $\hat{p} \geq p + z_\alpha \sigma_{\hat{p}} = 0.5 + 1.645(0.09) = 0.65$

which is identical with the result obtained by normal approximation to the binomial. In other words, we shall reject the null hypothesis that $p = 0.5$ in favor of the alternative hypothesis that $p = 0.8$ if the standardized normal deviate

$$Z = \frac{\hat{p} - 0.5}{0.09}$$

is equal to or greater than 1.645.

The basic concepts of error in statistical inference and of the power of a statistical test have now been briefly examined. The situations considered, however, are quite artificial. In actual research it is virtually nonexistent to call for a decision between two exact hypotheses as illustrated. What is widely practiced is testing two competing hypotheses, at least one of which is inexact. In the next section we shall consider the more realistic situations of testing inexact hypotheses.

Exercise 10-2

1 Distinguish between the following concepts:

a The probability of type I error α and the error of estimation α.
b Exact and inexact hypotheses.
c The power of a statistical test and the value of $1 - \alpha$.

2 Suppose that X is $B(30, p)$, and the following tests are conducted on the basis of an observation of X. Find the type I and type II errors:

	H_0	H_1	*Decision rule*
a	$p = 0.3$	$p = 0.5$	Reject H_0 if $X \geq 13$
b	$p = 0.5$	$p = 0.7$	Reject H_0 if $X \geq 18$
c	$p = 0.7$	$p = 0.5$	Reject H_0 if $X \leq 10$
d	$p = 0.1$	$p = 0.5$	Reject H_0 if $X \geq 5$
e	$p = 0.4$	$p = 0.3$	Reject H_0 if $X \leq 7$

3 Let X be $N(\mu, 100)$ and $\bar{X}$ be the mean of a random sample of 16 observations of X, Given that $\bar{X} = 14$, test the null hypothesis $H_0: \mu = 10$ against each of the following alternatives:

a $H_1: \mu = 13$ at $\alpha = 0.05$. *b* $H_1: \mu = 13$ at $\alpha = 0.01$.

4 Find the power of the test under each of the alternative hypothesis in Prob. 3.

5 Let X stand for the IQs of the high school students in a certain district and X be $N(\mu, 225)$. Suppose that a random sample of 25 students gives a mean $\bar{X} = 106$. Should we conclude that the true mean of the IQs of the students in the district is 100 against the alternative that it is 110 at $\alpha = 0.01$? Find the power of this test.

6 In Prob. 5, should we reject the null hypothesis at $\alpha = 0.05$? What is the power of this test?

7 Let X designate the defective parts produced by an automatic machine. From a randomly selected sample of 50 parts, 10 are defective. Let p be the true proportion of all the parts that are defective; test the null hypothesis $H_0: p = 0.1$ against the alternatives:

a $H_1: p = 0.2$ at $\alpha = 0.0094$. *b* $H_1: p = 0.2$ at $\alpha = 0.0579$.

8 In Prob. 7, what is the power of the test under each of the alternative hypotheses?

9 A supplier sells natural fibers to a leading manufacturing company. He claims that the fibers have a mean strength of 33 pounds and a variance of 64 pounds. A random sample of 25 fibers selected from a shipment yields a mean strength of 30 pounds. The buyer contends that the mean strength of all the fibers in the shipment is 30 pounds instead of 33 pounds as claimed. Do you reject the claim of the supplier at $\alpha = 0.05$? What is the power of this test? (*Hint:* H_0 is $\mu = 33$ pounds.)

10 A coin, the nature of which is unknown, is tossed 30 times, and 20 heads are obtained. Let p be the probability that the coin will turn up heads. Test the null hypothesis that $p = 0.5$ against the alternative that $p = 0.8$ at $\alpha = 0.0082$. Also, at $\alpha = 0.0495$.

11 Find the power of each of the two tests in Prob. 10.

12 Let p be the probability that a person who receives a certain vaccine will survive the winter without a cold. Let X be the number of persons who survive the winter without a cold in a sample of 100 persons who have received the vaccine. Test $H_0: p = 0.5$ against the alternative:

a That $p = 0.4$ at $\alpha = 0.05$ with $X = 45$.
b That $p = 0.4$ at $\alpha = 0.05$ with $X = 38$.
c That $p = 0.7$ at $\alpha = 0.05$ with $X = 67$.

13 The life length of a certain brand of fluorescent bulbs is believed to have a mean $\mu = 1{,}500$ hours and a standard deviation $\alpha = 160$ hours. The manufacturing company claims that the average length of service life of the bulbs is 1,600 hours. A random sample of 64 bulbs yields a sample mean $\bar{X} = 1{,}540$ hours. Does the evidence support the manufacturer's claim at $\alpha = 0.02$?

14 The proportion of TV viewers who watch a particular program on Saturday night was previously found to be 50 percent. It is suspected that the proportion has recently reduced to 40 percent. A sample of 100 viewers is randomly selected and interviewed, and 45 of them are found to watch the program. If the level of significance is specified at 0.025, shall we conclude that the proportion has indeed reduced to 40 percent?

10-3
Testing Inexact Hypotheses

Attention in this section turns to the test of inexact hypotheses. An *inexact hypothesis* is a statement that encompasses an interval of possible parameter values. For the present discussion, we shall continue to assume that the variance of the parent population is known.

Suppose that we wish to test the null hypothesis that a given parameter θ is equal to or less than a specific value θ_0 against the alternative hypothesis that θ is larger than θ_0; that is,

$$H_0\colon\ \theta \le \theta_0 \qquad H_1\colon\ \theta > \theta_0$$

Suppose also that a sample of n observations yields a sample statistic $\hat{\theta}$, on the basis of which a decision is made between H_0 and H_1 with α as the level of significance. Since H_0 is an inexact hypothesis, the sampling distribution of $\hat{\theta}$ is not defined. That being the case, how can we set up the region of rejection or choose the rule of decision?

In situations such as this, we may choose the decision rule *as if* the null hypothesis were an exact one, that is, $H_0\colon \theta = \theta_0$. We may take as the region of rejection all values of $\hat{\theta}$ equal to or greater than a certain value that cuts off the upper α of the sampling distribution, given $\theta = \theta_0$. As shown in II and III of Fig. 10-4, for any value of θ less than θ_0, the probability of rejecting the null hypothesis when it is true will be less than α. Thus if we reject $H_0\colon \theta = \theta_0$ with an error of α, we shall reject with even smaller error any hypothesis that $\theta < \theta_0$. This is to say that if we can be $1 - \alpha$ confident that we are not making any error in rejecting $H_0\colon \theta = \theta_0$, we should be more confident in rejecting other hypotheses that $\theta < \theta_0$. On the other hand, if $H_0\colon \theta \le \theta_0$ is false, that is, if the value of θ is actually larger than θ_0, the power of the test would not be smaller, and the type II error could not be larger, than when the null hypothesis

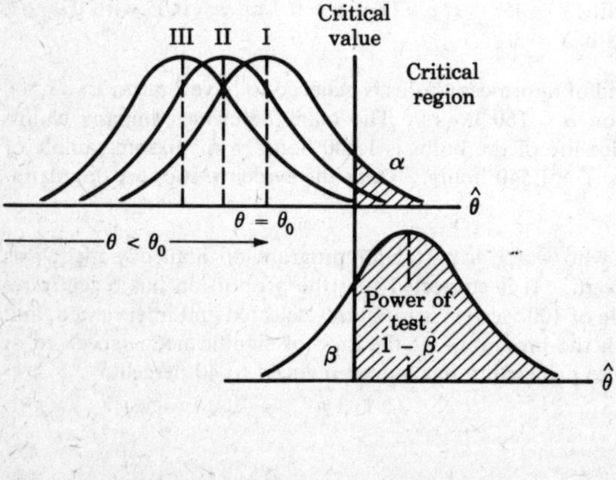

Figure 10-4
Distributions I, II, and III under $H_0\colon\ \theta = \theta_0$ showing that α is not increased by treating $\theta \le \theta_0$ as $\theta = \theta_0$ and that the power of the test is not reduced by doing so.

is true. For these reasons, we shall in this section consider only situations in which the null hypothesis is exact since a test for deciding between $H_0: \theta = \theta_0$ and $H_1: \theta > \theta_0$ (or $\theta < \theta_0$) can be used to replace another test for deciding between $H_0: \theta \leq \theta_0$ (or $\theta \geq \theta_0$) and $H_1: \theta > \theta_0$ (or $\theta < \theta_0$) without increasing the error or reducing the power of the test.

10-3-1 Test of hypotheses about the mean

Recall that in Sec. 10-2-1 the test was to decide between the null hypothesis that the mean of female students' IQs is 110 and the alternative hypothesis that the mean is 115. However, we may be interested in the question whether the mean of their IQs is more than 110 rather than whether it is exactly 115. That is, we may wish to decide between

$$H_0: \quad \mu = 110 \qquad H_1: \quad \mu > 110$$

The procedure of this test is identical with that in Sec. 10-2-1. The only difference is that when we decide to reject the null hypothesis, we accept the alternative hypothesis that $\mu > 110$ rather than $\mu = 115$. Given that $\alpha = 0.05$, $n = 64$, and $\sigma_X = 20$, the decision rule will remain

Reject H_0 if $\bar{X} \geq 110 + 1.645(2.5) = 114.1125$

or if the standardized normal deviate

$$Z = \frac{\bar{X} - 110}{2.5}$$

is larger than 1.645. When this happens, we shall conclude that the mean of female students' IQs is higher than the average 110 for all students, with the level of significance at 5 percent. Similarly, the decision rule for $\alpha = 0.01$ will be

Reject H_0 if $\bar{X} \geq 110 + 2.326(2.5) = 115.815$

or if the standardized normal deviate

$$Z = \frac{\bar{X} - 110}{2.5}$$

is larger than 2.326. If this happens, the mean of their IQs would be taken as higher than 110, with a significance level at only 1 percent.

In the general case, when we wish to decide between $H_0: \mu = \mu_0$ and $H_1: \mu > \mu_0$ on the basis of a sample mean $\bar{X}$, we shall reject H_0 if $\bar{X}$ is larger than or equal to the critical value c_1, a sufficiently large constant. If the test is at a significance level α, we should choose c_1 such that

$$P(\bar{X} \geq c_1 \mid \mu = \mu_0) = \alpha$$

In similar manner, when we wish to decide between $H_0: \mu = \mu_0$ and $H_1: \mu < \mu_0$, we should choose a small value for c_2 such that

$$P(\bar{X} \le c_2 | \mu = \mu_0) = \alpha$$

On the other hand, if we have a two-tailed test, that is, if the alternative hypothesis is stated as $\mu \ne \mu_0$, we should choose two critical values c_3 and c_4 such that

$$P(\bar{X} \ge c_3 \cup \bar{X} \le c_4 | \mu = \mu_0) = \alpha$$

Let X be a normal random variable, which has a known standard deviation σ_X. On the basis of the mean of a sample of size n, we wish to decide between $H_0: \mu = \mu_0$ and some alternative hypothesis such as any of the three mentioned above. Then the decision rule and the respective critical value will appear as in Table 10-1.

If the decison rule is based on the standardized normal score, the above tests for choosing between $H_0: \mu = \mu_0$ and each of the alternative hypotheses can be stated as in Table 10-2.

10-3-2 Test of hypotheses about the difference between two means

A frequently encountered problem in inferential statistics is the determination of whether two samples are taken from the same distribution. This kind of problem is dealt with by testing the hypothesis that the difference between the means of the two populations from which the two samples are drawn is 0.

Suppose, for example, that a certain manufacturing company employs both men and women doing the same type of work. We suspect that the average

Table 10-1

Tests of $H_0: \mu = \mu_0$ against H_1 when X is $N(\mu, \sigma_x^2)$

H_1	Decision rule: Reject H_0 if	Level of significance $\alpha = 0.05$	$\alpha = 0.01$
$\mu > \mu_0$	$\bar{X} \ge c_1$	$c_1 = \mu_0 + 1.645 \frac{\sigma_x}{\sqrt{n}}$	$c_1 = \mu_0 + 2.326 \frac{\sigma_x}{\sqrt{n}}$
$\mu < \mu_0$	$\bar{X} \le c_2$	$c_2 = \mu_0 - 1.645 \frac{\sigma_x}{\sqrt{n}}$	$c_2 = \mu_0 - 2.326 \frac{\sigma_x}{\sqrt{n}}$
$\mu \ne \mu_0$	$\bar{X} \ge c_3$	$c_3 = \mu_0 + 1.960 \frac{\sigma_x}{\sqrt{n}}$	$c_3 = \mu_0 + 2.575 \frac{\sigma_x}{\sqrt{n}}$
	or		
	$\bar{X} \le c_4$	$c_4 = \mu_0 - 1.960 \frac{\sigma_x}{\sqrt{n}}$	$c_4 = \mu_0 - 2.575 \frac{\sigma_x}{\sqrt{n}}$

Table 10-2

Tests of H_0: $\mu = \mu_0$ against H_1 when X is $N(\mu, \sigma_X^2)$

H_1	Decision rule: $\alpha = 0.05$ Reject H_0 if	Decision rule: $\alpha = 0.01$ Reject H_0 if
$\mu > \mu_0$	$\dfrac{\bar{X} - \mu_0}{\sigma_X/\sqrt{n}} \geq 1.645$	$\dfrac{\bar{X} - \mu_0}{\sigma_X/\sqrt{n}} \geq 2.326$
$\mu < \mu_0$	$\dfrac{\bar{X} - \mu_0}{\sigma_X/\sqrt{n}} \leq -1.645$	$\dfrac{\bar{X} - \mu_0}{\sigma_X/\sqrt{n}} \leq -2.326$
$\mu \neq \mu_0$	$\dfrac{\bar{X} - \mu_0}{\sigma_X/\sqrt{n}} \geq 1.960$ or $\dfrac{\bar{X} - \mu_0}{\sigma_X/\sqrt{n}} \leq -1.960$	$\dfrac{\bar{X} - \mu_0}{\sigma_X/\sqrt{n}} \geq 2.575$ or $\dfrac{\bar{X} - \mu_0}{\sigma_X/\sqrt{n}} \leq -2.575$

hourly output of women is higher than that of men. This problem can be handled by testing the hypothesis that the average hourly output is the same for both types of employees.

Let X_1 be the hourly output of a female employee, and X_2 be that of a male employee; also, μ_1 and μ_2 be the means and σ_1^2 and σ_2^2 the variances for X_1 and X_2, respectively. We wish to decide between the two hypotheses

$$H_0\colon\ \mu_1 - \mu_2 = 0 \qquad H_1\colon\ \mu_1 - \mu_2 > 0$$

Suppose that two separate samples are drawn at random, which result in two sample means $\bar{X}_1$ and $\bar{X}_2$. We shall try to determine whether the difference $\bar{D}$ between these sample means, or $\bar{X}_1 - \bar{X}_2$, is significantly deviant from 0.

To do this, we must first know the sampling distribution of $\bar{D}$. As we have repeatedly pointed out, the sampling distribution of $\bar{X}_1$ and that of $\bar{X}_2$ are normal or approximately so with means μ_1 and μ_2 and variances σ_1^2/n_1 and σ_2^2/n_2, respectively, regardless of the distributions of X_1 and X_2, as long as the sample sizes n_1 and n_2 are large. Thus $\bar{D}$, the linear combination of $\bar{X}_1$ and $\bar{X}_2$, must be normally distributed or nearly so. If the null hypothesis is true, $\bar{D}$ should have a mean

$$\mu_{\bar{D}} = \mu_1 - \mu_2 = 0$$

and, due to the independence of X_1 and X_2, a standard error

$$\sigma_{\bar{D}} = \sigma_{\bar{X}_1 - \bar{X}_2} = \sqrt{\frac{\sigma_1^2}{n_1} + \frac{\sigma_2^2}{n_2}}$$

The standardized normal deviate is therefore given by

$$Z = \frac{\bar{X}_1 - \bar{X}_2 - (\mu_1 - \mu_2)}{\sigma_D} = \frac{\bar{X}_1 - \bar{X}_2}{\sqrt{\sigma_1^2/n_1 + \sigma_2^2/n_2}} \tag{10-1}$$

on the basis of which we evaluate the probability of occurrence of the difference between the sample means. With α as the significance level, the decision rule will be

$$\text{Reject } H_0 \text{ if } \bar{D}, \text{ or } \bar{X}_1 - \bar{X}_2, \geq z_\alpha \sqrt{\frac{\sigma_1^2}{n_1} + \frac{\sigma_2^2}{n_2}}$$

Suppose that both X_1 and X_2 have the same variance $\sigma_1^2 = \sigma_2^2 = 32$. Suppose also that the two samples of 36 employees each yield $\bar{X}_1 = 65$ and $\bar{X}_2 = 58$; that is, the difference $\bar{D}$ is $65 - 58 = 7$. Can we conclude that at $\alpha = 0.05$ such a difference is not significant and the two samples of employees have been taken from the same distribution, so far as their hourly output is concerned? To answer this question, we compute the critical value, which is found to be $1.645\sqrt{\frac{32}{36} + \frac{32}{36}} = 2.193$. According to the decision rule, we shall reject H_0 if the observed difference $\bar{D}$ is larger than or equal to this number. We shall therefore reject H_0 since the obtained difference 7 is much larger than 2.193. Alternatively, by Eq. (10-1) we compute the standardized normal deviate as follows:

$$Z = \frac{65 - 58}{\sqrt{\frac{32}{36} + \frac{32}{36}}} = \frac{7}{\sqrt{\frac{64}{36}}} = \frac{7}{\frac{8}{6}} = \frac{42}{8} = 5.25$$

which is far greater than the standard score 1.645 above which lies the critical region. Accordingly, we reject the null hypothesis and accept the alternative that the average output of women is higher than that of men.

Table 10-3

Tests of H_0: $\mu_1 - \mu_2 = 0$ against H_1 when X_1 is $N(\mu_1, \sigma_1^2)$ and X_2 is $N(\mu_2, \sigma_2^2)$

H_1	Decision rule: Reject H_0 if	Level of significance $\alpha = 0.05$	$\alpha = 0.01$
$\mu_1 - \mu_2 > 0$	$\bar{X}_1 - \bar{X}_2 \geq c_1$	$c_1 = 1.645\sqrt{\frac{\sigma_1^2}{n_1} + \frac{\sigma_2^2}{n_2}}$	$c_1 = 2.326\sqrt{\frac{\sigma_1^2}{n_1} + \frac{\sigma_2^2}{n_2}}$
$\mu_1 - \mu_2 \neq 0$	$\bar{X}_1 - \bar{X}_2 \geq c_3$	$c_3 = 1.960\sqrt{\frac{\sigma_1^2}{n_1} + \frac{\sigma_2^2}{n_2}}$	$c_3 = 2.575\sqrt{\frac{\sigma_1^2}{n_1} + \frac{\sigma_2^2}{n_2}}$
	or		
	$\bar{X}_1 - \bar{X}_2 \leq c_4$	$c_4 = -1.960\sqrt{\frac{\sigma_1^2}{n_1} + \frac{\sigma_2^2}{n_2}}$	$c_4 = -2.575\sqrt{\frac{\sigma_1^2}{n_1} + \frac{\sigma_2^2}{n_2}}$

Table 10-4

Tests of $H_0: \mu_1 - \mu_2 = 0$ against H_1 when X_1 is $N(\mu_1, \sigma_1^2)$ and X_2 is $N(\mu_2, \sigma_2^2)$

	Decision rule	
H_1	$\alpha = 0.05$ *Reject H_0 if*	$\alpha = 0.01$ *Reject H_0 if*
$\mu_1 - \mu_2 > 0$	$\dfrac{\bar{X}_1 - \bar{X}_2}{\sqrt{\sigma_1^2/n_1 + \sigma_2^2/n_2}} \geq 1.645$	$\dfrac{\bar{X}_1 - \bar{X}_2}{\sqrt{\sigma_1^2/n_1 + \sigma_2^2/n_2}} \geq 2.326$
$\mu_1 - \mu_2 \neq 0$	$\dfrac{\bar{X}_1 - \bar{X}_2}{\sqrt{\sigma_1^2/n_1 + \sigma_2^2/n_2}} \geq 1.960$ or $\dfrac{\bar{X}_1 - \bar{X}_2}{\sqrt{\sigma_1^2/n_1 + \sigma_2^2/n_2}} \leq -1.960$	$\dfrac{\bar{X}_1 - \bar{X}_2}{\sqrt{\sigma_1^2/n_1 + \sigma_2^2/n_2}} \geq 2.575$ or $\dfrac{\bar{X}_1 - \bar{X}_2}{\sqrt{\sigma_1^2/n_1 + \sigma_2^2/n_2}} \leq -2.575$

There are situations in which a different alternative hypothesis is called for. We may, for instance, want to test $H_0: \mu_1 - \mu_2 = 0$ against $H_1: \mu_1 - \mu_2 < 0$. This is also a one-tailed test. For such a problem, we may simply apply the above procedure with X_1 and X_2 replaced by each other. We may also have a two-tailed test with the alternative hypothesis stated as $\mu_1 - \mu_2 \neq 0$.

In general, if X_1 and X_2 are two normal random variables with known variances σ_1^2 and σ_2^2 and if we wish to test the null hypothesis $H_0: \mu_1 - \mu_2 = 0$ against one of the above-mentioned alternative hypotheses on the basis of two samples of sizes n_1 and n_2 that yield two means $\bar{X}_1$ and $\bar{X}_2$, the decision rule and the critical values will be as in Table 10-3.

As in Sec. 10-3-1, the alternative hypotheses mentioned above can be stated in terms of the standard normal deviate, as shown in Table 10-4.

10-3-3 Test of hypotheses about the proportion

In Sec. 10-2-2 we examined the problem of testing the hypothesis $H_0: p = 0.50$ against $H_1: p = 0.80$ with $n = 30$. Here again we may be interested in choosing between $H_0: p = 0.50$ and the inexact hypothesis $H_1: p > 0.50$. Given the level α, the decision rule will remain the same as before. That is, we shall reject H_0 if X is larger than or equal to 20 at $\alpha = 0.0495$, or 22 at $\alpha = 0.0082$.

Of course, there are situations that call for testing an exact null hypothesis against some other alternative hypotheses. If p_0 is some value that a population proportion may assume, we may wish to test $H_0: p = p_0$ against $H_1: p < p_0$ or against $H_1: p \neq p_0$. Let X be a binomial random variable with parameters n and p. The tests for deciding between $H_0: p = p_0$ and the specified alternative hypotheses are shown in Table 10-5.

Table 10-5

Tests of $H_0: p = p_0$ against H_1 when X is $B(n,p)$

H_1	Decision rule: Reject H_0 if	Level of significance
$p > p_0$	$X \geq c_1$	$P(X \geq c_1 \mid p = p_0) = \alpha$
$p < p_0$	$X \leq c_2$	$P(X \leq c_2 \mid p = p_0) = \alpha$
$p \neq p_0$	$X \geq c_3$	$P(X \geq c_3 \mid p = p_0) = \alpha/2$
	or	
	$X \leq c_4$	$P(X \leq c_4 \mid p = p_0) = \alpha/2$

The expressions in the last column of Table 10-5 are referred to Appendix III for finding the c values. For any given values of n and p and a specified level of significance, the critical value c can be taken directly from such a table. For example, for $p_0 = 0.50$, $n = 30$, and $\alpha \approx 0.05$, we go to Appendix III and find that $c_1 = 20$, $c_2 = 10$, and c_3 and c_4 are 21 and 9, respectively.

Very few tables are available for n larger than 100, and the evaluation of probabilities for large n can be quite tedious and time consuming. Fortunately, for large n, we can find the probabilities for a binomial sampling distribution by the method of normal approximation. Also, the Poisson exponential distribution is found to be an excellent approximation in many cases.

If Y is a Bernoulli random variable with the possible values 0 and 1, and if a random sample of $Y_1, Y_2, \ldots, Y_n$ is taken, we have a random variable $\bar{Y} = \sum_{i=1}^{n} Y_i/n$, previously referred to as the sample proportion $\hat{p}$. When n is large, by the central limit theorem $\bar{Y}$, or $\hat{p}$, is approximately normal with mean p and variance pq/n. We may apply the method of normal approximation to choose between $H_0: p = p_0$ and $H_1: p > p_0$ on the basis of $\hat{p}$. When H_0 is true, or $p = p_0$, the standard error of $\hat{p}$ is $\sqrt{p_0 q_0/n}$. Given a level α, our decision rule will be

$$\text{Reject } H_0 \text{ if } \hat{p} \geq p_0 + z_\alpha \sqrt{\frac{p_0 q_0}{n}}$$

Table 10-6 lists the alternative hypotheses for testing $H_0: p = p_0$ with the decision rules for specified levels of significance. Alternatively, if we wish to state the critical value in terms of the standard normal score, the tests would appear as in Table 10-7.

Furthermore, if we want to conduct the tests in terms of the binomial random variable X, which for large n is approximately normally distributed with mean np and variance npq, the tests will be stated as in Table 10-8 or Table 10-9. Observe that the entries in Table 10-8 are obtained simply by multiplying each term in Table 10-6 by n. Similarly, by multiplying the numerator and the denominator of each of the fractions in Table 10-7 by n, we obtain Table 10-9.

Table 10-6

Tests of H_0: $p = p_0$ against H_1, with $\hat{p}$ approximately $N(p, p_0q_0/n)$ with large n

H_1	Decision rule: Reject H_0 if	Level of significance $\alpha = 0.05$	$\alpha = 0.01$
$p > p_0$	$\hat{p} \geq c_1$	$c_1 = p_0 + 1.645 \sqrt{\frac{p_0q_0}{n}}$	$c_1 = p_0 + 2.326 \sqrt{\frac{p_0q_0}{n}}$
$p < p_0$	$\hat{p} \leq c_2$	$c_2 = p_0 - 1.645 \sqrt{\frac{p_0q_0}{n}}$	$c_2 = p_0 - 2.326 \sqrt{\frac{p_0q_0}{n}}$
$p \neq p_0$	$\hat{p} \geq c_3$	$c_3 = p_0 + 1.960 \sqrt{\frac{p_0q_0}{n}}$	$c_3 = p_0 + 2.575 \sqrt{\frac{p_0q_0}{n}}$
	or		
	$\hat{p} \leq c_4$	$c_4 = p_0 - 1.960 \sqrt{\frac{p_0q_0}{n}}$	$c_4 = p_0 - 2.575 \sqrt{\frac{p_0q_0}{n}}$

Table 10-7

Tests of H_0: $p = p_0$ against H_1 with $\hat{p}$ approximately $N(p_0, p_0q_0/n)$ with large n

H_1	Decision rule $\alpha = 0.05$ Reject H_0 if	$\alpha = 0.01$ Reject H_0 if
$p > p_0$	$\frac{\hat{p} - p_0}{\sqrt{p_0q_0/n}} \geq 1.645$	$\frac{\hat{p} - p_0}{\sqrt{p_0q_0/n}} \geq 2.326$
$p < p_0$	$\frac{\hat{p} - p_0}{\sqrt{p_0q_0/n}} \leq -1.645$	$\frac{\hat{p} - p_0}{\sqrt{p_0q_0/n}} \leq -2.326$
$p \neq p_0$	$\frac{\hat{p} - p_0}{\sqrt{p_0q_0/n}} \geq 1.960$	$\frac{\hat{p} - p_0}{\sqrt{p_0q_0/n}} \geq 2.575$
	or	or
	$\frac{\hat{p} - p_0}{\sqrt{p_0q_0/n}} \leq -1.960$	$\frac{\hat{p} - p_0}{\sqrt{p_0q_0/n}} \leq -2.575$

Table 10-8

Tests of $H_0: p = p_0$ against H_1 when X is $B(n, p)$ with large n

H_1	Decison rule: Reject H_0 if	Level of significance $\alpha = 0.05$	$\alpha = 0.01$
$p > p_0$	$X \geq c_1$	$c_1 = np_0 + 1.645\sqrt{np_0q_0}$	$c_1 = np_0 + 2.326\sqrt{np_0q_0}$
$p < p_0$	$X \leq c_2$	$c_2 = np_0 - 1.645\sqrt{np_0q_0}$	$c_2 = np_0 - 2.326\sqrt{np_0q_0}$
$p \neq p_0$	$X \geq c_3$ or $X \leq c_4$	$c_3 = np_0 + 1.960\sqrt{np_0q_0}$ $c_4 = np_0 - 1.960\sqrt{np_0q_0}$	$c_3 = np_0 + 2.575\sqrt{np_0q_0}$ $c_4 = np_0 - 2.575\sqrt{np_0q_0}$

Table 10-9

Tests of $H_0: p = p_0$ against H_1 when X is $B(n, p)$ with large n

H_1	Decision rule: $\alpha = 0.05$ Reject H_0 if	$\alpha = 0.01$ Reject H_0 if
$p > p_0$	$\frac{X - np_0}{\sqrt{np_0q_0}} \geq 1.645$	$\frac{X - np_0}{\sqrt{np_0q_0}} \geq 2.326$
$p < p_0$	$\frac{X - np_0}{\sqrt{np_0q_0}} \leq -1.645$	$\frac{X - np_0}{\sqrt{np_0q_0}} \leq -2.326$
$p \neq p_0$	$\frac{X - np_0}{\sqrt{np_0q_0}} \geq 1.960$ or $\frac{X - np_0}{\sqrt{np_0q_0}} \leq -1.960$	$\frac{X - np_0}{\sqrt{np_0q_0}} \geq 2.575$ or $\frac{X - np_0}{\sqrt{np_0q_0}} \leq -2.575$

Exercise 10-3

1 Explain fully why a test about an exact null hypothesis can be used to replace a test about an inexact null hypothesis.

2 Suppose that X is $N(\mu, 100)$. Let $\bar{X}$ be the mean of a random sample of 16 observations. Given that $\bar{X} = 15$, test the null hypothesis $H_0: \mu = 10$ against each of the following alternative hypotheses at $\alpha = 0.01$:

a $H_1: \mu < 10$. *b* $H_1: \mu > 10$. *c* $H_1: \mu \neq 10$.

3 Do Prob. 2 for $\alpha = 0.05$.

4 Refer to Prob. 4 of Exercise 9-4. Let μ_1 and μ_2 be the true means of X_1 and X_2, respectively. Test the null hypothesis H_0: $\mu_1 = \mu_2$ against the alternative hypothesis at $\alpha = 0.05$ and $\alpha = 0.01$.

a H_1: $\mu_1 \neq \mu_2$. *b* H_1: $\mu_1 > \mu_2$.

5 Suppose that the variance of the IQs of the high school students in a certain district is 225. A random sample of 25 students has a mean IQ, or $\bar{X}$, of 106. If the level of significance is chosen at 0.01, should we conclude that the IQs of the high school students in this district are higher than 100? Find the 95 percent confidence interval for μ.

6 Let X designate the defective parts produced by an automatic machine. From a randomly selected sample of 20 parts, five parts are found to be defective. If p is the true proportion of all the parts produced that are defective, should we reject $H_0: p \leq 0.10$ in favor of $H_1: p > 0.10$ at $\alpha = 0.0431$?

7 Let X be $N(\mu, \sigma^2)$. A random sample of n observations of X yields a mean $\bar{X}$. Let e be the difference between $\bar{X}$ and μ, or $e = \bar{X} - \mu$. Find the sample size n:

a Given $e = +2$, $\sigma^2 = 100$, and $\alpha = 0.0455$.
b Given $e = -1$, $\sigma^2 = 150$, and $\alpha = 0.0027$.
c Given $e = +1.0$, $\sigma^2 = 100$, and $\alpha = 0.0455$.
d Given $|e| = 0.1$, $\sigma^2 = 200$, and $\alpha = 0.0455$.

8 Refer to Prob. 3 of Exercise 9-4. Let μ_1 and μ_2 be the true hourly outputs of the two departments, respectively. Test the null hypothesis H_0: $\mu_1 = \mu_2$ against the alternative hypothesis at $\alpha = 0.05$:

a H_1: $\mu_1 \neq \mu_2$. *b* H_1: $\mu_1 > \mu_2$.

9 Suppose that the mean μ of a random variable X is unknown, but the variance for X is known to be 144. Should we reject the null hypothesis H_0: $\mu = 15$ in favor of an alternative hypothesis H_1: $\mu \neq 15$ at $\alpha = 0.05$ if a random sample of 64 observations of X yields a mean $\bar{X} = 12$? What are the 95 percent confidence limits for μ?

10 Referring to Prob. 7 of Exercise 9-4, should we reject the null hypothesis that the weekly wages for machinists and carpenters are the same in favor of the alternative hypothesis that they are different at:

a $\alpha = 0.05$? *b* $\alpha = 0.01$?

11 In 30 tosses of a coin, the nature of which is unknown, 20 heads are obtained. Test the null hypothesis that the coin is well balanced against each of the following alternative hypotheses:

a The coin is unbalanced at $\alpha = 0.0430$.
b The probability of the coin's turning up heads is greater than 0.5 at $\alpha = 0.0495$.

12 In a poll of 10,000 voters selected at random from all the voters in a certain district, it is found that 5,180 voters are in favor of a particular candidate. Test the null hypothesis that the proportion of all the voters in the district who favor the candidate is equal to or less than 50 percent against the alternative that it is greater than 50 percent at $\alpha = 0.01$.

13 The Food and Drug Administration suspects that a newly developed medicine is less than 90 percent effective in curing a certain disease, while the drug manufacturing company believes that its effectiveness is at least 90 percent. In a random sample of 400 people who had the disease, 320 people were cured by using the medicine. Should we conclude that the medicine is at least 90 percent effective at $\alpha = 0.05$? At $\alpha = 0.01$? (*Hint*: $H_0: p \geq 0.9$.)

14 Prob. 7 above shows that one of the factors associated with a given sample size n is α. In fact, by increasing or decreasing the sample size, we can control not only α but also β. Suppose, for example, that a manufacturer of natural fibers wishes to control both types of errors. More specifically, he wants to control α by specifying that, given the mean strength of the fibers $\mu_X = \mu_1 = 30$ pounds, the probability of obtaining a sample mean $\bar{X} \geq C$ pounds is 0.05. He wants to control β by specifying that given the mean strength of the fibers, $\mu_X = \mu_2 = 33$ pounds, the probability of obtaining a sample mean $\bar{X} \leq C$ pounds is 0.01. Assume that X, the strength of natural fibers produced by this company, is normally distributed with a variance $\sigma_X^2 = 64$ pounds. What would be the sample size n required to satisfy these specifications? Derive a formula for computing n under similar circumstances.

10-4 The Power Function

As pointed out in Sec. 10-2-1, the power of a test is regarded as the ability of the test to detect the alternative hypothesis when it is, in fact, true. It is generally expressed as $1 - \beta$, the probability of rejecting the null hypothesis when it is false. Given the size of α, a test is considered *more powerful* than another if the value $1 - \beta$ of the test is greater than that of the other. However, the value $1 - \beta$ is known only if the sampling distribution is completely defined, which implies that the value of the parameter covered by the alternative hypothesis is specified. When the alternative hypothesis is inexact, the parameter value is not specified, and thus the sampling distribution is not defined. How can we then find the value of $1 - \beta$, or the power of the test? This question can be answered by introducing the concept of a power function, which is to be examined in the rest of this chapter.

10-4-1 The power function and the operating characteristic curve

First of all, let us define the concept of a *power function*:

> *A power function is a probability or probability density function that relates the probability of rejecting a null hypothesis to the various possible values of the parameter covered by the null and alternative hypotheses for a given level of significance.*

We shall start with the general case of testing the null hypothesis that a population parameter θ is equal to a certain value θ_0 against a hypothesis that θ is equal to any of the various alternative values. That is, we wish to decide between the two competing hypotheses

$$H_0\colon\ \theta = \theta_0 \qquad H_1\colon\ \theta \in V$$

where V is the set of all possible *alternative* values v for θ. Our decision between these two hypotheses will be based on a single statistic $\hat{\theta}$, which is calculated from a random sample of n observations. We shall reject H_0 if the observed $\hat{\theta}$ is sufficiently deviant from θ_0. The set of such values for $\hat{\theta}$, which causes us to reject H_0, is called the *critical region*, abbreviated CR. The level of significance α is therefore given by

$$\alpha = \sum P(\hat{\theta} \in CR \mid \theta = \theta_0) \tag{10-2}$$

Since $P(\hat{\theta} \mid \theta = \theta_0)$ is the probability of observing $\hat{\theta}$, given that the null hypothesis is true, the sum of these probabilities is the probability that $\hat{\theta}$ falls within the critical region that leads us to reject H_0 when it is, in fact, true.

If the null hypothesis is false, that is, if θ is actually equal to v, which belongs to V, then the sum of the probabilities of $\hat{\theta}$'s being in the critical region when v is the true value of the parameter is

$$\sum P(\hat{\theta} \in CR \mid \theta = v) = 1 - \beta \tag{10-3}$$

which is the probability of rejecting H_0 when H_0 is false. We should like this probability to be large, since the decision is a correct one. For a given test that has CR as its critical region, the likelihood function $PF(\theta)$ defined as

$$PF(\theta) = P(\hat{\theta} \in CR \mid \theta) \tag{10-4}$$

is referred to as the *power function* of the test. If $\theta \in V$, that is, if the null hypothesis is false, the power function becomes

$$PF(v) = P(\hat{\theta} \in CR \mid \theta = v)$$

which is equal to $1 - \beta$. If the null hypothesis is true, the power function becomes

$$PF(\theta_0) = P(\hat{\theta} \in CR \mid \theta = \theta_0)$$

which is identical with α.

The *complement* of the power function is known as the *operating characteristic function.* It is the likelihood function, designated $OC(\theta)$, and is defined as

$$OC(\theta) = P(\hat{\theta} \in AR \mid \theta) \tag{10-5}$$

where AR stands for the *acceptance region* for $\hat{\theta}$. It is the set of observed values that leads us to accept the null hypothesis. Note that

$$OC(\theta) = 1 - PF(\theta) \tag{10-6}$$

The operating characteristic function is also called the *performance function.* The widely known *operating characteristic curve*, or OCC, refers to the graph of this function. If the null hypothesis is true, the OC function becomes

$$OC(\theta_0) = P(\hat{\theta} \in AR \mid \theta = \theta_0)$$

which is identical with $1-\alpha$. It is the probability of accepting H_0 when H_0 is true—a correct decision. If the null hypothesis is false, or $\theta \in V$, the OC function becomes

$$OC(v) = P(\hat{\theta} \in AR \mid \theta = v)$$

which is identical with β. It is the probability of accepting H_0 when H_0 is false—an erroneous decision. Observe that

$$OC(\theta_0) = 1 - PF(\theta_0)$$

and

$$OC(v) = 1 - PF(v)$$

10-4-2 Examples

A few examples will help to make the above discussion clearer. Suppose that the McDonald Company is intending to open a new drive-in snack restaurant in a certain community. According to its experiences, the company will make less than normal profit if 20 percent or less of the adult population in the community likes the foods it serves. The greater this percentage, the more profitable the operation will be. Let p be the proportion of the adult population that actually likes the foods. We wish to decide between the following two competing hypotheses

$$H_0\colon\ p \leq 0.20 \qquad H_1\colon\ p > 0.20$$

As pointed out in the beginning of Sec. 10-3, a good test of $H_0\colon p = 0.20$ against $H_1\colon p > 0.20$ is also good for this test. Let X be the number of persons in a random sample of 30 people ($n = 30$) who say that they like the foods. On the basis of a single observed value of X, we shall decide whether to reject H_0. If H_0 is rejected in favor of H_1, the company will likely be successful in establishing the new restaurant. If α is chosen at 0.0094, the decision rule will be

Reject H_0 if $X \geq 12$

since

$$P(X \geq 12 \mid p = 0.20) = 1 - P(X \leq 11 \mid p = 0.2) = 1 - 0.9906 = 0.0094 \qquad \text{(see Appendix III)}$$

That is, the probability of X's being in the critical region when the null hypothesis is true is 0.0094, or approximately 1 percent. If an alternative value of p is specified, the power of the test can be evaluated. For instance, if $p = 0.3$, the value of the power can be obtained by applying Eq. (10-3) as follows:

$$PF(0.3) = P(X \geq 12 \mid p = 0.3) = 0.1594 \qquad \text{(see Appendix III)}$$

which is $1 - \beta$, the probability of rejecting H_0 when it is actually false. By applying Eq. (10-6) for the corresponding value of the operating characteristic function, we have

$$OC(0.3) = 1 - PF(0.3) = 1 - 0.1594 = 0.8406$$

which is β, the probability of accepting H_0 when the alternative hypothesis is true.

In general, for $n = 30$ and for the various possible alternative values of p, the corresponding values of the power function are given by

$$PF(v) = P(X \geq 12 | p = v)$$

where v stands for the possible alternative values of p. The corresponding values of the OC function can be obtained by applying Eq. (10-5) as follows:

$$OC(v) = P(X < 12 | p = v)$$

Observe that since $X \geq 12$ is the critical region, the acceptance region of the test is $X < 12$. Table 10-10 shows the values of the power function and of the operating characteristic function for specified values of the parameter p.

Figure 10-5 shows the sampling distributions for X when X is $B(30,p)$ for specified values of $p(p = 0.2, 0.3, 0.4, 0.5, 0.6, \text{and } 0.7)$. The interval of values for $\hat{p}$ to the right of the critical value c is the critical region, while that to the left of it is the region of acceptance. The sum of the probabilities to the right of c for each value of p larger than 0.2 is $1 - \beta$, the power of the test; the sum of the probabilities to the left of c for each value of p is β, the value of the operating

Table 10-10

Power and operating characteristic functions for the test of $H_0: p = 0.20$ and $H_1: p > 0.20$ at $\alpha = 0.0094$

Alternative values of p	*Value of power function,* $1-\beta$	*Value of OC function,* β
0.1	0.0000	1.0000
(0.2)	$(\alpha = 0.0094)$	$(1 - \alpha = 0.9906)$
0.3	0.1594	0.8406
0.4	0.5690	0.4310
0.5	0.8997	0.1003
0.6	0.9919	0.0081
0.7	0.9999	0.0001
0.8	1.0000	0.0000
0.9	1.0000	0.0000

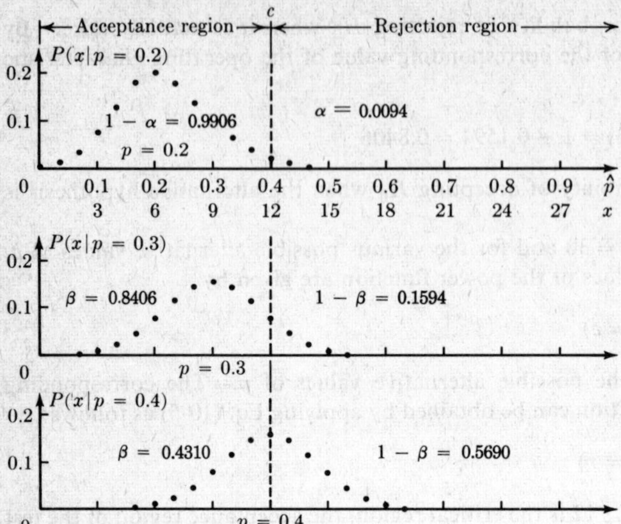

Figure 10-5

Sampling distributions for X when X is $B(30,p)$ for specified values of p.

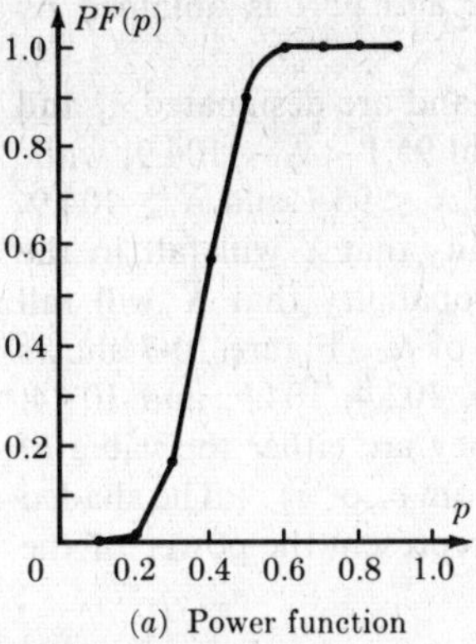

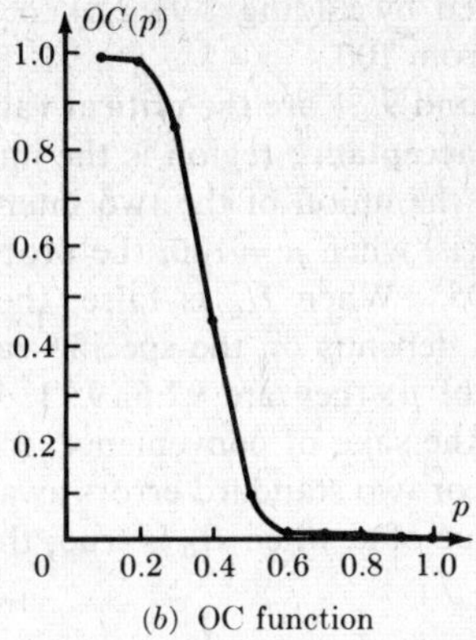

Figure 10-6
Power and operating characteristic functions for the test of H_0: $p = 0.20$ against H_1: $p > 0.20$ at $\alpha = 0.0094$ for specified values of p.

characteristic function. The latter value can be obtained simply by subtracting the corresponding value of the power from 1.

The values of the power function and of the operating characteristic function as in Table 10-10 are respectively represented by the points in Fig. 10-6*a* and *b*. The line segments connecting the adjacent points are added for visual assistance. If distributions for other values of p had been added to those in Fig. 10-5, there would have been more points plotted in each of the two curves in Fig. 10-6. If a sufficiently large number of points are plotted, or if the sample size n approaches infinity, the two functions would become smooth curves, known as the *power curve* and the *OC curve*.

As another example, suppose that we wish to test the hypothesis that a certain ethnic group in the United States has a mean IQ of 100 against the alternative hypothesis that it is not 100. That is,

$$H_0: \quad \mu = 100 \qquad H_1: \quad \mu \neq 100$$

Let the random variable X be the IQs of the members of this minority group. Our decision about the parameter μ will be made in the light of the observed mean $\overline{X}$ of a sample of 64 persons ($n = 64$) from this group. We assume that the standard deviation of X is known to be 20, and X is $N(\mu,400)$. Thus the sample mean $\overline{X}$ is normally distributed with mean μ and standard error 2.5. If the level of significance is set at 5 percent, our decision rule will be

Reject H_0 if $\overline{X} \geq 104.9$ or $\overline{X} \leq 95.1$

where 104.9 is obtained by adding 1.96(2.5) to 100, and 95.1 is obtained by subtracting 1.96(2.5) from 100.

The values 104.9 and 95.1 are the critical values and are designated c_1 and c_2 in Fig. 10-7. The acceptance region is the interval $95.1 < \overline{X} < 104.9$, while the rejection region is the union of the two intervals $\overline{X} \leq 95.1$ and $\overline{X} \geq 104.9$. When H_0 is true, that is, when $\mu = 100$, the probability that $\overline{X}$ will fall in the rejection region is 0.05. When H_0 is false, the probability that $\overline{X}$ will fall in the rejection region depends on the specific value of μ. Figure 10-7 shows six alternative values of μ; they are 92.6, 95.1, 97.6, 102.4, 104.9, and 107.4. They are selected for the sake of convenience, for they are either the value of c_1 or c_2 or exactly one or two standard errors away from c_1 or c_2. The shaded areas represent the value of α, when H_0 is true; they represent the power of the

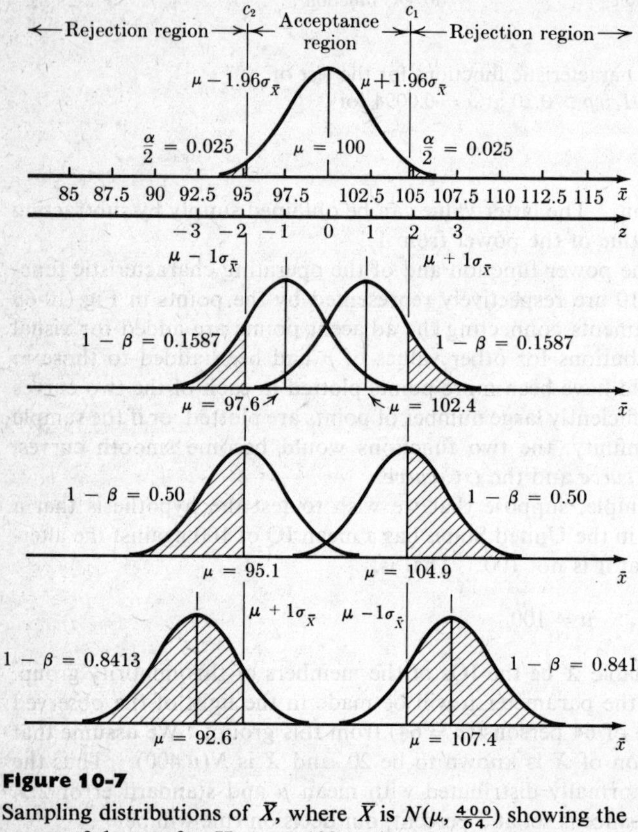

Figure 10-7

Sampling distributions of $\overline{X}$, where $\overline{X}$ is $N(\mu, \frac{400}{64})$ showing the power of the test for H_0: $\mu = 100$ against H_1: $\mu \neq 100$ for specified values of μ.

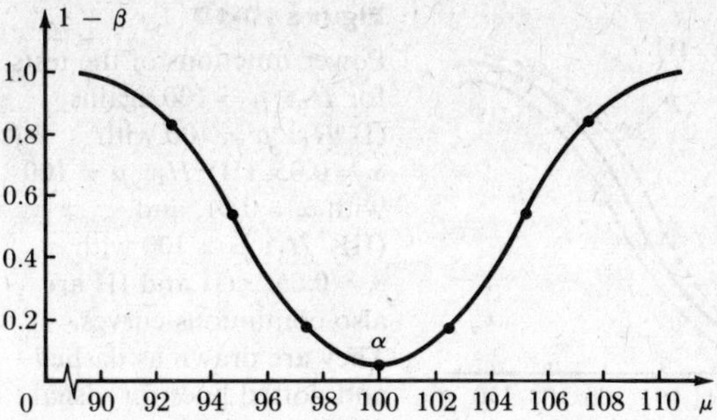

Figure 10-8

The power function of the test of H_0: $\mu = 100$ against H_1: $\mu \neq 100$ with $\alpha = 0.05$ and $\sigma_{\bar{x}} = 2.5$.

test, or $1 - \beta$, when H_0 is false. The areas lying under each curve between c_1 and c_2 represent $1 - \alpha$ or β, depending on whether the null hypothesis is true or false.

The power function of the test is shown in Fig. 10-8. The shaded areas under the various curves in Fig. 10-7 are represented by the points on the power curve in Fig. 10-8. Similarly, the unshaded areas under the various curves in Fig. 10-7 are represented by the points in the operating characteristic curve in Fig. 10-9.

10-4-3 Factors affecting the power of a test

The power of a test is affected by the level of significance and the size of the standard error. The standard error in turn depends on the standard deviation of the population and the sample size *n*. Thus the power of a test depends on three factors: the size of α, the sample size *n*, and the population standard deviation σ.

As pointed out previously, the value of β varies *inversely* with α. It follows that the power of a test, $1 - \beta$, varies *directly* with α. By setting α larger (or smaller), we can increase (or decrease) the power of a statistical test, other things being equal. For example, if in the above test for H_0: $\mu = 100$ against

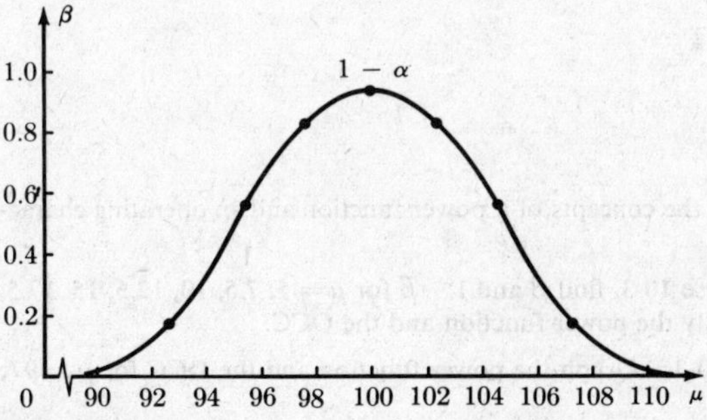

Figure 10-9

The operating characteristic function of the test of H_0: $\mu = 100$ against H_1: $\mu \neq 100$ with $\alpha = 0.05$ and $\sigma_{\bar{x}} = 2.5$.

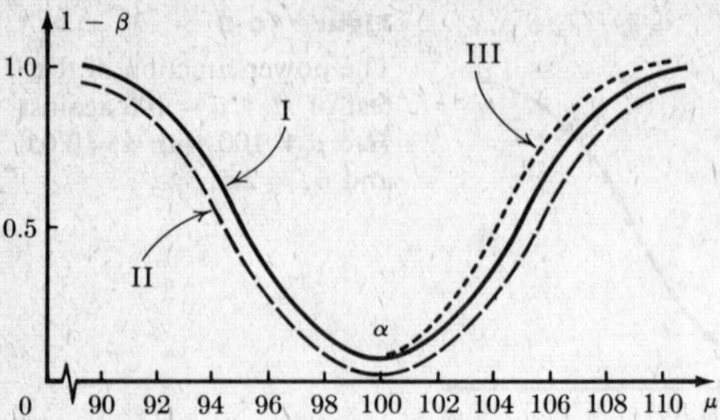

Figure 10-10
Power functions of the tests for H_0: $\mu = 100$ against (I) H_1: $\mu \neq 100$ with $\alpha = 0.05$, (II) H_1: $\mu \neq 100$ with $\alpha = 0.01$, and (III) H_1: $\mu > 100$ with $\alpha = 0.05$. (II and III are also continuous curves. They are drawn as dashed and dotted lines for visual aid purposes.)

$H_1: \mu \neq 100$, the size of α had been set at 1 percent rather than at 5 percent, the rejection region for each true value of the parameter μ would be smaller.

It follows that the original power curve I would lie above the new curve II, as shown in Fig. 10-10. (Note that the curve in Fig. 10-8 is the curve I in Fig. 10-10.) Since the probability of making the correct decision in rejecting the false null hypothesis is *always* greater in the original test, that test is said to be a *uniformly more powerful test*. The new test is then called *inadmissible*.

Moreover, if the alternative hypothesis had been $\mu > 100$, the critical region of this one-tailed test $\bar{X}$ would have been larger than the original test, even though the size of α remained unchanged. The power of the one-tailed test is shown as curve III in Fig. 10-10. A one-tailed test is always more powerful than a two-tailed test, given α and an exact null hypothesis.

The power of a statistical test also varies directly with the sample size n. The larger the value of n, the smaller will be the standard error, given the standard deviation of the population. When the standard error is smaller, the distribution of the sample statistic will be more concentrated about the mean, and the critical values will be closer to the mean. It follows that the rejection region, and so the power of the test, will be larger, given the size of α. Similarly, by reducing the standard deviation of the population, we can also decrease the standard error of the sampling distribution, and thus increase the power of the test.

Exercise 10-4

1 Use your own words to define the concepts of a power function and an operating characteristic function.

2 Referring to Prob. 2 of Exercise 10-3, find β and $1 - \beta$ for $\mu = 5$, 7.5, 10, 12.5, 15, 17.5, 20, 22.5. Represent graphically the power function and the OCC.

3 Refer to Prob. 5 of Exercise 10-3. Graph the power function and the OCC for $\mu = 97$, 100, 103, 106, 109, 112, 115.

4 Given the test in Prob. 6 of Exercise 10-3, graph the power function and the OCC for $p = 0.0, 0.1, 0.2, 0.3, \ldots, 0.9, 1.0$.

5 A coin is tossed 30 times to determine whether the probability that the coin turns up heads is 0.5. That is, it is desired to test the null hypothesis $H_0: p = 0.5$ against the alternative $H_1: p \neq 0.5$. Let α be 0.0429. Find the values of the power function for $p = 0.1, 0.2, 0.3, \ldots, 0.9$.

6 Refer to Prob. 9 of Exercise 10-3. Graph the power function and the OCC for $\mu = 7.56$, 9.06, 10.56, 12.06, 13.56, 15, 16.44, 17.94, 19.44, 20.94, and 22.44.

7 Explain fully the relation between the power function and the operating characteristic function.

8 Given the test in Prob. 12 of Exercise 10-3, graph the power function for p, the true proportion of all the voters who favor the candidate, with $\alpha = 0.05$. (For $p = 0.5000$, 0.5032, 0.5082, 0.5132, 0.5182, and 0.5232.)

9 Refer to Prob. 13 of Exercise 10-3. Let p be the true proportion of people who are cured by using the medicine. Graph the power function and the OCC for the level of significance at 5 percent. (For $p = 0.9$, 0.8903, 0.8753, 0.8603, 0.8453, and 0.8303.)

10 Refer to Prob. 9 of Exercise 10-2. If the buyer contends that the true mean strength is *less than 33 pounds*, what would be the value of $1 - \beta$ for the true mean strength of the fibers to be 26.6, 28.2, 29.8, 31.4, 33.0, 34.6, 36.2, 37.8, or 39.4?

11 What are the factors that determine the power of a test? Explain fully.

Chapter 11

The student's *t* distribution

Previous discussions have been based on the assumption that the population variance is known. This assumption, however, is usually unrealistic since in most applications the population variance is unknown and must be estimated from the sample variance. Consequently, the standard error of the sample mean $\overline{X}$, designated $\sigma_{\overline{X}}$, is not known and we cannot use the standardized normal deviate

$$Z = \frac{\overline{X} - \mu}{\sigma_{\overline{X}}}$$

for interval estimation or significance testing. If the denominator $\sigma_{\overline{X}}$ is replaced by its estimate $\hat{s}_{\overline{X}}$, we have the ratio $(\overline{X} - \mu)/\hat{s}_{\overline{X}}$, which, unlike the Z score, is not normally distributed, especially when the sample size n is small. This being the case, how can we draw inferences about the mean or the difference between two means when the standard error of $\overline{X}$ is unknown?

This problem was successfully tackled by William S. Gosset, who derived in 1908 a theoretical sampling distribution known as *Student's t distribution.* Gosset was then employed with the Guinness Brewery in Dublin, and the company did not permit him to use his real name for publishing the results of his theoretical work. Accordingly, he wrote under the pen name "Student," and so the name of the distribution. The Student's distribution is essentially similar to the normal distribution; it is in fact identical with the normal pattern when the sample size n approaches infinity.

In this chapter we shall discuss the nature of the Student's distribution. After that, we shall demonstrate how to make inferences about the mean, and about the difference between two means, when the variance of the population has to be estimated.

11-1 Nature of the t Distribution

The above-mentioned ratio with $\hat{s}_{\bar{X}}$ as the denominator is known as the t ratio; that is,

$$t = \frac{\bar{X} - \mu}{\hat{s}_{\bar{X}}} \tag{11-1}$$

the theoretical sampling distribution of which is the Student's t distribution. The symbol t is a random variable. It is called a *test statistic* to distinguish it from other sample statistics such as $\bar{X}$ and s^2. Unlike other statistics, however, the t ratio is not an estimator of any population parameter. Before we use the test statistic to make inferences about the mean, let us first look into the nature of the t distribution and the assumptions underlying it.

First of all, it must be assumed that the random variable X is *normally* distributed. The necessity of this assumption lies in the fact that in order to specify the distribution of t, its denominator $\hat{s}_{\bar{X}}$ and $\bar{X}$ in the numerator must be statistically independent. The mathematical explanation of this requirement is too advanced to be treated here. But unless the distribution of the population is normal, the sample statistics $\bar{X}$ and $\hat{s}_{\bar{X}}$ will not be statistically independent. Recall, for instance, that if X is a binomial random variable with mean np and variance npq, the corresponding sample statistics, $n\hat{p}$ and $n\hat{p}\hat{q}$, are obviously not independent of each other. Likewise, if X is a Poisson random variable, its mean and variance are identical, and thus are completely dependent on each other. It is in fact an important statistical principle that the *sample mean* $\bar{X}$ and *the sample variance* s^2 or $\hat{s}^2$ are statistically *independent* if the random variable X is normally distributed.

Under the assumption that X is normally distributed, the sample mean $\bar{X}$ will be normal. Thus, $\bar{X} - \mu$, the numerator of t, is normally distributed since μ is a constant and the subtraction of μ from $\bar{X}$ has no effect on the dispersion of the distribution but shifts the distribution μ units to the left. However, $\hat{s}_{\bar{X}}$, the denominator of t, unlike $\sigma_{\bar{X}}$, is not a constant. It is a statistic, the value of which changes from sample to sample. As a consequence, the t ratio, unlike the Z score, is usually not normal.

The distribution of t, nonetheless, is very close to that of the Z score, especially when the sample size n is large. It can be demonstrated that the mean of the t distribution is the same as the mean of the Z score; that is, $E(t) = 0$. The variance of the t distribution is $n/(n-2)$ for $n > 2$; it is slightly larger than 1, the variance of the Z score. For the exact nature of the sampling distribution of t, let us consider its density function.

11-1-1 The density function

The function rule for t is given as follows:

$$f(t) = \frac{k}{(1 + t^2/\nu)^{(\nu+1)/2}} \qquad \text{for } -\infty < t < \infty \tag{11-2}$$

where ν (a Greek letter read as "nu") stands for the number of *degrees of freedom* (the concept of which will be discussed shortly), and k is a constant dependent on the value of ν. Observe that ν is the sole parameter of the t distribution, and the function rule involves only the value of ν and the test statistic t. Although the above mathematical expression is different from the normal function rule, the shape of the t distribution is quite similar to that of the normal curve.

In the first place, we observe that the statistic t enters as a squared quantity. This means that two different values of t with the same absolute magnitude will have the same probability density. Accordingly, the t distribution must be *symmetric.* Furthermore, ν, the number of degrees of freedom, is a positive constant, and in the expression it always has a positive sign. Consequently, the exponent $(\nu + 1)/2$ must be positive. Thus the greater the absolute value of t, the smaller will be the probability density associated with it. That is, the tails of the curve have decreasing densities, while the center of it has the greatest density. This implies that the t distribution is *unimodal.* The mode of the distribution is $t = 0$ since at this value of t the density of the distribution reaches its maximum. Observe that this maximum value of the function is exactly k, since at $t = 0$ the denominator of the expression is equal to 1.

Since the statistic t has a variance greater than 1, its distribution can be expected to have a greater dispersion than the normal distribution, especially when the sample size n is small. The exact shape of the t distribution depends on the sample size, or the degrees of freedom, ν. The latter is equal to $n - 1$, or one less than the sample size, when a single sample is included in a statistical problem. Accordingly, the t distribution is not a single distribution such as the distribution of the standardized normal deviate, but rather a family of distributions, one for each number of degrees of freedom. If the number of degrees of freedom is small, that is, if the sample size n is small, the curve of the t distribution is more spread out in the tails and flatter around the center than is the normal curve. As the sample size increases, $\hat{s}^2$ becomes closer to σ^2; consequently, the t distribution approaches the standard normal distribution. As the sample size approaches infinity, $\hat{s}^2$ approaches σ^2; there is no difference between the two distributions.

Figure 11-1 shows the effect of an increase in the sample size on the shape of the t distribution. It is much closer to the normal distribution with $n = 10$

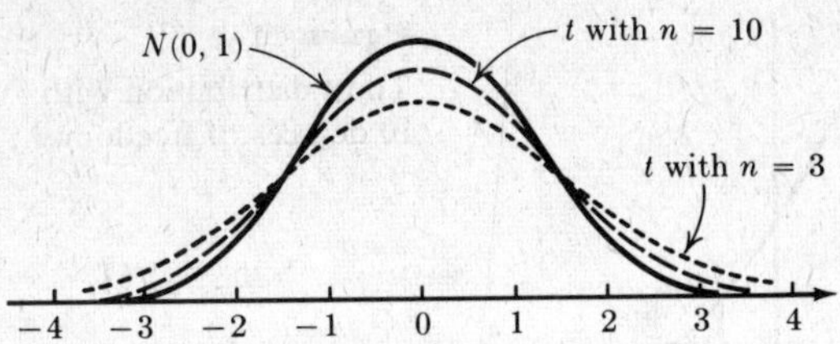

Figure 11-1

The t distribution, with $n = 3$ and $n = 10$, compared with the standard normal distribution.

than it is with $n = 3$. We can expect that as n continues to increase, the t distribution will approach the distribution of the normal deviate.[1]

The fact that with large n, the t distribution approaches the standard normal distribution can also be detected from the t table, Appendix VII. A t table of probabilities can be presented in the same way as the standard normal table. However, since for each different number of degrees of freedom there is a distribution of t, the t table would run impractically long if we wanted it to be in as detailed a form as the standard normal table is. For most practical purposes, this is unnecessary. Usually, the table presents only the most commonly used probabilities such as those listed in the top row of Appendix VII. Each of the probabilities listed in row $2Q$ is equal to the sum of the areas in the two tails, as indicated by the diagram at thc top of the table, while each of those listed in row Q is equal to either of the two-tail areas of a given distribution. For each of the values of v, which are listed in the left column of the table, the table entry of a specified column is the t value above which lies the probability listed in row Q. While only the positive values of t are entered in the table, its negative values are implied since the distribution is symmetrical around 0. Thus, for 10 degrees of freedom ($n = 11$), the probability is 0.10 that the values of t will be greater than 1.372, which is found in the intersection of row 10 and column 0.1 for Q (or 0.2 for $2Q$). For a one-tailed test, the probabilities listed in row Q are the values for α. For a two-tailed test, the values in row $2Q$ should be used as α. Thus the probability is 0.10 that the value of t will be greater than 1.812 or smaller than -1.812 for the distribution with 10 degrees of freedom. (See Fig. 11-2 for further clarification of the use of the t table.) Similarly, for 30 degrees of freedom, the t values 1.697 and -1.697 will respectively cut off the upper and lower 5 percent of the distribution. Observe that the value 1.697 is closer to the value 1.645 above which lies 5 percent of the standard normal distribution. When the number of degrees of freedom approaches infinity, as shown in the bottom row of Appendix VII, the t value is exactly 1.645. That is, the t distribution with $n \to \infty$ becomes identical with the standard normal distribution.

[1] The t curves are smooth and continuous. The dotted and broken curves in Fig. 11-1 are used for the purpose of visual aid.

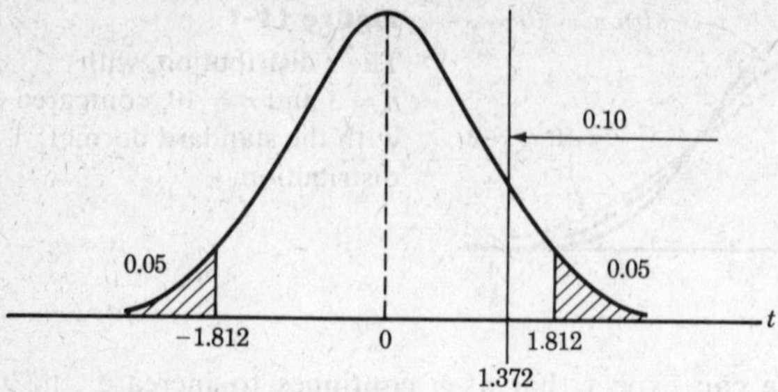

Figure 11-2
The t distribution with 10 degrees of freedom.

11-1-2 The concept of degrees of freedom

Reference has been made repeatedly to the concept of degrees of freedom. It is important that you understand this notion since it is the sole parameter of the t distribution and will be used frequently in this and the next few chapters.

The parameter v, or the degrees of freedom, reflects the fact that the statistic t uses a sample standard error $\hat{s}_{\bar{X}}$ in place of the population standard error $\sigma_{\bar{X}}$. As pointed out in Chap. 5, the sample variance, and thus the sample standard deviation, is derived from the sum of squared deviations from the sample mean. Moreover, the sum of deviations from the sample mean must be 0; that is,

$$\sum_{i=1}^{n} (X_i - \bar{X}) = 0$$

These two facts are sufficient to explain the concept of degrees of freedom.

Suppose, for example, that a sample of size 5 ($n = 5$) is taken and that we are asked to guess the five deviations from the sample mean. Let D_i stand for the deviation for $i = 1, 2, 3, 4$, and 5; we have

$$D_1 + D_2 + D_3 + D_4 + D_5 = 0$$

We have, of course, complete *freedom* in assigning values to any four of the five deviations. Once the four values are assigned, however, we no longer have any freedom in guessing the other value because the last is already determined. We may, for instance, assign *at will* to the first four deviations as follows:

$$D_1 = 9 \qquad D_2 = -7 \qquad D_3 = 0 \qquad D_4 = 4$$

The fifth deviation is then determined, and we cannot guess any value we please. The value of D_5 must be

$$D_5 = 0 - (D_1 + D_2 + D_3 + D_4) = 0 - 6 = -6$$

It is clear that with n observations in a sample there can be only $n - 1$ degrees of freedom in assigning values to the deviations from the sample mean. Accordingly, the sample variance, and so the standard error of the sample mean, has

$n - 1$ degrees of freedom. Thus the t distribution is not based on n, the sample size per se, but rather the degrees of freedom, or $n - 1$, in the sample standard deviation.

It ought to be pointed out that the sum of squared deviations from the population mean μ, or $\sum_{i=1}^{n}(X_i - \mu)^2$, has n, rather than $n - 1$, degrees of freedom because the sum of deviations, or $\sum_{i=1}^{n}(X_i - \mu)$, need not be 0, and there is freedom to guess the value of every deviation. In general whenever we use a sample statistic to estimate a population parameter, we lose 1 degree of freedom. Thus we may say that the sum of squared deviations from the sample mean, or $\sum_{i=1}^{n}(X_i - \bar{X})^2$, has only $n - 1$ degrees of freedom because $\bar{X}$ is used to estimate the population mean μ.

Exercise 11-1

1 Summarize the characteristic of the t distribution.

2 Explain how the t distribution is different from the normal distribution.

3 For a t distribution with 15 degrees of freedom, find the critical t value that cuts off each of the following areas under the curve:

a The upper 1 percent.
b The upper 5 percent.
c The upper 97.5 percent.
d The lower 5 percent.
e The lower 10 percent.

4 Do Prob. 3*a* through *e* for t distributions with 30 and 120 degrees of freedom, respectively. Compare the t values for each of the two distributions with the corresponding values for a normal distribution.

5 Explain the concept of degrees of freedom. Why is this concept so important to the t distribution?

6 Given that a random sample of n observations $X_1, X_2, \ldots, X_n$, where X is $N(5, \sigma^2)$, yields a ratio

$$\frac{\bar{X} - \mu}{\hat{s}/\sqrt{n}}$$

which is distributed as t with $n - 1$ degrees of freedom. Answer the following:

a If $n = 25$, $\bar{X} = 3$, and $\hat{s} = 2$, what is t?
b If $n = 9$, $\bar{X} = 2$, and $t = -2$, what is $\hat{s}$?
c If $n = 25$, $\hat{s} = 10$, and $t = 2$, what is $\bar{X}$?
d If $\hat{s} = 15$, $\bar{X} = 14$, and $t = 3$, what is n?

7 Find the t value such that:

a The probability of obtaining a smaller t value is 0.95 when ν, the degrees of freedom, is 20.
b The probability of obtaining a larger t value is 0.01 when $\nu = 30$.
c The probability of obtaining a larger t value is 0.95 when $\nu = 15$.
d The probability of obtaining a smaller t value is 0.05 when $\nu = 10$.

8 Some statisticians have mistakenly stated that in making an inference about a population mean μ on the basis of a sample mean $\overline{X}$, the choice between the normal model and the t model depends on whether the sample is "large" or "small." While this is technically incorrect, it works well as a rule-of-thumb in most circumstances. Explain.

11-2
Inference about the Mean

In this section we shall consider the methods for making inferences about the mean of a single population under the assumption that the *population variance is not known.* The procedures of estimation and hypothesis testing applied here are the same as those used in the preceding two chapters. However, since the population variance is unknown, we must use the substitute for σ^2, that is, its unbiased estimate, which is commonly designated $\hat{s}^2$. It can be expected that $\hat{s}^2$ will become closer to σ^2 as the sample size n is increased. When n is sufficiently large, there will be little error in replacing σ^2 by $\hat{s}^2$. Let us first take a look at the method of making inferences about the mean under the condition of a large sample.

11-2-1 Large sample

Since the sample variance for large n tends to be close to the population variance, the distribution of the normal deviate z is usually used as an approximation to the t distribution when the sample size is large. How large, then, must n be in order to apply the normal approximation? We know the critical z value for a two-tailed test with $\alpha = 0.05$ is 1.960, and the corresponding t value for 30 degrees of freedom is 2.042. Thus the risk of error is rather negligible if we consider a sample of 30 or more observations large and actually use the standard normal distribution for interval estimation and hypothesis testing. Of course, the larger the sample, the closer the approximation will be. With 120 degrees of freedom, the corresponding t value is 1.98, almost equal to the z value 1.96.

Accordingly, if we wish to test the null hypothesis

$$H_0: \quad \mu = \mu_0$$

against the alternative hypothesis

$$H_1: \quad \mu \neq \mu_0$$

with the level of significance at α, and if $\hat{s}^2$, the unbiased estimate of the population variance, is computed from a large sample ($n \geq 30$), our decision rule will be

$$\text{Reject } H_0 \text{ if } \overline{X} \geq \mu_0 + z_{\alpha/2}\frac{\hat{s}}{\sqrt{n}} \text{ or } \overline{X} \leq \mu_0 - z_{\alpha/2}\frac{\hat{s}}{\sqrt{n}}$$

where $\hat{s}/\sqrt{n}$ is the estimated standard error of the mean and $z_{\alpha/2}$ is the z value that cuts off $\alpha/2$ in each tail of the standard normal curve.

Suppose that the average weekly output turned out by workers in a particular trade has been 300 units. It is suspected that the weekly output is now significantly different from 300 units. Thus the hypotheses to be tested are

$$H_0:\quad \mu = 300 \quad H_1:\quad \mu \neq 300$$

A sample of 64 workers in this trade is randomly selected, and their weekly outputs during an ordinary week are recorded. The following statistics are obtained:

$$\hat{s} = 40 \qquad \overline{X} = 315$$

Should the null hypothesis be rejected in favor of the alternative with $\alpha = 0.05$? To answer this question, we set up the following decision rule:

$$\text{Reject } H_0 \text{ if } \overline{X} \geq 300 + 1.96\left(\frac{40}{8}\right) \text{ or } \overline{X} \leq 300 - 1.96\left(\frac{40}{8}\right)$$

that is,

$$\text{Reject } H_0 \text{ if } \overline{X} \geq 309.8 \text{ or } \overline{X} \leq 220.2$$

Since the observed value for $\overline{X}$ is 315, we should reject the null hypothesis in favor of the alternative that $\mu \neq 300$.

After a null hypothesis about the mean is rejected, we may wish to estimate the confidence limits for the true mean μ. When a sample mean $\overline{X}$ is computed from a large sample, and the probability that the limits will not include the population mean is specified at α, we have

$$P\left(\overline{X} - z_{\alpha/2}\frac{\hat{s}}{\sqrt{n}} < \mu < \overline{X} + z_{\alpha/2}\frac{\hat{s}}{\sqrt{n}}\right) = 1 - \alpha \tag{11-3}$$

In the above example, where the sample of 64 observations yields a standard deviation $\hat{s} = 40$ and a mean $\overline{X} = 315$, the 95 percent confidence interval of the population mean μ will be

$$315 - 1.960\frac{40}{\sqrt{64}} < \mu < 315 + 1.960\frac{40}{\sqrt{64}}$$

or

$$305.2 < \mu < 324.8$$

Observe that the procedures of estimation and hypothesis testing about the population mean are identical with those used in Secs. 9-3 and 10-2-1. The only difference is that the population standard deviation σ is replaced by its estimate $\hat{s}$.

11-2-2 Small sample

For small samples, the sample variance can be quite different from the population variance. There will be considerable error if we use the method of normal approximation and compute the standard error with σ replaced by $\hat{s}$. Thus, when the sample is small, we shall use the t distribution for inferences about the mean of the population.

If we wish to test the competing hypotheses

$$H_0: \ \mu = \mu_0 \qquad H_1: \ \mu \neq \mu_0$$

with the significance level at α, and if a small sample of less than 30 observations yields a mean $\bar{X}$ and a variance $\hat{s}^2$, our decision rule will be

$$\text{Reject } H_0 \text{ if } \bar{X} \geq \mu_0 + t_{(v,\,\alpha/2)} \frac{\hat{s}}{\sqrt{n}} \text{ or } \bar{X} \leq \mu_0 - t_{(v,\,\alpha/2)} \frac{\hat{s}}{\sqrt{n}}$$

Here $\hat{s}/\sqrt{n}$ is the estimated standard error of the mean and $t_{(v,\,\alpha/2)}$ is the t value that cuts off $\alpha/2$ in each tail of the distribution with v degrees of freedom. If the alternative hypothesis is stated as

$$H_1: \ \mu > \mu_0$$

we shall reject H_0 if $\bar{X} \geq \mu_0 + t_{(v/\,\alpha)}\, \hat{s}/\sqrt{n}$. Here $t_{(v/\,\alpha)}$ is the t value that cuts off the upper α of the t distribution with v degrees of freedom. Similarly, if the alternative is stated as

$$H_1: \ \mu < \mu_0$$

we shall reject H_0 if $\bar{X} \leq \mu_0 - t_{(v/\,\alpha)}\, \hat{s}/\sqrt{n}$. Here $t_{(v/\,\alpha)}$ is the t value that cuts off the lower α of the t distribution with v degrees of freedom.

The tests of the null hypothesis $H_0\colon \mu = \mu_0$ against each of the various alternative hypotheses are summarized in Table 11-1.

Suppose, for example, that a steel manufacturing company wishes to know whether the tensile strength of the steel wire has an overall average of 120 pounds. A sample of 25 units of steel wire produced by the company yields a mean strength of 110 pounds and a variance $\hat{s}^2 = 144$ pounds. Should the

Table 11-1

Tests of hypotheses H_0: $\mu = \mu_0$ against H_1 when X is $N(\mu, \sigma_X^2)$ with σ_X^2 unknown and $t = (\bar{X} - \mu_0)/(\hat{s}/\sqrt{n})$

H_1	*Decision rule: Reject H_0 if*	*Where at a level α and $v = n - 1$*
$\mu > \mu_0$	$t \geq c_1$	$c_1 = t_{(v,\alpha)}$
$\mu < \mu_0$	$t \leq c_2$	$c_2 = -t_{(v,\alpha)}$
$\mu \neq \mu_0$	$t \geq c_3$	$c_3 = t_{(v,\alpha/2)}$
	or	
	$t \leq c_4$	$c_4 = -t_{(v,\alpha/2)}$

company conclude that the average strength is *not* 120 pounds with $\alpha = 0.05$? This problem is one of deciding between the two competing hypotheses

H_0: $\mu = 120$ $\quad H_1$: $\mu \neq 120$

Since the level of significance is set at 0.05, with $v = 24$, the decision rule is

$$\text{Reject } H_0 \text{ if } \overline{X} \geq 120 + t_{(24,0.025)} \frac{\sqrt{144}}{\sqrt{25}} \text{ or } \overline{X} \leq 120 - t_{(24,0.025)} \frac{\sqrt{144}}{\sqrt{25}}$$

That is,

Reject H_0 if $\overline{X} \geq 120 + 2.064(2.4) = 124.9536$ or $\overline{X} \leq 120 - 2.064(2.4) = 115.0464$

Since the observed value for $\overline{X}$ is 110, we should reject H_0 in favor of H_1 that the true mean of the tensile strength is not equal to 120 pounds.

You should have no difficulty in testing the null hypothesis $H_0: \mu = 120$ against $H_1: \mu > 120$ or $H_1: \mu < 120$. This is left as an exercise.

When σ^2 is unknown and n is small, we use the t value rather than z for estimating the confidence interval for μ. If the probability that the interval will not include μ is set at α, the $1 - \alpha$ confidence interval is found within the parentheses of the following expression:

$$P\left(\overline{X} - t_{(v,\alpha/2)} \frac{\hat{s}}{\sqrt{n}} < \mu < \overline{X} + t_{(v,\alpha/2)} \frac{\hat{s}}{\sqrt{n}}\right) = 1 - \alpha \tag{11-4}$$

Observe that this expression differs from (11-3) only in that $z_{\alpha/2}$ is replaced by $t_{(v,\alpha/2)}$. Thus the $1 - \alpha$ confidence limits for μ are

$$\overline{X} \pm t_{(v,\alpha/2)} \frac{\hat{s}}{\sqrt{n}}$$

Suppose that the above-mentioned steel manufacturing company wants to find the 98 percent confidence interval for the true mean μ of the tensile strength. Since

$\overline{X} = 110 \qquad \alpha = 0.02 \qquad v = 24$

$t_{(v,\alpha/2)} = t_{(24,0.01)} = 2.492$

and

$$\sqrt{\frac{\hat{s}^2}{n}} = \sqrt{\frac{144}{25}} = 2.4$$

the 98 percent confidence interval for μ will be

$110 - 2.492(2.4) < \mu < 110 + 2.492(2.4)$

or

$$104.02 < \mu < 115.98$$

That is, the interval between 104.02 and 115.98 pounds obtained from a sample taken at random will include the true mean of the tensile strength of the steel wire with a probability of 0.98.

It should be pointed out that the technique of using an estimate of the population variance as discussed in this section applies only to testing hypotheses about the *mean* and not about the *proportion*. This is because the value of the standard error of the sample proportion $\hat{p}$ becomes known when an exact hypothesis about the population proportion p is specified. Consequently, it is not necessary to estimate, on the basis of sample information, the population variance for testing hypotheses about a population proportion.

Exercise 11-2

1 Suppose that X is $N(\mu, \sigma^2)$, where μ and σ^2 are not known, and that a sample of 100 ($n = 100$) observations yields a mean $\bar{X}$ and a variance $\hat{s}^2 = 25$. At $\alpha = 0.05$, set up the decision rule for testing the null hypothesis H_0 against the alternative hypothesis H_1.

	H_0	H_1
a	$\mu = 10$	$\mu > 10$
b	$\mu = 0$	$\mu < 0$
c	$\mu \geq -5$	$\mu < -5$
d	$\mu = 5$	$\mu \neq 5$

2 Suppose that X is $N(\mu, \sigma^2)$, where μ and σ^2 are not known, and that a sample of 16 ($n = 16$) observations yields a mean $\bar{X}$ and a variance $\hat{s}^2 = 64$. At $\alpha = 0.01$, set up the decision rule for testing H_0 against H_1.

	H_0	H_1
a	$\mu = 10$	$\mu > 10$
b	$\mu = 0$	$\mu < 0$
c	$\mu \geq -5$	$\mu < -5$
d	$\mu = 5$	$\mu \neq 5$

3 A truck was designed to carry 6.3 tons of earth per load. Exactly 100 sample loads were made on a test run. The sample yielded a mean $\bar{X} = 6.0$ tons and a variance $\hat{s}^2 = 2.25$ tons. On the basis of the sample information, will you reject the null hypothesis $H_0: \mu = 6.3$ in favor of the alternative hypotheses:

a $H_1: \mu < 6.3$ at $\alpha = 0.05$?

b $H_1: \mu < 6.3$ at $\alpha = 0.01$?

c $H_1: \mu \neq 6.3$ at $\alpha = 0.05$?

d $H_1: \mu \neq 6.3$ at $\alpha = 0.01$?

4 Suppose that in the production department of a manufacturing company, the average output per man-hour had been 100 units before an innovation device was introduced. After the innovation device was installed, a random sample of 36 man-hours yielded a mean $\bar{X} = 105$ units and a variance $\hat{s}^2 = 225$ units. On the basis of the sample information, will you reject the null hypothesis H_0: $\mu = 100$ in favor of the alternative hypothesis H_1: $\mu > 100$:

a At $\alpha = 0.05$? *b* At $\alpha = 0.01$?

5 Given the information in Prob. 4, if the true output per man-hour was 103 ($\mu = 103$) after the innovation device was installed, what would be the value of β, the size of the type II error:

a At $\alpha = 0.05$? *b* At $\alpha = 0.01$?

6 Given the following information, what is your conclusion in testing each of the indicated null hypotheses?

	n	$\bar{X}$	$\hat{s}^2$	α	H_0	H_1
a	9	12	36	0.05	$\mu = 10$	$\mu > 10$
b	16	13	64	0.05	$\mu = 10$	$\mu \neq 10$
c	16	11	81	0.01	$\mu \leq 10$	$\mu > 10$
d	25	8	64	0.01	$\mu \geq 10$	$\mu < 10$
e	25	9	49	0.02	$\mu = 10$	$\mu \neq 10$

7 Suppose that the average number of customers who entered a small discount house per day was 175 ($\mu = 175$). For 7 weeks, an extensive advertising campaign was conducted. During the campaign, the average number of customers increased to 181 per day ($\bar{X} = 181$) with a sample variance $\hat{s}^2 = 441$. Did the advertising campaign significantly increase the number of customers entering the store at $\alpha = 0.01$? At $\alpha = 0.05$?

8 A manufacturing company making automobile tires claims that the average life of its product is 35,000 miles. A random sample of 16 tires was selected; and it was found that the mean life was 34,000 miles with a standard deviation $\hat{s} = 2{,}000$ miles. Test the hypothesis H_0: $\mu = 35{,}000$ against the alternative H_1: $\mu < 35{,}000$ at $\alpha = 0.01$. Also, at $\alpha = 0.05$.

9 For a long period in the past, the length of a type of screw produced by a machine has been 2 inches. Suppose that it is desired to determine whether the machine is now functioning properly. A sample of 16 screws is selected at random, which gives an average length of 2.25 inches and a variance $\hat{s}^2 = 0.25$ inch. Test the hypothesis that the machine is functioning properly at:

a $\alpha = 0.01$. *b* $\alpha = 0.05$.

10 A manufacturing company claimed that the ropes it produces have a mean breaking strength of 6,500 pounds. A customer of the company, who was dubious about the company's claim, selected a sample of nine ropes at random and found that the mean breaking strength $\bar{X} = 6{,}200$ pounds and a variance $\hat{s}^2 = 3{,}600$ pounds. Should the customer accept the company's claim at a significant level of:

a Five percent? *b* One percent?

11-3
Inference about the Difference between Two Means

Statistical inference about the mean of a single population on the basis of a small sample represents an important application of the *t* distribution. Generally, however, we are more interested in making inferences about the difference between the means of two populations. Suppose that two samples are drawn, and to each of these samples a different treatment is applied. Each sample is considered representative of the potential population under the same treatment. On the basis of the difference between the sample means, inferences are made about the difference of the population means to determine the relative effects of the treatments. Or suppose that two experimental groups are selected. To one of the groups a certain treatment is applied, and to the other nothing is done. The latter is called the *control group* and is compared with the *treated group*. The two groups are samples from the two potential populations; one is treated and the other is not. On the basis of the difference between the sample means, inferences are made on whether the treatment has any significant effect.

When we infer about the difference between two means, our discussion should be conducted under two separate assumptions. One assumption is that the two population variances are different. Under this assumption, the computation of the number of degrees of freedom, ν, is quite involved. It is therefore more convenient to apply the normal approximation, and consequently, large samples are required. Another assumption is that the two population variances are equal. Under this assumption, the computation of the number of degrees of freedom is a simple process. We can then apply the *t* distribution, and small samples present no special problem. Accordingly, we shall proceed with our discussion covering both large and small samples.

11-3-1 Large samples

In the following discussion we shall identify one population with the subscript 1, and the other with the subscript 2. The mean and the variance relative to population 1 will be similarly identified with the subscript 1, and the mean and variance relative to population 2 will be identified with the subscript 2.

Let $\overline{X}_1$ and $\overline{X}_2$ be the means of two *independent* samples with sizes n_1 and n_2 taken respectively from populations 1 and 2. If $\overline{D}$ stands for the difference between $\overline{X}_1$ and $\overline{X}_2$, that is, $\overline{D} = \overline{X}_1 - \overline{X}_2$, and if the two populations have means μ_1 and μ_2 and variances ${\sigma_1}^2$ and ${\sigma_2}^2$, then we know from Sec. 9-4 that $\overline{D}$ is normally distributed with a mean

$$E(\overline{D}) = \mu_1 - \mu_2$$

and variance

$${\sigma_{\overline{D}}}^2 = \frac{{\sigma_1}^2}{n_1} + \frac{{\sigma_2}^2}{n_2}$$

provided that the two populations are normal. If the two populations are not normal, then by the central limit theorem, $\bar{D}$ will still be approximately normal as long as the samples are large. Thus, whenever the population variances are known and the samples are large, we can apply the normal distribution for inferences about the difference between two means regardless of the nature of the population distribution.

When the variances are not known, however, we have to estimate the population variances as in the last section. Consequently, the variance, and so the standard error, of the difference $\bar{D}$ must also be estimated. The best unbiased estimates for ${\sigma_1}^2$ and ${\sigma_2}^2$ are, of course, ${\hat{s}_1}^2$ and ${\hat{s}_2}^2$, the variances of the respective samples. The variances of the sample means are given by

$$\hat{s}_{\bar{X}_1}{}^2 = \frac{{\hat{s}_1}^2}{n_1}$$

and

$$\hat{s}_{\bar{X}_2}{}^2 = \frac{{\hat{s}_2}^2}{n_2}$$

Therefore the *standard error of the difference* between two sample means is estimated as

$$\hat{s}_{\bar{D}} = \sqrt{\frac{{\hat{s}_1}^2}{n_1} + \frac{{\hat{s}_2}^2}{n_2}} \tag{11-5}$$

Let δ (a Greek letter read as "delta") stand for the difference between the population means, that is, $\delta = \mu_1 - \mu_2$. Then the t ratio becomes

$$t = \frac{\bar{X}_1 - \bar{X}_2 - \delta}{\hat{s}_{\bar{D}}} = \frac{\bar{X}_1 - \bar{X}_2 - \delta}{\sqrt{{\hat{s}_1}^2/n_1 + {\hat{s}_2}^2/n_2}} \tag{11-6}$$

When samples are large, the value of t is close to the corresponding value of z. We can then use the normal approximation to the sampling distribution of the difference $\bar{D}$. This is permissible regardless of the distribution of the population from which each sample is taken. Thus, for estimating the difference between two population means and for hypothesis testing, the above ratio may be treated as the standardized normal score rather than a t value, just as we did in inferring about the mean of a single population in Sec. 11-2-1.

Suppose that we wish to test the hypothesis about the true difference between the average breaking strength of nylon cord and the average breaking strength of cotton cord. For this purpose, a random sample of 60 lengths of nylon cord ($n_1 = 60$) and another sample of 40 lengths of cotton cord ($n_2 = 40$) are tested, and the following means and variances are obtained:

	Nylon cord	*Cotton cord*
Mean	$\bar{X}_1 = 810$	$\bar{X}_2 = 835$
Variance	$\hat{s}_1^2 = 2{,}700$	$\hat{s}_2^2 = 1{,}440$

Substituting the values of n_1, n_2, s_1^2, and s_2^2 in Eq. (11-6), we have

$$\hat{s}_{\bar{D}} = \sqrt{\frac{2{,}700}{60} + \frac{1{,}440}{40}} = \sqrt{45 + 36} = \sqrt{81} = 9$$

We ask: Does the observed difference -25 significantly differ from 0? In other words, we want to test the hypothesis that $\mu_1 = \mu_2$, or $\delta = 0$, against the alternative hypothesis that $\mu_1 \neq \mu_2$; that is,

$$H_0\colon \quad \mu_1 - \mu_2 = 0 \qquad H_1\colon \quad \mu_1 - \mu_2 \neq 0$$

With the level of significance at 0.05, the critical t value is approximated as 1.96. Since $\hat{s}_{\bar{D}}$ is found to be 9, the decision rule will be

Reject H_0 if $\bar{D} \leq -1.96(9) = -17.64$ or $\bar{D} \geq +1.96(9) = +17.64$

Since the observed difference is -25, we shall reject the null hypothesis in favor of the alternative that the average breaking strength of the nylon cord is not equal to the average breaking strength of the cotton cord.

If a 95 percent confidence interval for δ is desired, the confidence limits for δ can be estimated by referring to the standard normal distribution for the t value ($t \approx 1.96$) as follows:

$$-25 - 1.96(9) < \delta < -25 + 1.96(9)$$

or

$$-42.64 < \delta < -7.36$$

That is, the interval between -42.64 and -7.36 obtained through a process of random sampling would include the true difference between the means of the breaking strengths of the nylon cord and cotton cord with a probability of 0.95.

11-3-2 Small samples

When samples are small, the sample variances $\hat{s}_1^2$ and $\hat{s}_2^2$ should not be used as estimates of σ_1^2 and σ_2^2, and consequently Eq. (11-5) should not be used to compute the standard error of the difference. This is so because when samples are small, we have to use the t distribution, which in turn requires that we know the degrees of freedom involved. When the two population variances are not equal, however, the number of degrees of freedom is not readily known. As a result, we must further assume that the two variances are *equal* when samples are small and the t distribution is used for inferring about the difference between two means. Making inferences about the difference between two means, we need further to distinguish between two cases that have rather different applications. In one case the two samples are *independent*; in the other they are *dependent* because observations in one sample are paired and related to those of the other sample.

Independent samples Two samples are known to be independent if items in one are not in any way associated with the items in the other. When two independent samples are drawn from two populations that have identical variances, the standard error of the difference $\bar{D}$ is derived as follows: Since $\sigma_1{}^2 = \sigma_2{}^2 = \sigma^2$, the population standard error of the difference between the means becomes $\sigma_D = \sigma\sqrt{1/n_1 + 1/n_2}$, and the sample standard error of the difference between the means will be

$$\hat{s}_D = \hat{s}\sqrt{\frac{1}{n_1} + \frac{1}{n_2}} = \hat{s}\sqrt{\frac{n_1 + n_2}{n_1 n_2}} \tag{11-7}$$

where $\hat{s}$ is the estimate of the population standard deviation. Our job is now reduced to finding the value for $\hat{s}$. It is a fact that we can improve the estimate of σ if $\hat{s}$ is computed by pooling the data in the two samples together rather than by using either sample separately. Recall that the estimate of the population standard deviation on the basis of a single sample is

$$\hat{s} = \sqrt{\frac{\sum_{i=1}^{n} (X_i - \bar{X})^2}{n - 1}}$$

The pooled estimate derived from two samples taken from the same population must be

$$\hat{s} = \sqrt{\frac{\sum_{i=1}^{n_1} (X_{1i} - \bar{X}_1)^2 + \sum_{i=1}^{n_2} (X_{2i} - \bar{X}_2)^2}{n_1 + n_2 - 2}}$$

Since by Eq. (5-2) we have

$$\sum_{i=1}^{n} (X_i - \bar{X})^2 = ns^2$$

the pooled estimate of the population standard deviation becomes

$$\hat{s} = \sqrt{\frac{n_1 s_1{}^2 + n_2 s_2{}^2}{n_1 + n_2 - 2}} \tag{11-8}$$

Substituting this quantity for $\hat{s}$ in Eq. (11-7), we obtain the estimated standard error of the difference between two sample means as follows:

$$\hat{s}_{\bar{D}} = \sqrt{\frac{n_1 s_1{}^2 + n_2 s_2{}^2}{n_1 + n_2 - 2}}\sqrt{\frac{n_1 + n_2}{n_1 n_2}} \tag{11-9}$$

This is the denominator of the t ratio used, whenever samples are small and population variances are assumed to be identical, with $n_1 + n_2 - 2$ degrees of freedom.

Suppose, for example, that an identical examination is administered to two classes of students. The first class consists of 18 students from college 1, and the second consists of 12 students from college 2. Relevant test data for these two classes are shown below:

	College 1	*College 2*
Sample size	$n_1 = 18$	$n_2 = 12$
Sample mean (score)	$\bar{X}_1 = 85$	$\bar{X}_2 = 80$
Sample variance	$s_1{}^2 = 240$	$s_2{}^2 = 340$

The estimated standard error of the difference between the two sample means is obtained by substituting the values of n_1, n_2, $s_1{}^2$, and $s_2{}^2$ in Eq. (11-9):

$$\hat{s}_D = \sqrt{\frac{18(240) + 12(340)}{18 + 12 - 2}}\sqrt{\frac{18 + 12}{18(12)}} = \sqrt{300}\sqrt{\frac{5}{36}} = \frac{38.74}{6} = 6.46$$

For testing the null hypothesis that the mean of the test scores for students from college 1 is equal to that for students from college 2 against the alternative hypothesis that they are not equal at significance level 0.05, the critical t ratio for 28 degrees of freedom, $18 + 12 - 2 = 28$, is 2.048. Thus the decision rule is

Reject H_0 if $\bar{D}$, or $\bar{X}_1 - \bar{X}_2$, $\leq -2.048(6.46) = -13.23$ or $\bar{D} \geq 2.048(6.46) = 13.23$

Since the observed difference is $85 - 80 = 5$, which is greater than -13.23 and smaller than 13.23, the null hypothesis shall not be rejected.

The same conclusion can be reached by comparing the computed t ratio with the critical t value taken from Appendix VII. The computed t ratio is

$$t = \frac{85 - 80}{6.46} = \frac{5}{6.46} = 0.77$$

Since the computed t ratio is smaller than 2.048 and greater than -2.048, the critical t values for this test, we should not reject the null hypothesis that the two means of the test scores are equal. In other words, we should conclude that as far as their performances in these tests are concerned, the students from the two colleges belong to the same population.

You are asked to verify that we should not reject the null hypothesis even if a one-tailed test is intended with the level of significance at 0.05.

The tests of the null hypothesis $H_0: \mu_1 = \mu_2$ against the respective alternative hypotheses when X_1 and X_2 are independent, and X_1 is $N(\mu_1, \sigma_1^2)$ and X_2 is $N(\mu_2, \sigma_2^2)$, are summarized in Table 11-2. For testing $\mu_1 < \mu_2$, we simply interchange $\bar{X}_1$ and $\bar{X}_2$ and apply the first test shown in the table.

As briefly indicated in the beginning of this chapter, in order to apply the t test, it is necessary to assume that the population under consideration is normal. This assumption is comparatively unimportant because any departure from normality will not affect the validity of the test if samples are sufficiently large. The assumption of identical variances, on the other hand, is a more crucial one. This explains why the t test was, in earlier statistical texts, preceded by a test to determine whether the difference between the two sample variances was significantly different from 0. If not, then the t test was applied. However, since the t test is used when samples are small, and with small samples the test about the identity of variances is not good, more modern statisticians suggest a different approach. It is believed that if samples are of *equal size*, comparatively large differences in the population variances may have little effect on the result of any t test. Accordingly, whenever the identity of variances is questioned, we should try to select samples of equal size.

Dependent samples It frequently happens that observations are sampled in pairs. Each observation in one sample is matched with some observation in another sample. The two elements of a matched pair are *dependent* in the sense that knowing one would reduce the error of guessing the other. For instance, two tests, which are different but designed to measure similar characteristics, are given to a group of students. For a given student, the score in one test can be expected to match his score in another test. Naturally, knowing one score in a pair gives us some knowledge about the other. Pairing may also be achieved if subjects are matched as husband and wife or by age or other characteristics.

Table 11-2

Tests of hypotheses $H_0: \mu_1 = \mu_2$ against H_1 when

$$t = \frac{\bar{X}_1 - \bar{X}_2}{\sqrt{(n_1 s_1^2 + n_2 s_2^2)/(n_1 + n_2 - 2)}\sqrt{(n_1 + n_2)/n_1 n_2}}$$

H_1	*Decision rule: Reject H_0 if*	*Where at a level α and $v = n_1 + n_2 - 2$*
$\mu_1 > \mu_2$	$t \geq c_1$	$c_1 = t_{(v,\alpha)}$
$\mu_1 \neq \mu_2$	$t \geq c_2$	$c_2 = t_{(v,\alpha/2)}$
	or	
	$t \leq c_3$	$c_3 = -t_{(v,\alpha/2)}$

Since subjects are paired, the samples must be of equal size. Given two samples of size n,

Sample 1: $X_{11}, X_{12}, \ldots, X_{1n}$

Sample 2: $X_{21}, X_{22}, \ldots, X_{2n}$

the expected difference between their respective means must be equal to the difference between the means of the populations from which the samples are taken. That is,

$$E(\bar{D}) = E(\bar{X}_1 - \bar{X}_2) = \mu_1 - \mu_2$$

However, because of pairing and the *dependence* between subjects within each pair, the standard error of the difference is now no longer equal to $\sigma_1^2/n_1 + \sigma_1^2/n_2$, as indicated below:

$$\begin{aligned}\sigma_D^2 &= E(\bar{X}_1 - \bar{X}_2 - \mu_1 + \mu_2)^2 \\ &+ E[(\bar{X}_1 - \mu_1) - (\bar{X}_2 - \mu_2)]^2 \\ &= E(\bar{X}_1 - \mu_1)^2 + E(\bar{X}_2 - \mu_2)^2 - 2E(\bar{X}_1 - \mu_1)(\bar{X}_2 - \mu_2) \\ &= \sigma_1^2 + \sigma_2^2 - 2E(\bar{X}_1 - \mu_1)(\bar{X}_2 - \mu_2)\end{aligned}$$

The last term consists of two units of the covariance of $\bar{X}_1$ and $\bar{X}_2$, which is not 0, for the two sample means are not independent.

To avoid the difficulty of dealing with the covariance, we usually consider the matched samples a single sample of D's; that is,

$$D_1 = X_{11} - X_{21},\ D_2 = X_{12} - X_{22}, \ldots, D_n = X_{1n} - X_{2n}$$

Thus we have the mean of the differences

$$\bar{D} = \frac{\sum_{i=1}^{n} D_i}{n} = \frac{\sum_{i=1}^{n} X_{1i} - \sum_{i=1}^{n} X_{2i}}{n} = \bar{X}_1 - \bar{X}_2$$

and the estimated standard error for $\bar{D}$,

$$\hat{s}_{\bar{D}} = \sqrt{\frac{\hat{s}_D^{\,2}}{n}} = \frac{\hat{s}_D}{\sqrt{n}} \tag{11-10}$$

which is used as the denominator of the t ratio. We now have

$$t = \frac{\bar{D} - E(\bar{D})}{\hat{s}_D/\sqrt{n}} \tag{11-11}$$

with $n - 1$ degrees of freedom.

Note that a confidence interval for $E(\bar{D})$ is estimated just as for a single mean with $\bar{X}$ and $\hat{s}/\sqrt{n}$ replaced by $\bar{D}$ and $\hat{s}_D/\sqrt{n}$, respectively. Any hypothesis about $E(\bar{D})$ can be tested in the usual way by applying Eq. (11-11) with $n - 1$ degrees of freedom.

Suppose, for example, that we wish to determine the effect of a given training course on the efficiency of workers. For this purpose, 16 production workers are selected, and their efficiency scores both before and after the training are obtained and shown in Table 11-3.

The mean of the differences, or $\bar{D}$, is

$$\bar{D} = \frac{\sum D}{n} = \frac{16}{16} = 1$$

and the unbiased estimator of the population variance is computed by Eq. (9-7) as follows:

$$\hat{s}_D{}^2 = \frac{\sum D^2 - (\sum D)^2/n}{n-1} = \frac{458 - (16)^2/16}{15} = \frac{458 - 16}{15} = \frac{442}{15} = 29.47$$

Thus the standard deviation of D is

$$\hat{s}_D = \sqrt{29.47} = 5.43$$

Table 11-3

Efficiency scores of 16 production workers before and after a training course

	Efficiency score			
Workers	*After training* X_1	*Before training* X_2	*Difference* $D = X_1 - X_2$	D^2
A	80	75	5	25
B	90	83	7	49
C	92	96	−4	16
D	75	77	−2	4
E	86	81	5	25
F	90	90	0	0
G	81	82	−1	1
H	70	67	3	9
I	89	94	−5	25
J	88	85	3	9
K	82	78	4	16
L	79	82	−3	9
M	91	96	−5	25
N	90	80	10	100
O	78	87	−9	81
P	89	81	8	64
Total			$\sum D = 45 - 29 = 16$	$\sum D^2 = 458$

and the standard error of $\bar{D}$ is

$$\hat{s}_{\bar{D}} = \frac{5.43}{\sqrt{16}} = \frac{5.43}{4} = 1.36$$

Let μ_1 and μ_2 be the true means of X_1 and X_2, the efficiency scores after and before training, respectively; then the hypotheses to be tested are

H_0: $\mu_1 = \mu_2$ $\qquad H_1$: $\mu_1 \neq \mu_2$

Since the number of degrees of freedom is $16 - 1 = 15$, the critical t values with the level of significance specified at 0.02 are $\pm t_{(15,0.01)} = \pm 2.602$. Accordingly, the decision rule is

Reject H_0 if the computed t ratio is greater than 2.602 or smaller than -2.602.

Since the computed t ratio is

$$t = \frac{\bar{D}}{\hat{s}_{\bar{D}}} = \frac{1}{1.36} = 0.74$$

which is greater than -2.602 and smaller than 2.602, we do not reject H_0. We conclude that the training course does not have a significant effect upon the production efficiency at $\alpha = 0.02$.

Given that α is specified at 0.10, then the $1 - \alpha = 90$ percent confidence interval for the true difference δ can be constructed as follows:

$$\bar{D} - t_{(15,0.05)}\hat{s}_{\bar{D}} < \delta < \bar{D} + t_{(15,0.05)}\hat{s}_{\bar{D}}$$

That is,

$$1 - 1.753(1.36) < \delta < 1 + 1.753(1.36)$$

Consequently,

$$-1.384 < \delta < 3.384$$

Exercise 11-3

1 What is the basic reason for testing about the significance of difference between two sample means?

2 Suppose that two randomly selected samples yield the following information:

	Sample I	*Sample II*
Size	$n_1 = 82$	$n_2 = 41$
Mean	$\bar{X}_1 = 50$	$\bar{X}_2 = 55$
Variance	$\hat{s}_1^2 = 405$	$\hat{s}_2^2 = 324$

Test the null hypothesis that the two population means are equal, that is, $H_0: \mu_1 = \mu_2$, against the alternative hypothesis:

a $H_1: \mu_1 \neq \mu_2$ at $\alpha = 0.05$. *b* $H_1: \mu_1 < \mu_2$ at $\alpha = 0.01$.

3 A production process is designed to produce identical parts. Two operators who work two separate shifts are required to work 49 days each. On the average, the first operator turns out 45 parts per day with a variance $\hat{s}^2 = 109$ parts, and the second operator turns out 40 parts with a variance $\hat{s}^2 = 87$ parts. At the 5 percent level, can we conclude that a significant difference exists between the average daily outputs of the two operators?

4 An identical examination was administered to two classes of randomly selected students. The first class, which contained 144 boys, scored an average grade of 88 points with a variance $\hat{s}_1{}^2 = 47$ points; the second class, which contained 144 girls, scored an average grade of 90 points with a variance $\hat{s}_2{}^2 = 34$ points. Is there a significant difference in the performance of the boys and the girls at the 5 percent level? At the 1 percent level?

5 Two samples are randomly selected from two classes of students who have been taught by different methods. An examination is given, and the results are shown as follows:

	Class I	*Class II*
Sample size	$n_1 = 8$	$n_2 = 10$
Mean	$\bar{X}_1 = 95$	$\bar{X}_2 = 97$
Variance	$s_1{}^2 = 47$	$s_2{}^2 = 30$

On the assumption that the test scores of the two classes of students have identical variances, determine whether the two different methods of teaching are equally effective at $\alpha = 0.01$ and 0.05 respectively.

6 Suppose that an experiment designed to compare the average service lives of two kinds of light bulbs has found that the difference between the means of the two experimental groups is 1,200 hours. Furthermore, the pooled estimate of the variance for the difference is found to be 360,000 hours. Fifteen bulbs are included in each of the two groups. Are the service lives of the two kinds of bulbs significantly different from each other at $\alpha = 0.02$? Find the 95 percent confidence interval for the true difference between the average lives.

7 Two random samples taken independently from normal populations with an identical variance yield the following results:

	Sample I	*Sample II*
Size	$n_1 = 10$	$n_2 = 18$
Mean	$\bar{X}_1 = 10$	$\bar{X}_2 = 25$
Variance	$s_1{}^2 = 1{,}200$	$s_2{}^2 = 900$

Test the hypothesis that the true difference between the population means is 10, that is, that $\mu_2 - \mu_1 = 10$, against the alternative that $\mu_2 - \mu_1 > 10$ at the 5 percent level.

8 Suppose that we wish to test the hypothesis that the means of two normal populations are equal. Two samples taken respectively from the populations give the following paired results:

Item	Values in sample I	Values in sample II
1	13	18
2	19	13
3	14	17
4	12	17
5	21	16
6	14	19
7	17	22
8	20	24
9	17	19

If we cannot assume that the populations have equal variances or that the samples are independent, test the hypothesis that $\mu_1 = \mu_2$ against the alternative $\mu_1 \neq \mu_2$ at $\alpha = 0.05$. Find the 99 percent confidence interval for the true difference between the means of the two populations.

9 Suppose it is desired to determine whether the graduating class of 16 students in the statistics department of a certain university can do equally well in two different tests. The test scores of the 16 students are shown as follows:

Student number	First test	Second test
1	90	84
2	90	84
3	80	82
4	90	94
5	92	90
6	88	85
7	90	89
8	63	62
9	70	65
10	54	52
11	78	72
12	86	90
13	99	98
14	84	89
15	56	58
16	85	85

Test the null hypothesis that $\mu_1 = \mu_2$ against the alternative hypothesis that $\mu_1 \neq \mu_2$ at the 5 and the 1 percent level, respectively. Also, find the 95 and 99 percent confidence limits for the true difference between the two population means.

Chapter 12

The chi-square distribution

Another well-known probability distribution is the chi-square distribution, or the distribution of χ^2 (square of the Greek letter read as "chi"). It was first described and named by Karl Pearson in about 1900. Like the t ratio, χ^2 is a random variable and test statistic. The distrubution of χ^2 has very broad applications, and ranks close to the normal distribution in general usefulness.

In the preceding chapters we examined the procedures of making inferences about measures of central tendency such as the mean and the proportion. Frequently, however, we encounter situations calling for inferences about the *dispersion* of a distribution, and thus a technique for inferring about the *variance* of a distribution is in order. In this chapter, we shall consider the method of hypothesis testing for the variance of a population, since the sample statistic computed for this purpose follows the pattern of the χ^2 distribution.

Another application of the χ^2 distribution is the test of *goodness of fit*. This is the test of a hypothesis concerning the *difference* between the *observed* frequencies of a sample and the corresponding *expected* frequencies based on a theoretical distribution. The sample statistic computed for this test is the sum of the squared differences each divided by the expected frequency. It has been found that such a sample statistic is distributed approximately as χ^2.

A closely related technique is the test of *independence*, also known as *Pearson's test for association*. This test is very similar to the test of goodness of fit, and some statisticians prefer to treat them as the same test with minor variations. It is more convenient, however, to discuss them separately. Although the procedure of an independence test is essentially the same comparison between observed and theoretical frequencies, its emphasis is on the classification of variables. Hypothesis testing is designed to determine whether there is any evidence of independence between the variables in the classification.

In short, we shall in this chapter concern ourselves primarily with three types of tests, all of which are applications of the χ^2 distribution. They include the test for the variance of a single population, the test for goodness of fit, and the test for independence. Before considering these tests, we shall look into the nature of the χ^2 distribution.

12-1
Nature of the χ^2 Distribution

Like the t ratio, the test statistic χ^2 is closely related to the standardized normal deviate Z. Also, the shape of the χ^2 distribution changes as the sample size n increases. We shall in this section examine the relationship between χ^2 and Z and the shape of the χ^2 distribution.

12-1-1 Relation of χ^2 to Z

If a random variable X is $N(\mu, \sigma^2)$, then the standardized variable

$$Z = \frac{X - \mu}{\sigma}$$

is normal with mean 0 and variance 1. We may raise the question: Is the distribution of Z^2 also normal? The answer is simply "no." It is interesting to note that when a single value of X is obtained, the resulting quantity Z^2 is exactly what we now refer to as χ^2. That is, the variable

$$Z^2 = \frac{(X - \mu)^2}{\sigma^2} \tag{12-1}$$

is distributed as χ^2 with 1 degree of freedom.

The variable χ^2 has a family of distributions, one corresponding to each number of degrees of freedom. The number of degrees of freedom is exactly equal to the number of observations. When a single observation is made, that is, when $n = 1$, the squared difference between the observed value and the population mean is simply

$$(X - \mu)^2$$

which we know from the last paragraph of Sec. 11-1-2 has 1 degree of freedom. Thus, using the number of degrees of freedom as a subscript of χ^2, we have

$$\chi^2_{(1)} = \frac{(X - \mu)^2}{\sigma^2} \tag{12-2}$$

Now suppose that two observations are made independently, or $n = 2$. Accordingly, two squared Z scores are obtained as follows:

$$Z_1{}^2 = \frac{(X_1 - \mu)^2}{\sigma^2} \qquad \text{and} \qquad Z_2{}^2 = \frac{(X_2 - \mu)^2}{\sigma^2}$$

Since the sum of the squared differences

$$\sum_{i=1}^{2} (X_i - \mu)^2$$

has 2 degrees of freedom, the sum of $Z_1{}^2$ and $Z_2{}^2$ has 2 degrees of freedom. It is in fact an important property of the χ^2 distribution that if two independent variables are each distributed as χ^2 with 1 degree of freedom, their sum is distributed as χ^2 with 2 degrees of freedom; that is,

$$Z_1{}^2 + Z_2{}^2 = \sum_{i=1}^{2} \frac{(X_i - \mu)^2}{\sigma^2} = \chi^2_{(2)} \tag{12-3}$$

It is now easy to see that in the general case when n observations are made, the sum of the squared differences

$$\sum_{i=1}^{n} (X_i - \mu)^2$$

must have n degrees of freedom. Therefore we have

$$\chi^2_{(n)} = \sum_{i=1}^{n} Z_i{}^2 = \sum_{i=1}^{n} \frac{(X_i - \mu)^2}{\sigma^2} \tag{12-4}$$

Since theoretically n can be any positive integer, there is conceivably an infinite number of χ^2 distributions.

The above discussion can be easily expanded to include two or more random variables. Suppose that X is normally distributed with mean μ_X and variance $\sigma_X{}^2$, and Y is also normally distributed with mean μ_Y and variance $\sigma_Y{}^2$, and X and Y are independent. Then a sample of $X_1, X_2, \ldots, X_n$ will yield

$$\sum_{i=1}^{n} Z_i{}^2 = \sum_{i=1}^{n} \frac{(X_i - \mu_X)^2}{\sigma_X{}^2}$$

which is distributed as χ^2 with n degrees of freedom. Similarly, a sample of $Y_1, Y_2, \ldots, Y_m$ will yield

$$\sum_{i=1}^{m} Z_i{}^2 = \sum_{i=1}^{m} \frac{(Y_i - \mu_Y)^2}{\sigma_Y{}^2}$$

which is distributed as χ^2 with m degrees of freedom. Adding these two sums, we get $n + m$ squared Z scores. Consequently, $\sum_{i=1}^{n} Z_i{}^2 + \sum_{i=1}^{m} Z_i{}^2$ must be distributed as χ^2 with $n + m$ degrees of freedom; that is,

$$\chi^2_{(n+m)} = \chi^2_{(n)} + \chi^2_{(m)} \tag{12-5}$$

This is to say that if one statistic has a χ^2 distribution with n degrees of freedom and another statistic, independent of the first, also has a χ^2 distribution with m

degrees of freedom, the new statistic obtained by adding the two statistics together is itself distributed as χ^2 with $n + m$ degrees of freedom.

Similarly, it can be easily demonstrated that

$$\chi^2_{(n-m)} = \chi^2_{(n)} - \chi^2_{(m)} \tag{12-6}$$

12-1-2 Pattern of the χ^2 distribution

First, let us take a cursory look at the function rule of the χ^2 distribution. For each value of χ^2, the probability density is given as follows:

$$f(\chi^2) = k(\chi^2)^{(\nu/2)-1}e^{-\chi^2/2} \qquad \text{for } \chi^2 > 0 \tag{12-7}$$

where $e = 2.71828$
ν = number of degrees of freedom
k = constant depending only on ν

Thus, like the t distribution, the χ^2 distribution has only one parameter ν.

It is rather difficult to detect the shape of the χ^2 distribution by examining the above function rule. However, since the number of degrees of freedom is the only parameter of the distribution, its exact shape must be dictated by the value of ν. In fact, for small ν the distribution is *positively skewed,* or skewed to the right. As ν increases, the curve becomes less skewed to the right and approaches symmetry rapidly. As indicated in Fig. 12-1, the distribution becomes almost symmetric when ν is only 10. For large values of ν, χ^2 becomes approximately normally distributed. In any case the distribution is unimodal and continuous.

The mean of the χ^2 distribution can be easily determined. From Eq. (12-4) we have

$$E[\chi^2_{(\nu)}] = E\left(\sum_{i=1}^{n} Z_i^2\right)$$
$$= \sum_{i=1}^{n} E(Z_i^2)$$

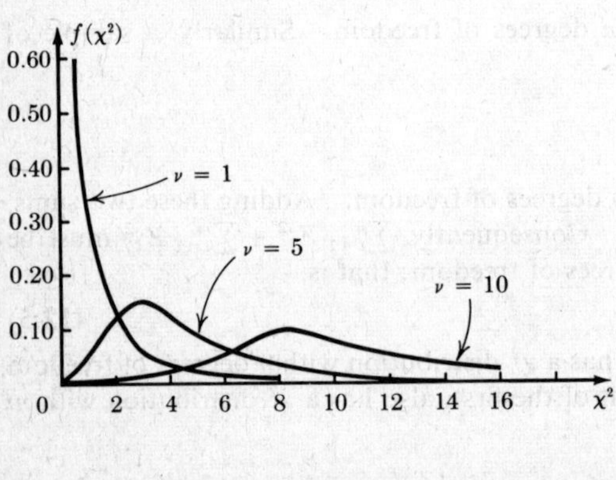

Figure 12-1
Chi-square distributions with 1, 5, and 10 degrees of freedom.

Since the variance of Z is

$$\sigma_Z^2 = E(Z^2) - [E(Z)]^2$$
$$= E(Z^2) - 0^2 = E(Z^2) = 1$$

we get

$$E[\chi^2_{(\nu)}] = \sum_{i=1}^{n} 1 = n = \nu \qquad (12\text{-}8)$$

That is, the mean of the χ^2 distribution is equal to the number of degrees of freedom. The variance of $\chi^2_{(\nu)}$ is simply equal to 2ν. This again illustrates that, given the number of degrees of freedom, the χ^2 distribution is completely defined.

As with other density functions, the area under the χ^2 curve represents probability. A table of areas, or probabilities, for a particular χ^2 distribution can be constructed in the same detailed form as the table for the standard normal distribution. However, since there is a distinct χ^2 distribution for each value of ν, and a full page is needed for each distribution, it would require as many pages for the χ^2 table as for the values of ν. To construct a χ^2 table in such a detailed fashion is obviously quite impractical. Actually, we seldom need such a detailed table; a condensed form such as that shown in Appendix VIII is sufficient for most practical purposes.

Appendix VIII gives the values of χ^2 that cut off the upper (right) tail of the distribution specified by the number of degrees of freedom (see the figure on top of the table). The top row shows the column heads, which are percentage points constituting the *right* tail of the curve. Each of the other rows lists the corresponding values of χ^2 for a particular distribution with its degrees of freedom indicated in the first column.

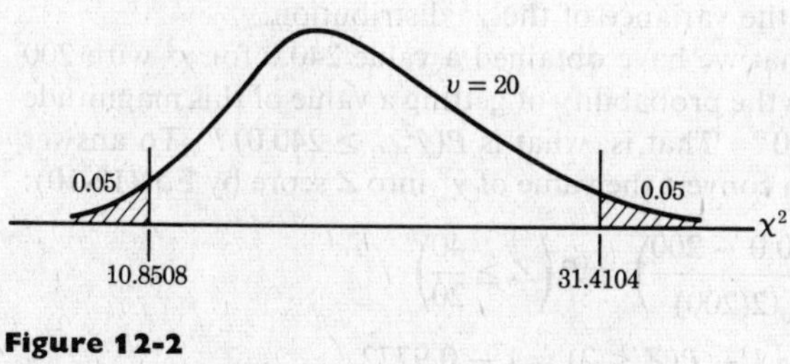

Figure 12-2
The χ^2 distribution with 20 degrees of freedom.

Suppose, for example, that we are dealing with a random variable possessing the χ^2 distribution with 10 degrees of freedom. We find that the value of χ^2 in the column headed 0.05 for 10 degrees of freedom is 18.307. That is, the probability that a random sample yields a χ^2 value 18.307 *or higher* is 0.05. In the distribution with 20 degrees of freedom, a χ^2 value 28.412 *or higher* will occur with a probability of 0.10. A χ^2 value that cuts off the lower (left) 5 percent of a distribution is the value that cuts off the upper 95 percent of the same distribution. Thus, as shown in Fig. 12-2, the χ^2 values that cut off the upper and lower 5 percent of the distribution with 20 degrees of freedom are, respectively, 31.4104 and 10.8508.

For values of v greater than 100, the χ^2 value may be approximated by using

$$\chi^2_{(v,\alpha)} = \tfrac{1}{2}(z_\alpha + \sqrt{2v - 1})^2 \tag{12-9}$$

where z_α is the standard normal score corresponding to the α level of significance. For instance, if we are dealing with a χ^2 distribution with 113 degrees of freedom, the χ^2 value that cuts off the upper 5 percent of the curve ($\alpha = 0.05$) is

$$\begin{aligned}\chi^2_{(113,0.05)} &= \tfrac{1}{2}(1.645 + \sqrt{2(113) - 1})^2 \\ &= 138.5\end{aligned}$$

That is, random samples of 113 observations each showing a χ^2 value of 138.5 or higher should occur with a probability of 0.05.

For extremely large v, normal approximation may be used to obtain probabilities for intervals of χ^2 values. This is permissible because the χ^2 distribution approaches the normal distribution as the number of degrees of freedom approaches infinity. Recall that by the central limit theorem, the statistic $\bar{X}$ computed from a *large* sample is normal or approximately normal regardless of the population distribution. This is also the case with the statistic χ^2. For sufficiently large v, the probability of any interval of χ^2 values may be approximated by using the standard normal table with the Z score given by

$$Z = \frac{\chi^2_{(v)} - v}{\sqrt{2v}} \tag{12-10}$$

since v is the mean and $2v$ is the variance of the χ^2 distribution.

Suppose, for instance, that we have obtained a value 240.0 for χ^2 with 200 degrees of freedom. What is the probability of getting a value of this magnitude or greater for χ^2 with $v = 200$? That is, what is $P(\chi^2_{200} \geq 240.0)$? To answer this question, we need only to convert the value of χ^2 into Z score by Eq. (12-10):

$$\begin{aligned}P(\chi^2_{200} \geq 240.0) &= P\left(Z \geq \frac{240.0 - 200}{\sqrt{2(200)}}\right) = P\left(Z \geq \frac{40}{20}\right) \\ &= P(Z \geq 2) = 1 - P(Z \leq 2) = 1 - 0.9772 \\ &= 0.0228\end{aligned}$$

Thus the probability of obtaining a value of χ^2 equal to or greater than 240.0 for a distribution with 200 degrees of freedom is approximately 0.0228.

Exercise 12-1

1 Define the following concepts:

a χ^2 with 1 degree of freedom. *b* χ^2 distribution.

2 Explain fully the "additivity" of χ^2.

3 In testing hypotheses, we reject the null hypothesis if a computed χ^2 value is too large. Is it possible that the computed χ^2 value is too small? Under what conditions might we be interested in the lower tail of the χ^2 distribution?

4 Given that X is a χ^2 variable with ν degrees of freedom, find the *approximate* values for the following probabilities:

a $P(X \geq 44)$, $\nu = 30$. *d* $P(X \leq 8.3)$, $\nu = 20$.
b $P(X \leq 18.5)$, $\nu = 30$. *e* $P(X \geq 11)$, $\nu = 5$.
c $P(X \geq 35)$, $\nu = 20$.

5 Given that a χ^2 distribution has 15 degrees of freedom, find the critical χ^2 value that cuts off each of the following areas under the χ^2 curve:

a The upper 5 percent. *d* The lower 5 percent.
b The upper 1 percent. *e* The lower 1 percent.
c The upper 90 percent.

6 At $\alpha = 0.01$, find the critical χ^2 value for:

a $\nu = 221$. *b* $\nu = 145$.

7 What is the approximate probability of obtaining a χ^2 value equal to or greater than 152.8 for a distribution with 128 degrees of freedom?

12-2
Inference about a Single Variance

We shall now examine the sampling distribution of the sample variance and the procedure of testing hypotheses about the population variance.

12-2-1 Sampling distribution of the sample variance

The basic assumption underlying the discussion in this section is that the distribution of the random variable X is *normal*; that is, X is $N(\mu, \sigma^2)$.

In Sec. 5-2 the variance of a sample of size n for X was defined as

$$s^2 = \frac{\sum_{i=1}^{n} (X_i - \bar{X})^2}{n}$$

If we divide both sides of this equation by σ^2, we obtain

$$\frac{s^2}{\sigma^2} = \frac{1}{n} \sum_{i=1}^{n} \frac{(X_i - \bar{X})^2}{\sigma^2}$$

Multiplying both sides by n, we have

$$\frac{ns^2}{\sigma^2} = \sum_{i=1}^{n} \frac{(X_i - \bar{X})^2}{\sigma^2}$$

The expression on the right side of this equation is similar to the expression on the right side of Eq. (12-4). The only difference between these two is that the one involves $\bar{X}$, while the other involves μ. We may wonder whether the expression

$$\sum_{i=1}^{n} \frac{(X_i - \bar{X})^2}{\sigma^2}$$

is also distributed as χ^2 with n degrees of freedom. We now proceed to demonstrate that this is *almost* the case if X is normally distributed.

Given the population mean μ, the sum of squared deviations from μ is

$$\sum_{i=1}^{n} (X_i - \mu)^2$$

Since subtracting and adding the same value to any quantity does not change the quantity, we have

$$\begin{aligned}\sum_{i=1}^{n} (X_i - \mu)^2 &= \sum_{i=1}^{n} [(X_i - \bar{X}) + (\bar{X} - \mu)]^2 \\ &= \sum_{i=1}^{n} (X_i - \bar{X})^2 + \sum_{i=1}^{n} (\bar{X} - \mu)^2 + 2\sum_{i=1}^{n} (X_i - \bar{X})(\bar{X} - \mu)\end{aligned}$$

However, $\bar{X} - \mu$ is a constant and $\sum_{i=1}^{n} (X_i - \bar{X}) = 0$, therefore

$$\begin{aligned}2\sum_{i=1}^{n} (X_i - \bar{X})(\bar{X} - \mu) &= 2(\bar{X} - \mu)\sum_{i=1}^{n} (X_i - \bar{X}) \\ &= 0\end{aligned}$$

Hence

$$\begin{aligned}\sum_{i=1}^{n} (X_i - \mu)^2 &= \sum_{i=1}^{n} (X_i - \bar{X})^2 + \sum_{i=1}^{n} (\bar{X} - \mu)^2 \\ &= \sum_{i=1}^{n} (X_i - \bar{X})^2 + n(\bar{X} - \mu)^2\end{aligned}$$

Dividing each term of this equation by σ^2, we get

$$\frac{\sum_{i=1}^{n}(X_i - \mu)^2}{\sigma^2} = \frac{\sum_{i=1}^{n}(X_i - \bar{X})^2}{\sigma^2} + \frac{n(\bar{X} - \mu)^2}{\sigma^2}$$

If X is normally distributed with mean μ and variance σ^2, $\bar{X}$ will be normally distributed with mean μ and variance σ^2/n. Thus

$$\frac{n(\bar{X} - \mu)^2}{\sigma^2} = \frac{(\bar{X} - \mu)^2}{\sigma^2/n} = Z^2 = \chi^2_{(1)}$$

Furthermore, as indicated above,

$$\frac{\sum_{i=1}^n (X_i - \bar{X})^2}{\sigma^2} = \frac{ns^2}{\sigma^2}$$

and

$$\frac{\sum_{i=1}^n (X_i - \mu)^2}{\sigma^2} = \chi^2_{(n)} \qquad \text{from (12-4)}$$

Consequently,

$$\chi^2_{(n)} = \frac{ns^2}{\sigma^2} + \chi^2_{(1)}$$

and by Eq. (12-6) we get

$$\frac{ns^2}{\sigma^2} = \chi^2_{(n-1)} \tag{12-11}$$

That is, the sample variance multiplied by the sample size n and divided by the population variance is distributed as χ^2 with $n - 1$ degrees of freedom. Thus the substitution of $\bar{X}$ for μ in Eq. (12-4) reduces the number of degrees of freedom from n to $n - 1$. This is so because 1 degree of freedom is lost whenever a parameter is replaced by a statistic used as its estimate.

Moreover, if the unbiased estimate $\hat{s}^2$ is used, then, because of the relationship

$$ns^2 = \sum_{i=1}^{n} (X_i - \bar{X})^2 = (n - 1)\hat{s}^2$$

we have

$$\frac{(n - 1)\hat{s}^2}{\sigma^2} = \chi^2_{(n-1)} \tag{12-12}$$

Thus, when inference about a single variance is desired, the test statistic computed from the sample is not s^2 (or $\hat{s}^2$) itself but s^2 (or $\hat{s}^2$) multiplied by n (or $n - 1$) and divided by σ^2. These two additional operations make it possible to apply the χ^2 distribution for inferring about the population variance on the basis of the sample variance.

12-2-2 Test of hypothesis about a single variance

Suppose that we are interested in determining whether a population variance has changed over time or whether the variance of a distribution is different from a specified value. For this purpose, we test the null hypothesis

$$H_0: \quad \sigma^2 = \sigma_0^2$$

where σ_0^2 is the known variance or some hypothetical value. The alternative hypothesis depends on the problem under consideration. If X is $N(\mu, \sigma^2)$ and a sample of size n for X yields a variance s^2, then given the significance level at α, the test statistic

$$\frac{ns^2}{\sigma_0^2}$$

is referred to the χ^2 distribution with $n - 1$ degrees of freedom for determining whether the variance of the random variable X is equal to σ_0^2.

Imagine, for example, that the variability in the hourly output of the workers of a large manufacturing company is suspected of having increased after the introduction of a job enlargement program in which several simple jobs are combined into a single job. The program is intended to reduce monotony of work and thus to increase production. It is suspected, however, that the program may favor the good workers and penalize the poor ones so that the hourly output of the good workers tends to increase and that of the poor workers tends to decrease. If this is the case, we can expect the dispersion of the hourly output to be greater after the program has been instituted. Let X be the average hourly output of any worker. It is assumed that X is $N(\mu, \sigma^2)$. The variance for X before the new program is known to be 80, and a random sample of 26 observations for X yields a variance $s^2 = 90$. Shall we conclude that the program of job enlargement does cause greater output variation?

The null and alternative hypotheses to be tested in this case are

$$H_0: \quad \sigma^2 = 80 \qquad H_1: \quad \sigma^2 > 80$$

We shall reject H_0 in favor of H_1 if the value of s^2 is sufficiently large; that is, if the test statistic

$$\chi^2 = \frac{26s^2}{80}$$

falls within the region of rejection. If the significance level is set at 0.05, the critical value for χ^2 with 25 degrees of freedom is 37.6525; that is,

$$P\left(\frac{26s^2}{80} \geq 37.6525\right) = 0.05$$

$$P\left[s^2 \geq \frac{80}{26}(37.6525)\right] = 0.05$$

hence

$P(s^2 \geq 115.85) = 0.05$

In other words, the critical value for s^2 is 115.85. We shall reject H_0 if s^2 is equal to or larger than 115.85. Since the observed value for s^2 is 90, we shall not reject the null hypothesis.

An equivalent, but more commonly used, test is to compare the value of ns^2/σ^2 with the critical value for χ^2. The null hypothesis will be rejected if $ns^2/\sigma^2 \geq \chi^2$ with $n - 1$ degrees of freedom. In the above example, we shall not reject the null hypothesis since

$$\frac{ns^2}{80} = \frac{26(90)}{80} = 29.25$$

which is smaller than the critical χ^2 value, 37.6525.

The null hypothesis $\sigma^2 = {\sigma_0}^2$ may be tested against the alternative hypothesis $H_1: \sigma^2 < {\sigma_0}^2$ or $H_1: \sigma^2 \neq {\sigma_0}^2$. If the alternative is $H_1: \sigma^2 < {\sigma_0}^2$, then the rejection region relates to the lower tail of the distribution. We shall reject H_0 if $ns^2/{\sigma_0}^2$ is equal to or less than the critical value that cuts off the upper $1 - \alpha$ of the χ^2 distribution with $n - 1$ degrees of freedom. Similarly, if the alternative is $H_1: \sigma^2 \neq {\sigma_0}^2$, the rejection region relates to both tails of the distribution. We shall reject H_0 if $ns^2/{\sigma_0}^2$ is equal to or greater than the χ^2 value cutting off the upper $\alpha/2$ of the distribution with $n - 1$ degrees of freedom, or if it is equal to or smaller than the χ^2 value cutting off the upper $1 - \alpha/2$ of the same distribution. The level of significance is divided equally for the two tails since the distribution becomes almost symmetric with even a moderate number of degrees of freedom.

The various tests for alternative hypotheses are summarized in Table 12-1.

Table 12-1

Tests of H_0: $\sigma^2 = {\sigma_0}^2$ against H_1 when X is $N(\mu, \sigma^2)$ and

$$s^2 = \frac{\sum_{i=1}^{n} (X_i - \bar{X})^2}{n}$$

H_1	*Decision rule: Reject H_0 if*	*Where at a level α and with $v = n - 1$*
$\sigma^2 > {\sigma_0}^2$	$\frac{ns^2}{{\sigma_0}^2} \geq c_1$	$c_1 = \chi^2_{(v,\,\alpha)}$
$\sigma^2 < {\sigma_0}^2$	$\frac{ns^2}{{\sigma_0}^2} \leq c_2$	$c_2 = \chi^2_{(v,\,1-\alpha)}$
$\sigma^2 \neq {\sigma_0}^2$	$\frac{ns^2}{{\sigma_0}^2} \geq c_3$	$c_3 = \chi^2_{(v,\,\alpha/2)}$
	or	
	$\frac{ns^2}{{\sigma_0}^2} \leq c_4$	$c_4 = \chi^2_{(v,\,1-\alpha/2)}$

The reader should be aware that in the case $H_1: \sigma^2 > \sigma_0{}^2$, the test is also good if the null hypothesis is $H_0: \sigma^2 \le \sigma_0{}^2$, and that in the case $H_1: \sigma^2 < \sigma_0{}^2$, the test is also good if the null hypothesis is $H_0: \sigma^2 \ge \sigma_0{}^2$.

12-2-3 Estimation of σ^2

As indicated in Sec. 9-1-2, the unbiased estimator of σ^2 is $\hat{s}^2$, that is, $E(\hat{s}^2) = \sigma^2$. The sample variance s^2 is not an unbiased estimator of σ^2 since

$$E(s^2) = E\left[\frac{1}{n}\sum_{i=1}^{n}(X_i - \bar{X})^2\right] = \frac{1}{n}E[(n-1)\hat{s}^2] = \frac{n-1}{n}E(\hat{s}^2)$$

and

$$E(s^2) = \frac{n-1}{n}\sigma^2 \tag{12-13}$$

Thus, if a point estimate of σ^2 is desired, we shall use $\hat{s}^2$ instead of s^2. The difference is rather small, especially when the sample size is large.

When an interval estimation is desired, we need, first of all, to choose the value of α, the probability that the interval constructed from the sample variance will not include the population variance. That is,

$$P\left[\chi^2_{(n-1,\,1-\alpha/2)} \ge \frac{ns^2}{\sigma^2}\right] + P\left[\chi^2_{(n-1,\,\alpha/2)} \le \frac{ns^2}{\sigma^2}\right] = \alpha$$

where $\chi^2_{(n-1,1-\alpha/2)}$ is the value that cuts off the lower $\alpha/2$, and $\chi^2_{(n-1,\alpha/2)}$ is the value that cuts off the upper $\alpha/2$ of the χ^2 distribution with $n-1$ degrees of freedom. Thus we have

$$P\left[\chi^2_{(n-1,\,1-\alpha/2)} < \frac{ns^2}{\sigma^2} < \chi^2_{(n-1,\,\alpha/2)}\right] = 1-\alpha$$

The inequality signs within the bracket will be reversed as each term is replaced by its reciprocal. Then, multiplying all three terms by ns^2, we obtain

$$P\left[\frac{ns^2}{\chi^2_{(n-1,\,\alpha/2)}} < \sigma^2 < \frac{ns^2}{\chi^2_{(n-1,\,1-\alpha/2)}}\right] = 1-\alpha \tag{12-14}$$

That is, an interval between

$$\frac{ns^2}{\chi^2_{(n-1,\,\alpha/2)}} \quad \text{and} \quad \frac{ns^2}{\chi^2_{(n-1,\,1-\alpha/2)}}$$

will include the true value of σ^2 with a probability of $1-\alpha$.

Suppose that we wish to estimate the upper and lower limits of the 98 percent confidence interval for σ^2 for the example discussed in Sec. 12-2-2. From Appendix VIII we find that

$$\chi^2_{(25,0.01)} = 44.3141 \quad \text{and} \quad \chi^2_{(25,0.99)} = 11.5240$$

Since $n = 26$ and $s^2 = 90$, by Eq. (12-14) we have

$$\frac{26(90)}{44.3141} < \sigma^2 < \frac{26(90)}{11.524}$$

or

$$52.8 < \sigma^2 < 203.1$$

That is, on the basis of the sample observations the 98 percent confidence limits for the true variance σ^2 are estimated to be 52.8 and 203.1.

Exercise 12-2

1 Assume that the random variable X is $N(\mu, \sigma^2)$. Three random samples for X provide the following information:

Size	*Variance*
$n_1 = 3$	$s^2 = 25$
$n_2 = 5$	$s^2 = 16$
$n_3 = 7$	$s^2 = 9$

Given that these sample values are correctly combined, what are the 95 percent confidence limits of σ^2?

2 The standard variance of a certain production process is equal to 4 inches. It is suspected that the variance of the process has become too large. A sample of 9 parts produced by this process are measured at 5, 7, 2, 4, 8, 9, 8, 6, and 5 inches, respectively. Test the hypothesis that the process still has the standard variance at the 1 percent level. Also, test the hypothesis at the 5 percent level.

3 From a sample of 10 observations, we find that the variance s^2 is equal to 15. Find the 90 percent confidence limits of σ^2.

4 Suppose we wish to determine whether the variance of the IQs of the students in a school district is larger than 250. A random sample of 30 students selected from the district yields a variance of 275. Test the hypothesis that the variance of the IQs remains 250 at the 1 percent level.

5 Suppose that in Prob. 4 we wish to conduct a two-tailed test of the null hypothesis at the 5 percent level. Shall we reject the null hypothesis? Find the 95 percent confidence limits for the true variance.

6 An experiment was performed five times to test the stability of a population variance. The five χ^2 values obtained from the experiments were 2.85, 3.05, 3.20, 2.65, and 3.50, each of which corresponds to 1 degree of freedom. Would the null hypothesis be rejected in any of the five experiments at $\alpha = 0.05$? Would it be rejected if the results of the five experiments were combined?

7 In the light of the answers to Prob. 6, what can you say about the relation between the size of a sample and the power of the test?

8 Suppose that the quality of a manufactured product has shown a standard deviation $\sigma = 8$. A random sample of 30 units of the product gives a standard deviation $s = 10$. Is there any justification for concluding that the variability of the quality has increased at $\alpha = 0.05$?

12-3 Test of Goodness of Fit

Another application of the χ^2 distribution is the test of *goodness of fit*. It involves a comparison of the frequency distribution *observed* in the sample with the *expected* frequency distribution based on some theoretical model. From the comparison, inferences are made about the distribution of the population from which the sample is taken. If the difference between the observed and the expected frequencies is so great that it can hardly happen by chance alone, we shall conclude that the sample is not taken from the population that gave rise to the expected frequencies.

Suppose that X is a random variable with a sample space S, and $C_1, C_2, \ldots, C_J$ are J event classes or subsets that partition the sample space so that $C_1 \cup C_2 \cup \cdots \cup C_J = S$. Suppose further that a sample of n observations for X is taken and $f_0(C_j)$ is the observed frequency in the jth class for $j = 1, 2, \ldots, J$. If $p_j = P(X \in C_j)$ is the probability that the jth event will occur, np_j will be the expected frequency for C_j, denoted as $f_e(C_j)$. Then the sum of squared differences between the observed and the expected frequencies as a fraction of the expected frequency, namely,

$$\sum_{j=1}^{J} \frac{[f_0(C_j) - f_e(C_j)]^2}{f_e(C_j)}$$

is commonly used as a measure of the difference between the observed distribution and the expected distribution. As a matter of fact, when sample size n is large, this quantity is distributed *approximately* as χ^2 with $J - 1$ degrees of freedom. For the sake of simplicity, let us use f_{0j} and f_{ej} to replace $f_0(C_j)$ and $f_e(C_j)$, respectively; we have

$$\chi^2_{(J-1)} = \sum_{j=1}^{J} \frac{(f_{0j} - f_{ej})^2}{f_{ej}} \tag{12-15}$$

The larger the sample size, the better the approximation will be. For most practical purposes, however, the approximation will be close enough if n is such that $np_j \geq 5$. When this condition is not met, we may pool the small classes into larger ones so that $np_j \geq 5$ for all j is satisfied. This will be illustrated later. In situations where np_j is less than 5 and the classes are not pooled, the tests should be based on the exact *multinomial* probabilities (discussed in Prob. 12 of Exercise 7-3) rather than the χ^2 approximation.

We can demonstrate the rationale of this approximation by considering the frequency distribution with only *two* classes, that is, $J = 2$. For a two-class distribution, the various quantities in Eq. (12-15) are given as follows:

Class j	*Observed frequency*	*Expected frequency*	*Sample proportion*	*Expected proportion*
1	$f_{01} = n\hat{p}$	$f_{e1} = np$	$\hat{p}$	p
2	$f_{02} = n(1 - \hat{p})$	$f_{e2} = n(1 - p)$	$1 - \hat{p}$	$1 - p$
Total	n	n	1	1

Substituting these quantities in the expression on the right side of Eq. (12-15), we obtain

$$\sum_{j=1}^{2} \frac{(f_{0j} - f_{ej})^2}{f_{ej}} = \frac{(f_{01} - f_{e1})^2}{f_{e1}} + \frac{(f_{02} - f_{e2})^2}{f_{e2}}$$

$$= \frac{(n\hat{p} - np)^2}{np} + \frac{[n(1 - \hat{p}) - n(1 - p)]^2}{n(1 - p)}$$

$$= \frac{(n\hat{p} - np)^2(1 - p) + (-n\hat{p} + np)^2 p}{np(1 - p)}$$

$$= \frac{(n\hat{p} - np)^2[(1 - p) + p]}{np(1 - p)}$$

$$= \frac{(n\hat{p} - np)^2}{np(1 - p)}$$

Observe that $n\hat{p}$ is a binomial random variable with mean np and variance $np(1-p)$. The above expression is exactly the square of the standardized normal score Z, which is used, when n is large, in the normal approximation to the binomial. Thus

$$\sum_{j=1}^{2} \frac{(f_{0j}-f_{ej})^2}{f_{ej}} = Z^2 \tag{12-16}$$

is distributed approximately as χ^2 with $2-1=1$ degree of freedom.

You should not, however, generalize that the number of degrees of freedom in a test of goodness of fit is always $J-1$ with J as the number of classes. Actually, the number of degrees of freedom is also dependent on the number of parameters that have to be estimated for computing the expected frequencies. If no parameter is replaced by its sample estimate in the computation, the number of degrees of freedom will be $J-1$. If one or two parameters are replaced by the sample estimates, the number of degrees of freedom will be $J-2$ or $J-3$, respectively. We shall illustrate this point with two examples. In the first no replacement is made; in the other the sample mean is used to replace the mean of the population.

Suppose, for example, that we wish to test the hypothesis that a die is perfectly balanced. For this purpose, the die is tossed 150 times, and the results of the tosses are noted and recorded. Let X be the number of spots shown on the face of the die after it is thrown. The sample space of X is $S = \{1,2,3,4,5,6\}$. Let us partition this sample space into six mutually exclusive and exhaustive event classes. (Note that we could, if we wished, partition it into two or three events.) That is, each C_j is a one-number subset of S. Here X is *uniformly* distributed since the probability of C_j is identical, or $p_j = \frac{1}{6}$, for all j. The results of the 150 tosses are tabulated in the second column of Table 12-2. The table also shows other quantities for computing the value of χ^2.

When the value of χ^2 is small, we say that the fit is good. Note that this is a one-tailed test since χ^2 can only take on a large value if the fit is bad. Appendix VIII shows that with 5 degrees of freedom, the value of χ^2 that cuts off the upper 5 percent is 11.0705. Thus, at a 5 percent significance level, the observed value 6.32 for χ^2, which is substantially less than 11.07, will not cause us to reject the null hypothesis. In other words, the evidence from this experiment shows that the fit is good and the die used in this experiment is not unbalanced at $\alpha = 0.05$.

Observe that the number of degrees of freedom is the number of distinct classes, or J, minus 1. This is the case because given any $J-1$ differences between the observed and the expected frequencies, the remaining difference is already determined. There is no additional loss of degrees of freedom since it is not required to replace the population parameter by its sample estimate.

As another example, suppose that a manufacturing company wants to

Table 12-2

Computation of χ^2 for the experiment of tossing a die 150 times to compare observed with expected distributions

Event class $C_j = x$	*Observed frequencies* f_{oj}	*Probabilities* p_j	*Expected frequencies* f_{ej}	$(f_{oj} - f_{ej})^2$	$\frac{(f_{oj} - f_{ej})^2}{f_{ej}}$
1	24	$\frac{1}{6}$	25	1	0.04
2	20	$\frac{1}{6}$	25	25	1.00
3	20	$\frac{1}{6}$	25	25	1.00
4	22	$\frac{1}{6}$	25	9	0.36
5	33	$\frac{1}{6}$	25	64	2.56
6	31	$\frac{1}{6}$	25	36	1.36
Total	150	1.0	150		6.32

$$\chi^2 = \sum_{j=1}^{6} \frac{(f_{oj} - f_{ej})^2}{f_{ej}} = 6.32$$

determine whether the number of defective parts used in a certain assembly follows the Poisson distribution. For this purpose, the quality control engineer inspected a sample of 200 boxes, each containing 100 finished parts. In this example, the random variable X is the number of defective parts found in each box and has the following set of possible values:

$\{0,1,2,3, \ldots, 100\}$

From each of the 200 boxes, the engineer found 0, 1, 2, 3, . . . defectives with the frequencies shown in the second column of Table 12-3. From these data, the value of χ^2 is computed to determine whether the random variable has a Poisson distribution.

Observe that in order to satisfy the requirement $np_j \geq 5$ for each p_j, we have to consider all the events "5 or more defective parts" as a single class. Thus we have $J = 6$, the number of classes used in computing the value of χ^2.

The mean of the Poisson distribution is not known, and has to be estimated from the sample mean. There are altogether 400 defective parts found in the 200 boxes, and so the mean of the sample is $400/200 = 2$. The probabilities shown in the third column are taken from Appendix V, and multiplying the probabilities by 200 gives the expected frequencies. The rest of the computation is self-explanatory. The obtained value for χ^2 is 7.01. Will this lead us to reject the null hypothesis that X has a Poisson distribution?

Table 12-3

Computation of χ^2 for comparing the observed with the expected frequencies based on Poisson distribution

Number of defective parts x	*Observed frequencies* f_{oj}	*Poisson probabilities* $\mu = 2$	*Expected frequencies* f_{ej}	$(f_{oj} - f_{ej})^2$	$\frac{(f_{oj} - f_{ej})^2}{f_{ej}}$
0	21	0.13534	27.068	36.8206	1.36
1	62	0.27067	54.134	61.8740	1.14
2	50	0.27067	56.134	17.0900	0.32
3	40	0.18045	36.090	15.2881	0.42
4	22	0.09022	18.044	15.6499	0.87
5	0				
6	5	0.05265	10.530	30.5809	2.90
7 or greater	0				
Total	200	1.00000	200.000		7.01

$$\chi^2 = \sum_{j=1}^{J} \frac{(f_{oj} - f_{ej})^2}{f_{ej}} = 7.01$$

From Appendix VIII, we know that the χ^2 value that cuts off the upper 5 percent of the distribution with $6 - 2 = 4$ degrees of freedom is 9.48773. Since the observed χ^2 value is less than this value, we should not reject the hypothesis that X, the number of defective parts per box, follows a Poisson distribution at $\alpha = 0.05$.

The number of degrees of freedom is $6 - 2 = 4$ rather than $6 - 1 = 5$ because the mean of the sample is used to estimate the mean of the Poisson distribution. As indicated in Sec. 11-1-2, 1 degree of freedom is lost whenever a population parameter is replaced by its estimate, a sample statistic.

The χ^2 test for goodness of fit has quite broad applications. It may be used to test the hypothesis that the sample is taken from a normal distribution. In order to do this, it is necessary to replace the parameters μ by $\overline{X}$ and σ^2 by $\hat{s}^2$, since μ and σ^2 are not known. The computed χ^2 value is then referred to the χ^2 distribution with $J - 3$ degrees of freedom. Two more degrees of freedom are lost because their sample estimates are substituted for both parameters.

Generally, if the expected frequencies in a test of goodness of fit are computed through estimating k parameters by the sample statistics, and J classes are used, the obtained test statistic will have a χ^2 distribution with $J - k - 1$ degrees of freedom.

Exercise 12-3

1 Given that a random digit table of 200 digits shows the following frequency distribution:

Digits	*Frequency*
0	18
1	23
2	22
3	17
4	20
5	25
6	14
7	19
8	22
9	20

Is the observed distribution significantly different from the expected distribution at the 1 percent level?

2 Assume that the number of fatal accidents on a certain freeway during a given week is distributed as follows: Sunday, 19; Monday, 14; Tuesday, 12; Wednesday, 12; Thursday, 15; Friday, 20; Saturday, 13; Total, 105. At a significance level 0.01, test the hypothesis that fatal accidents on that freeway are uniformly distributed over the 7 days of the week.

3 The number of people killed per week by traffic accidents on a certain freeway in a given year (52 weeks) is as follows:

Number of people killed:	0	1	2	3	4	5	Total
Frequency:	6	10	20	10	6	0	52

At the 5 percent level, test the hypothesis that the observed frequency follows the Poisson distribution.

4 Fit a normal curve to the frequency distribution of the data in Prob. 12 of Exercise 4-1. (The mean and the standard deviation are given in Prob. 12 of Exercise 4-2 and Prob. 12 of Exercise 5-2, respectively. For simplicity's sake, let $\bar{X} = 53$ and $\hat{s} = 16$.)

5 Suppose that six coins are tossed simultaneously 640 times and the following frequency distribution is observed:

Number of heads:	0	1	2	3	4	5	6	*Total*
Observed frequency:	13	70	137	210	145	56	9	640

Test the null hypothesis that the coins are well balanced with the level of significance at 0.01.

6 One thousand voters were selected at random from a state and were asked whether they would vote for a gubernatorial candidate. Of the 1,000 voters, 450 answered "yes" and 550 answered "no." At $\alpha = 0.05$, do you reject the hypothesis that 50 percent of the voters would vote for the candidate?

7 The number of calls received at a certain switchboard from 8:00 to 8:01 in the morning during a 100-day period are given as follows:

Number of calls:	0	1	2	3	4	5	6	7	*Total*
Frequency:	3	10	25	30	15	12	5	0	100

At the 5 percent level, test the hypothesis that the observed frequency follows a Poisson distribution. (Since f_{ej} for $x = 0$ is very close to 5, leave it as a separate class.)

8 The heights of 200 employees are distributed as follows:

Heights in inches	*Class boundaries*	*Observed frequency*
	57.5	
57.5–60.5		9
	60.5	
60.5–63.5		20
	63.5	
63.5–66.5		45
	66.5	
66.5–69.5		55
	69.5	
69.5–72.5		43
	72.5	
72.5–75.5		17
	75.5	
75.5–78.5		11
	78.5	
Total		200

Test the hypothesis that the observed frequency follows the normal pattern at $\alpha = 0.05$.

12-4 Test of Independence

The procedure discussed in the preceding section can be extended to tests of *statistical independence* between *two* variables, each of which is classified into a number of types or attributes. Frequently, we encounter problems that call for cross classification of the observed data. Suppose that n observations are made and that they are distributed jointly with respect to two variables C and R, each being classified into several classes. Let $C_1, C_2, \ldots, C_K$ be the K classes of the variable C, and $R_1, R_2, \ldots, R_J$ be the J classes of the variable R. The C's and the R's are mutually exclusive and exhaustive, respectively. Each observation represents the occurrence of one joint event (R_j, C_k) for $j = 1, 2, \ldots, J$ and $k = 1, 2, \ldots, K$. The sample of n observations can then be distributed over the $C \times R$ cells as follows:

	C				
R	C_1	C_2	$\cdots$	C_K	Total
R_1	f_{11}	f_{12}	$\cdots$	f_{1K}	f_1
R_2	f_{21}	f_{22}	$\cdots$	f_{2K}	f_2
$\cdots$	$\cdots$	$\cdots$	$\cdots$	$\cdots$	$\cdots$
R_J	f_{J1}	f_{J2}	$\cdots$	f_{JK}	f_J
Total	f_1'	f_2'	$\cdots$	f_K'	n

Here f_{jk} is the frequency with which the joint event (R_j, C_k) has occurred, and f_j and f_k' are frequencies with which the events R_j and C_k have occurred, respectively. Such a table is called a *contingency table* or, in particular, a $J \times K$ contingency table.

The null hypothesis to be tested is that R and C are *independent*; that is, there is no association or relation between the two variables. If this hypothesis is true, then, by Eq. (3-6), we have

$$P(R_j, C_k) = P(R_j)P(C_k)$$

That is, the probability of a joint event (R_j, C_k) is equal to the marginal probability of R_j multiplied by the marginal probability of C_k. Since the marginal probabilities are unknown, we have to replace them by their estimates, the sample marginal proportions. Let f_{ejk} be the expected frequency of the joint event (R_j, C_k); it follows that

$$P(C_j, C_k) = \frac{f_{ejk}}{n} = \frac{f_j}{n} \times \frac{f_k'}{n}$$

Thus we get

$$f_{ejk} = \frac{f_j f_k'}{n} \tag{12-17}$$

which is the expected frequency for the cell in the jth row and the kth column under the hypothesis that R and C are independent. We shall test this hypothesis by comparing f_{ejk} with f_{Ojk}, the observed frequency for the same cell, for all j and k. The test statistic is the sum of the squared differences between the observed and the expected frequencies each divided by the latter, and is distributed as χ^2 with the number of degrees of freedom equal to $(J - 1)(K - 1)$ for $J > 1$ and $K > 1$; that is,

$$\chi^2_{(J-1)(K-1)} = \sum_{j=1}^{J} \sum_{k=1}^{K} \frac{(f_{Ojk} - f_{ejk})^2}{f_{ejk}} \tag{12-18}$$

Here again a one-tailed test is applied since any significant association or relation between the two variables must be reflected in a large computed value for χ^2.

Suppose, for example, that it is desired to find out whether there is any significant association between educational attainment and job performance for a group of 200 employees. Educational attainment is classified into three classes, secondary school or lower, college, and graduate school, while job performance is classified as excellent, good, and fair. The joint frequency distribution of the 200 observations is represented in a 3 × 3 contingency table shown below:

	Educational attainment			
Job rating	*Secondary school or lower*	*College*	*Graduate school*	*Total*
Excellent	10	40	10	60
Good	30	30	20	80
Fair	10	30	20	60
Total	50	100	50	200

The expected frequency for each of the nine cells can be computed by applying Eq. (12-17), and the results are shown as follows:

	Educational attainment			
Job rating	*Secondary school or lower*	*College*	*Graduate school*	*Total*
Excellent	15	30	15	60
Good	20	40	20	80
Fair	15	30	15	60
Total	50	100	50	200

Given the quantities in the above two tables, we get the following value for χ^2:

$$\chi^2 = \frac{(10-15)^2}{15} + \frac{(40-30)^2}{30} + \frac{(10-15)^2}{15} + \frac{(30-20)^2}{20} + \frac{(30-40)^2}{40} + \frac{(20-20)^2}{20} + \frac{(10-15)^2}{15} + \frac{(30-30)^2}{30} + \frac{(20-15)^2}{15}$$

$$= \frac{25}{15} + \frac{100}{30} + \frac{25}{15} + \frac{100}{20} + \frac{100}{40} + \frac{0}{20} + \frac{25}{15} + \frac{0}{30} + \frac{25}{15}$$

$$= \frac{200 + 400 + 200 + 600 + 300 + 200 + 200}{120} = \frac{2{,}100}{120}$$

$$= 17.5$$

which has $(3-1)(3-1)=4$ degrees of freedom. From Appendix VIII, we find that the χ^2 value that cuts off the upper 5 percent of the distribution with 4 degrees of freedom is 9.48773, and the upper 1 percent is 13.2767. The observed value is substantially higher than these values. Thus, even if we set the significance level at 0.01, we can still reject the null hypothesis that there is no significant relation between the educational attainment of the employees and their job performance.

A special case in the test of independence is one that uses a 2×2 contingency table. If such a fourfold table is used, a simplified formula can be applied for computing the value of χ^2. Suppose that the observed frequencies in a 2×2 contingency table are designated *a*, *b*, *c*, and *d* as follows:

	A	*B*	*Total*
X	*a*	*b*	$a+b$
Y	*c*	*d*	$c+d$
Total	$a+c$	$b+d$	*n*

The value of χ^2 can then be computed from the following formula:

$$\chi^2 = \frac{n(ad-bc)^2}{(a+b)(a+c)(c+d)(b+d)} \tag{12-19}$$

which has $(2-1)(2-1)=1$ degree of freedom.

Frequently, the *Yates' correction for continuity*, which is analogous to the continuity correction for normal approximation to the binomial, is applied for better approximation to the exact *multinomial* probability. The corrected χ^2 value is computed as follows:

$$\chi^2 = \frac{n(|ad-bc|-n/2)^2}{(a+b)(a+c)(c+d)(b+d)} \tag{12-20}$$

This correction is used, however, only when the number of degrees of freedom is 1.

As already indicated, the χ^2 test in Secs. 12-3 and 12-4 is only a method of approximation. The larger the sample size, the better the approximation will be. In the test of goodness of fit, the expected frequency of each class must be *at least* 5 in order to have a good fit. Similarly, in a test of independence the expected frequency should also be 5 *or more* when the number of degrees of freedom is larger than 1. When the number of degrees of freedom is exactly 1, or when a 2×2 contingency table is used, the sample size should be large enough so that no expected frequency is smaller than 10.

Exercise 12-4

1 What is the number of degrees of freedom for a contingency table containing four rows and three columns?

2 Suppose that the following is a random sample of 1,000 voters classified according to party affiliation and voting preference regarding a certain issue:

Preference	*Democrats*	*Republicans*
For	400	150
Against	250	200

Test the hypothesis that party affiliation is not related to voting preference at $\alpha = 0.01$.

3 A certain company wishes to determine whether absenteeism is related to age. A random sample of 200 employees is selected and classified according to age and cause of absenteeism as follows:

	Age		
Cause	*Under 30*	*30–50*	*Over 50*
Illness	40	28	52
Other	20	36	24

Is age related to absenteeism at $\alpha = 0.01$?

4 An automobile manufacturing company wishes to find out whether the sex of its prospective customers is unrelated to model preference. A random sample of 2,000 prospective buyers is selected and classified as follows:

	Model		
Sex	*I*	*II*	*III*
Male	350	270	380
Female	340	400	260

Test the hypothesis that sex is unrelated to model preference at $\alpha = 0.01$.

5 Two groups, I and II, consist of 50 patients each who were attacked by the same disease. Group I (a control group) received identical treatment as group II except that it did not receive a vaccine given to each patient in group II. It was found that 38 people in group II and 32 people in group I recovered from the disease. Test the hypothesis that the vaccine had no effect on the recovery of the patients at $\alpha = 0.05$.

6 The management of a certain firm conducted a survey to determine whether the type of employees is related to preference for a particular insurance policy. A random sample of 300 employees was interviewed, and data collected were classified as follows:

	Insurance Policy		
Type of employees	*I*	*II*	*III*
Supervisors	18	6	12
Office workers	42	24	30
Factory workers	36	72	60

Test the hypothesis that the type of employment is independent of insurance policy preference at $\alpha = 0.01$.

7 For a 2×2 contingency table shown below:

	A	*B*	*Total*
X	a	b	$a+b$
Y	c	d	$c+d$
Total	$a+c$	$b+d$	n

prove that

$$\chi^2 = \frac{n(ad - bc)^2}{(a+b)(a+c)(c+d)(b+d)}$$

Chapter 13

The *F* distribution

Another important distribution is the *F distribution*, named in honor of R. A. Fisher, who first developed and described it. The random variable F is also a test statistic, and is defined as the ratio of the unbiased estimates of two population variances; that is,

$$F = \frac{\hat{s}_1^{\,2}}{\hat{s}_2^{\,2}} \tag{13-1}$$

This ratio has several applications, two of which are widely used. They are the test of equality between two variances and the test of equality among three or more means.

In the preceding chapter we examined the method of testing hypotheses about a *single* population variance. Often, however, we encounter a situation that calls for comparison between *two* population variances, that is, for determining whether the variability of one population differs from that of the other population. The F distribution is used for inferences of this kind.

In Chap. 11 we considered the procedure of making inferences about the difference between two population means. Our primary interest then was to determine, on the basis of two sample means, whether the two population means were equal. Frequently, however, we are interested in deciding whether *three or more* population means are equal. In the latter case we are required to test a set of sample means to see whether their differences are *statistically* significant. It is interesting to note that the F test can also be applied to such a situation. The test of homogeneity of three or more sample means is usually conducted by an analysis of the variability among these means. This technique is known as the *analysis of variance*, which represents an important application of the F distribution.

In this chapter, we shall first examine the nature of the F distribution. Then we shall review the procedure for testing hypotheses about the difference between two population variances. After that, attention will be directed toward the analysis of variance. In this connection, two linear models, namely, the one-way analysis of variance and two-way analysis of variance, will be considered.

13-1 Nature of the *F* Distribution

Suppose that two independent random variables X_1 and X_2 are each normally distributed; that is, X_1 is $N(\mu_1, \sigma_1^2)$ and X_2 is $N(\mu_2, \sigma_2^2)$. If two samples for X_1 and X_2 are taken with sizes n_1 and n_2, respectively, from which two unbiased estimates $\hat{s}_1^2$ and $\hat{s}_2^2$ for the two population variances are obtained, by Eq. (12-12) the following relationships hold:

$$\hat{s}_1^2 = \frac{\sigma_1^2 \chi^2_{(n_1-1)}}{n_1 - 1}$$

and

$$\hat{s}_2^2 = \frac{\sigma_2^2 \chi^2_{(n_2-1)}}{n_2 - 1}$$

Since $F = \hat{s}_1^2/\hat{s}_2^2$, under the hypothesis that $\sigma_1^2 = \sigma_2^2$, we have

$$F = \frac{\chi^2_{(n_1-1)}/(n_1 - 1)}{\chi^2_{(n_2-1)}/(n_2 - 1)} \tag{13-2}$$

That is, the *F ratio* is actually the ratio of two independent χ^2 statistics each divided by its degrees of freedom.

The density function for the variable F is found to be

$$f(F) = kF^{(\nu_1/2)-1}\left(1 + \frac{\nu_1 F}{\nu_2}\right)^{-(\nu_1+\nu_2)/2} \qquad \text{for } F > 0 \tag{13-3}$$

where $\nu_1 = n_1 - 1$ and $\nu_2 = n_2 - 1$, respectively, the degrees of freedom for the numerator and the denominator of the F ratio, and k is a constant depending only on the values of ν_1 and ν_2. Thus the F distribution has two parameters, namely, ν_1 and ν_2, and is completely defined if the values of these two parameters are given.

The derivation of the density rule for F need not concern us here. The mathematical expression of this rule is so complicated that we are unable to detect the characteristics of this distribution. All we can say here is that the variable F takes on only *nonnegative* values since both the numerator and the denominator of the F ratio are squared quantities. The range of the F values is from zero to infinity. The shape of the F curve depends on the parameters ν_1 and ν_2.

Generally, it is nonsymmetric and skewed to the right. However, when one or both of the parameter values increase, the F distribution tends to become more and more symmetrical. This tendency is demonstrated in Fig. 13-1, which presents three F curves with specified degrees of freedom. The first subscript of F is the value of v_1, and the second is the value of v_2, for each of the distributions.

We shall mention in passing that the expected value of F is given by

$$E(F) = \frac{v_2}{v_2 - 2} \tag{13-4}$$

and the variance of F is expressed as follows:

$$E\{[F - E(F)]^2\} = \frac{2{v_2}^2(v_1 + v_2 - 2)}{v_1(v_2 - 2)^2(v_2 - 4)} \tag{13-5}$$

It is obvious that for $v_2 \leq 2$ there is no mean, and for $v_2 \leq 4$ there is no variance, for the F distribution.

Since there are two sets of degrees of freedom for the F distribution, it would take a considerable amount of work if the F table were presented in as detailed a form as the t or the χ^2 table. Consequently, the F table is usually presented in a much condensed fashion. It gives only the frequently used α values such as 1, 2.5, 5, and 10 percentage points, the F values corresponding to the significance level, and the degrees of freedom v_1 and v_2. Such a table of F is presented as Appendix IX. Given a specified value of α, the upper row of the table lists the values of v_1, the degrees of freedom associated with the *numerator*; the first column lists the values of v_2, the degrees of freedom associated with the *denominator*; and the table entry is the F value, commonly designated $F_{(\alpha; v_1, v_2)}$, which cuts off the *upper* α of the distribution with v_1 and v_2 degrees of freedom.

The table is constructed under the assumption that the larger variance is placed in the numerator and the smaller variance in the denominator. Accordingly, the F value must always be greater than 1, and it approaches 1 only when both v_1 and v_2 approach infinity. Under the hypothesis that ${\sigma_1}^2 = {\sigma_2}^2$, that is,

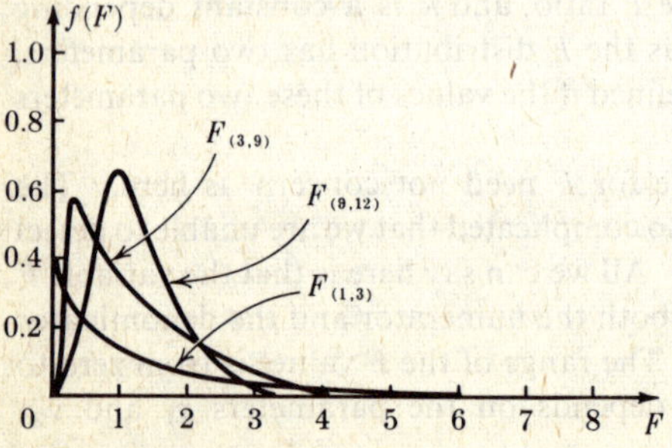

Figure 13-1
F distribution with specified degrees of freedom.

that the sample variances $\hat{s}_1^2$ and $\hat{s}_2^2$ are taken from populations with an identical variance, the F ratio is *expected* to be equal to 1. As the difference between $\hat{s}_1^2$ and $\hat{s}_2^2$ becomes greater, the value of the F statistic will rise above 1 by greater magnitudes. Accordingly, the hypothesis is rejected only when the value of the F statistic is sufficiently larger than 1. This means that the entire region of rejection relates to the *upper* tail of the distribution, and the F test, as conducted by direct application of Appendix IX, is basically a one-tailed test.

The practice of considering $\hat{s}_1^2$ the larger variance, however, is rather arbitrary, and we adhere to it simply because a consistent procedure has to be followed in applying the F table. Under the hypothesis that the sample variances are computed from two samples taken from two populations with the same variance, $\hat{s}_1^2$ can be equal to, or larger than, as well as smaller than, $\hat{s}_2^2$. Accordingly, the F ratio can be equal to, or greater than, as well as smaller than, 1. The expected value of the F ratio, as indicated above, is close to 1. Thus the alternative hypothesis can be nondirectional, and need not always be $\sigma_1^2 > \sigma_2^2$. How can we conduct a two-tailed test about the hypothesis that $\sigma_1^2 = \sigma_2^2$ by using Appendix IX?

The use of the F table for finding a proportion in the lower tail of the distribution requires a method for finding the F value that cuts off that proportion of the distribution. It is a fact that:

The critical value of F that cuts off a specified proportion in the lower tail of the distribution turns out to be the reciprocal of the F value that cuts off the upper tail of the same magnitude with the degress of freedom v_1 and v_2 interchanged; that is,

$$F_{(1-\alpha;\, v_1,\, v_2)} = \frac{1}{F_{(\alpha;\, v_2,\, v_1)}} \tag{13-6}$$

Suppose, for example, that we wish to find the critical F value that cuts off the *lower* 5 percent of the distribution with $v_1 = 5$ and $v_2 = 10$. We obtain this value by first finding the value that cuts off the *upper* 5 percent of the distribution with $v_1 = 10$ and $v_2 = 5$. From Appendix IX, we get $F_{(0.05;10,5)} = 4.74$. Thus we have

$$F_{(0.95;5,10)} = \frac{1}{F_{(0.05;10,5)}} = \frac{1}{4.74} = 0.21$$

In other words, the critical value is 0.21, *below* which the F statistic with $v_1 = 5$ and $v_2 = 10$ will lie with a probability of 0.05. Figure 13-2 shows the F values that cut off the upper and the lower 5 percent of the distribution with $v_1 = 5$ and $v_2 = 10$.

Observe that Appendix IX presents only four one-tailed significance levels, namely, 0.01, 0.025, 0.05, and 0.10. If a two-tailed test is desired, the table provides only significance levels 0.02, 0.05, 0.10, and 0.20 for the F test.

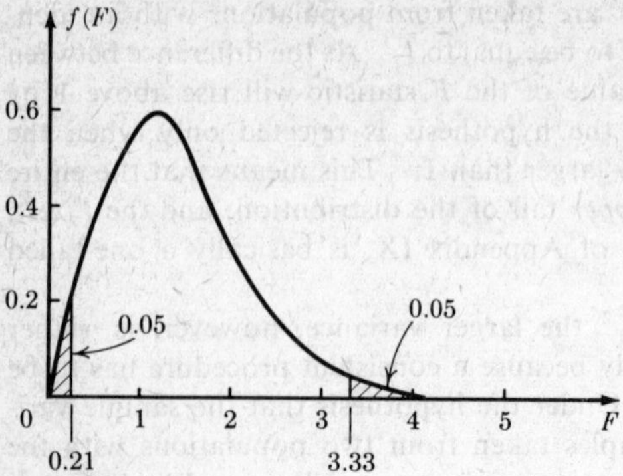

Figure 13-2
The F distribution with $\nu_1 = 5$ and $\nu_2 = 10$.

Exercise 13-1

1 Define or explain the following concepts:

a The test statistic F.
b F distribution.
c The degrees of freedom ν_1 and ν_2.

2 Explain the relation between the F and χ^2 distributions.

3 Explain how the F distribution is related to normal distribution.

4 Describe some of the important characteristics of the F distribution.

5 For the specified areas under the F distributions with ν_1 and ν_2 as the degrees of freedom, find the values for k:

	ν_1	ν_2	*Area*
a $P(F \geq k)$	6	8	0.01
b $P(F \geq k)$	10	12	0.01
c $P(F \geq k)$	20	10	0.05
d $P(F \geq k)$	10	20	0.05
e $P(F \leq k)$	10	20	0.01
f $P(F \leq k)$	20	10	0.05

6 Find the values of k for the following probabilities:

a $P(F \geq k) = 0.05$, given $\nu_1 = 12$ and $\nu_2 = 13$.
b $P(F \geq k) = 0.01$, given $\nu_1 = 5$ and $\nu_2 = 40$.
c $P(F \geq k) = 0.025$, given $\nu_1 = 40$ and $\nu_2 = 10$.
d $P(F \geq k) = 0.95$, given $\nu_1 = 10$ and $\nu_2 = 20$.
e $P(F \geq k) = 0.99$, given $\nu_1 = 20$ and $\nu_2 = 10$.

7 How are the parameters of the F distribution related to the size of the random samples?

8 Suppose that all the values in each of two samples are multiplied by a constant b and then another constant, a, is added to each product thus obtained. Will the F ratio be changed? Explain.

9 What are some of the basic assumptions underlying the F distribution?

10 Use an equation to express the relation between the statistics t and F.

13-2
Inference about the Difference between Two Variances

As pointed out above, the F statistic is used, among other applications, for inferences about the difference between two population variances. We shall now illustrate the procedure of making this type of inference.

In Sec. 11-3-1 we tested the hypothesis about the difference between the average breaking strength of nylon cord and that of cotton cord. It was found that the difference between the two means was significantly different from 0, and that the average breaking strength of the nylon cord is lower than that of the cotton cord. Now we wish to compare the variability of the breaking strength of nylon cord with that of cotton cord. In other words, we wish to determine whether σ_1^2, the variance of the breaking strength of nylon cord, is equal to or smaller than σ_2^2, the variance of the breaking strength of cotton cord. For this purpose we set up two competing hypotheses as follows:

H_0: $\sigma_1^2 \leq \sigma_2^2$ $\quad$ H_1: $\sigma_1^2 > \sigma_2^2$

Recall that a test about the null hypothesis H_0: $\sigma_1^2 = \sigma_2^2$ is also good for H_0: $\sigma_1^2 \leq \sigma_2^2$. We shall therefore proceed with the test as if the null hypothesis were H_0: $\sigma_1^2 = \sigma_2^2$.

Suppose that two samples are randomly selected and the relevant data are shown below:

	Nylon cord	*Cotton cord*
Sample size	$n_1 = 61$	$n_2 = 41$
Degrees of freedom	$\nu_1 = 60$	$\nu_2 = 40$
Variance	$\hat{s}_1^2 = 2{,}745$	$\hat{s}_2^2 = 1{,}451$

Accordingly, the F ratio is 2,745/1,451 = 1.89. The critical F value that cuts off the upper 5 percent ($\alpha = 0.05$) of the distribution with $\nu_1 = 60$ and $\nu_2 = 40$ is 1.64, and the F value that cuts off the upper 1 percent ($\alpha = 0.01$) is 2.02. Thus, at $\alpha = 0.05$, we shall reject the null hypothesis that the variability in the breaking strength of nylon cord is equal to or less than that of cotton cord. We shall not reject this hypothesis, however, at the 1 percent significance level.

For a two-tailed test, on the other hand, we have the hypotheses

H_0: $\sigma_1^2 = \sigma_2^2$ $\quad$ H_1: $\sigma_1^2 \neq \sigma_2^2$

As usual, we have to find two critical values cutting off the upper and lower $\alpha/2$ of the distribution with ν_1 and ν_2 degrees of freedom. In the above example, with the significance level still set at $\alpha = 0.05$, these two critical values turn out to be

$$F_{(0.025;\,60,40)} = 1.80$$

and

$$F_{(0.975;60,40)} = \frac{1}{F_{(0.025;\,40,60)}} = \frac{1}{1.74} = 0.57$$

Since the observed F value is 1.89, which lies in the upper region of rejection, we should conclude that the variability of the breaking strength of nylon cord is different from the variability for cotton cord.

Exercise 13-2

1 A random variable X_1 is $N(\mu_1, \sigma_1^2)$, and another random variable X_2, independent of X_1, is $N(\mu_2, \sigma_2^2)$. Two samples of $n_1 = 25$ and $n_2 = 25$ are randomly selected for X_1 and X_2, and yield two variances $\hat{s}_1^2 = 225$ and $\hat{s}_2^2 = 100$, respectively. Test the null hypothesis that $\sigma_1^2 = \sigma_2^2$ against the alternative that $\sigma_1^2 > \sigma_2^2$ at $\alpha = 0.05$.

2 It is suspected that the weights of junior high school students have greater variation than the weights of senior high school students. A sample of 25 students randomly selected from junior high schools yields a standard deviation $\hat{s}_1 = 20$ pounds; another sample of 20 students selected from senior high schools gives a standard deviation $\hat{s}_2 = 15$ pounds. Should we conclude that the variability of the weights of junior high students is greater than that of senior high students at $\alpha = 0.01$? What is your assumption when you test the null hypothesis that the two variances are equal?

3 Two different machines, A and B, are used to produce identical bolts that are supposed to be 2 inches in length. Two samples of 25 bolts each are randomly selected from the output of each of the machines, and they yield two variances $\hat{s}_1^2 = 0.03$ inch for machine A and $\hat{s}_2^2 = 0.04$ inch for machine B. At $\alpha = 0.05$, test the null hypothesis that the population variances are equal against the alternative that they are unequal.

4 In the preceding problem, if the alternative hypothesis is $\sigma_1^2 < \sigma_2^2$, what would be your conclusion for a computed F ratio $0.03/0.04 = 0.75$?

5 The ABC Company seeks to determine whether its female customers differ from its male customers in the variability of their yearly purchases. A sample of 31 female customers and another of 25 male customers for a given year are selected for comparison, and they yield two variances $\hat{s}_1^2 = 196$ and $\hat{s}_2^2 = 170$ in units sold, respectively. Test the null hypothesis that the variabilities of annual purchases of the two types of customers are equal against the alternative that they are not equal with the level of significance at 0.05.

6 X and Y are two normal and independent variables with variances σ_X^2 and σ_Y^2. Design a test between $H_0\colon \sigma_X^2 = \sigma_Y^2$ and $H_1\colon \sigma_X^2 \neq \sigma_Y^2$ at α level of significance, on the basis of two samples of n_1 observations for X and n_2 observations for Y, which yield two variances $\hat{s}_X^2$ and $\hat{s}_Y^2$.

7 Suppose that two samples of $n_1 = 25$ and $n_2 = 16$ are taken from two different types of salesmen whose annual volumes of sales have variances $\sigma_1{}^2$ and $\sigma_2{}^2$, respectively. Suppose further that the two samples yield two variances $\hat{s}_1{}^2 = \$60{,}000$ and $\hat{s}_2{}^2 = \$50{,}000$. At $\alpha = 0.05$, test the null hypothesis H_0: $\sigma_1{}^2 = \sigma_2{}^2$ against:

a $H_1: \sigma_1{}^2 > \sigma_2{}^2$. *b* $H_1: \sigma_1{}^2 \neq \sigma_2{}^2$.

13-3 One-Way Analysis of Variance

Since we know the nature of the F distribution and the procedure of testing about the difference between two sample variances, we are now ready to examine the procedure of testing hypotheses about the differences among *three or more* population means. This is known as the method of *analysis of variance*, which will be dealt with in the rest of the chapter. In the present section consideration is given to the *one-way* analysis of variance, in which only one type of treatment is involved in the experiment.

In Sec. 11-3 we discussed the method of comparing the means of two samples. Two samples known as *experimental groups* are selected. A treatment is administered to one group, and no treatment is applied to the other group, called the *control group*. Or, to each of the two groups a different treatment is administered, and each group is regarded as representative of the potential population under the same treatment. On the basis of the group means, inferences are made about the means of the potential populations by the use of the t test. When more than two experimental groups are involved, however, the t test is of little use. For one thing, the t test is rather impractical, especially when the number of experimental groups is large. For example, in an experiment with only five groups, there will be $\binom{5}{2} = 5!/(2!\,3!) = 10$ pairs of means, or 10 t tests to perform. For another, the various t tests cannot be considered *independent* since it is impossible to avoid redundance, or overlapping, in the tests. Consequently, they make it very difficult to determine the overall levels of significance.

Fortunately, when we wish to test the hypothesis about the overall difference among three or more population means, we can apply the F test by finding the ratio of two variances computed from the experimental data. For the rationale of this approach, we first consider the model of linear relationship.

13-3-1 The model

We now examine the linear model on which the method of one-way analysis of variance is based. Let X_{ik} be the ith observation of the kth experimental group. Then, X_{ik} is nothing more than the sum of three component quantities:

$$X_{ik} = \mu + \gamma_k + e_{ik} \tag{13-7}$$

where μ = grand mean of all K treatment populations

γ_k = *effect* of treatment for population k (γ is a Greek letter read as "gamma")

e_{ik} = random error associated with ith observation of treatment population k

Among these three components, the grand mean μ is self-explanatory. The *treatment effect* γ_k is the difference between the grand mean μ and the mean μ_k of the treatment population k; that is,

$$\gamma_k = \mu_k - \mu \tag{13-8}$$

It is assumed that inference about treatment effect is limited to the treatment populations actually under consideration. We have no intention of inferring about the effects of other treatments that might conceivably be included in the experiment. Accordingly, if K treatments are included in an experiment, the sum of all K treatment effects must be 0,

$$\sum_{k=1}^{K} \gamma_k = \sum_{k=1}^{K} (\mu_k - \mu) = \sum_{k=1}^{K} \mu_k - K\mu = 0 \tag{13-9}$$

since the grand mean μ is the mean of all the treatment population means.

The last term e_{ik} reflects the variability within each of the treatment populations. Its occurrence is attributed to the process of randomness, or acts of nature. It is the *residual* after subtracting the mean of the treatment population from the observed score; that is,

$$e_{ik} = X_{ik} - \mu_k \tag{13-10}$$

The expectation of e_{ik} is 0:

$$E(e_{ik}) = 0 \tag{13-11}$$

When a sample of K groups with n_k observations in group k is taken, certain relationships in connection with this model become apparent. First, the mean $\bar{X}_k$ of the treatment group k can be stated as follows:

$$\bar{X}_k = \sum_{i=1}^{n_k} \frac{\mu + \gamma_k + e_{ik}}{n_k} = \frac{n_k\mu + n_k\gamma_k + \sum_{i=1}^{n_k} e_{ik}}{n_k} = \mu + \gamma_k + \bar{e}_k \tag{13-12}$$

Also, the grand sample mean $\bar{X}$ can be written as

$$\bar{X} = \sum_{i=1}^{n_k} \sum_{k=1}^{K} \frac{(\mu + \gamma_k + e_{ik})}{n_k K} = \mu + \sum_{k=1}^{K} \frac{\gamma_k}{K} + \bar{e} = \mu + \bar{e} \tag{13-13}$$

since the sum of the treatment effects is equal to 0. Observe that $\bar{e}_k$ is the error mean of treatment group k, and $\bar{e}$ is the error mean for the entire sample.

A simple numerical example will help you to better understand the relationships as expressed in Eqs. (13-7) through (13-13). Suppose that an experiment involves three treatment groups with four observations each. Suppose also that the grand population mean is 10, the effects of the treatment populations are respectively 1,2, and −3, and the random errors are shown in the third column under each treatment group in Table 13-1. We shall then have the various means shown in the bottom row of the table. You are advised to verify these before you proceed.

This model is based on a number of assumptions. First, it is assumed that the random errors e_{ik} are $N(0,\sigma^2)$ for each treatment population k. It is also assumed that e_{ik} are independently distributed both within and between treatment populations. Finally, we assume that the error variance σ^2 remains the same for each of the populations.

13-3-2 Partition of the variability

In testing hypotheses about the difference among the means of several treatment populations, the various treatment groups constitute the total sample. The variance of the total sample is partitioned into the variance *within* the treatment groups and the variance *between* the groups. Since the sample variance as an unbiased estimator of the population variance is the sum of squared deviations from the sample mean divided by the number of degrees of freedom, partition of the sample variance implies the *partition of the sum of squares.* We shall, for the moment, restrict our attention to partitioning the sum of squares for the total sample, SST, into the sum of squares within the treatment groups, SSW, and the sum of squares between groups, SSB.

Suppose that an experiment includes K treatment groups, and each group consists of n_k observations for $k = 1, 2, \ldots, K$. The total number of observations in the K groups is N; that is,

$$N = n_1 + n_2 + \ldots + n_K$$

Table 13-1
Numerical illustrations of the linear model for one-way analysis of variance

	Group 1 $\mu+\gamma_1+e_{i1}=X_{i1}$	Group 2 $\mu+\gamma_2+e_{i2}=X_{i2}$	Group 3 $\mu+\gamma_3+e_{i3}=X_{i3}$	
$i=1$	$10+1+(-2)=9$	$10+2+(+1)=13$	$10+(-3)+(+2)=9$	
$i=2$	$10+1+(+1)=12$	$10+2+(-2)=10$	$10+(-3)+(+3)=10$	
$i=3$	$10+1+(+3)=14$	$10+2+(+3)=15$	$10+(-3)+(-2)=5$	
$i=4$	$10+1+(+2)=13$	$10+2+(0)\ =12$	$10+(-3)+(+3)=10$	
Treatment effect	$\gamma_1=1$	$\gamma_2=2$	$\gamma_3=-3$	$\sum_{i=1}^{3}\gamma_i=0$
Means	$\bar{e}_1=1$ $\bar{X}_1=12.0$ $\mu_1=11$	$\bar{e}_2=0.5$ $\bar{X}_2=12.5$ $\mu_2=12$	$\bar{e}_3=1.5$ $\bar{X}_3=8.5$ $\mu_3=7$	$\bar{e}=1$ $\bar{X}=11$ $\mu=10$

Suppose also that the total sample yields a grand sample mean $\bar{X}$, and the group k yields a group mean $\bar{X}_k$. The total sum of squares of deviations of X_{ik} from $\bar{X}$ for $i = 1, 2, \ldots, n_k$ is given as follows:

$$\sum_{i=1}^{n_k}\sum_{k=1}^{K}(X_{ik}-\bar{X})^2 = \sum_{i=1}^{n_k}\sum_{k=1}^{K}[(X_{ik}-\bar{X}_k)+(\bar{X}_k-\bar{X})]^2$$

$$= \sum_{i=1}^{n_k}\sum_{k=1}^{K}(X_{ik}-\bar{X}_k)^2 + \sum_{i=1}^{n_k}\sum_{k=1}^{K}(\bar{X}_k-\bar{X})^2$$

$$+ 2\sum_{i=1}^{n_k}\sum_{k=1}^{K}(X_{ik}-\bar{X}_k)(\bar{X}_k-\bar{X})$$

The last term on the right side of this equation vanishes since the sum of $X_{ik} - \bar{X}_k$ is 0 and the quantity $\bar{X}_k - \bar{X}$ is the same for all i in each group k. Moreover, because of the fact that $\bar{X}_k - \bar{X}$ is a constant for all i in group k, we get

$$\sum_{i=1}^{n_k}\sum_{k=1}^{K}(\bar{X}_k-\bar{X})^2 = \sum_{k=1}^{K} n_k(\bar{X}_k-\bar{X})^2$$

Consequently, we have the relationship

$$\sum_{i=1}^{n_k}\sum_{k=1}^{K}(X_{ik}-\bar{X})^2 = \sum_{i=1}^{n_k}\sum_{k=1}^{K}(X_{ik}-\bar{X}_k)^2 + \sum_{k=1}^{K} n_k(\bar{X}_k-\bar{X})^2 \tag{13-14}$$

That is, the *sum of squares total* is equal to the *sum of squares within* groups plus the *sum of squares between* groups. Symbolically,

$$SST = SSW + SSB \tag{13-15}$$

which is generally referred to as the *partition of the sum of squares.*

13-3-3 Expectations of sum of squares and mean square

We shall first examine the expectations of the sum of squares and the mean square *within* groups. After that, the expectations of the sum of squares and the mean square *between* groups will be considered.

Within groups It must be pointed out that the sum of squares within groups reflects only the random errors since the treatment applied does not affect the variations among the observations within any given group. We shall demonstrate that this is actually the case. From Eqs. (13-7) and (13-12), we have

$$X_{ik} = \mu + \gamma_k + e_{ik}$$

and

$$\bar{X}_k = \mu + \gamma_k + \bar{e}_k$$

The sum of squares of the differences between X_{ik} and $\overline{X}_k$ therefore becomes

$$\begin{aligned} SSW &= \sum_{i=1}^{n_k} \sum_{k=1}^{K} (X_{ik} - \overline{X}_k)^2 \\ &= \sum_{i=1}^{n_k} \sum_{k=1}^{K} (\mu + \gamma_k + e_{ik} - \mu - \gamma_k - \bar{e}_k)^2 \\ &= \sum_{i=1}^{n_k} \sum_{k=1}^{K} (e_{ik} - \bar{e}_k)^2 \end{aligned}$$

Taking the expectation of SSW, we get

$$\begin{aligned} E(SSW) &= E\sum_{i=1}^{n_k} \sum_{k=1}^{K} (e_{ik} - \bar{e}_k)^2 = \sum_{k=1}^{K} E\sum_{i=1}^{n_k} (e_{ik} - \bar{e}_k)^2 \\ &= \sum_{k=1}^{K} (n_k - 1)\sigma^2 \end{aligned}$$

This is the case because for any treatment group k, the unbiased estimate of σ^2 is $\hat{s}_k^{\,2}$, the error variance of the sample group k, defined as

$$\hat{s}_k^{\,2} = \frac{\sum_{i=1}^{n_k} (e_{ik} - \bar{e}_k)^2}{n_k - 1}$$

therefore

$$E(\hat{s}_k^{\,2}) = \frac{E\sum_{k=1}^{n_k} (e_{ik} - \bar{e}_k)^2}{n_k - 1} = \sigma^2$$

and

$$E\sum_{k=1}^{n_k} (e_{ik} - \bar{e}_k)^2 = (n_k - 1)\sigma^2$$

We know that the sum of squares within groups has $N - K$ degrees of freedom. Thus the expected *mean square within groups*, or $E(MSW)$, is simply the error variance, as indicated below:

$$E(MSW) = \frac{\sum_{k=1}^{K} (n_k - 1)\sigma^2}{N - K} = \frac{(N - K)\sigma^2}{N - K} = \sigma^2 \tag{13-16}$$

Since the sum of squares for each group k has $n_k - 1$ degrees of freedom, and the sum of squares within all groups has $\sum_{k=1}^{K} (n_k - 1) = N - K$ degrees of freedom, the mean square within groups is in fact the *pooled* error variance within sample groups.

Between groups Let us now turn our attention to the second term on the right side of Eq. (13-14), the sum of squares between groups. We know from Eqs. (13-12) and (13-13) that

$$\overline{X}_k = \mu + \gamma_k + \bar{e}_k$$

and

$\overline{X} = \mu + \bar{e}$

Thus the expected sum of squares of difference between $\overline{X}_k$ and $\overline{X}$ for all k is

$$\begin{aligned} E(SSB) &= E\sum_{k=1}^{K} n_k(\overline{X}_k - \overline{X})^2 = E\sum_{k=1}^{K} n_k[\gamma_k + (\bar{e}_k - \bar{e})]^2 \\ &= E\sum_{k=1}^{K} n_k[\gamma_k^{\,2} + (\bar{e}_k - \bar{e})^2 + 2\gamma_k(\bar{e}_k - \bar{e})] \\ &= E\sum_{k=1}^{K} n_k\gamma_k^{\,2} + E\sum_{k=1}^{K} n_k(\bar{e}_k - \bar{e})^2 + 2E\sum_{k=1}^{K} n_k\gamma_k(\bar{e}_k - \bar{e}) \end{aligned}$$

The last term vanishes because of independence between random errors and treatment effect. Furthermore, the second term appears to be

$$\begin{aligned} E\sum_{k=1}^{K} n_k(\bar{e}_k - \bar{e})^2 &= E\sum_{k=1}^{K} (n_k\bar{e}_k^{\,2} - 2n_k\bar{e}_k\bar{e} + n_k\bar{e}^2) \\ &= E\left(\sum_{k=1}^{K} n_k\bar{e}_k^{\,2} - 2\bar{e}\sum_{k=1}^{K} n_k\bar{e}_k + \bar{e}^2\sum_{k=1}^{K} n_k\right) \\ &= E\left(\sum_{k=1}^{K} n_k\bar{e}_k^{\,2} - 2\bar{e}N\bar{e} + N\bar{e}^2\right) \\ &= E\left(\sum_{k=1}^{K} n_k\bar{e}_k^{\,2} - N\bar{e}^2\right) \\ &= \sum_{k=1}^{K} n_k E(\bar{e}_k^{\,2}) - NE(\bar{e}^2) \end{aligned}$$

Since $E(\bar{e}_k) = E(\bar{e}) = 0$, $E(\bar{e}_k^{\,2}) = \sigma^2/n_k$, and $E(\bar{e}^2) = \sigma^2/N$, we have

$$\begin{aligned} E(SSB) &= \sum_{k=1}^{K} n_k\gamma_k^{\,2} + \sum_{k=1}^{K} n_k\frac{\sigma^2}{n_k} - N\frac{\sigma^2}{N} \\ &= \sum_{k=1}^{K} n_k\gamma_k^{\,2} + K\sigma^2 - \sigma^2 \\ &= \sum_{k=1}^{K} n_k\gamma_k^{\,2} + (K-1)\sigma^2 \end{aligned}$$

We know that the expected *mean square between groups*, or $E(MSB)$, is the expected sum of squares between, or $E(SSB)$, divided by the number of degrees of freedom, $K - 1$. Thus we get

$$E(MSB) = \frac{\sum_{k=1}^{K} n_k\gamma_k^{\,2}}{K-1} + \sigma^2 \tag{13-17}$$

That is, the expected mean square between groups is equal to the error variance σ^2 plus a term that must not be negative because γ_k enters as a squared quantity.

13-3-4 Test of hypotheses

Keep in mind that MSW is the pooled variance within sample groups, or $\hat{s}_w^2$. Thus, by Eq. (12-12), $(N - K)\hat{s}_w^2/\sigma^2$ is distributed as χ^2 with $N - K$ degrees of freedom. Similarly, MSB is the variance between sample groups, or $\hat{s}_b^2$. Hence $(K - 1)\hat{s}_b^2/\sigma^2$ is distributed as χ^2 with $K - 1$ degrees of freedom. Consequently, the following important relationship holds:

$$\frac{\hat{s}_b^2}{\hat{s}_w^2} = \frac{\hat{s}_b^2/\sigma^2}{\hat{s}_w^2/\sigma^2} = \frac{\chi^2_{(K-1)}/(K-1)}{\chi^2_{(N-K)}/(N-K)}$$

By Eq. (13-2), we have

$$F = \frac{MSB}{MSW} \tag{13-18}$$

That is, the ratio of the mean square between to the mean square within sample groups is distributed as F with $v_1 = K - 1$ and $v_2 = N - K$.

If the null hypothesis is true, that is, if there exists no treatment effect, the first term on the right side of Eq. (13-17) would vanish. Then the expected mean square between groups would be equal to the expected mean square within groups. Accordingly,

$$\frac{E(MSB)}{E(MSW)} = \frac{\sigma^2}{\sigma^2} = 1$$

On the other hand, if treatment effects exist, the expected mean square between groups will be gretaer than the expected mean square within groups; consequently,

$$\frac{E(MSB)}{E(MSW)} > 1$$

Thus, if treatment effects exist, they are expected to show up in an F ratio larger than 1. If it is significantly larger than 1, it reflects the existence of large treatment effects; and accordingly, we shall reject the null hypothesis that the means of the various treatment populations are equal. On the other hand, if the computed F value is not significantly larger than 1, we shall conclude that treatment effects do not exist, or no conclusion is reached pending further investigation. The F value may show up as less than 1. When this happens it may indicate nothing but sampling errors, or it may be a reflection that something has gone wrong in the design or execution of the experiment.

Recall that the sum of squares for all items in the sample is equal to the sum of squares within groups plus the sum of squares between groups. Similarly, the degrees of freedom for the sum of squares total are equal to the degrees of freedom within groups plus the degrees of freedom between groups. These relationships may serve as checks in the computation process.

13-3-5 Computation process

Let X_{ik} be the ith observation of the kth sample group, and the total sample of size N be made up of K treatment groups each consisting of n_k subjects for $k = 1, 2, \ldots, K$. Also, let C_k designate the treatment group k, and T_k the sum of the observed values for group k. Then the sample data will appear in the following form:

C_1	C_2	$\cdots$	C_K
X_{11}	X_{12}	$\cdots$	X_{1K}
X_{21}	X_{22}	$\cdots$	X_{2K}
$\cdots$	$\cdots$	$\cdots$	$\cdots$
$X_{n_1 1}$	$X_{n_2 2}$	$\cdots$	$X_{n_K K}$
T_1	T_2	$\cdots$	T_K

Also, let T stand for the grand total of all observed values, that is,

$$T = T_1 + T_2 + \cdots + T_K$$

For computational purposes, the sum of squares of total deviations is expressed below:

$$
\begin{aligned}
SST &= \sum_{i=1}^{n_k} \sum_{k=1}^{K} (X_{ik} - \overline{X})^2 \\
&= \sum_{i=1}^{n_k} \sum_{k=1}^{K} (X_{ik}{}^2 - 2X_{ik}\overline{X} + \overline{X}^2) \\
&= \sum_{i=1}^{n_k} \sum_{k=1}^{K} X_{ik}{}^2 - 2\overline{X} \sum_{i=1}^{n_k} \sum_{k=1}^{K} X_{ik} + \sum_{i=1}^{n_k} \sum_{k=1}^{K} \overline{X}^2 \\
&= \sum_{i=1}^{n_k} \sum_{k=1}^{K} X_{ik}{}^2 - 2\overline{X}N\overline{X} + N\overline{X}^2 \\
&= \sum_{i=1}^{n_k} \sum_{k=1}^{K} X_{ik}{}^2 - N\overline{X}^2
\end{aligned}
$$

Since

$$\overline{X}^2 = \left(\frac{\sum_{i=1}^{n_k} \sum_{k=1}^{K} X_{ik}}{N} \right)^2$$

and

$$N\overline{X}^2 = \left(\frac{\sum_{i=1}^{n_k} \sum_{k=1}^{K} X_{ik}}{N} \right)^2 = \frac{T^2}{N}$$

we have

$$SST = \sum_{i=1}^{n_k} \sum_{k=1}^{K} X_{ik}^2 - \frac{T^2}{N} \tag{13-19}$$

Similarly, the sum of squares between sample groups can be expressed as follows:

$$\begin{aligned} SSB &= \sum_{k=1}^{K} n_k(\bar{X}_k - \bar{X})^2 = \sum_{k=1}^{K} n_k(\bar{X}_k^2 - 2\bar{X}\bar{X}_k + \bar{X}^2) \\ &= \sum_{k=1}^{K} n_k \bar{X}_k^2 - 2\bar{X} \sum_{k=1}^{K} n_k \bar{X}_k + \sum_{k=1}^{K} n_k \bar{X}^2 \\ &= \sum_{k=1}^{K} n_k \left(\frac{\sum_{i=1}^{n_k} X_{ik}}{n_k} \right)^2 - 2\bar{X}N\bar{X} + N\bar{X}^2 \\ &= \sum_{k=1}^{K} \frac{T_k^2}{n_k} - N\bar{X}^2 \\ &= \sum_{k=1}^{K} \frac{T_k^2}{n_k} - \frac{T^2}{N} \end{aligned} \tag{13-20}$$

The sum of squares within sample groups is the difference between the two sums of squares shown above; that is,

$$SSW = \sum_{i=1}^{n_k} \sum_{k=1}^{K} X_{ik}^2 - \sum_{k=1}^{K} \frac{T_k^2}{n_k} \tag{13-21}$$

These quantities are usually summarized in an analysis of variance table such as Table 13-2.

Table 13-2

Summary table for one-way analysis of variance

Source of variation	*Sum of squares SS*	*Degrees of freedom,* ν	*Mean square MS*	*F ratio*
Between groups, B	$\sum_{k=1}^{K} \frac{T_k^2}{n_k} - \frac{T^2}{N}$	$K-1$	$\frac{SSB}{K-1}$	$\frac{MSB}{MSW}$
Within groups, W	$\sum_{i=1}^{n_k} \sum_{k=1}^{K} X_{ik}^2 - \sum_{k=1}^{K} \frac{T_k^2}{n_k}$	$N-K$	$\frac{SSW}{N-K}$	
Total T	$\sum_{i=1}^{n_k} \sum_{k=1}^{K} X_{ik}^2 - \frac{T^2}{N}$	$N-1$		

13-3-6 An example

We shall illustrate the method of one-way analysis of variance with a simple example. Suppose that we wish to determine whether the daily outputs, for the same kind of work, of three groups of workers with different ethnic backgrounds are significantly different. That is, we wish to test the hypothesis that the three population means are the same against the alternative that they are different. For the sake of simplicity, we assume that the three sample groups are of equal size n. (The procedure remains the same if the sample groups are of different size.) Let $n = 5$, and the three randomly selected independent samples yield the following results:

Group 1	*Group 2*	*Group 3*
56	48	55
60	61	60
50	48	44
65	52	46
64	46	55

The various quantities required for computing the F ratio are given below:

$$n_1 = 5 \qquad n_2 = 5 \qquad n_3 = 5 \qquad N = 15$$

$$T_1 = 295 \qquad T_2 = 255 \qquad T_3 = 260 \qquad T = 810$$

$$T_1^2 = 87{,}025 \qquad T_2^2 = 65{,}025 \qquad T_3^2 = 67{,}600 \qquad T^2 = 656{,}100$$

$$\frac{T^2}{N} = \frac{810^2}{15} = \frac{656{,}100}{15} = 43{,}740$$

$$\sum_{i=1}^{5} \sum_{k=1}^{3} X_{ik}^2 = 56^2 + 60^2 + \cdots + 55^2 = 44{,}408$$

$$SST = 44{,}408 - 43{,}740 = 668$$

$$SSB = \frac{87{,}025}{5} + \frac{65{,}025}{5} + \frac{67{,}600}{5} - 43{,}740 = 190$$

$$SSW = SST - SSB = 668 - 190 = 478$$

Source of variation	*Sum of squares*	*Degrees of freedom*	*Mean square*	*F ratio*
Between groups	190	$\nu_1 = 3 - 1 = 2$	$\frac{190}{2} = 95$	$\frac{95}{39.8} = 2.4$
Within groups	478	$\nu_2 = 15 - 3 = 12$	$\frac{478}{12} = 39.8$	
Total	668	$15 - 1 = 14$		

The computed F ratio 2.4 is to be compared with the critical F ratio from Appendix IX. The critical F ratio that cuts off the upper 5 percent of the distribution with $\nu_1 = 2$ and $\nu_2 = 12$ is 3.89. Thus, at $\alpha = 0.05$, we conclude that no significant differences exist in the outputs for the three different ethnic groups of workers. The same conclusion is reached even at $\alpha = 0.10$, since the critical F ratio that cuts off the upper 10 percent of the distribution with $\nu_1 = 2$ and $\nu_2 = 12$ is 2.81, greater than the observed F ratio.

Exercise 13-3

1 What are the basic assumptions in the one-way analysis of variance discussed in this section? Do you see the reasons why these assumptions are made? Explain.

2 As mentioned in the text, when more than two means are compared, the F rather than the t test is used. Explain the advantages of the F test when three or more means are compared.

3 Define the following concepts relative to the linear model discussed in this section:

a Random error.
b Treatment effect.
c Sum of squares between groups.
d Sum of squares within groups.
e Sum of squares total.

4 An English test is given to three freshman classes, I, II, and III. Suppose that we wish to test whether the mean grades of the three classes are equal. Three samples of size $n_1 = n_2 = n_3 = 6$ are randomly selected, and they give the following grades:

I	*II*	*III*
76	79	94
68	79	84
64	77	96
78	73	88
79	81	78
73	79	86

Test the hypothesis that the mean grades of the three classes are equal at $\alpha = 0.05$, and again at $\alpha = 0.01$.

5 Four kinds of fertilizers, A, B, C, and D, are used on four identical plots of land in each of five farms growing wheat. Let the wheat yields be as follows:

A	*B*	*C*	*D*
2	2	3	5
4	5	5	5
5	5	6	7
3	7	6	7
1	1	5	6

Test the hypothesis that the fertilizers have the same effect on the yields at $\alpha = 0.01$.

6 Explain the rationale underlying the F test of the equality of population means.

7 Given that the sample size, the arithmetic mean, and the variance $\hat{s}^2$ for each of three independent samples are known, how do you compute the SSW and SSB to carry out an analysis of variance for these three samples?

8 Suppose that a company makes four kinds of light bulbs, A, B, C, and D, and it is desired to test whether there are any differences in the durabilities of the bulbs. Random samples of sizes $n_1 = 5$, $n_2 = 10$, $n_3 = 7$, and $n_4 = 5$ are selected and the following results (service lives in thousands of hours) are obtained:

A	*B*	*C*	*D*
12	30	13	27
18	10	15	25
10	28	18	30
14	26	10	15
16	29	26	13
	29	21	
	27	16	
	26		
	28		
	27		

Can we say that the service lives of the four kinds of bulbs do not differ from one another at $\alpha = 0.01$?

9 Three machines, A, B, and C, are designed to produce identical parts. We want to test whether the hourly outputs of the machines are different. Random samples of a specified number of hours for the machines are selected, and they give the following results:

A	*B*	*C*
2	4	6
3	5	5
4	7	7
5	3	4
	4	6
		8

Test the hypothesis that no differences exist among the hourly outputs of the three machines at $\alpha = 0.05$.

10 Suppose that X_1, X_2, X_3, and X_4 are four variables each known to be approximately normal and to have an identical variance. A sample of five observations for each variable is drawn, and the sum of squares total and the sum of squares within are as follows: $SST = 98.6$ and $SSW = 56.0$. Test the hypothesis that the population means of the four random variables are equal at $\alpha = 0.05$.

11 In a one-way analysis of variance, if hypothesis testing shows a nonsignificant value for F, can we conclude that all the effects of the treatments are actually equal to 0? Explain.

12 Design an experiment in your own field of specialization to illustrate the technique of a one-way analysis of variance in which the linear model discussed in the text would be most appropriate.

13-4 Two-Way Analysis of Variance

The one-way analysis of variance discussed in Sec. 13-3 represents the simplest model of variance analysis. A comprehensive account of the other models is far beyond the scope of this book. We shall, however, include in our discussion one more model, the somewhat more complex two-way analysis of variance.

Frequently, we are interested in examining the effects of two types of factors or treatments. For instance, in marketing research, salesmen may be classified by two factors, age and educational attainment. The purpose is to determine whether age and education level have any significant effect on the volume of sales. We may be interested not only in the individual effect of each of these two factors but also in the joint effect, called *interaction*, since certain age groups may work better when combined with particular educational attainment. As another example, in education research it may be desired to investigate the effect of different teaching methods and of the subjects taught on students' performance. Interaction may also exist here because particular teaching methods may be especially effective when applied to certain subjects.

13-4-1 The model

The discussion in this section is limited to differences about the *specified* classes of each type of factors or treatments. No consideration is given to other classes that might be included. Furthermore, it is assumed that two or more observations are made for each *joint* treatment, that is, a combination of two treatments, each belonging to one of the two types. This requirement is made in order that a test about the existence of the interaction effect can be performed. If only a single observation is made per cell, or per joint treatment, the interaction effect will be combined with the random error and cannot be identified. When two or more observations per cell are made, it is possible to compute the error sum of squares, or to estimate the error variance, and thus separate the interaction effect from the random error.

Suppose that an experiment includes two types of factors: one type designated C (suggesting column) consists of K different treatments, and the other, designated R (suggesting row), consists of J different treatments. It is assumed that under the jth treatment of R and the kth treatment of C for $j = 1, 2, \ldots, J$ and $k = 1, 2, \ldots, K$, there are n observations for $n \geq 2$. (That is, in the cell corresponding to the jth row and the kth column two or more values are entered.) The ith observation for each combination of treatments j and k, or X_{ijk}, is thus the sum of five components as follows:

$$X_{ijk} = \mu + \rho_j + \gamma_k + \iota_{jk} + e_{ijk} \tag{13-22}$$

where μ = grand mean of X regardless of treatment
ρ_j = effect of treatment j (ρ is a Greek letter read as "rho")
γ_k = effect of treatment k
ι_{jk} = effect of interaction between treatments j and k (ι is a Greek letter read as "iota")
e_{ijk} = random error associated with ith observation under the combination of treatments j and k

It is assumed that

$$\sum_{j=1}^{J} \rho_j = 0 \qquad \sum_{k=1}^{K} \gamma_k = 0 \qquad \sum_{j=1}^{J} \iota_{jk} = 0 \qquad \sum_{k=1}^{K} \iota_{jk} = 0$$

Other assumptions underlying the two-way analysis of variance are similar to those made under the one-way model. We assume that e_{ijk} are $N(0,\sigma^2)$ and are independently distributed both within and between treatments or joint treatments. The error variance σ^2 is taken as the same for the cell corresponding to j and k for all j and k.

13-4-2 Partition of the variability

According to the model discussed here, the total sample will consist of $n \times J \times K$ observations. Putting the data in tabular form, we shall have J rows, K columns, and n items in each cell corresponding to the jth row and the kth column. Accordingly, the total variance of the sample can be partitioned into the variance among the rows, variance among the columns, variance among the $J \times K$ cells, and the variance of the random error. Since the sample variance as an unbiased estimator of the population variance is the sum of squared deviations from the grand mean of the sample divided by its degrees of freedom, partition of the sample variance entails the partition of the sum of squares. We shall proceed to see how the sum of squares for the total sample in a two-way classification can be partitioned into the various components.

Let $\overline{X}$ be the grand mean of the sample, and $\overline{X}_j$, $\overline{X}_k$, and $\overline{X}_{jk}$ be respectively the means of row j, column k, and the cell corresponding to row j and column k. Since adding and subtracting the same value will not change the original quantity, the sum of squared deviations of X_{ijk} from the grand mean of the sample can be stated as follows:

$$\sum_{i=1}^{n}\sum_{j=1}^{J}\sum_{k=1}^{K}(X_{ijk}-\bar{X})^2 = \sum_{i=1}^{n}\sum_{j=1}^{J}\sum_{k=1}^{K}(X_{ijk}+\bar{X}_{jk}-\bar{X}_{jk}+\bar{X}_j-\bar{X}_j + \bar{X}_k-\bar{X}_k+\bar{X}-\bar{X}-\bar{X})^2$$

$$= \sum_{i=1}^{n}\sum_{j=1}^{J}\sum_{k=1}^{K}[(X_{ijk}-\bar{X}_{jk})+(\bar{X}_j-\bar{X}) + (\bar{X}_k-\bar{X})+(\bar{X}_{jk}-\bar{X}_j-\bar{X}_k+\bar{X})]^2$$

It can be demonstrated that the right side of this equation is equal to

$$\sum_{i=1}^{n}\sum_{j=1}^{J}\sum_{k=1}^{K}(X_{ijk}-\bar{X}_{jk})^2+\sum_{j=1}^{J}nK(\bar{X}_j-\bar{X})^2+\sum_{k=1}^{K}nJ(\bar{X}_k-\bar{X})^2 + \sum_{j=1}^{J}\sum_{k=1}^{K}n(\bar{X}_{jk}-\bar{X}_j-\bar{X}_k+\bar{X})^2$$

Thus we get

$$\sum_{i=1}^{n}\sum_{j=1}^{J}\sum_{k=1}^{K}(X_{ijk}-\bar{X})^2 = \sum_{i=1}^{n}\sum_{j=1}^{J}\sum_{k=1}^{K}(X_{ijk}-\bar{X}_{jk})^2+\sum_{j=1}^{J}nK(\bar{X}_j-\bar{X})^2 + \sum_{k=1}^{K}nJ(\bar{X}_k-\bar{X})^2 + \sum_{j=1}^{J}\sum_{k=1}^{K}n(\bar{X}_{jk}-\bar{X}_j-\bar{X}_k+\bar{X})^2 \quad (13\text{-}23)$$

The four terms on the right side of this equation are respectively called the *sum of squares of error, SSE*; *sum of squares between rows, SSR*; *sum of squares between columns, SSC*; and *sum of squares for interaction, SSI*. That is,

$$SSE = \sum_{i=1}^{n}\sum_{j=1}^{J}\sum_{k=1}^{K}(X_{ijk}-\bar{X}_{jk})^2 \quad (13\text{-}24)$$

$$SSR = \sum_{j=1}^{J}nK(\bar{X}_j-\bar{X})^2 \quad (13\text{-}25)$$

$$SSC = \sum_{k=1}^{K}nJ(\bar{X}_k-\bar{X})^2 \quad (13\text{-}26)$$

$$SSI = \sum_{j=1}^{J}\sum_{k=1}^{K}n(\bar{X}_{jk}-\bar{X}_j-\bar{X}_k+\bar{X})^2 \quad (13\text{-}27)$$

and the *sum of squares for the total sample, SST*, is

$$SST = SSE + SSR + SSC + SSI \quad (13\text{-}28)$$

For the sake of simplicity, we shall not go into the rationale of the two-way analysis of variance since in principle it is similar to that of the one-way model. Suffice it to say here that the sum of squares for error has $JK(n-1)$ degrees of freedom and, accordingly, the mean square error MSE becomes

$$MSE = \frac{SSE}{JK(n-1)} \tag{13-29}$$

Similarly, the sum of squares between rows, between columns, and for interaction have $J-1$, $K-1$, and $(J-1)(K-1)$ degrees of freedom, respectively. Hence we have the mean square between rows, MSR, between columns, MSC, and for interaction, MSI, as follows:

$$MSR = \frac{SSR}{J-1} \tag{13-30}$$

$$MSC = \frac{SSC}{K-1} \tag{13-31}$$

and

$$MSI = \frac{SSI}{(J-1)(K-1)} \tag{13-32}$$

13-4-3 Test of hypotheses

For testing the null hypothesis that there is no interaction effect, we compute the ratio

$$F = \frac{MSI}{MSE} \tag{13-33}$$

which has $(J-1)(K-1)$ and $JK(n-1)$ degrees of freedom. The null hypothesis that no row effect exists can be tested by

$$F = \frac{MSR}{MSE} \tag{13-34}$$

with $J-1$ and $JK(n-1)$ degrees of freedom. Similarly, the hypothesis that no column effect exists can be tested by

$$F = \frac{MSC}{MSE} \tag{13-35}$$

with $K-1$ and $JK(n-1)$ degrees of freedom.

13-4-4 Computation process

In a two-way analysis of variance, the entire sample of N observations is classified into J rows and K columns with n observations in each row-column combination. As indicated in the table below, there are nk observations in row j,

	C				
R	C_1	C_2	$\cdots$	C_K	*Total*
R_1	X_{111}	X_{112}	$\cdots$	X_{11K}	
	X_{211}	X_{212}	$\cdots$	X_{21K}	
	$\cdots$	$\cdots$	$\cdots$	$\cdots$	
	X_{n11}	X_{n12}	$\cdots$	X_{n1K}	$T_1.$
R_2	X_{121}	X_{122}	$\cdots$	X_{12K}	
	X_{221}	X_{222}	$\cdots$	X_{22K}	
	$\cdots$	$\cdots$	$\cdots$	$\cdots$	
	X_{n21}	X_{n22}	$\cdots$	X_{n2K}	$T_2.$
	$\cdots$	$\cdots$	$\cdots$	$\cdots$	
R_J	X_{1J1}	X_{1J2}	$\cdots$	X_{1JK}	
	X_{2J1}	X_{2J2}	$\cdots$	X_{2JK}	
	$\cdots$	$\cdots$	$\cdots$	$\cdots$	
	X_{nJ1}	X_{nJ2}	$\cdots$	X_{nJK}	$T_J.$
Total	$T._1$	$T._2$	$\cdots$	$T._K$	T

yielding a total score $T_{j.}$ and nJ observations in column k yielding a total score $T_{.k}$ for $j = 1, 2, \ldots, J$ and $k = 1, 2, \ldots, K$. (Observe that a dot "•" is used as one of the subscripts of the marginal total T. If the second subscript of T is a dot, it indicates a row total, whereas a dot used as the first subscript of T indicates a column total.) The grand total is equal to the sum of the row totals, or the sum of the column totals; that is,

$$\begin{aligned} T &= T_1. + T_2. + \cdots + T_J. \\ &= T_{.1} + T_{.2} + \cdots + T_{.K} \end{aligned}$$

For computation purposes, the sums of squares can be written as follows:

$$SST = \sum_{i=1}^{n} \sum_{j=1}^{J} \sum_{k=1}^{K} (X_{ijk}^2 - \bar{X})^2 = \sum_{i=1}^{n} \sum_{j=1}^{J} \sum_{k=1}^{K} X_{ijk}^2 - \frac{T^2}{N} \tag{13-36}$$

$$SSR = \sum_{j=1}^{J} nK(\bar{X}_j - \bar{X})^2 = \sum_{j=1}^{J} \frac{T_j.^2}{nK} - \frac{T^2}{N} \tag{13-37}$$

$$SSC = \sum_{k=1}^{K} nJ(\bar{X}_k - \bar{X})^2 = \sum_{k=1}^{K} \frac{T._k^2}{nJ} - \frac{T^2}{N} \tag{13-38}$$

$$SSE = \sum_{i=1}^{n} \sum_{j=1}^{J} \sum_{k=1}^{K} (X_{ijk} - \bar{X}_{jk})^2$$

$$= \sum_{i=1}^{n} \sum_{j=1}^{J} \sum_{k=1}^{K} X_{ijk}^2 - \sum_{j=1}^{J} \sum_{k=1}^{K} \frac{(\sum_{i=1}^{n} X_{ijk})^2}{n} \tag{13-39}$$

$$SSI = SST - SSC - SSR - SSE$$

$$= \frac{T^2}{N} - \sum_{j=1}^{J} \frac{T_{j\cdot}^{\ 2}}{nK} - \sum_{k=1}^{K} \frac{T_{\cdot k}^{\ 2}}{nJ} + \sum_{j=1}^{J} \sum_{k=1}^{K} \frac{(\sum_{i=1}^{n} X_{ijk})^2}{n} \tag{13-40}$$

These apparently tedious computations can be made quite simple and easy if an electronic computer is available. The computation process for a two-way analysis of variance is summarized in Table 13-3.

13-4-5 An example

A manufacturing company uses three different machines operated by three different workmen. The company wishes to determine whether the differences in the number of defective products turned out by the workmen and the differences in the number of defective products turned out by the machines are significant. Also, it desires to test whether there is any interaction effect between the workmen and the machines. For testing the null hypothesis that there is no difference

Table 13-3

Summary table for two-way analysis of variance

Source of variation	*Sum of squares SS*	*Degrees of freedom* ν	*Mean square MS*	*F ratio*
Row R	$\sum_{j=1}^{J} \frac{T_{j\cdot}^{\ 2}}{nK} - \frac{T^2}{N}$	$J-1$	$\frac{SSR}{J-1}$	$\frac{MSR}{MSE}$
Column C	$\sum_{k=1}^{K} \frac{T_{\cdot k}^{\ 2}}{nJ} - \frac{T^2}{N}$	$K-1$	$\frac{SSC}{K-1}$	$\frac{MSC}{MSE}$
Interaction I	$\frac{T^2}{N} - \sum_{j=1}^{J} \frac{T_{j\cdot}^{\ 2}}{nK} - \sum_{k=1}^{K} \frac{T_{\cdot k}^{\ 2}}{nJ} + \sum_{j=1}^{J} \sum_{k=1}^{K} \frac{(\sum_{i=1}^{n} X_{ijk})^2}{n}$	$(J-1(K-1)$	$\frac{SSI}{(J-1)(K-1)}$	$\frac{MSI}{MSE}$
Error E	$\sum_{i=1}^{n} \sum_{j=1}^{J} \sum_{k=1}^{K} X_{ijk}^2 - \sum_{j=1}^{J} \sum_{k=1}^{K} \frac{(\sum_{i=1}^{n} X_{ijk})^2}{n}$	$JK(n-1)$	$\frac{SSE}{JK(n-1)}$	
Total T	$\sum_{i=1}^{n} \sum_{j=1}^{J} \sum_{k=1}^{K} X_{ijk}^2 - \frac{T^2}{N}$	$N-1$		

in the performance of the workmen, no difference in the performance of the machines, and no interaction effects, each workman is required to operate each machine on two different days. The results are shown in Table 13-4. The various quantities required to test the hypotheses are computed below:

$$\frac{T^2}{N} = \frac{282^2}{18} = \frac{79{,}524}{18} = 4{,}418$$

$$\sum_{j=1}^{J}\sum_{k=1}^{K}\frac{(\sum_{i=1}^{n} X_{ijk})^2}{n} = \frac{(10+13)^2}{2} + \frac{(13+16)^2}{2} + \cdots + \frac{(14+17)^2}{2}$$

$$= \frac{9{,}284}{2} = 4{,}642$$

$$\sum_{i=1}^{n}\sum_{j=1}^{J}\sum_{k=1}^{K} X_{ijk}^2 = 10^2 + 13^2 + \cdots + 14^2 + 17^2 = 4{,}736$$

$$SST = 4{,}736 - 4{,}418 = 318$$

$$SSR = \sum_{j=1}^{J}\frac{T_{j\cdot}^{\,2}}{nK} - \frac{T^2}{N} = \frac{8{,}649}{6} + \frac{11{,}449}{6} + \frac{6{,}724}{6} - 4{,}418$$

$$= 4{,}470.33 - 4{,}418 = 52.33$$

$$SSC = \sum_{k=1}^{K}\frac{T_{\cdot k}^{\,2}}{nJ} - \frac{T^2}{N} = \frac{5{,}625}{6} + \frac{10{,}816}{6} + \frac{10{,}609}{6} - 4{,}418$$

$$= 4{,}508.33 - 4{,}418 = 90.33$$

$$SSI = \frac{T^2}{N} - \sum_{j=1}^{J}\frac{T_{j\cdot}^{\,2}}{nK} - \sum_{k=1}^{K}\frac{T_{\cdot k}^{\,2}}{nJ} + \sum_{j=1}^{J}\sum_{k=1}^{K}\frac{(\sum_{i=1}^{n} X_{ijk})^2}{n}$$

$$= 4{,}418 - 4{,}470.33 - 4{,}508.33 + 4{,}642 = 81.34$$

$$SSE = 4{,}736 - 4{,}642 = 94$$

Table 13-4

Number of defective products turned out by three workmen and three machines in 2 days

Machines	*Workmen* C_1	C_2	C_3	*Total*
R_1	10 13	14 16	18 22	$T_{1\cdot} = 93$
R_2	13 16	19 27	14 18	$T_{2\cdot} = 107$
R_3	9 14	11 17	14 17	$T_{3\cdot} = 82$
Total	$T_{\cdot 1} = 75$	$T_{\cdot 2} = 104$	$T_{\cdot 3} = 103$	$T = 282$

The analysis of variance table for this example is given as follows:

Source of variation	*Sum of squares SS*	*Degrees of freedom v*	*Mean square MS*	*F ratio*
Row R	52.33	2	26.17	$\frac{26.17}{10.44} = 2.51$
Column C	90.33	2	45.17	$\frac{45.17}{10.44} = 4.34$
Interaction I	81.34	4	20.34	$\frac{20.34}{10.44} = 1.95$
Error E	94.00	9	10.44	
Total	318.00	17		

For testing the null hypothesis that there is no interaction effect, we compute the ratio of the mean square interaction to the mean square error

$$F = \frac{MSI}{MSE} = 1.95$$

which has 4 and 9 degrees of freedom. The critical ratio for this F distribution at $\alpha = 0.05$ is 3.63, and at $\alpha = 0.01$ it is 6.42. Thus we shall not reject the null hypothesis at the 5 percent significance level and at the 1 percent level as well.

For testing the hypothesis of no column effect, we compute the ratio

$$F = \frac{MSC}{MSE} = 4.34$$

which has 2 and 9 degrees of freedom. The critical F value for this distribution is 4.26 at $\alpha = 0.05$, and 8.02 at $\alpha = 0.01$. The result barely leads us to reject the null hypothesis that there is no difference in the performance of the workmen at $\alpha = 0.05$. Since the F ratio for testing about the difference in the performance of the machines is only 2.51 with the same degrees of freedom, we shall *not* reject the null hypothesis of no difference in the performance of the machines at both the 5 and the 1 percent levels of significance.

When there is no significant interaction effect, a pooled mean square error may be used as shown in Prob. 10, Exercise 13-4.

Exercise 13-4

1 Define the following concepts relative to the linear model for a two-way analysis of variance discussed in this section:

a Interaction effect.
b Sum of square error.
c Sum of square interaction.
d Random error.

2 In the light of the discussion in Sec. 13-3, explain the rationale for the two-way analysis of variance.

3 In a two-way analysis of variance, if the tests of hypotheses show that there exist no significant differences in the row, column, or interaction effects, will you conclude that all the row, column, and interaction effects are actually equal to 0? Explain.

4 Design an experiment in your own field of specialization to illustrate the technique of a two-way analysis of variance in which the linear model discussed in this section would be most appropriate.

5 Suppose that there are three kinds of rice, R_1, R_2, and R_3, each planted on six identical plots of land and that three kinds of fertilizers, F_1, F_2, and F_3 are used. The yields of rice in bushels are cross classified as follows:

	F_1	F_2	F_3
R_1	3	8	12
	5	7	10
R_2	5	10	13
	4	9	13
R_3	5	8	10
	6	10	12

At $\alpha = 0.05$, test the null hypothesis that:

a There are no differences in the mean yields for the different kinds of rice.
b There are no differences in the effects of the fertilizers on the rice yields.
c There are no differences in the interaction effects.

6 What are the basic assumptions underlying the two-way analysis of variance discussed in this section? Do you see the reasons why these assumptions are made? Explain.

7 Explain each of the terms on the right side of Eq. (13-22).

8 Suppose that in an experiment there are four row factors and three column factors, and that there are five observations in each of the row-column combinations. Suppose further that the experiment yields the following results using the same symbols as in the text:

$$\sum_{i=1}^{n}\sum_{j=1}^{J}\sum_{k=1}^{K} X_{ijk}^2 = 4{,}110 \qquad \sum_{j=1}^{J}\sum_{k=1}^{K} \frac{(\sum_{i=1}^{n} X_{ijk})^2}{n} = 3{,}150$$

$$\frac{T^2}{N} = 720 \qquad \sum_{j=1}^{J} \frac{T_{j\cdot}{}^2}{nK} = 990 \qquad \sum_{k=1}^{K} \frac{T_{\cdot k}{}^2}{nJ} = 1{,}080$$

Test the following null hypotheses at $\alpha = 0.05$:

a There are no differences in the interaction effects.
b There are no differences in the column effects.
c There are no differences in the row effects.

9 Suppose that a certain company divides the territory it serves into five regions and classifies its salesmen into four age groups. It wants to determine whether there are differences in the annual sales of the salesmen in different age brackets and in the annual mean sales of the regions. It also wants to find out whether there are differences in the interaction effects between the ages of salesmen and the regions covered. Sales data (in thousands of dollars) for a two-year period are collected and tabulated as follows:

	C_1	C_2	C_3	C_4
R_1	32	45	42	29
	25	33	34	34
R_2	30	36	36	23
	21	14	18	30
R_3	22	30	33	20
	12	22	20	19
R_4	28	25	34	34
	29	40	22	31
R_5	38	37	36	34
	24	35	26	28

Test the following null hypotheses at $\alpha = 0.05$:

a There are no differences in sales among the age groups.
b There are no differences in the average sales among the regions.
c There are no differences in the interaction effects.

10 In Prob. 9, since there are no significant differences in the interaction effects, we may compute the *pooled* mean square error by adding *SSE* to *SSI* and then dividing the sum by the sum of the degrees of freedom for *SSE* and *SSI*. Test the hypotheses in the preceding problem by using the pooled mean square error.

Chapter 14

Simple regression and correlation

So far we have been primarily concerned with the problems of statistical inference relative to a *univariate distribution*, that is, the distribution of only a single variable, say, X. A single observation is made on each random trial; on the basis of a statistic resulting from sample observations, we try to infer about a parameter of the distribution of X. Our attention now turns to problems with a *bivariate distribution*, a distribution that contains two variables, say, X and Y. The sample space of a bivariate experiment is some set of ordered pairs of measurements. In other words, two observations are made on a single trial. For example, for each student, we may observe his intelligence quotient and grade average. Intelligence quotient may be designated X, and grade average Y. Or, X and Y may denote the heights and weights of enlisted men, the gross national product and new construction of any period of time in the United States, etc.

The central problem of statistical inference involving a bivariate distribution is to determine the true relationship between X and Y; that is, how the two variables behave in relation to each other. College students with high IQs are likely to have high grades, and those with low IQs are likely to have low grades. This relationship is true only on the average because occasionally we may observe that rather low grades are earned by a student with a very high intelligence quotient, and vice versa. Our task is to establish this type of relationship in the form of some equation so that on the basis of someone's IQ, or X, we can find an estimate of his grade average, or Y. Such an equation may be referred to as an *estimation relation.*

The procedure of estimation is also a techniqué for prediction, which is the basic function of science, natural as well as social. In the natural sciences, prediction is often made possible because a *cause-and-effect* relation exists between two or more variables. For instance, with the *mass* held constant, a certain increase in the *speed* at which the mass is moving will cause a predictable amount of rise in the kinetic energy, or KE, since it is well known that

$$\mathrm{KE} = \frac{1}{2} mv^2$$

where m is the mass, and v is the speed. It is also apparent, to use another example, that a cause-and-effect relation exists between the amount of precipitation and the amount of wheat raised from a given acreage.

However, prediction does not necessarily require the existence of such a relationship. In the social sciences we frequently encounter variables that are merely *functionally* related; that is, they are associated in some functional way. Although a functional relation between two variables implies nothing about cause and effect, it nonetheless enables us to predict the value of one variable on the condition that we have prior information about the other. This leads us to an important topic in statistics, namely, *regression analysis.* If two variables are involved,

> *The variable that is the basis of estimation is conventionally called the independent variable and is designated X, and the variable whose value is to be estimated is called the dependent variable and is designated Y. When an equation is formulated for estimating Y from X, such an equation is called a regression of Y on X.*

The technique of regression analysis is nothing more than a procedure of estimation, or prediction. The term regression is used here simply because it is well established in statistics. It was first introduced in 1877 by Sir Francis Galton, who found in his studies of heredity that tall fathers tended to have tall sons and short fathers tended to have short sons. However, the average height of the sons of very tall fathers was less than the average height of their fathers, while the sons of very short fathers were, on the average, taller than their fathers. Galton referred to this tendency toward the average height of all men as a regression. Similar associations were observed in several other phenomena by Galton, who generalized them as a universal law of behavior between two or more associated variables.

Regression analysis is generally classified into two kinds, simple and multiple. *Simple regression* involves only two variables, such as the regression of Y on X mentioned above. *Multiple regression* involves three or more variables, one of which is the dependent variable that is to be associated with the values of all others. Only simple regression is considered in this chapter.

Furthermore, our discussion will be restricted to *linear regression* only. That is, the equation describing the relation between X and Y is linear and is graphically represented by a straight line. Sometimes, a regression equation that can best describe the relation between variables is *curvilinear*; its geometrical representation is a curve rather than a straight line. This chapter will not be concerned with problems of curvilinear regression.

When variables are found to be related, we often want to know how close the relationship is. The degree or closeness of the relationship is commonly referred to as the *correlation* between the variables. The problem of correlation is intimately associated with that of regression and is an integral part of bivariate analysis.

In short, we shall in this chapter consider only simple regression and correlation problems on the assumption that the two variables have a linear relation.

14-1 The Population Regression Line

In our previous discussions on univariate distributions we proceeded by first assuming some population model and then using sample statistics as estimates of population parameters. Among the possible models we chose one that appeared to fit the problem in hand. In bivariate analysis we are concerned with a particular model, the *population regression line*, and we shall use a sample regression line as an estimate of the population regression line. As with a univariate population model, which requires assumptions about the population distribution, the linear regression model is assumed to have the following properties:

Firstly, the possible values of the independent variable X are fixed in advance. They are arbitrarily selected *constants*, and thus have no observation errors associated with them. The values of the dependent variable Y are assumed to be dependent on the value of X. The variable Y possesses a random property; it is left free to take on any value that may be possibly associated with a given value of X.

Secondly, it is assumed that the true relationship between X and Y can be described by the equation

$$Y_i = \alpha + \beta X_i + \epsilon_i \tag{14-1}$$

where α and β are parameters of the population and ϵ_i (a Greek letter read as "epsilon") is the difference between Y_i and the expected value of Y given X_i, designated $\mu_{Y|X_i}$; that is,

$$\epsilon_i = Y_i - \mu_{Y|X_i} \tag{14-2}$$

the sum and mean of which are 0. Thus in general we obtain the population regression line as

$$\mu_{Y|X} = \alpha + \beta X \tag{14-3}$$

which represents the average relationship between X and Y in the population. It is referred to as the *line of regression of Y on X*, and its properties are shown in Fig. 14-1. The parameter α is the average value of Y corresponding to 0 for X, and is called the *Y intercept*. The parameter β is the *slope* of the population regression line, also known as the *population regression coefficient*. It

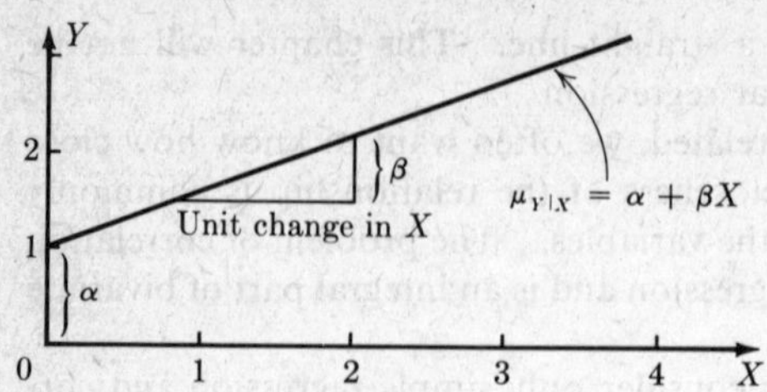

Figure 14-1
Properties of a regression line.

represents the amount of increase in Y for each unit increase in X. When the line slopes downward to the right, the value of β will be negative; it then represents the amount of decrease in Y for each unit increase in X. Since a line is completely defined by the Y intercept and the slope, the task of estimating a population regression line is nothing more than estimating these two parameters.

For each fixed value of X, the height of the regression line represents the arithmetic mean of a theoretically infinite number of Y values. This explains why the regression line is often referred to as the *line of conditional means.*

Thirdly, we assume that the distribution of the Y values corresponding to a predetermined X value is *normal*, or Y is $N(\mu_{Y|X}, \sigma^2_{Y|X})$. Such a normal distribution is a *probability* distribution just like any other distribution discussed in the previous chapters. It differs from the others in that it is *conditional*; that is, it relates to a fixed value of X. Also, Y is assumed to be independently distributed.

Finally, the *conditional probability distribution* is assumed to have the same variance $\sigma^2_{Y|X}$, and thus the same standard deviation $\sigma_{Y|X}$ for each of the possible values of X. The value of $\sigma_{Y|X}$ is given by

$$\sigma_{Y|X} = \sqrt{E((Y - \mu_{Y|X})^2)} \tag{14-4}$$

The assumption that this conditional standard deviation for Y remains the same for each and every value of X may not be a realistic one. It is made, however, in order to simplify our discussion and to make it easier to understand the basic principles in regression analysis.

Since the values of Y are normally distributed, we can make use of the normal distribution to determine the probability of occurrence for Y. We can say, for instance, that about 68.27 percent of the Y's associated with a fixed value of X will fall within one standard deviation from $\mu_{Y|X}$, and that 95.45 percent of them will fall within two standard deviations from $\mu_{Y|X}$. The conditional probability distribution of Y about the conditional mean $\mu_{Y|X}$ is illustrated in Fig. 14-2.

With these properties of the population regression line in mind, we can now turn our attention to the task of inferring about the conditional means in the light of empirical sample information.

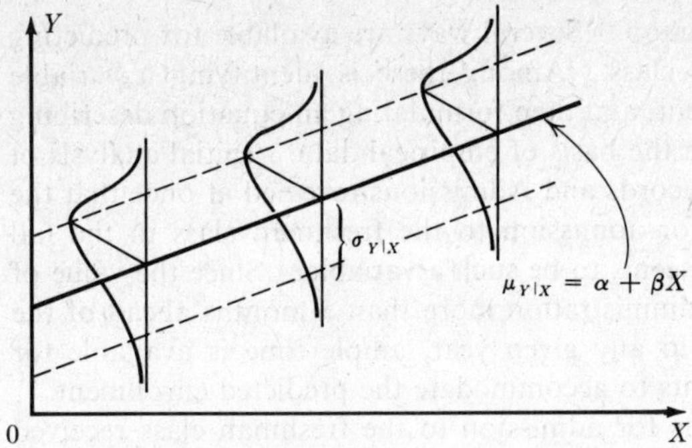

Figure 14-2
Normal curves showing conditional probability distributions of *Y* at given values of *X*.

Exercise 14-1

1 Distinguish between the following concepts:

a Univariate and bivariate distribution.
b Simple and multiple regression.
c Linear and curvilinear regression.

2 List and explain the assumptions underlying the linear regression model.

3 Why is the regression line often referred to as the line of conditional means?

4 By means of a graph, identify or define each of the following symbols from Eqs. (14-1) and (14-3).

a Y_i. *d* β.
b $\mu_{Y|X}$. *e* ϵ_i.
c α.

5 Suppose that the true relation between *X* and *Y* is expressed by the equation

$Y = 8 + 2X + \epsilon$

and the conditional variance for *Y*, given *X*, has the identical value 6.25. Graph the line of the conditional means and the 95.45 percent confidence limits for *Y*.

6 What are the parameters of the simple linear regression model? What do you propose to do for estimating these parameters?

14-2
Point Estimation of the Population Regression Line

In order to explain the basic notions of estimating the regression line by the use of sample data, let us describe a typical situation to which the regression analysis is usually applied.

Suppose that the University of Southland wishes to forecast the size of its freshman class so that adequate preparation can be made to ensure efficiency of operation and quality of instruction. Several ways are available for predicting the enrollment in the freshman class. Among these is identifying a variable closely associated with enrollment, and then formulating an equation describing the relation between the two on the basis of empirical data. Initial analysis of the problem by the Office of Records and Admissions revealed at once that the total number of applications for admission to the freshman class in the fall semester received as of April 1 seems to be such a variable. Since the value of this variable is known to the administration more than 5 months ahead of the beginning of the fall semester in any given year, ample time is available for making the necessary adjustments to accommodate the predicted enrollment.

The number of applications for admission to the freshman class received before April 1 (hereafter referred to as applications) is the independent variable denoted by X, and the number of freshmen enrolled in the fall semester of the same year (hereafter referred to as enrollments) is the dependent variable denoted by Y. Data for these two variables for the 20 years from 1953 to 1972 are shown in Table 14-1. We are considering the problem of predicting the value of Y on

Table 14-1

Number of freshmen enrolled in September and number of applications received as of April 1 for the University of Southland, 1953–1972

Year	*Number of applications in hundreds* X	*Number of freshmen enrolled in hundreds* Y
1953	2	3
1954	4	6
1955	5	6
1956	5	7
1957	7	9
1958	6	9
1959	9	11
1960	11	13
1961	12	12
1962	13	14
1963	13	15
1964	13	13
1965	14	13
1966	14	15
1967	15	15
1968	16	18
1969	17	19
1970	17	20
1971	18	20
1972	19	22

the basis of the predetermined value of X. We shall first consider the technique of using a scatter diagram and the freehand method for estimation, and then examine the method of least squares in fitting a regression line.

14-2-1 The scatter diagram

In analyzing the relation between two variables it is often desirable to present the sample data in a diagram called the *scatter diagram.* It gives us a visual impression of the relations possible and suggests the type of model that may best fit the data. The conventional procedure in constructing such a diagram is to have the independent variable X scaled on the horizontal axis and the dependent variable Y on the vertical axis. A point representing a pair of observations of X and Y is plotted; the resulting graph of all the points thus plotted for all the pairs of X and Y values in the sample is the scatter diagram. A hypothetical example of a scatter diagram is shown in Fig. 14-3. It would appear that the regression relation between X and Y is not linear. This observation is based on the fact that while an increase in the value of the independent variable X would be accompanied by an increase in the value of the dependent variable Y, the rate of increase in the value of Y is not a constant but is rising relative to that of X. A linear equation seems inappropriate for fitting this kind of data.

The scatter diagram for the data in Table 14-1 is shown in Fig. 14-4. It clearly indicates that there is a positive relation between the number of freshmen enrolled and the number of applications received. That is, there is a general tendency for the former to rise with an increase in the latter. It further suggests that the average relation between the two variables seems to be linear since a straight line would result if we trace along the center of the scatter from the lower left to the upper right.

Once we are reasonably assured that a linear relation exists between the two variables, our next task is to estimate the true relationship. The simplest and crudest way of doing this is the so-called *freehand method.* It is drawing a straight line freehand near or through the points on the scatter diagram so that the line appears to describe the relation. The principal drawback of such a method is, of course, the absence of precision measurement of any prediction based on such a line.

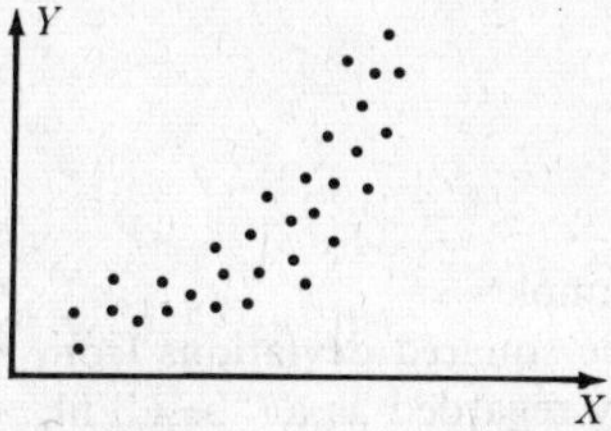

Figure 14-3
A scatter diagram.

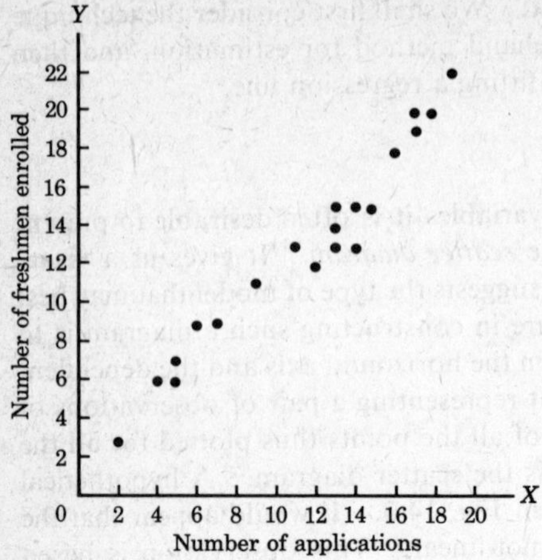

Figure 14-4

Scatter diagram of applications and enrollments (in hundreds) for the University of Southland. From data of Table 14-1.

14-2-2 The least-squares method

There are many lines that could possibly be drawn on a given scatter diagram. Which one then should we choose as the *estimating* line? Of course we wish to choose a line that can fit the sample data in hand. The most commonly used method for selecting such a line is the *least-squares method*, and the resulting line is called the *least-squares line*. The term least squares is used because the sum of squares of the vertical deviations of the points from this line is less than the sum of squares of such deviations from any other line. Let Y_c be the computed value for Y; that is, the vertical coordinate at the point on the line that has the horizontal coordinate X_i. We know that an estimating line

$$Y_c = a + bX_i$$

is completely defined if the statistics a, the Y intercept, and b, the slope of the line, are known. Let Y_i be the ith *observation* of the variable Y associated with X_i, the ith observation on X. Then the least-squares lines is the line that minimizes

$$\sum_{i=1}^{n} e_i^2 = \sum_{i=1}^{n} [Y_i - (a + bX_i)]^2$$

where e_i = deviation of Y_i from Y_c

n = number of pairs of observations in the sample

Since the least-squares line minimizes the sum of the squared deviations from the conditional means, the least-squares fit is usually regarded as a "best" fit.

The values of a and b that make the sum of squares of the e's as small as possible are obtained by solving the following two normal equations simultaneously:

$$\sum_{i=1}^{n} Y_i = na + b\sum_{i=1}^{n} X_i \tag{14-5}$$

$$\sum_{i=1}^{n} X_i Y_i = a\sum_{i=1}^{n} X_i + b\sum_{i=1}^{n} X_i^2 \tag{14-6}$$

Besides a and b, all the quantities in these two equations can be directly obtained from the sample data as shown in Table 14-2, and the values of a and b can be computed from these quantities. Solving for b, we simply multiply Eq. (14-5) by $\sum_{i=1}^{n} X_i$ and Eq. (14-6) by n and then subtract one from the other so that the term containing a is eliminated. Consequently, we have

$$b = \frac{(\sum_{i=1}^{n} X_i Y_i)/n - (\sum_{i=1}^{n} X_i \sum_{i=1}^{n} Y_i)/n^2}{(\sum_{i=1}^{n} X_i^2)/n - [(\sum_{i=1}^{n} X_i)/n]^2} = \frac{(\sum_{i=1}^{n} X_i Y_i)/n - \overline{X}\,\overline{Y}}{(\sum_{i=1}^{n} X_i^2)/n - \overline{X}^2} \tag{14-7}$$

where the numerator is the covariance of X and Y, or cov(X, Y), and the denominator is the variance of X [see Eqs. (5-16) and (5-4)]. That is,

$$b = \frac{\text{cov}(X, Y)}{s_X^{\ 2}} \tag{14-8}$$

After the value of b has been obtained, we can compute the value of a by substituting the b value in either Eq. (14-5) or (14-6). Since

$$\sum_{i=1}^{n} Y_i = na + b\sum_{i=1}^{n} X_i$$

we have

$$a = \frac{\sum_{i=1}^{n} Y_i}{n} - b\frac{\sum_{i=1}^{n} X_i}{n} \tag{14-9}$$

or

$$a = \overline{Y} - b\overline{X} \tag{14-10}$$

Thus, for obtaining the values of a and b, we need to know the quantities of n, $\sum_{i=1}^{n} X_i$, $\sum_{i=1}^{n} Y_i$, $\sum_{i=1}^{n} X_i Y_i$, and $\sum_{i=1}^{n} X_i^2$. ($\sum_{i=1}^{n} Y_i^2$ also needed for later computations.) They are computed in Table 14-2. The slope and the Y

Table 14-2

Computation for regression of number of freshmen enrolled, Y, on number of applications, X, for the University of Southland

	Number of applications in hundreds X	Number of freshmen enrolled in hundreds Y	XY	X^2	Y^2
	2	3	6	4	9
	4	6	24	16	36
	5	6	30	25	36
	5	7	35	25	49
	7	9	63	49	81
	6	9	54	36	81
	9	11	99	81	121
	11	13	143	121	169
	12	12	144	144	144
	13	14	182	169	196
	13	15	195	169	225
	13	13	169	169	169
	14	13	182	196	169
	14	15	210	196	225
	15	15	225	225	225
	16	18	288	256	324
	17	19	323	289	361
	17	20	340	289	400
	18	20	360	324	400
	19	22	418	361	484
Total	230	260	3,490	3,144	3,904

$n = 20$

$\sum_{r=1}^{n} X_i = 230 \qquad \bar{X} = \frac{230}{20} = 11.5$

$\sum_{i=1}^{n} Y_i = 260 \qquad \bar{Y} = \frac{260}{20} = 13$

$\sum_{i=1}^{n} X_i Y_i = 3{,}490 \qquad \sum_{i=1}^{n} X_i^2 = 3{,}144 \qquad \sum_{i=1}^{n} Y_i^2 = 3{,}904$

Source: Table 14-1.

intercept of the regression line based on the sample data are then obtained as follows:

$$b = \frac{\frac{3,490}{20} - 11.5(13)}{\frac{3,144}{20} - (11.5)^2}$$

$$= \frac{25}{24.95}$$

$$\approx 1.00$$

and

$$a = 13 - 1(11.5) = 1.50$$

Our regression equation for the sample data is therefore

$$Y_c = 1.5 + X$$

which is graphically represented in Fig. 14-5. It is our estimate of the population regression equation

$$\mu_{Y|X} = \alpha + \beta X$$

That is, our estimate of α is 1.5 and that of β is 1. In this instance the estimate of α is of little significance. Its only importance lies in the fact that it locates the regression line at the point when X is 0. However, when X is 0, it is usually not necessary to consider the value of a in Y_c. On the other hand, the slope b is of great importance. It represents an estimate of the average change in the value of the dependent variable Y for each unit change in the value of the independent variable X. In this particular example the value of positive 1 for b

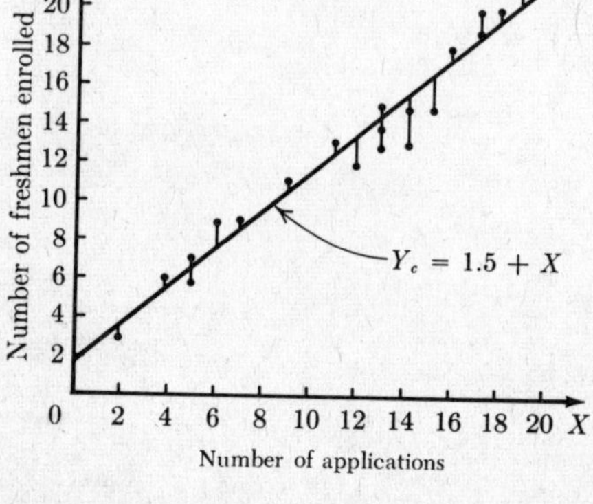

Figure 14-5
Estimated regression line of number of enrollment on number of applications (in hundreds) and deviations of observed values from the line. From data of Table 14-1.

means that the enrollment is estimated to change in about the same magnitude and direction as the change in the number of applications.

The linear regression equation based on sample data is widely used for prediction purposes. Given a value of X, we can predict what the associated Y value will be, on the average. For instance, if during the year the number of applications by April 1 is 20 (in hundreds), the predicted enrollment in the freshman class in the following fall would be

$$Y_c = 1.5 + 20 = 21.5$$

which is a *point estimate* of the conditional mean $\mu_{Y|X}$. How good, then, is this estimate? Or, what is the degree of precision we can put on the predicted value? In order to answer this question, we need to consider interval estimation of the conditional mean $\mu_{Y|X}$.

Exercise 14-2

1 Show that the least-squares line $Y = a + bX$ can also be expressed as $Y = \bar{Y} + b(X - \bar{X})$.

2 Suppose that we wish to study the distribution of the weights of college men with relation to their heights. We divide the population of college men into groups of approximately equal heights, and for any selected height there is a distribution of weights. Let X be the measurement of heights and Y be the measurement of weights. Suppose further that the values of X are chosen at 60, 62, 64, 66, 68, and 70 inches, and 20 men having these heights are randomly selected. The observations are shown as follows:

Heights in inches X	*Weights in pounds* Y
60	115
60	120
60	130
60	125
62	130
62	140
62	120
64	155
64	130
64	145
66	135
66	170
66	140
66	155
68	150
68	160
68	175
70	180
70	160
70	175

a Plot the data as a scatter diagram.
b Draw a freehand line on the diagram to fit these data, and then construct an equation to represent the freehand line.
c How many such freehand lines can you possibly draw on the diagram?
d Would you use the line you have just drawn as an estimate of the true regression line? Explain.

3 Fit a least-squares line to the data of the preceding problem with X as the independent variable. If the height of a college man is 67 inches, what would be his predicted weight?

4 Given the following data on the amount of fertilizer X and the yield Y of wheat:

Fertilizer in pounds X	*Wheat yield in bushels* Y
2	8
4	9
5	11
7	11
10	12
11	14
12	15
15	16

a Plot the data as a scatter diagram.
b Find the regression line of Y on X.
c Estimate the wheat yield when 13 pounds of fertilizer is used.

5 Show that a least-squares line always passes through the point $(\bar{X}, \bar{Y})$.

6 Suppose that it is desired to determine the relation between the length of sales experience X and the volume of sales Y for each salesman from a group of 10 salesmen of an insurance company. Years of sales experience and sales volume are shown as follows:

Sales experience in years X	*Volume of sales in \$10,000* Y
1	3
2	2
3	5
4	4
5	6
6	8
7	9
8	9
9	12
10	10

a Plot the data as a scatter diagram.
b Construct a regression line of Y on X.
c Estimate the amount of sales for a salesman who has had 12 years of sales experience.

7 What are the similarities and differences between point estimation of a population mean and point estimation of a population regression line?

8 Show that the least-squares line of Y on X can be expressed as:

a $y = \dfrac{\sum xy}{\sum x^2} x$ when $x = X - \overline{X}$ and $y = Y - \overline{Y}$.

b $Y = \overline{Y} + \dfrac{\sum XY}{\sum X^2} X$ when $\overline{X} = 0$.

14-3 Interval Estimation of the Population Regression Line

Just as the degree of precision in point estimation of the parameter of a univariate distribution is measured by constructing a confidence interval for that parameter, the preciseness of a point estimate of the conditional mean $\mu_{Y|X}$ of a bivariate population is measured by constructing a confidence interval for $\mu_{Y|X}$. The construction of a confidence interval for $\mu_{Y|X}$ requires, of course, a knowledge of the standard deviation of the estimate of $\mu_{Y|X}$, that is, the standard deviation of Y_c, or σ_{Y_c}. When σ_{Y_c} is known, the confidence limits are easily determined as

$$Y_c \pm z_{\alpha/2}\,\sigma_{Y_c} \tag{14-11}$$

where $z_{\alpha/2}$ is the standard normal deviate corresponding to the desired level of significance α. When σ_{Y_c} is not known, we have to estimate it in order to establish the confidence limits.

In addition to interval estimation for the population regression line, we shall also in this section try to construct a confidence interval for the *next* Y value. As will be seen in Figs. 14-6 and 14-7, the confidence intervals to be discussed in this section actually form a *confidence belt* which is symmetric in width around the value of $X = \overline{X}$. For this reason they are often referred to as *symmetric confidence intervals.*

14-3-1 Estimation of the standard deviation σ_{Y_c}

The standard deviation σ_{Y_c} measures the dispersion of the dependent variable Y_c around the conditional mean $\mu_{Y|X}$ of the population. The value of σ_{Y_c} is dependent on the value of $\sigma_{Y|X}$, the conditional standard deviation of Y at a given value of X, which is called the *standard deviation of the error term* in the linear regression model. Thus estimating the value of σ_{Y_c} requires, first of all, an estimation of the value of $\sigma_{Y|X}$ if the latter is not known. Before finding the estimate for $\sigma_{Y|X}$, let us first review the standard deviation of the variable Y in a univariate distribution. When $\hat{s}_Y$ is an estimator of σ_Y for a univariate distribution, by Eq. (9-4) we have

$$\hat{s}_Y = \sqrt{\frac{\sum_{i=1}^{n} (Y_i - \bar{Y})^2}{n-1}}$$

Similarly, for a bivariate distribution, we have the conditional standard deviation $\hat{s}_{Y|X}$ for the sample as an estimator of the population conditional standard deviation $\sigma_{Y|X}$ as follows:

$$\hat{s}_{Y|X} = \sqrt{\frac{\sum_{i=1}^{n} (Y_i - Y_c)^2}{n-2}} \qquad (14\text{-}12)$$

where Y_c, as before, is the observed conditional mean of Y at a given value of X. The reason why $n - 2$ is used as the denominator is that 2 degrees of freedom are lost when both α and β have to be estimated by a and b (calculated from the sample data) in order to estimate the conditional mean $\mu_{Y|X}$. The estimate $\hat{s}_{Y|X}$ is the standard deviation of the *observed* Y values *from* the computed regression line corresponding to the various values of X.

Table 14-3 shows the process of computing the conditional standard deviation from the sample data. Using the sum of squares of the deviations from the computed value of Y as the numerator of the expression on the right side of Eq. (14-12), we obtain

$$\hat{s}_{Y|X} = \sqrt{\frac{23}{18}} = \sqrt{1.2778} = 1.13$$

Note that the deviations of observed Y values from the computed conditional means sum to 0. This is the basic property of the least-squares line, and it can be used to check the accuracy of the computations in Table 14-3 and in Table 14-2 as well, about the quantities from which the Y intercept a and the slope b are calculated. The sum of deviations, however, may not be exactly 0 because of rounding.

Equation (14-12), however, is quite cumbersome to work with, especially when the sample is large. It requires the computation of Y_c for each observed value of Y, and then squaring the difference between the two values. Much simpler methods of computing the value of $\hat{s}_{Y|X}$ are available, and one of them is given as follows:

$$\hat{s}_{Y|X} = \sqrt{\frac{\sum_{i=1}^{n} Y_i^2 - a\sum_{i=1}^{n} Y_i - b\sum_{i=1}^{n} X_i Y_i}{n-2}} \qquad (14\text{-}13)$$

where a is the Y intercept and b is the slope already determined. Other quantities in the numerator are all given in Table 14-2. Substituting these quantities in Eq. (14-13), we obtain

$$\hat{s}_{Y|X} = \sqrt{\frac{3{,}904 - 1.5(260) - 1(3{,}490)}{20-2}} = \sqrt{\frac{24}{18}} = 1.15$$

Table 14-3
Computation of the estimate of the population conditional standard deviation

	Applications X	*Enrollment* Y	*Conditional mean* Y_c	*Deviations* $Y - Y_c$	*Squares of deviations* $(Y - Y_c)^2$
	2	3	3.5	−0.5	0.25
	4	6	5.5	0.5	0.25
	5	6	6.5	−0.5	0.25
	5	7	6.5	0.5	0.25
	7	9	8.5	0.5	0.25
	6	9	7.5	1.5	2.25
	9	11	10.5	0.5	0.25
	11	13	12.5	0.5	0.25
	12	12	13.5	−1.5	2.25
	13	14	14.5	−0.5	0.25
	13	15	14.5	0.5	0.25
	13	13	14.5	−1.5	2.25
	14	13	15.5	−2.5	6.25
	14	15	15.5	−0.5	0.25
	15	15	16.5	−1.5	2.25
	16	18	17.5	0.5	0.25
	17	19	18.5	0.5	0.25
	17	20	18.5	1.5	2.25
	18	20	19.5	0.5	0.25
	19	22	20.5	1.5	2.25
Total	230	260	252.00	0	23.00

Source: Table 14-1.

which represents a slight error of 0.02 as compared with the value obtained by the earlier tedious inethod. The slight discrepancy is due to rounding; the slope b is actually about 1.002 rather than 1 as used here. The above two methods are expected to yield the same result.

As already noted, to find the confidence limits for the population regression line, what we need is the standard deviation *of* the sample regression line, or σ_{Y_c}, not just the standard deviation *from* the sample regression line. When σ_{Y_c} is not known, it has to be estimated from the sample data. Let $\hat{s}_{Y_c}$ be an estimate of σ_{Y_c}. The formula for obtaining $\hat{s}_{Y_c}$ is given as follows:[1]

$$\hat{s}_{Y_c} = \hat{s}_{Y|X}\sqrt{\frac{1}{n} + \frac{(X - \overline{X})^2}{\sum_{i=1}^{n} (X_i - \overline{X})^2}}$$

[1] See Supplement IX for derivation.

For computation convenience, the denominator of the second term under the radical is better expressed as

$$\sum_{i=1}^{n} X_i^2 - \frac{(\sum_{i=1}^{n} X_i)^2}{n}$$

which is equal to $\sum_{i=1}^{n}(X_i - \overline{X})^2$. Thus we have

$$\hat{s}_{Y_c} = \hat{s}_{Y|X}\sqrt{\frac{1}{n} + \frac{(X-\overline{X})^2}{\sum_{i=1}^{n} X_i^2 - (\sum_{i=1}^{n} X_i)^2/n}} \tag{14-14}$$

In the above example, when X is equal to $\overline{X}$, then

$$\hat{s}_{Y_c} = \frac{\hat{s}_{Y|X}}{\sqrt{n}} = \frac{1.13}{\sqrt{20}} = 0.25$$

and when X is 18.5, by Eq. (14-14) we obtain

$$\hat{s}_{Y_c} = 1.13\sqrt{\frac{1}{20} + \frac{(18.5-11.5)^2}{3{,}144-(230)^2/20}} = 1.13\sqrt{0.05 + \frac{49}{3{,}144-2{,}645}}$$

$$= 1.13\sqrt{0.05 + \tfrac{49}{499}} = 1.13(0.385) = 0.435$$

In general it can be said that the more deviant the value of X from its mean $\overline{X}$, the larger the value of $\hat{s}_{Y_c}$ will be.

14-3-2 The confidence interval for $\mu_{Y|X}$

The confidence interval of the population regression line is, of course, constructed on the basis of the estimated regression line. When the sample size is not sufficiently large, the ratio

$$t = \frac{Y_c - \mu_{Y|X}}{\hat{s}_{Y_c}} \tag{14-15}$$

follows the t distribution with $n-2$ degrees of freedom. Since

$$P(Y_c - t_{\alpha/2}\hat{s}_{Y_c} < \mu_{Y|X} < Y_c + t_{\alpha/2}\hat{s}_{Y_c}) = 1-\alpha$$

the $1-\alpha$ symmetrical confidence limits for $\mu_{Y|X}$ are given by

$$Y_c - t_{\alpha/2}\hat{s}_{Y_c} < \mu_{Y|X} < Y_c + t_{\alpha/2}\hat{s}_{Y_c} \tag{14-16}$$

The $t_{\alpha/2}$ value is read from Appendix VII with $n-2$ degrees of freedom, and the value of $\hat{s}_{Y_c}$ is obtained from Eq. (14-14). For the above example, the 95 percent symmetric confidence limits for the population regression line are

$$Y_c \pm (2.101)(1.13)\sqrt{\frac{1}{20} + \frac{(X-11.5)^2}{499}}$$

the graph of which is shown in Fig. 14-6. At $X = \bar{X} = 11.5$, we have $Y_c = 13$ and $\hat{s}_{Y_c} = 1.13\sqrt{20} = 0.25$. Thus we have

$$13 - 2.101(0.25) < \mu_{Y|X} < 13 + 2.101(0.25)$$

or

$$12.47 < \mu_{Y|X} < 13.53$$

At $X = 18.5$, then $Y_c = 20$ and $\hat{s}_{Y_c} = 0.435$. Consequently, we get the 95 percent confidence interval for $\mu_{Y|X}$ at $X = 18.5$ as follows:

$$20 - 2.101(0.435) < \mu_{Y|X} < 20 + 2.101(0.435)$$

or

$$19.086 < \mu_{Y|X} < 20.914$$

You are advised to verify the 95 percent confidence limits at some other values of X.

It is interesting to note that the length of the confidence interval is dependent on the value of X. As the value of X deviates farther from the mean $\bar{X}$, the confidence interval becomes wider; that is, the estimate of the true conditional mean becomes less precise. This is shown algebraically in the expression of $\hat{s}_{Y_c}$. Equation (14-14) shows that $\hat{s}_{Y_c}$ had its minimum value when X equals its mean $\bar{X}$. Its value becomes greater as X moves farther from $\bar{X}$. Thus we have greater assurance about the relationship between the two variables toward the mean of the independent variable than at its extreme values.

We have now considered the interval estimation of the conditional mean of the population. When we are interested in the mean value of Y for a given value of X, we construct a confidence interval for the conditional mean $\mu_{Y|X}$, which shows the range within which we believe the conditional mean lies. Such

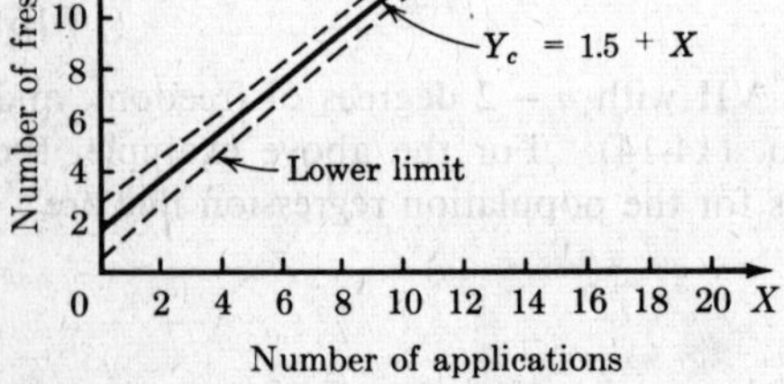

Figure 14-6
Ninety-five percent confidence intervals for the population regression line, $\mu_{Y|X}$ (in hundreds). From data of Table 14-1.

a range indicates the precision with which the conditional mean is estimated, and is not the spread of the distribution of the individual values of Y. On the other hand, when we are interested in predicting the *next* observation for Y, we construct a confidence interval for that specific event. The next observation will seldom coincide with the conditional mean. We should be concerned with the range within which the next Y value is expected to fall and shall now turn our attention to interval estimation of the next Y value.

14-3-3 The confidence interval for Y_a

In practical work prediction of a specific event is frequently required. For instance, management may wish to forecast what the company sales will be in the next period; the University of Southland may want to know the size of the freshman class of the next academic year, that is, the value of Y for the next year. This is a simple task since we just substitute a fixed X value in the regression equation once such an equation has been developed. Such a value for Y is an estimate of the actual value of Y for the next year. Let Y_a be the next value of Y. How can we construct a confidence interval for Y_a? In other words, what is the precision measure for the estimate Y_c calculated from a fixed value of X?

The construction of a confidence interval for the next value of Y requires, of course, an estimation of the standard deviation of Y_c from Y, which we shall designate as σ_{Y_a}. Let $\hat{s}_{Y_a}$ be an estimate of σ_{Y_a}. As demonstrated in Supplement IX, $\hat{s}_{Y_a}$ is expressed as follows:

$$\hat{s}_{Y_a} = \hat{s}_{Y|X}\sqrt{1 + \frac{1}{n} + \frac{(X - \overline{X})^2}{\sum_{i=1}^{n} X_i^2 - (\sum_{i=1}^{n} X_i^2)/n}} \tag{14-17}$$

The ratio

$$t = \frac{Y_a - Y_c}{\hat{s}_{Y_a}} \tag{14-18}$$

is distributed as t with $n - 2$ degrees of freedom.

When the sample size is small, the $1 - \alpha$ symmetric confidence intervals for the actual event Y_a are found as follows:

$$Y_c - t_{\alpha/2}\hat{s}_{Y_a} < Y_a < Y_c + t_{\alpha/2}\hat{s}_{Y_a} \tag{14-19}$$

since

$$P(Y_c - t_{\alpha/2}\hat{s}_{Y_a} < Y_a < Y_c + t_{\alpha/2}\hat{s}_{Y_a}) = 1 - \alpha$$

The value of $t_{\alpha/2}$ is taken from Appendix VII with $n - 2$ degrees of freedom.

The 95 percent symmetrical confidence limits for Y_a in the above example are shown in Fig. 14-7. Here also, the confidence interval is at its minimum when $X = \overline{X}$, and becomes wider when X moves farther from $\overline{X}$. When X is equal to $\overline{X}$, or 11.5, $Y_c = 13$, and

$$\hat{s}_{Y_a} - 1.13\sqrt{1 + 1/20} = 1.15$$

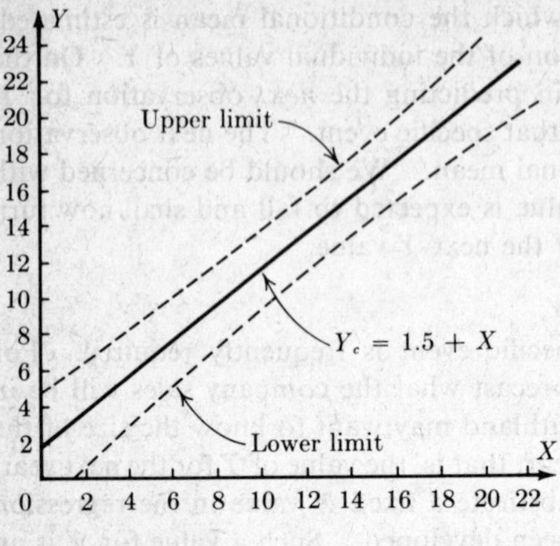

Figure 14-7 Ninety-five percent confidence intervals for Y_a, the actual value of Y. From Table 14-1.

Thus the 95 percent confidence limits for Y_a at $X = \bar{X} = 11.5$ are

$13 \pm 2.101(1.15)$

or

$10.58 < Y_a < 15.42$

When X is equal to 18.5, then $Y_c = 20$, and

$\hat{s}_{Y_a} = 1.13\sqrt{1 + 1/20 + (18.5 - 11.5)^2/499} = 1.21$

Consequently, the 95 percent confidence limits for Y_a at $X = 18.5$ are

$20 \pm 2.101(1.21)$

or

$17.46 < Y_a < 22.54$

You are again advised to verify the 95 percent confidence limits for Y_a at some other values of X.

Observe that the confidence intervals for Y_a are wider than those for $\mu_{Y|X}$. It demonstrates the fact that the estimate of the *actual* Y value is less precise than the estimate of the *mean* value.

However, in our example the regression equation was obtained from sample data that did not include any X value less than 2 or more than 19. It can be considered an estimate only of the population regression equation that has as its domain the same X values as used in our example. Consequently, extreme caution must be exercised if we want to extrapolate the regression line outside the domain, that is, to predict the size of enrollment when the number of applications lies beyond the limits of past experience. Extrapolation is especially hazardous for X values considerably beyond these limits. The reason for this is that beyond the range of the data in hand there is no assurance that the regression equation remains linear. The more deviant the X value is from the existing data, the more doubtful will be the reliability of the regression equation developed from these data.

Exercise 14-3

1 Explain the rationale that in computing the confidence interval for Y_a or $\mu_{Y|X}$, the t distribution rather than the normal distribution is applied.

2 Explain the differences between the confidence interval for a specific event Y_a and that for the conditional mean $\mu_{Y|X}$.

3 Suppose that a large discount house in a metropolitan area wishes to improve its ability to forecast weekly sales of dishwashers. For this purpose, data are collected on the number X of inquiries about dishwashers received during a given week and the number Y of dishwashers sold during the following week for a period of 52 weeks ($n = 52$). It is believed that the number of inquiries received during any week would be a reliable basis for forecasting the number of sales during the following week. The data collected are summarized as follows:

$\sum X = 9{,}710$ $\qquad \sum Y = 2{,}365$

$\sum XY = 700{,}000$ $\qquad \sum X^2 = 3{,}187{,}700$

$\sum Y^2 = 188{,}247$

a Using the least-squares method, estimate the population regression line.
b For $X = 400$, find the 99 percent confidence interval for the true mean of Y.
c If the number of inquiries received this week is 400, what would be the predicted number of sales during the next week? Find the 99 percent confidence interval for the actual sales for the next week.

4 Using the data given in Prob. 4 of Exercise 14-2, do the following:

a Compute $\hat{s}_{Y|X}$, the standard deviation of the error term, by the method of Eq. (14-12).
b Check your answer to part *a* by using the method of Eq. (14-13).
c Compute $\hat{s}_{Y_c}$, the standard deviation of the sample regression line.
d Find the 95 percent confidence interval for the population regression line $\mu_{Y|X}$, and plot the results as in Fig. 14-6.
e Find the 95 percent confidence interval for Y_a, the actual wheat yield in bushels, and plot the results as in Fig. 14-7.

5 Explain the differences between point and interval estimation of the conditional mean $\mu_{Y|X}$.

6 Using the data in Prob. 6 of Exercise 14-2, do the following:

a Compute $\hat{s}_{Y|X}$ by the method of Eq. (14-12).
b Check your answer to part *a* by using the method of Eq. (14-13).
c Compute $\hat{s}_{Y_c}$, the standard deviation of the sample regression line.
d Find the 95 percent confidence interval for the population regression line $\mu_{Y|X}$, and plot the results as in Fig. 14-6.
e Find the 95 percent confidence interval for Y_a, the actual volume of sales, and plot the results as in Fig. 14-7.

14-4
Test of Hypothesis

For a regression problem, hypothesis testing is not about the regression line, which is a set of conditional means, or about $\mu_{Y|X}$, but rather, about the crucial value of β, the slope of the population regression line. As already noted, the estimator of β is b, which is obtained from sample data by the least-squares method. The statistic b is subject to sampling variability; its value will vary with different samples. That is, b may take on values other than 0 even if β actually equals 0, indicating that no linear relation exists between the two variables. Because of such chance variations, we must know something about the sampling distribution of b if we want to use the estimate of the population regression line with assurance. In other words, if we want to construct a confidence interval for β, we must know the standard deviation of the sampling distribution of b, or σ_b, called the *standard error of b*. For the model considered in this chapter, it is assumed that b is normally distributed with a mean β and a standard error σ_b, or b is $N(\beta, \sigma_b^2)$.

With σ_b^2 unknown, however, we must find its estimate. The estimate of σ_b, denoted by $\hat{s}_b$, is obtained as follows:[1]

$$\hat{s}_b = \frac{\hat{s}_{Y|X}}{\sqrt{\sum_{i=1}^{n} X_i^2 - (\sum_{i=1}^{n} X_i)^2/n}} \tag{14-20}$$

Since the standard error of b has to be estimated, the ratio

$$t = \frac{b - \beta}{\hat{s}_b} \tag{14-21}$$

is distributed as t with $n - 2$ degrees of freedom because $\hat{s}_b$ varies with $\hat{s}_{Y|X}$, which is the root mean of squared deviations from regression that has $n - 2$ degrees of freedom. Of course, when the sample size is large, the ratio in Eq. (14-21) is distributed approximately as the standardized normal score Z.

The null hypothesis to be tested about β is usually stated such that no relation exists between the two variables. The test may be either one-tailed or two-tailed. If it is two-tailed, the test between the two hypotheses will appear as follows:

H_0: $\beta = 0$ $\quad$ H_1: $\beta \neq 0$

Given α, the level of significance, the decision rule will be

Reject H_0 if $b \geq t_{\alpha/2}\hat{s}_b$ or $b \leq -t_{\alpha/2}\hat{s}_b$

For the above example, the estimated standard error of b is obtained by substituting the various quantities in Eq. (14-20) as follows:

$$\hat{s}_b = \frac{1.13}{\sqrt{3{,}144 - (230)^2/20}} \approx 0.05$$

[1] The proof of this formula is given in Supplement IX.

With $\alpha = 0.05$, and $20 - 2 = 18$ degrees of freedom, the value of $t_{\alpha/2}$ is read from Appendix VII as 2.101. Thus the decision rule is

Reject H_0 if $b \geq 0.105$ or $b \leq -0.105$

Since our observed value for b is 1.00, we must reject the null hypothesis and conclude that there is a positive relation between the two variables X and Y.

The confidence interval for β is estimated as usual. Since for small n we have

$$P(b - t_{\alpha/2}\hat{s}_b < \beta < b + t_{\alpha/2}\hat{s}_b) = 1 - \alpha$$

the $1 - \alpha$ confidence interval for β is

$$b - t_{\alpha/2}\hat{s}_b < \beta < b + t_{\alpha/2}\hat{s}_b \qquad (14\text{-}22)$$

The value for $t_{\alpha/2}$ is taken from Appendix VII with $n - 2$ degrees of freedom. Thus, for the above example, the 95 percent confidence interval for β is

$$1.00 - 2.101(0.05) < \beta < 1.00 + 2.101(0.05)$$

or

$$0.895 < \beta < 1.105$$

Rejection of the null hypothesis merely indicates that there exists a relation between the two variables. It does not reveal how strong the relation is. For measuring the strength of relation between the variables, we turn our attention to the problem of correlation.

Exercise 14-4

1 Explain why the t test is applied when we test the hypothesis about the slope of the population regression line, and why the number of degrees of freedom is $n - 2$ for such a test.

2 Under what condition are we interested in testing the null hypothesis $\beta = 0$ against the alternative $\beta < 0$? What are the characteristics of the sampling distribution of b, the slope of the sample regression line?

3 Test the hypothesis that the slope of the population regression line is 0 against the alternative hypothesis that it is not 0 for the data in Probs. 2 and 3 of Exercise 14-2 at $\alpha = 0.10$. Also, find the 95 percent confidence limits for β.

4 Given the data in Prob. 4 of Exercise 14-2, test the following hypotheses at $\alpha = 0.05$:

H_0: $\beta = 0$ $\quad H_1$: $\beta \neq 0$

Determine the 95 percent confidence interval for β.

5 Given the following information: X = the independent variable, Y = the dependent variable, $n = 9$, and

$\sum X = 45 \qquad \sum XY = 168$

$\sum X^2 = 285 \qquad \sum Y^2 = 285$

$\sum Y = 45$

a Determine the slope *b* for the linear regression line.
b Test the null hypothesis $\beta = 0$ against the alternative hypothesis $\beta < 0$ at $\alpha = 0.02$.
c Find the 99 percent confidence limits for β.

6 Given the data in Prob. 6 of Exercise 14-2, test the null hypothesis that the population regression coefficient β is equal to 0 against each of the following alternatives at $\alpha = 0.05$:
a $\beta > 0$. *b* $\beta \neq 0$.

7 In Prob. 6 above, find the 95 percent confidence limits for β.

8 Using the data in Prob. 3 of Exercise 14-3, test the following hypotheses at $\alpha = 0.05$:

H_0: $\beta = 0$ H_1: $\beta \neq 0$

What is the point estimate for β? What is the 95 percent confidence interval for β?

14-5 Correlation Analysis

The second major part of our bivariate analysis concerns the problem of *correlation*. In a regression problem, the independent variable X is not a random variable since its values are fixed or preassigned; the dependent variable Y is a random variable because its observation is randomly selected from the probability distribution on the condition that X has occurred. In correlation analysis, however, there is no distinction between the two variables. Both X and Y are random variables; neither the values of X nor those of Y are predetermined. For example, if a college man is selected at random, and both his height and weight are left completely free to take on any possible values, we have a situation of correlation.

Other assumptions underlying the correlation problem considered in this section are similar to those assumed for the regression model. We assume that the relation between X and Y is linear. Not only do the conditional means $\mu_{Y|X}$ all fall on a regression line described by

$$\mu_{Y|X} = \alpha + \beta X$$

but also the conditional means $\mu_{X|Y}$ all fall on a regression line described by a similar equation

$$\mu_{X|Y} = \alpha + \beta Y$$

Although the parameters α and β in the first regression equation shown above could be the same as those in the second equation, the two parameters with X as the inbependent variable are ordinarily different from those with X as the dependent variable.

For correlation problems, the probability distribution of X's, given a fixed Y, is also assumed *normal*. This means that the sample is drawn from a *joint normal distribution;* that is, both the conditional probability distributions of Y's and those of X's are assumed to be normal. Also, the standard deviation of X's given each value of Y is assumed to be the same, just as the standard deviation of Y's, given each value of X, is the same.

Under these assumptions, we proceed to consider the measurement of the closeness of the relation between the variables X and Y.

14-5-1 Coefficients of determination and correlation

In estimation or prediction, the standard deviation of the sampling distribution plays a crucial role. In general, the smaller the standard deviation, the closer the observed value will be to the predicted, and the more precise the estimate is.

Let us first assume that we wish to estimate the Y value without any reference to the value of X. For this purpose, we need to know the standard deviation of Y, or σ_Y, which can be estimated by

$$\hat{s}_Y = \sqrt{\frac{\sum_{i=1}^{n} (Y_i - \bar{Y})^2}{n-2}}$$

The estimated standard deviation of Y, of course, varies directly with the sum of squared deviations from the sample mean $\bar{Y}$. Here there are $n-2$ degrees of freedom since the parameters α and β are now "estimated" as $\bar{Y}$ and 0, respectively.

Figure 14-8 shows the deviations of the observed Y values from their mean for the above example. A comparison of this graph with that in Fig. 14-5 immediately reveals that the deviations in Fig. 14-8 are much greater. The root mean of squares of these deviations is

$$\hat{s}_Y = \sqrt{\frac{\sum_{i=1}^{n} Y_i^2 - n\bar{Y}^2}{n-2}} = \sqrt{\frac{3{,}904 - 20(13^2)}{18}} = \sqrt{29.1111} = 5.4$$

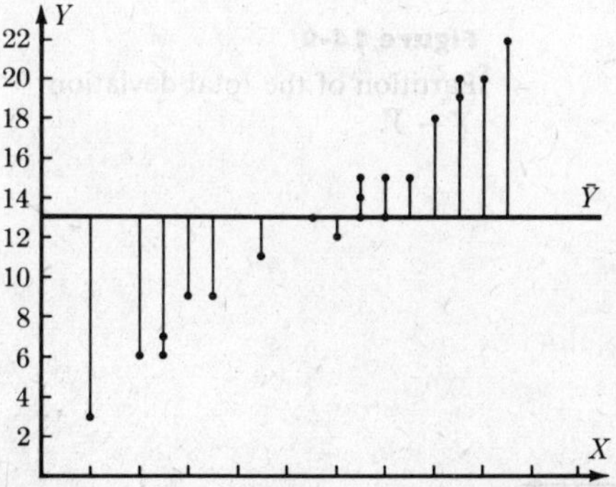

Figure 14-8

Deviations of the Y values from the mean $\bar{Y}$. From data of Table 14-2.

which is substantially greater than 1.13, the previously computed standard deviation of the error term. Consequently, the estimate of the true Y value will be more precise if it is made on the condition that X is known than if made without reference to X.

Thus, if we can find a device to measure the reduction in the variability of the distribution of Y, given a regression of Y on X, we shall be able to measure the closeness, or strength, of the relation between X and Y. We shall now proceed to identify such a device.

The difference between the individual Y value and $\bar{Y}$, the estimate of the true mean of Y, when no regression line is fitted, is called the *total deviation.* As indicated in Fig. 14-9, the total deviation, or $Y - \bar{Y}$, can be partitioned into the *explained deviation*, or $Y_c - \bar{Y}$, and the *unexplained deviation*, or $Y - Y_c$. That is,

$$Y - \bar{Y} = (Y - Y_c) + (Y_c - \bar{Y}) \tag{14-23}$$

The deviation $Y_c - \bar{Y}$ is called "explained" because it is regarded as the amount of error that is removed by fitting the regression line to the data. The deviation $Y - Y_c$ is called "unexplained" since it is the amount of error that still remains after the regression line has been fitted. Each of the total deviations shown in Fig. 14-8 can be partitioned in this manner.

Let us now demonstrate how the *sum of squares total* is partitioned into the *sum of squares explained* and the *sum of squares unexplained.* Since

$$Y_{ci} = \bar{Y} + b(X_i - \bar{X}) \qquad \text{(See Prob. 1 of Exercise 14-2.)}$$

we have

$$\begin{aligned}\sum_{i=1}^{n} (Y_i - Y_{ci})^2 &= \sum_{i=1}^{n} [Y_i - \bar{Y} - b(X_i - \bar{X})]^2 \\ &= \sum_{i=1}^{n} (Y_i - \bar{Y})^2 + b^2 \sum_{i=1}^{n} (X_i - \bar{X})^2 - 2b \sum_{i=1}^{n} (X_i - \bar{X})(Y_i - \bar{Y})\end{aligned}$$

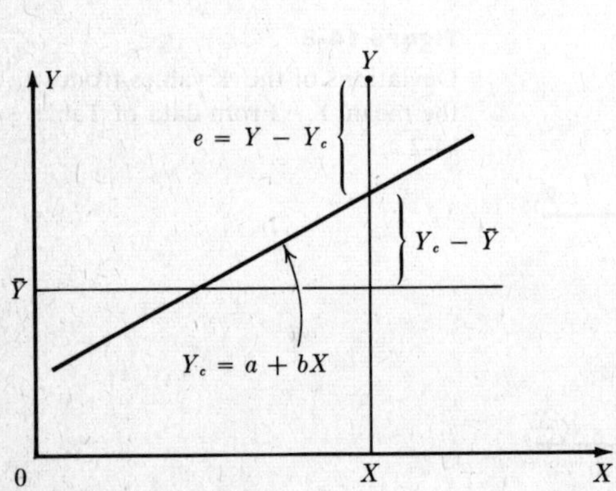

Figure 14-9
Partition of the total deviation $Y - \bar{Y}$.

From Eq. (14-7), we have

$$b = \frac{(\sum_{i=1}^{n} X_i Y_i)/n - \bar{X}\bar{Y}}{(\sum_{i=1}^{n} X_i^2)/n - \bar{X}^2}$$

$$= \frac{\sum_{i=1}^{n} X_i Y_i - n\bar{X}\bar{Y}}{\sum_{i=1}^{n} X_i^2 - n\bar{X}}$$

$$= \frac{\sum_{i=1}^{n} (X_i - \bar{X})(Y_i - \bar{Y})}{\sum_{i=1}^{n} (X_i - \bar{X})^2}$$

and consequently,

$$b \sum_{i=1}^{n} (X_i - \bar{X})^2 = \sum_{i=1}^{n} (X_i - \bar{X})(Y_i - \bar{Y})$$

Accordingly,

$$\sum_{i=1}^{n} (Y_i - Y_{ci})^2 = \sum_{i=1}^{n} (Y_i - \bar{Y})^2 + b^2 \sum_{i=1}^{n} (X_i - \bar{X})^2 - 2b^2 \sum_{i=1}^{n} (X_i - \bar{X})^2$$

$$= \sum_{i=1}^{n} (Y_i - \bar{Y})^2 - b^2 \sum_{i=1}^{n} (X_i - \bar{X})^2$$

From $Y_{ci} = \bar{Y} + b(X_i - \bar{X})$, we have

$$Y_{ci} - \bar{Y} = b(X_i - \bar{X}) \qquad \text{and} \qquad \sum_{i=1}^{n} (Y_{ci} - \bar{Y})^2 = b^2 \sum_{i=1}^{n} (X_i - \bar{X})^2$$

Thus

$$\sum_{i=1}^{n} (Y_i - Y_{ci})^2 = \sum_{i=1}^{n} (Y_i - \bar{Y})^2 - \sum_{i=1}^{n} (Y_{ci} - \bar{Y})^2$$

or

$$\sum_{i=1}^{n} (Y_{ci} - \bar{Y})^2 = \sum_{i=1}^{n} (Y_i - \bar{Y})^2 - \sum_{i=1}^{n} (Y_i - Y_{ci})^2 \tag{14-24}$$

That is, the sum of squares explained is equal to the sum of squares total minus the sum of squares unexplained. Dividing each side of this equation by $\sum_{i=1}^{n} (Y_i - \bar{Y})^2$, we obtain

$$\frac{\sum_{i=1}^{n} (Y_{ci} - \bar{Y})^2}{\sum_{i=1}^{n} (Y_i - \bar{Y})^2} = \frac{\sum_{i=1}^{n} (Y_i - \bar{Y})^2 - \sum_{i=1}^{n} (Y_i - Y_{ci})^2}{\sum_{i=1}^{n} (Y_i - \bar{Y})^2}$$

$$= \frac{\hat{s}_Y{}^2 - \hat{s}_{X|Y}^2}{\hat{s}_Y{}^2} \tag{14-25}$$

which is known as the *coefficient of determination* and is commonly designated r^2. The square root of r^2 is called the *correlation coefficient*; that is,

$$r = \sqrt{\frac{\sum_{i=1}^{n} (Y_i - \bar{Y})^2 - \sum_{i=1}^{n} (Y_i - Y_{ci})^2}{\sum_{i=1}^{n} (Y_i - \bar{Y})^2}} = \sqrt{\frac{\hat{s}_Y{}^2 - \hat{s}_{Y|X}^2}{\hat{s}_Y{}^2}} \tag{14-26}$$

The coefficient r^2 or r is therefore a ratio designed to measure the proportional reduction in the variability of Y, given the regression of Y on X. The numerator of the ratio r^2 is the variance explained, or accounted for, by linear regression, and the ratio itself can be said to be the proportion of the total variance accounted for by linear regression.

Observe that r^2 and r are statistics computed from the sample. Their population counterparts, ρ^2 and ρ (a Greek letter read as "rho"), are expressed as follows:

$$\rho^2 = \frac{\sigma_Y{}^2 - \sigma_{Y|X}^2}{\sigma_Y{}^2} \tag{14-27}$$

and

$$\rho = \sqrt{\frac{\sigma_Y{}^2 - \sigma_{Y|X}^2}{\sigma_Y{}^2}} \tag{14-28}$$

The variance $\hat{s}_{Y|X}^2$, or the sum of squares unexplained, may be considered representing the "leftover" variability. If the variability in the conditional probability distribution were as large as that of the unconditional probability distribution, that is, if $Y_c = \bar{Y}$ and $\hat{s}_{Y|X}^2 = \hat{s}_Y{}^2$, it follows that knowing X would be of no help in the prediction of Y. This being the case, the coefficient of determination would be 0, indicating no relation between X and Y. Accordingly, all the conditional probability distributions would have the same mean $\bar{Y}$ and the same variance $\hat{s}_Y{}^2$. This is, of course, an extreme case, which is nonexistent in practice.

On the other hand, if the variance $\hat{s}_{Y|X}^2 = 0$, which signifies $Y = Y_c$, the coefficient of determination would be equal to 1. This means that no variability existed in the conditional probability distribution, and thus a perfect prediction of Y could be made from a knowledge of X. This being the case, all the points would fall on the regression line. This is the other extreme case, which is similarly nonexistent.

Always, the coefficient of determination lies between 0 and 1. If it is close to 0, it indicates that the prediction of Y is not much improved by knowing X. On the other hand, as it moves away from 0 toward 1, knowing X will be increasingly helpful in the prediction of Y.

The correlation coefficient r is a better-known and more widely used index. The value of r takes the same sign as b and ranges from -1 to $+1$. A minus sign indicates *negative* correlation, and a plus sign indicates *positive* correlation.

The meaning of the correlation coefficient is not as clear as that of the coefficient of determination, though the former is more frequently referred to than the latter.

For the example considered in this chapter, the correlation coefficient is obtained as

$$r = \sqrt{\frac{29.1111 - 1.2778}{29.1111}} = \sqrt{0.96} = 0.98$$

which is positive since the value of b obtained earlier takes a positive sign.

In practice, r is frequently computed by using the following formula:

$$r = \frac{\operatorname{cov}(X, Y)}{s_X s_Y} = \frac{(\sum_{i=1}^{n} X_i Y_i)/n - \bar{X}\bar{Y}}{s_X s_Y} \tag{14-29}$$

or

$$r = \frac{n \sum_{i=1}^{n} X_i Y_i - \sum_{i=1}^{n} X_i \sum_{i=1}^{n} Y_i}{\sqrt{[n\sum_{i=1}^{n} X_i^2 - (\sum_{i=1}^{n} X_i)^2]\,[n \sum_{i=1}^{n} Y_i^2 - (\sum_{i=1}^{n} Y_i)^2]}} \tag{14-30}$$

Substituting the quantities from Table 14-2 in Eq. (14-30), we get

$$r = \frac{20(3,490) - 230(260)}{\sqrt{[20(3,144) - (230)^2][20(3,904) - (260)^2]}}$$

$$= \frac{10,000}{\sqrt{(9,980)(10,480)}} = 0.98$$

which is the same as the value obtained by using Eq. (14-26).

14-5-2 Test of the hypothesis $\rho = 0$

The value of r is subject to sampling error since it is computed from sample data. Even if the population correlation coefficient ρ is actually 0, that is, if no linear relation exists between the two variables, r may assume a negative or positive value as a matter of chance. How can we tell, then, whether a given value of r is significantly different from 0? This situation requires the testing of the hypothesis $\rho = 0$.

As pointed out earlier, the sample from which r is computed is taken from a joint normal distribution. Consequently, r tends to be normally distributed as the size of the sample increases. This tendency is much stronger when $\rho = 0$ than when ρ is close to -1 or $+1$. This is because when $\rho = 0$, the distribution of r is symmetric, while for a value of ρ different from 0, the distribution of r is skewed. It is skewed to the left when ρ is positive, and to the right when ρ is negative. The property that the distribution of r is symmetric when ρ is 0 makes it possible for us to test the hypotheses

H_0: $\rho = 0$ $\quad$ H_1: $\rho \neq 0$

by using the t distribution. The ratio

$$t = r\sqrt{\frac{n-2}{1-r^2}} \tag{14-31}$$

is distributed as t with $n - 2$ degrees of freedom. For the above example, with 18 degrees of freedom and at $\alpha = 0.05$, we should reject $H_0: \rho = 0$ if the t value is less than -2.101 or larger than $+2.101$. For the value of r obtained above, we have

$$t = 0.98\sqrt{\frac{20-2}{1-0.98^2}} \approx 20.7$$

which is substantially greater than 2.101. It is therefore quite safe to reject H_0 and state that there is a correlation between the two variables. That is, the r value as large as 0.98 is not likely to occur by chance alone; it is significantly different from 0.

The value of t as obtained here should be identical with the t value implied in the test about β in Sec. 14-4. Any discrepancy between the two values is due to rounding.

It must be pointed out that this test is used only for the hypothesis that $\rho = 0$. If this hypothesis is rejected, we simply accept the alternative hypothesis that some relation exists between the two variables. For testing the hypothesis that ρ assumes some value other than 0, we need to resort to a different technique not discussed in this chapter.

Exercise 14-5

1 Explain fully the differences between regression and correlation analyses.

2 What are the basic assumptions underlying correlation analysis?

3 Do the following for the data in Prob. 2 of Exercise 14-2:

a Compute r, the coefficient of correlation between X and Y.

b Test the hypothesis that the population correlation coefficient ρ is 0 against the alternative that it is not, at $\alpha = 0.05$.

4 Do the following for the data in Prob. 4 of Exercise 14-2:

a Compute the correlation coefficient r.

b Test the hypothesis that the population correlation coefficient ρ is 0 against the alternative that it is not, at $\alpha = 0.05$.

5 Let s_X and s_Y be the standard deviation of X and Y, respectively, and b the slope of the sample regression line. Prove

$$b = \frac{s_Y}{s_X} r$$

6 Use the data in Prob. 5 of Exercise 14-4 to verify the relation between b and r as expressed in the preceding problem.

7 Use the data in Prob. 3 of Exercise 14-3 to verify the relation between b and r as expressed in Prob. 5 above.

8 Define as fully as you can each of the following concepts:

a Coefficient of determination.
b Coefficient of correlation.
c Joint normal distribution.

9 Do the following for the data in Prob. 6 of Exercise 14-2:

a Compute r, the coefficient of correlation between X and Y.
b Test the hypothesis that the population correlation coefficient ρ is 0 against the alternative that it is not, at $\alpha = 0.05$ and $\alpha = 0.01$.

10 Do the following for the data in Prob. 3 of Exercise 14-3:

a Compute r, the coefficient of correlation between X and Y.
b Test the hypothesis that the population correlation coefficent ρ is 0 against the alternative that it is not, at $\alpha = 0.05$ and $\alpha = 0.01$.

11 Prove that if $r = 0$,

$$s_{X+Y}^2 = s_X{}^2 + s_Y{}^2$$

Chapter 15

Time-series analysis

Any institution, whether it be the family, the school, the business firm, or the government, must plan for the future if it is to survive and make progress. The family must make future plans with respect to income and expenditures in order to make its financial ends meet. A school system must plan for future expansion if it is to provide adequate educational services for the ever-growing population of school-age children. Business firms must make future plans in order to meet changes in the demand for its products. They have to make plans for production, finance, personnel, marketing, and other phases of management. It is hardly necessary to point out that the government, Federal or local, must plan for the future if it is to provide the required services to maintain peace and justice on the one hand and to promote economic progress on the other.

Rational planning requires *forecasting* future events that are likely to happen. Forecasting, in turn, is usually based on what has occurred in the past. Thus we have another type of statistical inference, which is made about the future for some variable or composite of variables on the basis of past events. The most important techniques of making inferences about the future on the basis of what has happened in the past is the *analysis of time series*, which may be defined as follows:

> *A time series is a set of quantitative readings of some variable or composite of variables arranged in chronological order of their occurrences.*

The term applies, for example, to such economic indicators as gross national product, index of industrial production, housing starts, and sales of motor vehicles. However, it must be recognized that time series and the related techniques of analysis are not limited to the field of economics. They apply also to other variables such as enrollment in an educational institution, population of the nation or of a region, and rainfall in a certain locality. A time series describes the variations in the values of the variable through time; such variations are results of the systematic as well as the random behavior of the variable. If a series has shown some trend or persistent pattern in its variations for a long period of time in the past, it would be sensible to assume that such patterns or regularities will continue to exist in the future. The possibility that the past performance of a variable may continue into the future provides a reasonable basis for forecasting, or inference, which is the principal objective of time-series analysis.

The analysis of time series constitutes a major portion of conventional statistics. A comprehensive treatment of the subject is far beyond the scope intended for this text. All we plan to do in this chapter is to review the general nature of time-series analysis, to describe some of the techniques of measuring the various types of variations embodied in the series, and to introduce a method of summarizing numerical data, known as index numbers, for facilitating comparisons of the variations from one time period or locality to another.

15-1 Nature of Time Series

The variation of time series can be attributed to a number of factors. Some are natural or institutional, others are socioeconomic in nature. Some of them are responsible for short-run variation, while others produce long-run variation. Thus a time series is made up of a variety of elements, or components, which account for the observed changes over a period of time.

A cursory look at some time series would probably lead you to think that it can hardly be possible to find some order from the fluctuating movement. If we make some simplifying, yet realistic, assumptions, however, we shall be able to identify, measure, and interpret the variations contained in the series. It is assumed that there are *four* basic types of components in a time series. They are secular trend, seasonal variations, cyclical fluctuations, and random, or irregular, changes. In this section we shall briefly examine the nature of each of these four components and their relationships.

15-1-1 Components of time series

In analyzing time series we may take the observed composite series as a whole for study, or study, one by one, respectively, the components in their own right. When some components are singled out for analysis, we have to decompose the series into its parts, and this necessitates an assumption about the relation existing among the various components. Before considering the methods of decomposition of time series, we shall try to acquaint ourselves with the characteristic of each of the components.

Secular trend A *secular*, or long-term, *trend* refers to the smooth and regular movement of a series, reflecting continuous growth, stagnation, or decline over a rather long period of time. While the exact length of the period is not defined, the consensus is that, for a time series in economics, the period must be long enough to include two or more business cycles in order to obtain any reasonable results. What the trend measures is the *average* change in the variable per unit of time. It characterizes the gradual and general pattern of developments, which is often described by a straight line or some type of smooth curve. Figure 15-1, for example, shows the shipments of mobile homes in the United States for the period from 1958 to 1972. Apparently, there is a general sweep, or trend, in its movement, which is suggested roughly by the broken curve.

In Fig. 15-2 the data for farm employment in the United States are plotted. In spite of the irregular movements in the series, there is a general downward trend in farm employment, and this downward movement is suggested by the broken straight line.

Seasonal variations The word "seasonal" as used here has a rather broad connotation. By *seasonal variations*, we mean *periodic variations* that recur with some degree of regularity within a specific period of 1 year or shorter. The most frequently studied periodic variation is one that recurs over a 12-month period as in the production of motor vehicles, sales of greeting cards, and so forth. Since these are related to the seasons of the year, the name seasonal

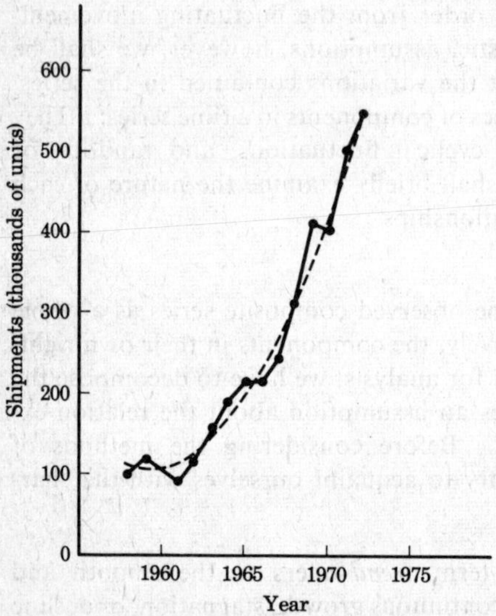

Figure 15-1
Shipments of mobile homes (in thousands of units) in the United States, 1958 to 1972. (*Source*: U.S. Department of Commerce, *Survey of Current Business and Business Statistics*, 1971 Biennial edition.)

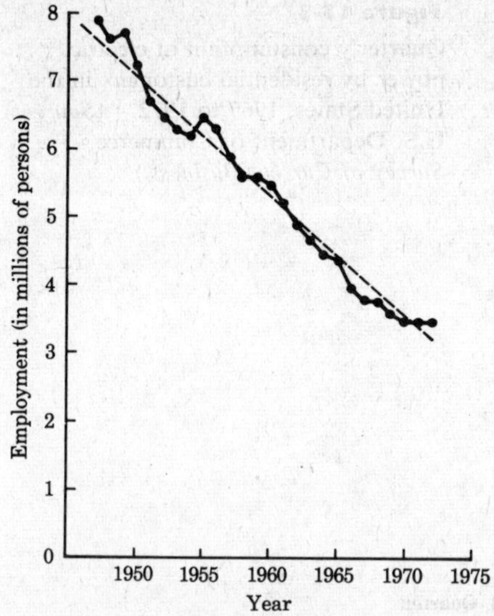

Figure 15-2

Farm Employment in the United States (in millions of persons), 1947 to 1972. (*Source*: U.S. Department of Commerce, *Survey of Current Business* and *Business Statistics*, 1971 biennial edition.)

variation is used. Actually, it is used to indicate any kind of periodic movement, daily, weekly, or otherwise, with a year as its maximum period of recurrence.

Among the most important factors responsible for seasonal variations are *climatic conditions*, *social customs*, and *religious festivities*. Climatic conditions are obviously the most important cause for seasonal fluctuation in farm production, construction activities, and tourist business. The demand for gifts before Valentine's Day, Easter, Mother's Day, Father's Day, Christmas, etc., has its source in the customs and religions of the population. Figure 15-3 illustrates the highly seasonal behavior of consumption of electricity in the United States. As shown in the graph, the consumption of electric power by residential customers is always highest in the first quarter, and lowest in the second quarter.

Cyclical variations Cyclical variation is characterized by recurring up-and-down movements, which are different from seasonal fluctuations in that they extend over longer periods of time, usually 2 or more years. They are attributed to a variety of factors, and a tremendous amount of effort has been expended to identify and appraise these factors. Consequently, many theories have been proposed to explain cyclical fluctuations, yet few of them have achieved the status of universal acceptance. We shall not attempt to study the forces that result in this kind of fluctuation. In general, they are economic in nature and reflect the business status from time to time. The interested reader is referred to any textbook on business cycles for such information.

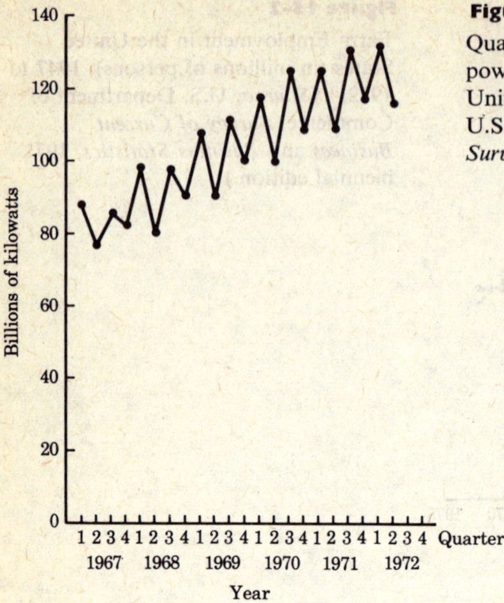

Figure 15-3

Quarterly consumption of electric power by residential customers in the United States, 1967 to 1972. (*Source*: U.S. Department of Commerce, *Survey of Current Business.*)

Figure 15-4 provides an example of cyclical variation in the data for the index of industrial production from 1954 to 1972. Peaks can be noted in 1957 and 1969, while troughs occurred in 1958 and 1970 to 1971. The peak and trough of a cycle are often referred to as the *downturn* and *upturn*, respectively. There are in general four phases in a business cycle. The downturn or *prosperity* is followed by a period of *contraction* that culminates in a *recession*, or *depression*. Following that, *recovery*, or *expansion*, ensues and eventually develops into a period of prosperity. A cycle is measured either from downturn to downturn, or from upturn to upturn.

Irregular variations Irregular movements of time series are either random or caused by some sporadic forces such as war, earthquake, flood, droughts, and other natural catastrophes. Such fluctuations are nonrecurring, and therefore completely unpredictable. However, these unpredictable events can be easily recognized and identified, and thus can be easily eliminated from the data when we try to measure the secular trend, seasonal variation, and cyclical fluctuations. As for random, or chance, variations, they often are comparatively unimportant, and are usually considered a part of the seasonal or cyclical variations or simply ignored.

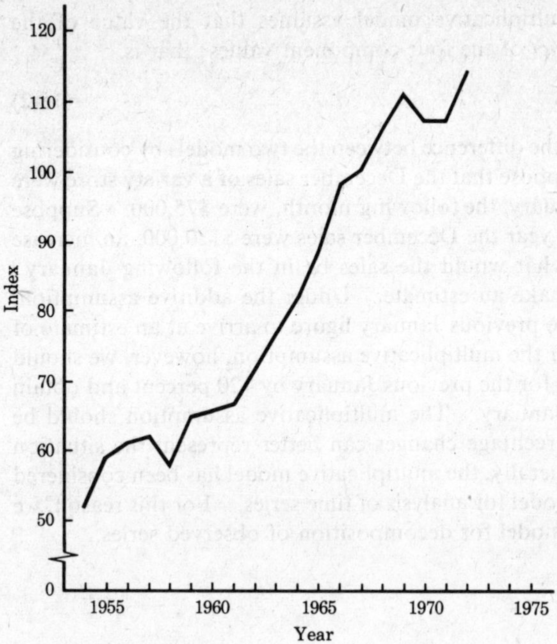

Figure 15-4
Index of industrial production, 1954 to 1972; 1967 = 100. (*Source*: Board of Governors of the Federal Reserve System, *Federal Reserve Bulletin.*)

15-1-2 Models of time series

Two time-series models are generally accepted as good *approximations* to the true relationships among the components of the observed data. They are the *additive* and *multiplicative* models, and are the most commonly assumed relationships between time series and its elements. Let

Y = original observed value

T = value of secular trend

S = value of seasonal component

C = value of cyclical component

I = value of irregular component

The additive model assumes that the value of the composite series is the *sum* of the four components; that is,

$$Y = T + S + C + I \tag{15-1}$$

On the other hand, the multiplicative model assumes that the value of the composite series is the *product* of the four component values; that is,

$$Y = T \times S \times C \times I \tag{15-2}$$

We shall now illustrate the difference between the two models by considering the seasonal element S. Suppose that the December sales of a variety store were $100,000 and the sales in January, the following month, were $75,000. Suppose further that in the following year the December sales were $120,000, an increase of $20,000 or 20 percent; what would the sales be in the following January? Two ways can be used to make an estimate. Under the additive assumption, we should add $20,000 to the previous January figure to arrive at an estimate of $95,000 for January. Under the multiplicative assumption, however, we should multiply the sales of $75,000 for the previous January by 120 percent and obtain an estimate of $90,000 for January. The multiplicative assumption should be used in such a case since percentage changes can better represent the situation than absolute changes. Generally, the multiplicative model has been considered the standard conventional model for analysis of time series. For this reason, we shall use the multiplicative model for decomposition of observed series.

Exercise 15-1

1 Define the following terms in your own words as fully as you can:

a Time series.
b Secular trend.
c Seasonal variations.
d Cyclical fluctuations.
e Irregular movements.

2 Distinguish between the additive and multiplicative models of time series.

3 Give an example to show each of the following patterns of seasonal variation:

a Daily pattern.
b Weekly pattern.
c Monthly pattern.
d Yearly pattern.

4 What are some of the most important factors that affect each of the following components of time series?

a Secular trend.
b Seasonal variations.
c Cyclical fluctuations.

5 Which component of time series do we usually use in each of the following problem situations? Why?

a A forecast of the demand for electronic computers during the next 5 years is called for in order to make plans for the expansion of plant facilities.
b An airline company must schedule its flights for the next month.

c The management of a manufacturing company must estimate its manpower and material needs throughout the coming year.

d A mutual fund must decide whether to purchase a certain common stock for long-run price appreciation.

6 Explain fully the importance of time-series analysis.

15-2 Decomposition of Time Series

As indicated above, the original data for the variable Y can be broken down into four components. Under the multiplicative model, the relation of Y and the four components is described by Eq. (15-2). According to this formula, only the trend element is viewed as absolute values, whereas the other three elements are all expressed in percentages. Thus a seasonal index of 89 percent would indicate that the actual value is expected to be 11 percent lower than it would be without the seasonal influence.

We shall in this section use one example to illustrate how the fourfold breakdown of a time series can be achieved. In doing so, a number of limitations must be recognized. First, the four components are obviously *not independent* of one another. An extremely unusual seasonal movement, for instance, may precipitate, or at least aggravate, the cyclical development, while a cyclical movement may greatly influence the seasonal. Similarly, a severe cyclical variation may seriously affect the trend, and random variations caused by war or earthquake may substantially alter any or all of the other components. Another limitation relates to the difficulty in decomposing the seasonal influence from cyclical or irregular elements. This is the case because seasonal variations are usually not uniform in timing and amplitude, which makes it extremely difficult to identify the seasonal with complete certainty. In short, the fourfold breakdown is by no means an accurate account of the relationships among the components. Rather, it is a technique of *approximation.* With these limitations in mind, we shall now study each of the four components resulting from decomposition.

The time series we plan to analyze is the series of housing starts in the United States from the third quarter of 1964 to the second quarter of 1972. We realize that for trend analysis and for the study of cyclical movement, the period under consideration should be longer. However, because our primary purpose is to illustrate the *method* of decomposition and the *techniques* of making inferences from the decomposed elements, the insufficiency of data should not unduly bother us.

In analyzing seasonal variations, most texts use a series of monthly data. While there is some advantage in going into such detailed analysis, for illustra-

Table 15-1
New housing units started in the United States, third quarter of 1964 to second quarter of 1972 (in thousands of units)

Year	I	II	III	IV	Yearly total
1964			398	352	
1965	283	454	392	345	1,474
1966	274	392	290	210	1,166
1967	218	382	382	340	1,322
1968	298	452	423	372	1,545
1969	336	468	387	309	1,500
1970	264	399	408	396	1,467
1971	389	604	579	513	2,085
1972	510	661			

Source: U.S. Department of Commerce, *Survey of Current Business.*

tion purposes we find it quite sufficient to use quarterly data. The quarterly data for dwelling units started in the United States from the third quarter of 1964 to the second quarter of 1972 are given in Table 15-1.

15-2-1 Detrending

Under the multiplicative assumption, decomposition is of course achieved by division. First of all, we shall try to eliminate the trend from the observed series. The process of eliminating the trend is referred to as *detrending*, and the trend itself can be represented either as a straight line or some type of smooth curve. We shall now review the procedure of fitting the observed data with a straight line and the technique of curve fitting by a second-degree equation.

Linear trend In Chap. 14 we examined two trend-fitting techniques, namely, the freehand method and the least-squares method. We shall again use the least-squares procedure for trend fitting to the data in Table 15-1.

Let X designate each of the 32 quarters from 1964 to 1972; that is, $X = 1$ for the third quarter of 1964, $X = 2$ for the fourth quarter, and so on. Thus the domain of X is the set of integers from 1 to 32 inclusive. Let Y designate the quarterly housing starts. The values of X and Y are shown in Table 15-2. For computing the values of a and b in the trend line

$$Y_t = a + bX$$

where the subscript t suggests trend, we obtain the following quantities from the data in Table 15-2:

$$n = 32 \qquad \sum X = 528 \qquad \sum Y = 12{,}480$$
$$\sum XY = 223{,}229 \qquad \sum X^2 = 11{,}440$$

Table 15-2

Computation of trend for housing starts in the United States, third quarter of 1964 to second quarter of 1972

Year	Quarter	X	Y	Trend $Y_t = 285.39 + 6.34X$
1964	3	1	398	291.73
	4	2	352	298.07
1965	1	3	283	304.41
	2	4	454	310.75
	3	5	392	317.09
	4	6	345	323.43
1966	1	7	274	329.77
	2	8	392	336.11
	3	9	290	342.45
	4	10	210	348.79
1967	1	11	218	355.13
	2	12	382	361.47
	3	13	382	367.81
	4	14	340	374.15
1968	1	15	298	380.49
	2	16	452	386.83
	3	17	423	393.17
	4	18	372	399.51
1969	1	19	336	405.85
	2	20	468	412.19
	3	21	387	418.53
	4	22	309	424.87
1970	1	23	264	431.21
	2	24	399	437.55
	3	25	408	443.89
	4	26	396	450.23
1971	1	27	389	456.57
	2	28	604	462.91
	3	29	579	469.25
	4	30	513	475.59
1972	1	31	510	481.93
	2	32	661	488.27

Source: Table 15-1.

By Eq. (14-7) we have

$$b = \frac{223{,}229/32 - 528(12{,}480)/32(32)}{11{,}440/32 - 528(528)/32(32)}$$

$$= \frac{32(223{,}229) - 528(12{,}480)}{32(11{,}440) - 528(528)}$$

$$= \frac{7{,}143{,}328 - 6{,}589{,}440}{366{,}080 - 278{,}784}$$

$$= \frac{553{,}888}{87{,}296} = 6.34$$

and by Eq. (14-9) we obtain the value of a as follows:

$$a = \frac{12{,}480}{32} - 6.34(528/32)$$

$$= 390.0 - 104.61 = 285.39$$

Thus the trend line appears as follows:

$$Y_t = 285.39 + 6.34X \tag{15.3}$$

The computed trend values are shown in the last column of Table 15-2.

Figure 15-5 shows graphically the trend line fitted to the quarterly data for new dwelling units started in the United States from 1964 to 1972. The dotted line beyond 1972 represents projections (discussed in Sec. 15-2-5).

Curvilinear trends The above analysis is, of course, based on the assumption that the variables X and Y have a linear relation; that is, the dwelling units started each quarter during the 8-year period are assumed to have a linear relation. Many time series, however, contain trends that are nonlinear in nature. Some trends can be represented by second-degree curves, while others may be portrayed by logarithmic, exponential, or other types of curves. Figure 15-6 shows some of the more frequently applied curvilinear trends. It should be kept in mind that when a higher-degree polynomial equation should be used for fitting to the data at hand, we are almost certain to be mixing cycles with trend.

For actual applications of the equations shown in Fig. 15-6, you are referred to any of the many standard descriptive statistics textbooks, in which illustrations of their computation can be found. When such a curvilinear trend is applied, it must be remembered that only a *portion* of the curve is usually used to represent the trend.

A comprehensive treatment of the curve-fitting techniques for the various equations is far beyond the scope for which this discussion is planned. We shall, however, present one example to illustrate their application.

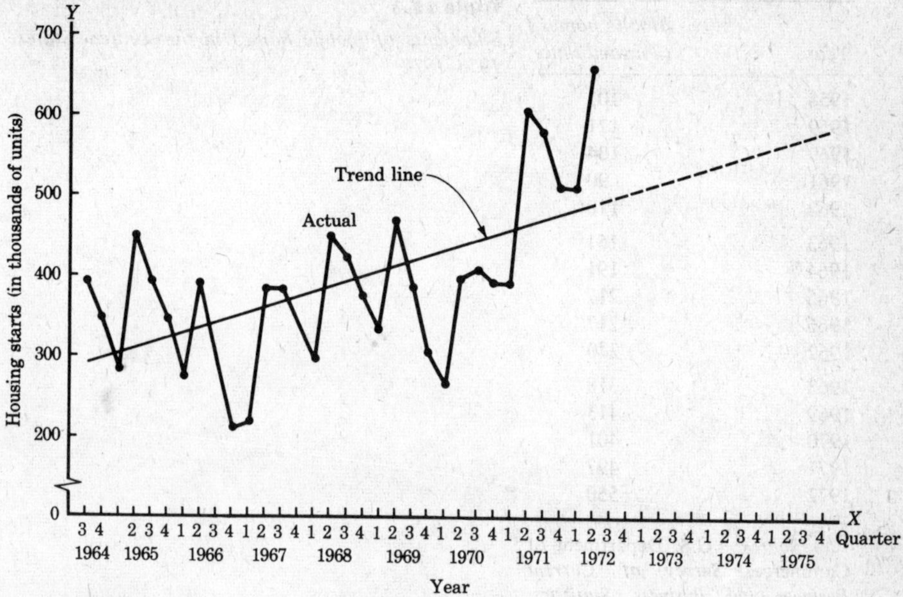

Figure 15-5

New housing units started in the United States, third quarter of 1964 to second quarter of 1972. From data of Table 15-2.

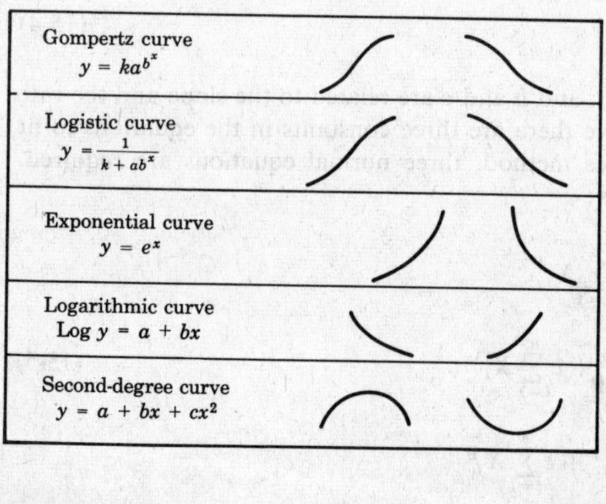

Figure 15-6

Some forms of nonlinear trend functions.

Table 15-3
Shipments of mobile homes in the United States, 1958–1972

Year	*Mobile homes, thousand units*
1958	102
1959	121
1960	104
1961	90
1962	118
1963	151
1964	191
1965	217
1966	217
1967	240
1968	318
1969	413
1970	401
1971	497
1972	550

Source: U.S. Department of Commerce, *Survey of Current Business* and *Business Statistics*, 1971 Biennial ed.

The mobile-home industry has shown a rate of rapid growth during recent decades. The trend of this series is definitely not linear. We shall show how these data can be fitted by a second-degree curve. The data for shipments of mobile homes for the 15 years from 1958 to 1972 are presented in Table 15-3.

The general form of a second-degree equation is

$$Y_t = a + bX + cX^2 \tag{15-4}$$

where a is still the Y intercept, and b and c are related to the slope and the rate of change of the curve. Since there are three constants in the equation, to fit the series by the least-squares method, three normal equations are required. These are obtained as follows:

$$\begin{aligned}
\sum_{i=1}^{n} Y_i &= na + b\sum_{i=1}^{n} X_i + c\sum_{i=1}^{n} X_i^2 \\
\sum_{i=1}^{n} X_iY_i &= a\sum_{i=1}^{n} X_i + b\sum_{i=1}^{n} X_i^2 + c\sum_{i=1}^{n} X_i^3 \\
\sum_{i=1}^{n} X_i^2Y_i &= a\sum_{i=1}^{n} X_i^2 + b\sum_{i=1}^{n} X_i^3 + c\sum_{i=1}^{n} X_i^4
\end{aligned} \tag{15-5}$$

When the midpoint of the periods covered by the series is taken as 0, or the origin, all the sums of X raised to odd powers will become 0. Consequently, the above three equations will be reduced to

$$\sum_{i=1}^{n} Y_i = na + c\sum_{i=1}^{n} x_i^2$$

$$\sum_{i=1}^{n} x_i Y_i = b\sum_{i=1}^{n} x_i^2$$

$$\sum_{i=1}^{n} x_i^2 Y_i = a\sum_{i=1}^{n} x_i^2 + c\sum_{i=1}^{n} x_i^4$$

where x stands for the difference between X and the midpoint. Accordingly, the three constants can be computed by the following equations, respectively:

$$a = \frac{\sum_{i=1}^{n} Y_i - c\sum_{i=1}^{n} x_i^2}{n} \tag{15-6}$$

$$b = \frac{\sum_{i=1}^{n} x_i Y_i}{\sum_{i=1}^{n} x_i^2} \tag{15-7}$$

$$c = \frac{n\sum_{i=1}^{n} x_i^2 Y_i - \sum_{i=1}^{n} x_i^2 \sum_{i=1}^{n} Y_i}{n\sum_{i=1}^{n} x_i^4 - (\sum_{i=1}^{n} x_i^2)^2} \tag{15-8}$$

With 1965 taken as the origin, the computation process is shown in Table 15-4. The values of $\sum Y$, $\sum xY$, $\sum x^2$, $\sum x^2 Y$, and $\sum x^4$ are shown in the bottom row of Table 15-4. Thus we have the values of the constants computed as follows:

$$a = \frac{3{,}730 - 2.699(280)}{15} = \frac{2{,}974.28}{15} = 198.29$$

$$b = \frac{8{,}973}{280} = 32.05$$

and

$$c = \frac{15(80{,}765) - 280(3{,}730)}{15(9{,}352) - (280)^2} = \frac{1{,}211{,}475 - 1{,}044{,}400}{140{,}280 - 78{,}400} = \frac{167{,}075}{61{,}880} = 2.699$$

The second-degree trend curve is therefore

$$Y_t = 198.29 + 32.05x + 2.699x^2 \tag{15-79}$$

and the computed trend values are shown in the last column of Table 15-4.

Table 15-4

Computation of second-degree trend curve fitted to shipments of mobile homes in the United States, 1958 to 1972 (in thousands of units)

Year X	x	x^2	x^4	Y	xY	x^2Y	Trend $Y_t = a + bx + cx^2$
1958	−7	49	2,401	102	−714	4,998	106.24
1959	−6	36	1,296	121	−726	4,356	103.19
1960	−5	25	625	104	−520	2,600	105.54
1961	−4	16	256	90	−360	1,440	113.29
1962	−3	9	81	118	−354	1,062	126.44
1963	−2	4	16	151	−302	604	144.99
1964	−1	1	1	191	−191	191	168.94
1965	0	0	0	217	0	0	198.29
1966	1	1	1	217	217	217	233.04
1967	2	4	16	240	480	960	273.19
1968	3	9	81	318	954	2,862	318.74
1969	4	16	256	413	1,652	6,608	369.69
1970	5	25	625	401	2,005	10.025	426.04
1971	6	36	1,296	497	2,982	17,892	487.79
1972	7	49	2,401	550	3,850	26,950	554.94
Total	0	280	9,352	3,730	8,973	80,765	

Source: Table 15-3.

Figure 15-7 presents the trend parabolic curve fitted to the data. Projections beyond 1972 will be considered, as previously noted, in Sec. 15-2-5.

A secular trend can also be obtained by the *method of moving averages.* By properly selecting moving averages, the seasonal and cyclical variations may be eliminated, and thus a trend pattern is approximated. The technique of moving averages, however, is more frequently used for computing seasonal indexes, to which we shall now direct our attention.

15-2-2 Computation of seasonal indexes

A knowledge of the seasonal pattern is important, especially in business and economic analysis. In forecasting the sales of a company or an industry, for instance, it is essential that the seasonal variations throughout the year are known. For this and other reasons, it is necessary to compute *seasonal indexes,* which are measures of seasonal variations in the behavior of any variable. By using such indexes, we can deseasonalize the time series; deseasonalized data are known as *seasonally adjusted.*

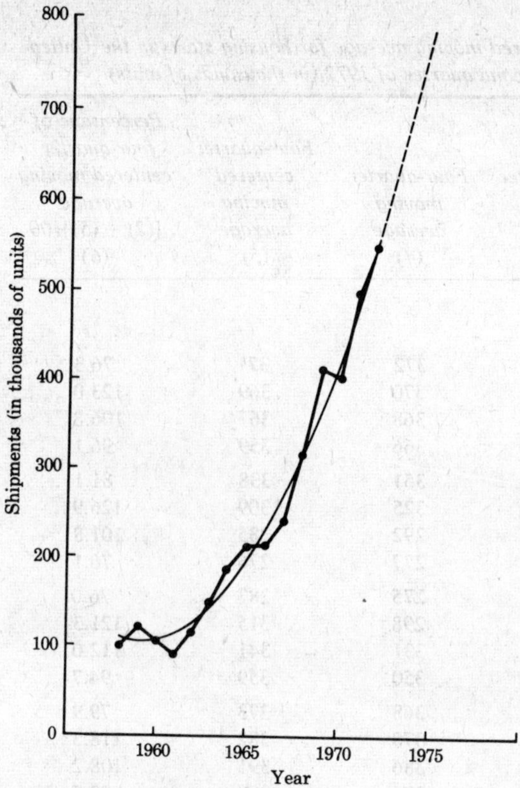

Figure 15-7

Second-degree trend curve fitted to shipments of mobile homes in the United States, 1958 to 1973 (in thousands of units). From data of Table 15-3.

We shall use the data for housing starts presented in Table 15-1 to compute the seasonal indexes. In order to do this, we must first take out the seasonal fluctuations contained in the series. For this purpose, we calculate a four-quarter *moving average* for each period, since the yearly data are made up of four quarters.

A moving average is computed by adding the values for a certain number of successive periods and then dividing the sum thus obtained by the number of periods included. In Table 15-5, for example, the four-quarter moving average for the first quarter of 1965 is obtained by adding the values for the third and fourth quarters of 1964 and the first and second quarters of 1965 and then dividing the sum by 4. The average for the second quarter of 1965 is obtained by adding the values for the fourth quarter of 1964 and the first, second, and third quarters of 1965 and then dividing the sum by 4. Thus, for each successive average, the earliest quarter is subtracted and the next later quarter is added.

Table 15-5

Computation of four-quarter centered moving average for housing starts in the United States, third quarter of 1964 to second quarter of 1972 (in thousands of units)

Year, by quarters (*1*)	*Original data* *Y* (*2*)	*Four-quarter moving total* (*3*)	*Four-quarter moving average* (*4*)	*Four-quarter centered moving average* (*5*)	*Percentage of four-quarter centered moving average* [(*2*) ÷ (*5*)]*100* (*6*)
1964: 3	398				
4	352				
1965: 1	283	1,487	372	371	76.3
2	454	1,481	370	369	123.0
3	392	1,474	368	367	106.8
4	345	1,465	366	359	96.1
1966: 1	274	1,403	351	338	81.1
2	392	1,301	325	309	126.9
3	290	1,166	292	285	101.8
4	210	1,110	277	276	76.1
1967: 1	218	1,100	275	287	76.0
2	382	1,192	298	315	121.3
3	382	1,322	331	341	112.0
4	340	1,402	350	359	94.7
1968: 1	298	1,472	368	373	79.9
2	452	1,513	378	382	118.3
3	423	1,545	386	391	108.2
4	372	1,583	396	398	93.5
1969: 1	336	1,599	400	395	85.1
2	468	1,563	391	383	122.2
3	387	1,500	375	366	105.7
4	309	1,482	357	349	88.5
1970: 1	264	1,359	340	343	77.0
2	399	1,380	345	356	112.1
3	408	1,467	367	383	106.5
4	396	1,592	398	424	93.4
1971: 1	389	1,797	449	471	82.6
2	604	1,968	492	507	119.1
3	579	2,085	521	537	107.8
4	513	2,206	552	559	91.8
1972: 1	510	2,263	566		
2	661				

Source: Table 15-1.

Column 4 of Table 15-5 shows the four-quarter moving averages derived in this way for the data on housing starts for 1964 to 1972. The moving average does not eliminate the very extreme fluctuations of the series, but it does substantially reduce the amplitude of the variations of the original data.

If an odd number of periods is included in the computation of a moving average, the process would be simpler since the number of periods before and after the period for which the average is computed are equal. When an even number of periods is included, as is true in the above example, the same number of periods after and before a specified period cannot be employed. Consequently, the moving average must fall midway between the values for two consecutive periods, and it is not related to any one period. This problem can be solved by computing a *centered* moving average on the series. It is achieved by first obtaining a moving average of an even number of periods as has been done above. Then we calculate a centered two-quarter moving average of the already obtained moving averages. The first centered moving average is the mean of the first two four-quarter moving averages, the second centered moving average is the mean of the second and the third four-quarter moving averages, etc. In so doing, there will be an equal number of periods after and before the specified period for which the centered moving average is computed. The four-quarter centered moving averages are shown in column 5 of Table 15-5.

Column 6 of Table 15-5 shows the original data as percentages of the centered moving averages. They are obtained by dividing the original data by the centered moving averages and then multiplying by 100. As pointed out previously, the time series is assumed to consist of $T \times S \times C \times I$. The four-quarter centered moving average is an approximate estimate of $T \times C$ since the process of averaging roughly smooths out the seasonal fluctuations and, for the most part, irregular fluctuations as well. The latter is possible because irregular movements are by and large of short duration and small amplitude, and the irregular component will be canceled out by averaging. Thus the centered moving averages in column 5 are an *estimate* of the *trend* and *cyclical movements combined*, or $T \times C$. The original data divided by $T \times C$ and then multiplied by 100 yield an estimate of the *seasonal* and *irregular* movement combined; that is,

$$\frac{T \times S \times C \times I}{T \times C} 100 = S \times I \tag{15-10}$$

which is the same as in column 6 of Table 15-5. This is tantamount to saying that this last operation is designed to eliminate the trend and cyclical elements from the original series.

The next operation is to eliminate the irregular variations as best we can in order to obtain the seasonal index. To that end, we rearrange the percentages in column 6 of Table 15-5 as in Table 15-6. When the seasonal-irregular percentages are averaged for each quarter, it can be expected that any remaining

Table 15-6

Computation of seasonal index

Year	*Quarters* 1	2	3	4	*Total*
1964					
1965	76.3	123.0	106.8	96.1	
1966	81.1	126.9	101.8	76.1	
1967	76.0	121.3	112.0	94.7	
1968	79.9	118.3	108.2	93.5	
1969	85.1	122.2	105.7	88.5	
1970	77.0	112.1	106.5	93.4	
1971	82.6	119.1	107.8	91.8	
1972					
Total	558.0	842.9	748.8	634.1	
Average	79.714	120.414	106.971	90.586	397.685
Seasonal index	80.18	121.12	107.59	91.11	400.000

Source: Table 15-5.

irregular fluctuations will be smoothed out in the process of averaging. The four averages will then be an estimate of the seasonal indexes, and the sum of the four values should equal 400. Since the total is 397.685 and not 400, an adjustment must be made. This is done by multiplying each of the quarterly averages by the ratio of 400 to 397.685, or $\frac{400}{397.685}$. The final values of the seasonal index, which must total 400, are shown in the bottom row of Table 15-6. A graphic representation of the seasonal indexes and the percentages for each year are given in Fig. 15-8.

Many published time series are seasonally adjusted data; that is, they are deseasonalized series. The process of eliminating seasonal fluctuation, or deseasonalization of data, consists of dividing each value in the original series by the corresponding value of the seasonal index. Table 15-7 shows the process of deseasonalization of the series of housing starts for 1966. The deseasonalized series shown in the last column is obtained by dividing each of the quarterly values of housing starts by the seasonal index from Table 15-6.

An important and interesting fact can be discovered from Table 15-7. For example, there is an increase of 118,000 dwelling units from the first to the second quarter of 1966. The increase, however, was actually much less than the 139,905 units [274,000 $(\frac{121.12}{80.18})$ − 274,000] expected on the basis of typical seasonal patterns. Thus the increase of 118,000 units should be viewed as a

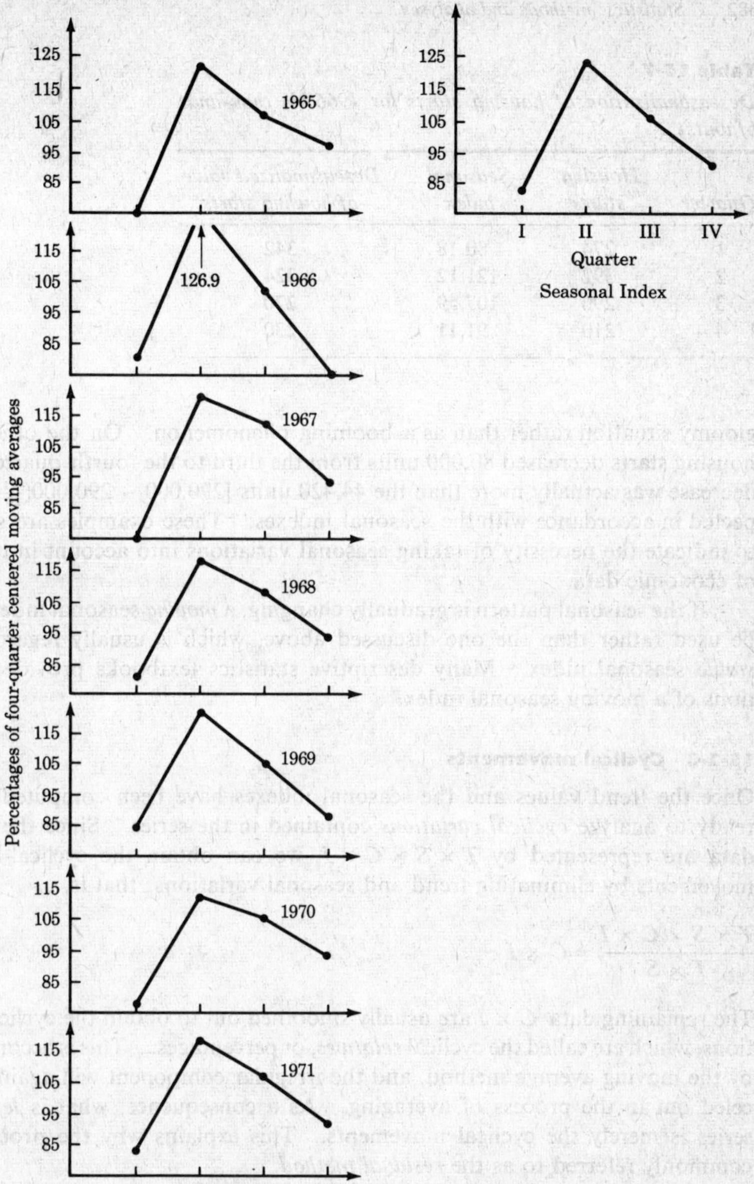

Figure 15-8

Seasonal index and the percentages of four-quarter centered moving averages plotted for each year. From data of Table 15-6.

Table 15-7

Deseasonalization of housing starts for 1966 (in thousands of units)

Quarter	*Housing starts*	*Seasonal index*	*Deseasonalized value of housing starts*
1	274	80.18	342
2	392	121.12	324
3	290	107.59	270
4	210	91.11	230

gloomy situation rather than as a booming phenomenon. On the other hand, housing starts decreased 80,000 units from the third to the fourth quarter. The decrease was actually more than the 44,420 units $[290{,}000 - 290{,}000(\frac{91.11}{107.59})]$ expected in accordance with the seasonal indexes. These examples are sufficient to indicate the necessity of taking seasonal variations into account in the study of economic data.

If the seasonal pattern is gradually changing, a *moving* seasonal index should be used rather than the one discussed above, which is usually regarded as a *stable* seasonal index. Many descriptive statistics textbooks provide illustrations of a moving seasonal index.

15-2-3 Cyclical movements

Once the trend values and the seasonal indexes have been computed, we are ready to analyze *cyclical variations* contained in the series. Since the original data are represented by $T \times S \times C \times I$, we can obtain the cyclical-irregular movements by eliminating trend and seasonal variations; that is,

$$\frac{T \times S \times C \times I}{T \times S} = C \times I \qquad (15\text{-}11)$$

The remaining data $C \times I$ are usually smoothed out to obtain the cyclical variations, which are called the cyclical *relatives*, or percentages. This is accomplished by the moving average method, and the irregular component will again be canceled out in the process of averaging. As a consequence, what is *left* in the series is merely the cyclical movements. This explains why the procedure is commonly referred to as the *residual method.*

The procedure of computing the cyclical movements according to the residual method consists of first multiplying the trend values by the seasonal index; the product thus obtained is divided into the original series, resulting in the combination of the cyclical and irregular movements. Table 15-8 shows the process of computing the cyclical-irregular movements.

Table 15-8

Computation of cyclical movements for data on housing starts in the United States, third quarter of 1964 to second quarter of 1972

Year, by quarters (1)	Original data $T \times S \times C \times I$ (2)	Trend value T (3)	Seasonal index S (4)	Trend-seasonal $T \times S \div 100$ (5)	Cyclical-irregular percentages $C \times I$ (6)	Three-quarter moving total (7)	Cyclical percentages C (8)
1964: 3	398	291.73	107.59	313.872	126.80		
4	352	298.07	91.11	271.572	129.62	372.37	124.12
1965: 1	283	304.41	80.18	244.076	115.95	366.19	122.06
2	454	310.75	121.12	376.380	120.62	351.47	117.16
3	392	317.09	107.59	341.157	114.90	352.58	117.53
4	345	323.43	91.11	294.677	117.08	335.61	111.87
1966: 1	274	329.77	80.18	264.410	103.63	317.00	105.67
2	392	336.11	121.12	407.096	96.29	278.63	92.88
3	290	342.45	107.59	368.442	78.71	241.08	80.36
4	210	348.79	91.11	317.783	66.08	221.35	73.78
1967: 1	218	355.13	80.18	284.743	76.56	229.89	76.63
2	382	361.47	121.12	437.812	87.25	260.34	86.78
3	382	367.81	107.59	395.727	96.53	283.52	94.50
4	340	374.15	91.11	340.888	99.74	293.95	97.98

Table 15-8 (*Continued*)

Year, by quarters (1)	Original data $T \times S \times C \times I$ (2)	Trend value T (3)	Seasonal index S (4)	Trend-seasonal $T \times S \div 100$ (5)	Cyclical-irregular percentages $C \times I$ (6)	Three-quarter moving total (7)	Cyclical percentages C (8)
1968: 1	298	380.49	80.18	305.077	97.68	293.89	97.96
2	452	386.83	121.12	468.528	96.47	294.15	98.05
3	423	393.17	107.59	423.012	100.00	298.67	99.56
4	372	399.51	91.11	363.994	102.20	305.45	101.82
1969: 1	336	405.85	80.18	325.411	103.25	299.19	99.73
2	468	412.19	121.12	499.245	93.74	282.93	94.31
3	387	418.53	107.59	450.296	85.94	259.50	86.50
4	309	424.87	91.11	387.099	79.82	242.12	80.71
1970: 1	264	431.21	80.18	345.744	76.36	231.47	77.16
2	399	437.55	121.12	529.961	75.29	237.08	79.03
3	408	443.89	107.59	477.581	85.43	257.26	85.75
4	396	450.23	91.11	410.205	96.54	288.23	96.08
1971: 1	389	456.57	80.18	366.078	106.26	310.53	103.51
2	604	462.91	121.12	560.677	107.73	328.67	109.56
3	579	469.25	107.59	504.866	114.68	340.80	113.60
4	513	475.59	91.11	433.310	118.39	365.05	121.68
1972: 1	510	481.93	80.18	386.411	131.98	362.14	120.71
2	661	488.27	121.12	591.383	111.77		

Source: Tables 15-1, 15-4, and 15-6.

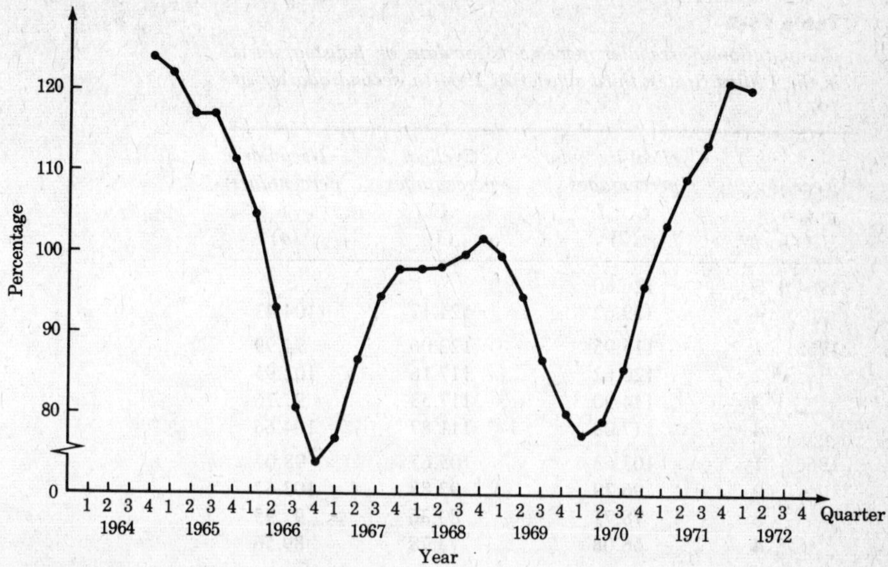

Figure 15-9

Percentages of cyclical movements of housing starts in the United States, fourth quarter of 1964 to first quarter of 1972. From data of Table 15-8.

The next step in computing the cyclical movement for the data on housing starts is the smoothing of irregular movements from $C \times I$ data. It has been found that most irregular variations are of short duration, 1 month or slightly longer. Since we have used quarterly data, a major portion of the irregular movements has probably already been canceled out. Occasionally, however, there may be irregular variations that last more than 3 months. Thus, for removing the irregular component from the cyclical-irregular movements, we plan to use a three-quarter moving average of the $C \times I$ data even though in so doing we might run into the danger of oversmoothing. The last column in Table 15-8 shows the cyclical percentages after the irregular movement has been eliminated.

The graphic representation of the cyclical movements of the data on housing starts for 1964 to 1972 is shown in Fig. 15-9.

15-2-4 Irregular movements

The *irregular movements* of a time series, or I, can be obtained by dividing the cyclical percentages into the combined cyclical-irregular variations; that is,

$$I = \frac{C \times I}{C} \tag{15-12}$$

For the data on housing starts, this means dividing the value in column 6 of

Table 15-9

Computation of irregular movements for data on housing starts in the United States, third quarter of 1964 to second quarter of 1972

Year, by quarters (*1*)		Cyclical-irregular percentages $C \times I$ (*2*)	Cyclical percentages C (*3*)	Irregular percentages I (*4*)
1964:	3	126.80		
	4	129.62	124.12	104.43
1965:	1	115.95	122.06	94.99
	2	120.62	117.16	102.95
	3	114.90	117.53	97.76
	4	117.08	111.87	104.66
1966:	1	103.63	105.67	98.07
	2	96.29	92.88	103.67
	3	78.71	80.36	97.95
	4	66.08	73.78	89.56
1967:	1	76.56	76.63	99.91
	2	87.25	86.78	100.54
	3	96.53	94.50	102.15
	4	99.74	97.98	101.80
1968:	1	97.68	97.96	99.71
	2	96.47	98.05	98.39
	3	100.00	99.56	100.44
	4	102.20	101.82	100.37
1969:	1	103.25	99.73	103.52
	2	93.74	94.31	99.40
	3	85.94	86.50	99.35
	4	79.82	80.71	98.90
1970:	1	76.36	77.16	98.96
	2	75.29	79.03	95.27
	3	85.43	85.75	99.63
	4	96.54	96.08	100.49
1971:	1	106.26	103.51	102.66
	2	107.73	109.56	98.33
	3	114.68	113.60	100.95
	4	118.39	121.68	97.30
1972:	1	131.98	120.71	109.34
	2	111.77		

Source: Table 15-8.

Table 15-10

Decomposition of quarterly series of housing starts in the United States by multiplicative model, third quarter of 1964 to second quarter of 1972

Year, by quarters *(1)*	*Original data* $T \times S \times C \times I$ *(2)*	*Trend value* *T* *(3)*	*Seasonal percentages* *S* *(4)*	*Cyclical percentages* *C* *(5)*	*Irregular percentages* *I* *(6)*
1964: 3	398	291.73	107.59		
4	352	298.07	91.11	124.12	104.43
1965: 1	283	304.41	80.18	122.06	94.99
2	454	310.75	121.12	117.16	102.95
3	392	317.09	107.59	117.53	97.76
4	345	323.43	91.11	111.87	104.66
1966: 1	274	329.77	80.18	105.67	98.07
2	392	336.11	121.12	92.88	103.67
3	290	342.45	107.59	80.36	97.95
4	210	348.79	91.11	73.78	89.56
1967: 1	218	355.13	80.18	76.63	99.91
2	382	361.47	121.12	86.78	100.54
3	382	367.81	107.59	94.50	102.15
4	340	374.15	91.11	97.98	101.80
1968: 1	298	380.49	80.18	97.96	99.71
2	452	386.83	121.12	98.05	98.39
3	423	393.17	107.59	99.56	100.44
4	372	399.51	91.11	101.82	100.37
1969: 1	336	405.85	80.18	99.73	103.52
2	468	412.19	121.12	94.31	99.40
3	387	418.53	107.59	86.50	99.35
4	309	424.87	91.11	80.71	98.90
1970: 1	264	431.21	80.18	77.16	98.96
2	399	437.55	121.12	79.03	95.27
3	408	443.89	107.59	85.75	99.63
4	396	450.23	91.11	96.08	100.49
1971: 1	389	456.57	80.18	103.51	102.66
2	604	462.91	121.12	109.56	98.33
3	579	469.25	107.59	113.60	100.95
4	513	475.59	91.11	121.68	97.30
1972: 1	510	481.93	80.18	120.71	109.34
2	661	488.27	121.12		

Source: Summary from Tables 15-8 and 15-9.

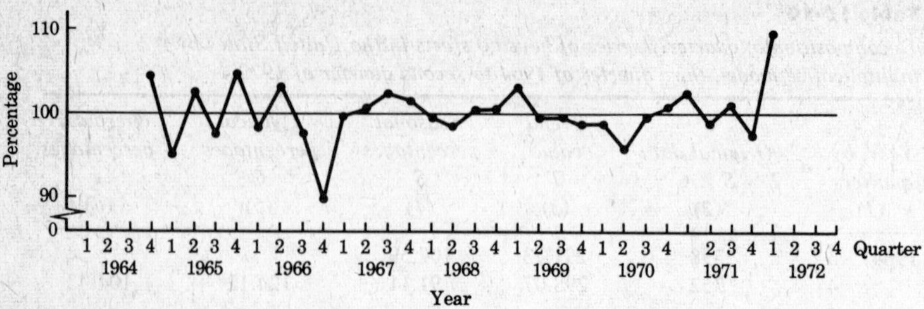

Figure 15-10

Percentages of irregular movements of housing starts in the United States, fourth quarter of 1964 to first quarter of 1972. From data of Table 15-9.

Table 15-8 by the corresponding values in column 8 and then multiplying the quotients by 100. The results are shown in Table 15-9. The graphic representation of the irregular movements of the data is shown in Fig. 15-10.

Thus the original observed data for housing starts in the United States from 1964 to 1972 have been decomposed into the four components, which are recapitulated in Table 15-10.

15-2-5 Forecasting

We shall now proceed to see how time series are used for forecasting. As pointed out previously, the rationale of making predictions on the basis of time series and their components lies in the fact that there exists some regularity in the movements of the various components. In other words, forecasting is based on the belief that what has occurred in the past will, by and large, continue to occur in the future. We shall now demonstrate how the components of trend and seasonal indexes can be used for forecasting purposes.

In Chap. 14 and in Sec. 15-2-1, the least-squares method was discussed. A least-squares equation is often used for forecasting by extrapolation, that is, by extending into the future. The trend movement expressed in a least-squares line describes, on the average, what has happened in the past, and estimates values that lie ahead beyond the series on the basis of which the trend equation is derived.

Let us use Eq. (15-3) for illustration. The trend equation

$$Y_t = 285.39 + 6.34X$$

was computed by the least-squares method for the quarterly data on housing starts in the United States from 1964 to 1972. On the assumption that this trend movement will continue into the following few years, the forecast for quarterly values of housing starts in the United States up to 1975, as shown in Fig. 15-5, will appear as in Table 15-11.

The process of extrapolation is a simple one. As shown in Table 15-11, for forecasting the values for the third and fourth quarters of 1972, we simply substitute 33 and 34 in the trend equation. A similar procedure is followed for forecasting the quarterly values for 1973 through 1975.

Since the housing industry is strongly influenced by seasonal factors, it seems unrealistic to expect that the housing starts will be 507,290 units for the first quarter and 526,310 units for the fourth quarter of 1973. As a matter of fact, the seasonal index computed from the same series shown in Sec. 15-2-2 indicates that the housing starts for the first quarter are 80.18 percent of those for the average quarter, while those for the fourth quarter are 91.11 percent of the average. Multiplying the quarterly trend values by the corresponding seasonal indexes, we obtain the forecasted housing starts of 406,745 units and 479,521 units for the first and fourth quarters of 1973, respectively. This process may be called the *reverse* of deseasonalization. Applying the same procedure, we arrive at a seasonally adjusted forecast for each quarter through 1975 as shown in Table 15-12.

If we forecast the shipments of mobile homes on the basis of the second-degree parabolic trend equation, Eq. (15-9), we obtain the following projected values for 1973 to 1978:

Year	Y_t
1973	627,490
1974	705,440
1975	788,790
1976	877,540
1977	971,690
1978	1,071,240

Observe that the 1973 figure is computed by substituting the value $x = 8$ in Eq. (15-9). For the figure for 1974, we use the value $x = 9$, and so forth. You are advised to verify the values for Y_t shown above.

The forecasting techniques discussed above are elementary and rudimentary. They serve only as an introduction to the more realistic and deeper econometric treatment of the subject. It would hardly be necessary to point out that extrapolation into the future on the basis of what has happened in the past can be precarious. It is well known that any projection from past experience into the future, especially the uncertain distant future, is highly speculative and hazardous. However, business and other institutions must make future plans in order to operate rationally and successfully. With refined and more sophisticated techniques, forecasts can be improved and more intelligent decisions can be made possible.

Table 15-11
Forecasts of quarterly trend values of housing starts in the United States, third quarter of 1972 to fourth quarter of 1975 (in thousands of units)

Year, by quarters		X	Projected housing starts Y
1972:	3	33	494.61
	4	34	500.95
1973:	1	35	507.29
	2	36	513.63
	3	37	519.97
	4	38	526.31
1974:	1	39	532.65
	2	40	538.99
	3	41	545.33
	4	42	551.67
1975:	1	43	558.01
	2	44	564.35
	3	45	570.69
	4	46	577.03

Table 15-12
Seasonally adjusted housing starts in the United States, third quarter of 1972 to fourth quarter of 1975 (in thousands of units)

Year, by quarters (1)		Trend value T (2)	Seasonal index percentage S (3)	Predicted quarter housing starts (4)
1972:	3	494.61	107.59	532.151
	4	500.95	91.11	456.416
1973:	1	507.29	80.18	406.745
	2	513.63	121.12	622.109
	3	519.97	107.59	559.436
	4	526.31	91.11	479.521
1974:	1	532.65	80.18	427.079
	2	538.99	121.12	652.825
	3	545.33	107.59	586.721
	4	551.67	91.11	502.627
1975:	1	558.01	80.18	447.412
	2	564.35	121.12	683.541
	3	570.69	107.59	613.962
	4	577.03	91.11	525.732

Our discussion of forecasting is based on trend and seasonal patterns only. Cyclical and irregular movements are extremely difficult, if not impossible, to predict. Both the Federal government and many private institutions furnish statistics relative to business-cycle developments; it is not the purpose of this book to go into further discussion of that topic.

Exercise 15-2

1 Explain the limitations of the techniques usually used in time-series analysis.

2 Define the following terms in your own words as fully as you can:

a Detrending.
b Deseasonalization.
c Moving average.
d Centered moving average.
e Seasonal index.
f Seasonally adjusted data.

3 Data presented below are quarterly sales Y of a certain product during the 12 years from 1962 through 1973; X represents time coded in quarters beginning in the first quarter of 1962.

Year	*Quarter*	*X*	*Y*	*Year*	*Quarter*	*X*	*Y*
1962	1	1	14	1968	1	25	22
	2	2	17		2	26	30
	3	3	13		3	27	27
	4	4	10		4	28	21
1963	1	5	9	1969	1	29	24
	2	6	10		2	30	31
	3	7	9		3	31	30
	4	8	7		4	32	22
1964	1	9	10	1970	1	33	27
	2	10	12		2	34	33
	3	11	11		3	35	29
	4	12	6		4	36	27
1965	1	13	8	1971	1	37	32
	2	14	15		2	38	36
	3	15	16		3	39	32
	4	16	19		4	40	29
1966	1	17	21	1972	1	41	31
	2	18	30		2	42	37
	3	19	28		3	43	34
	4	20	28		4	44	33
1967	1	21	27	1973	1	45	37
	2	22	33		2	46	40
	3	23	32		3	47	38
	4	24	29		4	48	37

a Plot the data on a two-dimensional plane.

b Determine the trend line by the least-squares method. Plot the line in the scatter diagram.

c On the basis of the least-squares line and the multiplicative model, compute the following:

1 Seasonal indexes for the four quarters by the method of four-quarter centered moving averages. Plot the results as in Fig. 15-8.
2 Deseasonalized values for *Y* for the year 1973.
3 Cyclical and irregular variations by the methods shown in Tables 15-8 and 15-9. Plot the results as in Figs. 15-9 and 15-10.
4 Predicted quarterly sales for 1974 by applying the least-squares line and the seasonal indexes.

4 What is the rationale that time series should, in many cases, be seasonally adjusted?

5 Describe the residual method in time-series analysis. What are some of the assumptions underlying the residual method?

6 Show the derivation of Eq. (15-8).

7 Explain why it is more difficult to forecast cyclical movements than secular trend and seasonal variations.

8 Given the following data on production of the XYZ Company:

Month	*Production units*	*Seasonal indexes*
January	54	130
February	58	125
March	57	120
April	43	80
May	48	75
June	45	90
July	38	75
August	44	85
September	50	70
October	55	120
November	50	110
December	58	120

What is your conclusion regarding the production activities of this company after you have analyzed the data?

9 Explain fully the irregular movements of a time series.

10 The monthly shipments in units of the ABC Company for 6 years from 1968 to 1973 are listed as follows:

Month	*1968*	*1969*	*1970*	*1971*	*1972*	*1973*
January	7	13	12	13	18	26
February	9	15	13	12	22	27
March	10	16	11	13	24	29
April	14	19	13	16	26	30
May	15	18	14	18	28	33
June	13	19	16	19	29	34
July	15	18	14	17	30	34
August	14	20	16	17	28	33
September	15	19	15	16	29	32
October	14	17	13	15	28	32
November	12	18	12	15	27	31
December	11	16	13	16	29	30

a Plot the quarterly data on a two-dimensional plane.
b Determine the trend line by the least-squares method (use the quarterly data). Plot the line in the scatter diagram.
c On the basis of the least-squares line and the multiplicative model, compute the following:
1 Seasonal indexes for the four quarters by the method of four-quarter centered moving averages. Plot the results as in Fig. 15-8.
2 Deseasonalized values for the quarterly shipments for 1973.
3 Cyclical and irregular variations by the methods shown in Tables 15-8 and 15-9. Plot the results as in Figs. 15-9 and 15-10.
4 Predicted quarterly shipments for 1974 on the basis of the least-squares line and the seasonal indexes.

11 The method illustrated in Tables 15-5 and 15-6 is known as the *ratio to the moving average* technique of computing seasonal indexes. Explain how the secular trend and cyclical influences are removed by such a method.

12 Plot the data below as a scatter diagram. Fit a second-degree curve to the data, and then plot the curve in the diagram.

Year	*X*	*Y*
1957	1	10
1958	2	11
1959	3	6
1960	4	8
1961	5	4
1962	6	3
1963	7	6
1964	8	8
1965	9	7
1966	10	10
1967	11	15
1968	12	15
1969	13	22
1970	14	21
1971	15	25
1972	16	30
1973	17	35

15-3 Index Numbers

A time series may be simplified by reducing the data for the various periods to a series of *index numbers*, which are usually expressed as percentages, but could be any convenient number, relative to a specified period called the *base period.* The value for the base period is taken as 100 when the data are reduced to percentages. Index numbers are convenient devices for measuring relative changes or differences from time to time or from place to place. Just as the arithmetic mean is used to represent a set of values, an index number is used to represent the average change of a set of values over two or more different periods or localities.

Index numbers are examined here because some techniques for making forecasts, or inferences about the future, are applied in terms of index numbers. In regression analysis, for instance, either the independent or dependent variable or both may be in the form of index numbers. They are less unwieldy than large numbers, and are readily understandable. Often business firms and many other institutions, including the government, are reluctant to release information concerning their financial conditions but do not hesitate to give out the same data in the form of percentages. These and other functions of index numbers have made the device an important tool in statistical analysis.

It ought to be pointed out that index numbers are not only associated with business and economics as many people think they are, but are actually used in a variety of fields. Psychologists use index numbers to measure intelligence quotients; they also present test results in the form of percentiles, which are essentially index numbers comparing individual performances with established norms. Educational authorities use indexes to measure the effectiveness of different school systems, and to project enrollment changes as a basis for future planning. In short, index numbers find application in virtually all disciplines and walks of life.

In general index numbers are classified into two types: simple and composite. A *simple index* is one that is computed for a single variable, while a *composite index* is constructed from two or more variables. Most index numbers are composite in nature. We shall now briefly consider the construction of a simple index for the primary purpose of illustrating the basic concept and use of index numbers.

Table 15-13

Illustration of simple index numbers for enrollment at the University of Southland

Year	*Enrollment*	*Index I* (1964–1966 = 100)	*Index II* (1973 = 100)
1964	15,000	88	56
1965	16,500	97	61
1966	19.500	115	72
1967	19,800	117	74
1968	20,500	121	76
1969	21,500	126	80
1970	22,000	130	81
1971	24,000	141	89
1972	25,600	151	95
1973	27,000	160	100

Let us use the hypothetical data on enrollment at the University of Southland during the 10-year period from 1964 to 1973 to illustrate the computation of a simple index. As shown in Table 15-13, different indexes based on different base periods can be constructed from the same set of data. Index I is obtained by using the average enrollment for the 3 years from 1964 to 1966 as the base period. The index is computed by dividing each year's enrollment by the average enrollment for the 3-year period and multiplying by 100. The percentage for each year is interpreted in terms of the base period. The index for 1972, for example, signifies that enrollment in that year was 151 percent of the average

Table 15-14

Consumer Price Index of the United States, 1947 to 1972 (*1967* = 100)

Year	*Index*
1947	66.9
1948	72.1
1949	71.4
1950	72.1
1951	77.8
1952	79.5
1953	80.1
1954	80.5
1955	80.2
1956	81.4
1957	84.3
1958	86.6
1959	87.3
1960	88.7
1961	89.6
1962	90.6
1963	91.7
1964	92.9
1965	94.5
1966	97.2
1967	100.0
1968	104.2
1969	109.8
1970	116.3
1971	119.7
1972	123.1

Source: U.S. Department of Commerce, *Survey of Current Business* and *Business Statistics* 1971 Biennial ed.

enrollment for 1964 to 1966. Similarly, index II is obtained by selecting 1973 as the base period. The index is computed by dividing each year's enrollment by the base 27,000 and multiplying by 100. In this case the enrollment in 1965 was 61 percent of the enrollment in 1973. By the use of either index, the absolute numbers of the variable are transformed into relative values, and, consequently, the changes in the variable can be readily compared.

By far the most important index, however, is the composite index that includes two or more variables. The *index of industrial production* shown in Fig. 15-4, for instance, is a composite index published monthly by the Board of Governors of the Federal Reserve System. The oldest composite index published continuously for the United States is the *Wholesale Price Index*, which is compiled monthly by the Bureau of Labor Statistics. This series has been in publication since 1890. The Bureau of Labor Statistics also compiles the monthly *Consumer Price Index* (CPI), which has been in publication since 1913. The CPI includes individual indexes for apparel, food, housing, transportation, and other goods and services. The annual CPI of the United States for all items from 1947 to 1972 with 1967 = 100 is shown in Table 15-14.

In the remainder of this section the discussion will be limited to composite index only. Composite indexes are of several types. Generally, they are classified into *unweighted* and *weighted* indexes. Before we take up the methods of constructing the various types of composite indexes, let us first examine the problems usually encountered in index construction.

15-3-1 Problems in index construction

Among the problems faced in constructing index numbers are the selection of data, the choice of the base period, and the selection of an appropriate weighting system. Each of these is briefly discussed below.

Selection of data In selecting data and making computations it is important to understand the *purpose* for which the index is used. If the purpose is to measure the changes in the cost of education for college students, for instance, it would not do much good to include prices of household appliances or rates for airline and railroad fares. Likewise, if the purpose is to measure the changes in family medical costs, it would be quite irrelevant to gather the prices of potatoes and gasoline. Obviously, an index properly constructed is a useful tool, while an improperly designed index can be misleading and meaningless.

In constructing a simple index as illustrated in Table 15-13, or in comparing the price of a single commodity in two different periods, there is no question what data should be included. Typically, index numbers are constructed from *samples*, not necessarily random but selected according to the judgment of the index compiler. In a composite index that may possibly involve hundreds or even thousands of variables, the selection of a sample among the variables can be a difficult and troublesome task.

One rule that must be observed in gathering data is that the samples selected must be *representative* of the population from which they are taken. The much favored way to achieve this is through *stratification* of all the conceivable data into groups or subgroups, from each of which a representative subsample is drawn. Through this procedure, we can make sure that every group or subgroup will be represented in the index to be constructed.

Another important principle to keep in mind when gathering data for index construction is the *comparability* of data. Frequently, statisticians run into serious difficulty because the data they collected are not strictly comparable. For example, a television set made in 1950 can be quite different from a present-day set, and a 1976 automobile can hardly be compared with a 1920 model. Thus, when a cost or price index is desired, we may wonder whether the goods or services included are identical in quality over the various periods considered.

Selection of the base period The base period of an index should be carefully selected because it is used as a point of reference in comparing the various index numbers and in describing their individual behaviors. At least two rules should be observed in choosing a base period. The first is that it should be a "normal" period, that is, one of *relative stability*. In the construction of a price index, for example, the war year of 1917 or 1944 should be avoided as the base year because of the abnormal and disrupted economic conditions. Consumption patterns might be quite irregular due to shortage of goods and services and operations of the black market.

The other rule for choosing the base period is that it should be relatively *recent*. One of the reasons for this is that the more distant the base period is in the past, the more difficult it will be to find comparable data for the later periods. Furthermore, a more recent period is preferred because we are more concerned with changes relative to the present than with fluctuations related to the distant past.

Selection of weights It is necessary to point out that each variable in a composite index should have a reasonable influence on the index. That is, in constructing an index due consideration should be given to the relative importance of each variable, which relates to the purpose for which the index is to be used. Suppose, for instance, that we wish to compare the prices of foods used in the home for 1958 and 1974, and that milk and vinegar are the variables used to construct the index. If the price of milk increased from 20 cents in 1958 to 30 cents in 1974 while the price of vinegar decreased from 20 to 14 cents during the same period, we might claim that the price index for 1974 is 110 percent of 1958, or

$$\frac{0.30 + 0.14}{0.20 + 0.20} = \frac{0.44}{0.40} = 110$$

This is obviously misleading since milk is a much more important item than vinegar in the diet of the average American. Daily consumption of milk is many times higher than that of vinegar, and thus should be weighted with the quantity actually consumed. In so doing, the 1974 price index of these two commodities must be more than 10 percent higher than in 1958. This clearly indicates the necessity of weighting the variables in accordance with their relative importance or significance in order to get meaningful results. Although in this particular example the quantities of the two commodities consumed monthly or yearly appear to be reasonable weights, the problem of weighting different variables is not always easy, and frequently can be difficult and troublesome.

The above-mentioned considerations are only a few of the many problems faced by statisticians in compiling index numbers. Some of the other problems will be considered later when we illustrate the construction of indexes.

15-3-2 Unweighted index numbers

Unweighted indexes are generally classified into *simple aggregative indexes* and *simple averages of relatives*. We shall now use an example to illustrate the construction of both.

Simple aggregative indexes The construction of a simple aggregative index is a rather easy procedure. If a *price* index is desired, we first get the sum of the various prices for each of the periods under consideration, and then divide each sum by the sum for the base period. Let $\sum p_0$ be the sum of the base-period prices and $\sum p_n$ be the sum of the given-period prices; the ratio of the two sums multiplied by 100 gives the index P expressed as a percentage; that is,

$$P = \frac{\sum p_n}{\sum p_0} 100 \tag{15-13}$$

Keep in mind that the sum is over all the variables in the computation. Table 15-15 shows the computation of a simple aggregative index with 1967 as the base period.

A simple aggregative index is easy to understand, but it has a serious defect. Table 15-15 shows that the price of commodity X exerts much greater influence

Table 15-15
Computation of simple aggregative index

Commodities	*1967* p_0	*1970* p_1	*1973* p_2
X, per ton	$12.00	$15.00	$15.60
Y, per dozen	3.00	3.60	3.30
Z, per bushel	5.00	6.00	5.70
Aggregate	$20.00	$24.60	$24.60
Index	100.0	123.00	123.00

than the price of each of the other two commodities. This is, in part at least, due to the *units* used in quoting the prices of the commodities. If we had used pounds instead of tons as units for X, we should have obtained a quite different indicator of the changes in these prices. Since the units by which the commodities happen to be measured become the so-called "concealed weights," which are of no logical significance, simple aggregative indexes are seldom used in actual applications. Furthermore, the simple aggregative method cannot be used to compute a quantity index if the variables are expressed in divergent units such as tons and dozens.

Simple average of relatives One way to rectify the drawbacks of a simple aggregative index is to construct a simple average of relatives. To compute such an index, we convert the actual prices for each variable to percentages of the base period. These percentages are called *relatives* because they are relative to the value for the base period. A price relative, for instance, is obtained by dividing the price of a commodity for a given period, p_n, by the base-period price p_0. The sum of all the price relatives, or $\sum (p_n/p_0)$, divided by n, the number of commodities in the computation, and multiplied by 100, gives the simple average of price relatives, or P; that is,

$$P = \frac{\sum (p_n/p_0)}{n} 100 \qquad (15\text{-}14)$$

which is in fact the arithmetic mean of relatives. Table 15-16 shows the computation of a simple average of price relatives for the data in Table 15-15.

Since relatives are pure numbers, price index numbers computed from relatives will remain the same regardless of the units by which the prices are quoted. Thus a simple average of relatives is said to meet the criterion called the *units test.* Also, a quantity index can be constructed for a group of variables that are expressed in divergent units.

Table 15-16
Computation of simple average of price relatives

Commodities	1967 $(p_0/p_0)100$	1970 $(p_1/p_0)100$	1973 $(p_2/p_0)100$
X	100.0	125.0	130.0
Y	100.0	120.0	110.0
Z	100.0	120.0	114.0
Aggregate	300.0	365.0	354.0
Index	100.0	121.7	118.0

Source: Price relatives computed from data in Table 15-15.

The simple average of relatives, however, is still not a satisfactory index in spite of these improvements. It still has the drawback of assuming equal importance for each of the relatives. For instance, commodity *Y*, though a minor item, exerts a tremendous influence on the final outcome since a small absolute change in the price of *Y* will result in great change in its relative, and thus unduly affect the average. This again amounts to objectionable "concealed weighting," and limits the practical usefulness of the simple average of relatives.

Since concealed weighting is the principal defect of the unweighted indexes, improvement of the indexes depends on the introduction of an appropriate weighting system.

15-3-3 Weighted index numbers

Weighted indexes can also be classified into two types: *weighted aggregative indexes* and *weighted average of relatives.* We shall use an example to illustrate the construction of each.

Weighted aggregative indexes The selection of appropriate weights is a rather difficult task; only some simple illustrations are presented here. The most frequently used weights for an aggregative price index are base-period quantities of the commodities in the computation. Let q_0 be the base-period quantity. Multiplying the price of each commodity for each period by the base-period quantity of that commodity and then adding the products thus obtained, we get the sum $\sum p_0 q_0$ for the base period and $\sum p_n q_0$ for the given period. Then, the weighted aggregative price index with base-period quantity as weights, or P_b, is the ratio of $\sum p_n q_0$ to $\sum p_0 q_0$ multiplied by 100; that is,

$$P_b = \frac{\sum p_n q_0}{\sum p_0 q_0} 100 \tag{15-15}$$

which is in fact the ratio of weighted average of prices for the given period to the weighted average of prices for the base period, with the common denominator n of the two averages canceled. This formula is also known as *Laspeyres' method*, after the statistician who first recommended it.

The weighted aggregative price index may also be computed with given-period quantities as weights. Let q_n be a given-period quantity of a commodity. The formula for the weighted aggregative price index, or P_g, with the values of q_n as weights, appears as follows:

$$P_g = \frac{\sum p_n q_n}{\sum p_0 q_n} 100 \tag{15-16}$$

which is sometimes referred to as *Paasche's method.* Since the procedure of constructing the Laspeyres and Paasche indexes are similar, and since the Paasche index is rarely used because of the necessity of applying new weights for each new period, only one example, Table 15-17, illustrating the computation of the Laspeyres index is presented here.

Table 15-17

Computation of weighted aggregative price index with base-period quantities as weights

	Quantity	Price			Value of 1967 quantity at prices		
Commodity	q_0 1967	p_0 1967	p_1 1970	p_2 1973	p_0 1967	p_1 1970	p_2 1973
X	10	20	22	22	200	220	220
Y	20	10	12	15	200	240	300
Z	50	5	6	7	250	300	350
Aggregate					650	760	870
Index					100.0	118.5	133.8

Observe that if the base-period prices had been weighted with the base-period quantities, and if the given-period prices had been weighted with the given-period quantities, the resulting index would have been a *value index* rather than a price index. The total values of the commodities would have been compared instead of the weighted means of their prices.

Weighted average of relatives With the exception of weighting, the procedure of constructing a weighted average of relatives index is similar to that of computing the simple average of relatives. Customarily, weights used for computing an average of relatives are *value weights.* If the base-period values are used, the weights will be $p_0 q_0$; similarly, if the given-period values are used, the value will be $p_n q_n$. Also, $p_0 q_n$ or $p_n q_0$ may be used as weights. If the values of $p_0 q_0$ are used as weights, we shall have the Laspeyres average of price relatives, or P_0, defined as follows:

$$P_0 = \frac{\sum (p_n/p_0)(p_0 q_0)}{\sum p_0 q_0} 100 \tag{15-17}$$

Observe that this formula is identical with Eq. (15-15) since the p_0's in the numerator cancel and the numerator is actually $\sum p_n q_0$. Similarly, the Paasche average of price relatives, or P_n, appears to be

$$P_n = \frac{\sum (p_n/p_0)(p_0 q_n)}{\sum p_0 q_n} 100 \tag{15-18}$$

which is identical with Eq. (15-16). Table 15-18 illustrates the computation of a weighted average of price relatives index with the base-period values as weights, that is, the Laspeyres method.

Table 15-18

Computation of weighted average of price relatives index with base-period values as weights

	Weights	Price relatives		Weighted price relatives	
Commodity	p_0q_0 1967	(p_1/p_0) ·(100) 1970	(p_2/p_0) ·(100) 1973	$p_0q_0(p_1/p_0)$ ·(100) 1970	$p_0q_0(p_2/p_0)$ ·(100) 1973
X	200	110	110	22,000	22,000
Y	200	120	150	24,000	30,000
Z	250	120	140	30,000	35,000
Aggregate	650			76,000	87,000
Index				118.5	133.8

It should not be surprising that the index numbers obtained in Table 15-18 are the same as those obtained in Table 15-17. This being the case, what are the advantages, if any, of using the weighted average of relatives over using the weighted aggregative index? The weighted average of relatives is often used because, first of all, the relative for each single variable is itself a simple index and can be analyzed to yield useful information. Moreover, many variables, which are not comparable or cannot be readily reduced to dollars and cents, such as absenteeism and defective parts produced, can be made additive and comparable by being converted into relatives of some base.

15-3-4 Some important applications

Some uses of index numbers have already been noted in the foregoing discussions. In concluding this section we shall briefly mention some of the other most important applications of index numbers to our economic life.

One of the uses of index numbers is the so-called *statistical deflation.* It is a procedure by which the effects of price changes on the values of sales or production are removed. For statistical deflation, a certain appropriate price index called the *deflator* is used. For instance, if we wish to remove the effect of price changes on the value of the gross national product, or GNP, for a given period, we may divide the *GNP deflator*, which is a price index constructed by the Department of Commerce, into the *current* value of the GNP for the period to derive the *real* value of the GNP. The last is then expressed in *constant* dollars that have the same purchasing power as in the base period. Generally, the procedure is expressed in the following relationship:

$$\text{Real value of goods and services} = \frac{\text{current value of goods and services}}{\text{appropriate price index}} \tag{15-19}$$

For example, if the GNP for 1973 is \$1,200 billion, and the deflator (with 1967 as the base period) is 125 for 1973, then the 1973 GNP in constant 1967 dollars will be

$$\frac{\$1{,}200 \text{ billion}}{125} 100 = \$956 \text{ billion}$$

A similar application of index numbers is in the computation of *real wages*. In order to measure the purchasing power of what is paid by the company to the workers, the money wages can be *deflated* by some price index. For this purpose, the Bureau of Labor Statistics' *consumer price index* is commonly used as the deflator. Thus we have

$$\text{Real wages} = \frac{\text{money wages}}{\text{consumer price index}} 100 \tag{15-20}$$

which is the amount of wages expressed in constant dollars, or dollars with the same purchasing power as in the base period of the index. If the consumer price index increased 100 percent from one period to another but money wages increased only 50 percent, real wages actually declined 25 percent during the same period in spite of the enlarged pay checks. To protect workers against such adverse effects on their scale of living, many union-management contracts include the so-called *escalator clause* that provides for automatic increases in wage rates to match advances in the consumer price index.

Exercise 15-3

1 Distinguish between the following terms relative to index numbers:

a Base and given periods.
b Simple and composite indexes.
c Weighted and unweighted indexes.

2 Suppose that a price index is constructed by using a fixed quantity q as weight for each price. How should the formula be stated? What does such an index measure?

3 The prices and quantities of three commodities sold by a company during 1973 and 1974 are given below.

	Price, \$		*Quantity*	
Commodity	*1973*	*1974*	*1973*	*1974*
A	12.50	10.00	200 tons	225 tons
B	4.00	3.00	125 pounds	150 pounds
C	120.00	104.00	375 yards	450 yards

a Using 1973 as the base period, compute the simple aggregative price index and the simple average of price relatives for 1974.
b Using 1973 as the base period and the base-period quantities as weights, compute the weighted aggregative price index and the weighted average of price relatives for 1974.
c Using 1973 as the base period and 1974 quantities as weights, compute the weighted aggregative price index and the weighted average of price relatives for 1974.

4 List the problems of index-number construction, and explain what you would do about each of them.

5 Compare the following concepts:

a Index number and arithmetic mean.
b Laspeyres price index and Paasche price index.
c Current dollar and constant dollar.
d Weighted-aggregative price index and weighted average of price relatives.

6 In a certain Latin American country the average monthly salary of public school teachers and consumer price indexes are as follows:

Year	*Average monthly salary pesos*	*Consumer price index 1967 = 100*
1967	750	100
1968	785	110
1969	820	115
1970	850	130
1971	980	140
1972	1,100	160
1973	1,240	185
1974	1,400	205

a Using the consumer price index as deflator, find the averge monthly salary in 1967 pesos for each year from 1967 to 1974.
b Compare the *actual* percentage change in the salaries from 1967 to 1974 with the corresponding percentage change in *real* salaries.

7 Do the preceding problem with the base period shifted from 1967 to 1971. (The average monthly salary in *a* is to be stated in 1971 pesos.)

8 The prices and quantities of three commodities sold by the XYZ Company during 1967 and 1973 are shown below.

	Quantity		*Price, $*	
Commodities	*1967*	*1973*	*1967*	*1973*
X	10	5	30	40
Y	20	10	15	20
Z	50	40	10	15

a Compute the following indexes by Laspeyres' method:

1 Weighted aggregative price index.
2 Weighted average of price relatives.

b Compute the following indexes by Paasche's method:

1 Weighted aggregative price index.
2 Weighted average of price relatives.

9 In the computation of the weighted average of price relatives, why do you use values rather than quantities as weights?

10 Explain the importance of index numbers in statistical analysis.

Chapter 16

Utilities and Bayesian decision theory

Up to this point, the decision problems we have considered are such that each population parameter is regarded as having some specific value (or values). Although the parameter value is unknown, on the basis of sample information taken from the population a decision is made between accepting or rejecting the specified value (or values) of the population. Since a probability distribution can be attached to possible values of the parameter, the decision to accept or reject the specified value (or values), or the null hypothesis, is made in terms of probabilities. When the null hypothesis is true and is rejected, we have made an error, the probability of which is usually designated α. The probability of accepting the null hypothesis when it is false is designated β. Furthermore, when the null hypothesis is true (false), the probability of making the correct decision by accepting (rejecting) it is $1 - \alpha$ $(1 - \beta)$. The task is completed when the error probability is determined, and the decision is made accordingly.

In other words our previous discussions did not deal with the problems of *gains* or *losses* relative to a correct or erroneous decision. They were not concerned with the *payoff* of a decision. Bayesian statistics goes a step further since it is concerned not only with error probabilities but also with the payoff resulting from a decision.

In this chapter we shall give a brief account of the elementary Bayesian decision theory. We shall make extensive use of the concepts of probability function and the notion of expected value discussed in Chaps. 3 and 4, respectively. First of all, consideration is given to the concept of payoff, which is often referred to as the *value of utility*.

16-1 Concept of Utility

In everyday usage the word *utility* is interpreted as the quality or property of being useful. For example, we might say that there is a certain utility in having a remote-control tuner for changing channels on a TV set. We might also say that the utility of having a three-car garage cannot be underestimated. In this chapter, however, we shall interpret utility in a much more restricted manner. In particular,

> *The word utility will be used to denote a numerical measure of the relative worth to a decision maker of the various consequences that may result from a set of actions, or alternatives, available to him.*

If a decision—the words *decision* and *action* will be used interchangeably—results in one of a set of possible distinct consequences, and each consequence happens with a probability assumed to be known to the decision maker, then we shall say that the decision maker is operating in the realm under *risk*. Suppose, for example, that an ideal coin is tossed. If it turns up heads, you receive a reward of $1; if it turns up tails, you lose $1. This game, of course, has an uncertain outcome, or risk. Whether you would play this game depends on, among other things, your utility of money. In this chapter, the idea of utility is applied as we discuss decision making under risk, or *uncertainty*.

Moreover, the measure of utility is stated in terms of preference, in the subjective sense, of the decision maker. Suppose that a decision maker faces two alternatives, a_1 and a_2. He will prefer a_1 to a_2 if and only if a_1 would lead to greater utility, according to his subjective evaluation, irrespective of how great the difference between the utilities of a_1 and a_2 may be. Let $u(a_1)$ and $u(a_2)$ designate the values of utility for a_1 and a_2, respectively; then

$$u(a_1) > u(a_2)$$

means that a_1 is *preferred to*, or is *better than*, a_2. Likewise, the relation

$$u(a_1) \geq u(a_2)$$

means that a_1 is *not worse than* a_2 since the utility of a_1 is at least as great as that of a_2. In short the primary interest of our discussion centers in the so-called *ordinal* utilities, or "preferences."

An action may result in any of various possible consequences. The utility of such an action is the weighted average of the utilities of the respective consequences with the corresponding probabilities as weights. Let c_i be the ith consequence of an action a and $u(c_i)$ be the utility of c_i, which occurs with a probability p_i for $i = 1, 2, \ldots, n$; then the utility of a, or $u(a)$, is obtained as follows:

$$u(a) = p_1 u(c_1) + p_2 u(c_2) + \cdots + p_n u(c_n) \tag{16-1}$$

where the p's sum to unity. Under the circumstances, $u(a)$ is the *expected utility*, or expected payoff, for taking action a.

Suppose, for instance, that an investment company is engaged in a certain venture, which may result in one of three possible consequences: c_1, tremendous success; c_2, median degree of success; and c_3, a failure. If c_1, c_2, and c_3 will occur with the probabilities of 0.1, 0.2, and 0.7, and will yield payoffs, or utilities, of 20, 10, and -2, respectively, the expected utility from the venture, or action a will be

$$u(a) = 0.1(20) + 0.2(10) + 0.7(-2) = 2 + 2 - 1.4 = 2.6$$

The symbol u in the notations $u(a)$ is used to designate utility functions since to each action a there corresponds one and only one value as its utility. Before we examine the concept of a utility function, let us first briefly consider the assumptions underlying this discussion.

16-1-1 Assumptions

There are two basic postulates from which the concept of utility is developed. They are: *(1)* The set of all actions, and their consequences of a decision situation can be identified. *(2)* The preference relationship can be established among all the possible actions and consequences; that is, all the available actions and consequences in a particular decision situation are amenable to *ordering* according to their desirabilities, or to the *ranking* of their utility values, with ties not excluded. The following are some of the important assumptions relative to the concept of utility:

First, subjective preferences are assumed to be *transitive*. The transitivity of preferences can be best explained by an illustration. If an individual is faced with three alternatives, a_1, a_2, and a_3, and if a_1 is preferred to a_2, and a_2, in turn, is preferred to a_3, then a_1 is preferred to a_3. That is, if $u(a_1) > u(a_2)$, and $u(a_2) > u(a_3)$, then $u(a_1) > u(a_3)$.

Second, it is assumed that if the alternative a_1 is preferred to a_2, and a_3 is another alternative, then the combination of a_1 and a_3 (which is the prospect of facing a_1 with the probability p and a_3 with the probability $1 - p$) will be preferred to the combination of a_2 and a_3, regardless of the probability distribution between the two alternatives as long as the distribution remains the same for both combinations. Suppose, for example, that $u(a_1) = 5$, $u(a_2) = 2$, and $u(a_3) = 4$, and that the two alternatives of each combination are equally probable. Then, we have the utility of the combination of a_1 and $a_3 = \frac{1}{2}(5) + \frac{1}{2}(4) = 4.5$, which is greater than the utility of the combination of a_2 and $a_3 = \frac{1}{2}(2) + \frac{1}{2}(4) = 3$. You are advised to try different sets of probabilities to verify this assumption.

Finally, we assume that the preference over a combination depends on the probability associated with each of the alternatives included in the combination. For instance, suppose the three alternatives a_1, a_2, and a_3 have the transitive relation $u(a_1) > u(a_2) > u(a_3)$, that is, a_1 is preferred to a_2, which is preferred to a_3. Suppose further that $u(a_1) = 6$, $u(a_2) = 3$, and $u(a_3) = 2$. Then a_2 is preferred to the combination of a_1 and a_3 if the probabilities for a_1 and a_3 are $\frac{1}{5}$ and $\frac{4}{5}$, respectively, since

$$\frac{1}{5}u(a_1) + \frac{4}{5}u(a_3) = \frac{1}{5}(6) + \frac{4}{5}(2) = 2.8$$

which is smaller than 3, the utility of a_2. On the other hand, if the probabilities for a_1 and a_3 are $\frac{3}{5}$ and $\frac{2}{5}$, respectively, the utility of the combination of a_1 and a_3 will be

$$\frac{3}{5}u(a_1) + \frac{2}{5}u(a_3) = \frac{3}{5}(6) + \frac{2}{5}(2) = \frac{22}{5} = 4.4$$

which is greater than the utility of a_2, namely, 3. Accordingly, the combination of a_1 and a_3 is preferred to a_2.

A combination may consist of three or more alternatives; the utility of such a combination is then the weighted average of the utilities of the various alternatives with the corresponding probabilities as weights. That is,

$$u(A) = p_1 u(a_1) + p_2 u(a_2) + \cdots + p_n u(a_n) \tag{16-2}$$

where A is the combination of n alternatives $a_1, a_2, \ldots, a_n$, and $p_1 + p_2 + \cdots + p_n = 1$.

For instance, if the investment company cited above has the privilege of choosing either one of two available ventures a_1 and a_2, and if a_1 and a_2 are equally likely to be chosen and $u(a_1) = 2.6$ and $u(a_2) = 3.4$, then the expected utility of the combination A of these two alternatives will be

$$u(A) = 0.5(2.6) + 0.5(3.4) = 1.3 + 1.7 = 3.0$$

16-1-2 Utility functions

As stated, to each alternative a there corresponds a value $u(a)$ referred to as the utility of a. Thus a *utility function* is defined over the set of distinct actions, or consequences of actions, available to the decision maker. Since utility is a subjective measure of preference of any individual, not only different individuals may have different utility functions but also the same individual may have different utility functions under changed conditions or circumstances.

Let us use the utility of money to illustrate the above argument. It may be assumed that the utility of money is *monotonically increasing;* that is, regardless of how rich a person may be, additional money will always provide for him added utility.

Linear utility function Let us first consider a highly hypothetical case. Suppose that the *marginal utility* of money to an individual is a *constant;* that is, the utility of each additional unit of money is a constant irrespective of the amount of money he possesses; his utility function is therefore a straight line as shown in Fig. 16-1. It is assumed that the total amount of money he has, denoted by x_0, is \$10,000, the utility of which is 1.0, or $u(x_0) = 1.0$, as indicated on the vertical scale of the figure. Imagine, for a moment, that he is offered a *fair game*, which means that the *mathematical expectation of gain* from this game is 0. The game is set up so that he will win or lose \$5,000 with an identical probability. The expected gain of the game therefore is

$$\frac{1}{2}(\$5{,}000) + \frac{1}{2}(-\$5{,}000) = 0$$

If he loses, he will have only \$5,000, or x_1; if he wins, he will have \$15,000, or x_2. Since the weighted average of his possible monetary fortunes remains \$10,000, the *utility of the weighted average* of his monetary fortunes is still 1.[1] That is,

$$u\left[\frac{1}{2}(\$5{,}000) + \frac{1}{2}(\$15{,}000)\right] = u(\$10{,}000) = 1.0$$

which is equal to $u(x_0)$. The two alternatives, a_1 and a_2, available to him are "play the game" and "not play the game." Since $u(a_1) = u(a_2)$, he is *indifferent* to the game.

His indifference can be further demonstrated by the *weighted average of the utilities* corresponding to his possible monetary positions resulting from the game. Let us again assume that the probabilities of winning and losing are

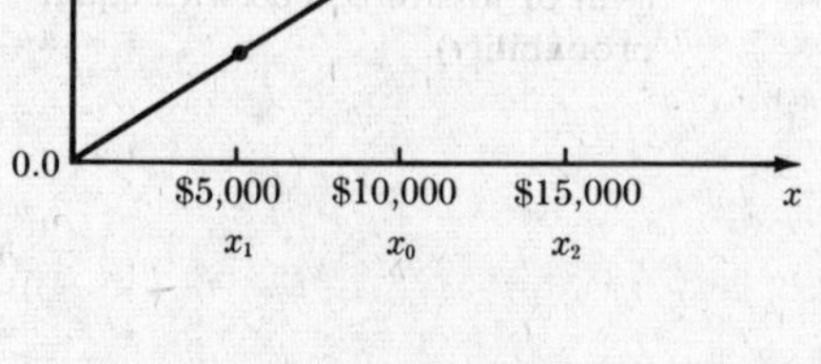

Figure 16-1
Constant utility per unit of money.

[1] Note that the horizontal scale of Fig. 16-1 (and also of Figs. 16-2 and 16-3) measures not only the *actual* amount of money possessed but also the *possible* amount of money possessed as a consequence of playing the game.

equal; his utility after winning, or $u(x_2)$, is 1.5, and that after losing, or $u(x_1)$, is 0.5. The weighted average of these utilities is

$$\frac{1}{2}u(\$15{,}000) + \frac{1}{2}u(\$5{,}000) = \frac{1}{2}(1.5 + 0.5) = 1.0$$

In short, for an indifferent man, or a man whose utility per unit of money is a constant, the utility of the weighted average of his monetary rewards is identical with the weighted average of the utilities of the same rewards.

The assumption that the utility function of money for an individual is a straight line is quite unrealistic since the utility of a dollar to any individual when he is, for example, trying to eke out a living by raising chickens on a 2-acre farm is evidently greater than that of a dollar to him when, for example, a substantial amount of oil is found in his land and his daily royalty amounts to thousands of dollars.

Concave-downward utility function As pointed out above, with a linear utility function the marginal utility of money is a constant. If the utility function of money is *curvilinear*, however, the marginal utility of money changes as the total amount of money possessed increases or decreases. Let x designate any amount of money that falls within the domain of the function. Then the function is called *concave downward* if the marginal utility of money at x is *decreasing*. Such a utility curve is shown in Fig. 16-2.

The downward concavity of a utility function can also be expressed in terms of the utility of the weighted average and the weighted average of utilities. Let x_0 stand for the amount of money a man currently possesses. He is asked to play a game. With a probability of p, he will win the game and his total monetary fortune will increase to x_2; the probability of his losing the game is

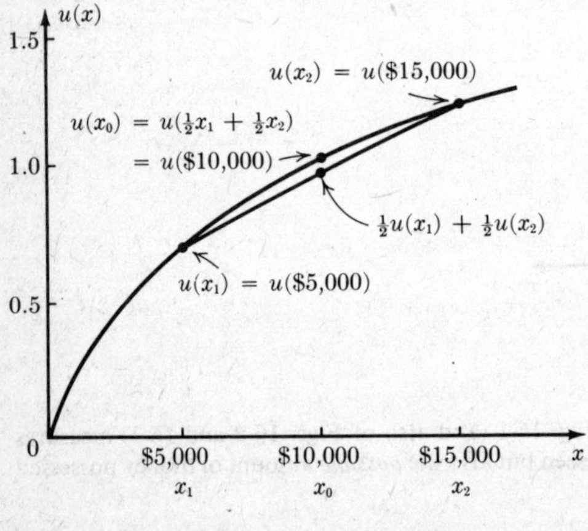

Figure 16-2
A concave downward utility function of money, showing reduced utility from a fair game (win or loss of $5,000 with equal probability).

$1 - p$, and then his monetary fortune will decrease to x_1. Suppose that this game is a *fair game*, which means that the mathematical expectation of gain from the game leaves him in the same monetary position as before, that is, the weighted average of x_1 and x_2 is x_0:

$$(1 - p)x_1 + px_2 = x_0$$

Our principal question now is: Will this person whose utility function of money is concave downward play this game?

To answer this question, we need only to find out whether playing the game would increase his utility. We shall now demonstrate that since his utility function of money is concave downward, he would refuse to play the game and would keep his money with certainty. Let $x_0 = \$10{,}000$, $x_1 = \$5{,}000$, and $x_2 = \$15{,}000$ and $p = 1 - p = \frac{1}{2}$. This is obviously a fair game since the mathematical expectation of gain is 0. That is,

$$\frac{1}{2}\ \$5{,}000 + \frac{1}{2}(-\ \$5{,}000) = 0$$

Without playing the game, his \$10,000 gives him a utility of 1, as shown in Fig. 16-2. The utility of the weighted average of \$5,000 and \$15,000, which amounts to \$10,000, is still 1. In general, for a fair game,

$$u[(1 - p)x_1 + px_2] = u(x_0) \tag{16-3}$$

On the other hand, the weighted average, or expected value, of the utilities of x_1 and x_2 is smaller than 1 in this case. From Fig. 16-2, we read that $u(x_1) = u(\$5{,}000) = 0.7$, and $u(x_2) = u(\$15{,}000) = 1.2$. The weighted average of these utilities is

$$\frac{1}{2}\,u(\$5{,}000) + \frac{1}{2}\,u(\$15{,}000) = \frac{1}{2}(0.7 + 1.2) = 0.95$$

which is smaller than 1. This explains why the man with a concave-downward utility function of money would not play a fair game; the utility of playing the game a_1 is smaller than the utility of keeping his money with assurance a_0. The utility of playing the game is generally expressed as

$$u(a_1) = (1 - p)u(x_1) + pu(x_2) \tag{16-4}$$

In short, if a person has a concave-downward utility function of money, he will not play a fair game because the utility of playing the game is smaller than the utility of keeping his money intact; that is,

$$(1 - p)u(x_1) + pu(x_2) < u[(1 - p)x_1 + px_2] \tag{16-5}$$

or

$$u(a_1) < u(a_0)$$

Such a person is often referred to as a *risk averter*. As shown above, a risk averter will not take a fair bet. This does not mean that he would not play a game if the "odds" were overwhelmingly in his favor. If a game were sufficiently *unfair* to his opponent so that the risk averter's utility of playing the game would be greater than the utility of keeping his money intact, he may risk his money to play the game.

Concave-upward utility function A concave-upward utility function is exactly the opposite of a downward concavity. If a person's utility function of money is concave upward, his marginal utility of money *increases* as his total amount of money increases. A concave-upward utility function is shown in Fig. 16-3.

The upward concavity of a utility function can also be expressed in terms of the utility of the weighted average and the weighted average of utilities. Suppose that a person possesses \$8,000, that is, $x_0 = \$8{,}000$. He is to choose between two alternatives: either he refuses to play the game, or he plays the game with the probability of $\frac{2}{3}$ to lose \$2,000 and $\frac{1}{3}$ to win \$4,000. This is again a fair game since the mathematical expectation of gain from this game leaves him in the same monetary position as before. That is, the weighted average of x_1, or \$6,000, and x_2, or \$12,000, is still x_0, or \$8,000:

$$\frac{2}{3}x_1 + \frac{1}{3}x_2 = \frac{2}{3}\,\$6{,}000 + \frac{1}{3}\,\$12{,}000 = \$8{,}000$$

As indicated in Fig. 16-3, the utility of keeping his \$8,000 intact is 0.3, which is also the utility of the weighted average, or expected value, of his possible monetary possessions.

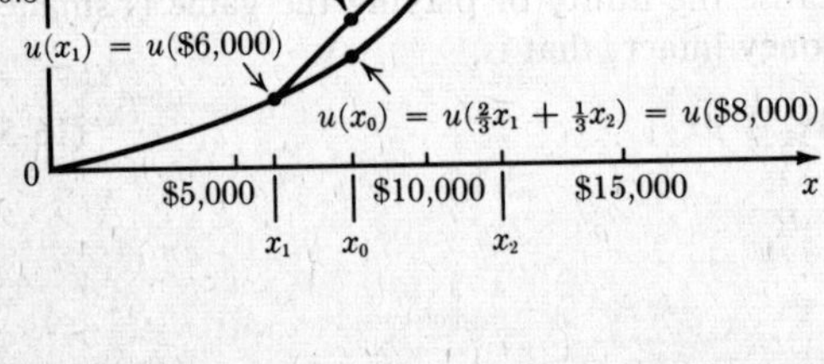

Figure 16-3
A concave-upward utility function of money, showing increased utility from a fair game (win of \$4,000 and loss of \$2,000 with probabilities $\frac{1}{3}$ and $\frac{2}{3}$, respectively).

On the other hand, the weighted average, or expected value, of the utilities of x_1 and x_2 is greater than 0.3. From Fig. 16-3, we read that $u(x_1) = u(\$6{,}000) = 0.2$ and $u(x_2) = u(\$12{,}000) = 0.8$. The weighted average of these utilities is

$$\frac{2}{3}(0.2) + \frac{1}{3}(0.8) = 0.4$$

which is higher than 0.3, the utility of keeping his \$8,000 with certainty. This explains why a person with a concave-upward utility function of money would prefer a fair fame; the utility of playing the game is greater than the utility of keeping his money intact. That is,

$$(1 - p)u(x_1) + pu(x_2) > u[(1 - p)x_1 + px_2] \tag{16-6}$$

or

$$u(a_1) > u(a_0)$$

Such a person is often referred to as a *risk lover*. As already pointed out, a risk lover would take a fair bet. This does not mean, however, that he would play any game if the odds were overwhelmingly against him. If the game is sufficiently unfair to him, so that the utility of playing is smaller than the utility of keeping his money with assurance, he would not risk his money to play.

In short the test for the decision to play or not to play a game is the subjective utility to the decision maker.

It is not necessarily true, however, that a person must be either a risk lover or a risk averter throughout his life, excluding the man with a linear utility function. A person can be both a risk lover and a risk averter depending on the amount of his monetary holding. For instance, a person's utility function may be similar to the curve shown in Fig. 16-4. When the money holding is small, he is a risk lover, willing to take a fair bet since the utility of playing a fair game is above the utility of keeping his money intact. On the other hand, when his money holding is large, he would not prefer a fair game because the utility of playing the game is smaller than keeping his money with certainty.

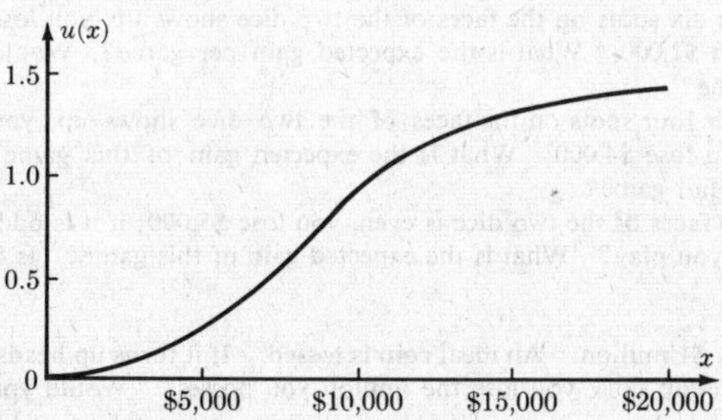

Figure 16-4
The utility function of money, showing that a fair game is preferred when money holding is small, not preferred when money holding is large.

Exercise 16-1

1 Use your own words to define the following terms as fully as you can:

a Utility.
b Utility function.
c Transitivity of preferences.
d Expected utility.
e Mathematical expectation of gain.
f Fair game.
g Monotonically increasing.
h Linear utility function.

2 What are the axioms on the basis of which the concept of utility is developed?

3 Suppose that your utility function is the same as the line in Fig. 16-1 and that you are offered a game that involves the roll of an ideal die. If a single spot of the die shows up, you collect $12; otherwise, you pay $3. Would you play the game? If you are obligated to play the game, what is the maximum amount of money you are willing to pay to avoid the game?

4 In the preceding problem, if the die shows four or five spots, you collect $10; otherwise, you pay $3.50. Would you play the game? If you are entitled to play the game, what is the minimum amount of money you would be willing to accept in order to give up the game?

5 Distinguish between the following concepts:

a Concave-downward and concave-upward utility functions.
b Risk averter and risk lover.
c Utility of weighted average and weighted average of utilities.

6 Suppose that your utility function is the same as the curve in Fig. 16-2, and that you possess $10,000. Suppose further that a game, which involves the toss of three idea coins, is offered to you:

a If the coins all turn up heads, you lose $5,000; otherwise, you win $5,000. What is your expected gain per game? Is this a fair game? Would you play?
b If two heads turn up, you lose $5,000; otherwise, you win $5,000. What is the expected gain per game? Would you play? Is it a fair game?

7 Suppose that your utility function is the same as the curve in Fig. 16-3, and that you possess $10,000. Suppose further that a game, which involves the roll of two perfect dice, is offered to you:

a If the sum of four, five, or six spots on the faces of the two dice shows up, you lose $4,000; otherwise, you win $2,000. What is the expected gain per game? Would you play? Is it a fair game?
b If the sum of two, three, or four spots on the faces of the two dice shows up, you win $2,000; otherwise, you lose $4,000. What is the expected gain of this game? Would you play? Is it a fair game?
c If the sum of spots on the faces of the two dice is even, you lose $5,000; if it is odd, you win $3,000. Would you play? What is the expected gain of this game? Is it a fair game?

8 Suppose that you have exactly $1 million. An ideal coin is tossed. If it turns up heads, you win $5 million; if it turns up tails, you lose the million you possess. Would you play? Why or why not?

9 Two actions are available to a decision maker. Each action may result in any of three possible consequences. The table below presents the payoffs of the various consequences and their corresponding probabilities for each of the two actions.

Action a_1			*Action a_2*		
Consequence	*Probability*	*Payoff*	*Consequence*	*Probability*	*Payoff*
c_1	$\frac{1}{6}$	2	c'_1	$\frac{1}{3}$	3
c_2	$\frac{2}{6}$	2	c'_2	$\frac{1}{3}$	2
c_3	$\frac{3}{6}$	4	c'_3	$\frac{1}{3}$	4

a Which action will the decision maker take and why?

b Suppose that the decision maker cannot make up his mind on which action to take. An ideal coin is tossed. If it turns up heads, he takes action a_1; if it turns up tails, he takes action a_2. What will be his expected payoff in this decision situation?

10 The utility function of money for Mr. Smith is given by $u(x) = x^2 - 2x + 3$ for $1 \leq x \leq 4$, where x is the amount of money in thousands of dollars. Mr. Smith now has $2,000. Will he take a fair bet?

11 Suppose that Mr. Smith of Prob. 10 has an opportunity to invest his money in some common stocks. The outcome of his investment venture would leave him in either of the following financial positions: $1,000, $2,000, $3,000, or $4,000 with probabilities 0.5, 0.3, 0.1, and 0.1, respectively. Should he invest?

12 Give an example (not used in the text) to illustrate decision making under uncertainty.

16-2
Some Decision Criteria

In this section we shall briefly consider the *maximax* and *maximin* criteria of decision making. Following these, we shall pay greater attention to the *minimax* criterion since this is often discussed in contrast with the Bayesian decision rule. Finally, the Bayesian rule for problems that are to be solved without data gathering will be reviewed.

16-2-1 Maximax and maximin criteria

An aggressive and optimistic individual or institution may take the approach that it should endeavor to obtain the maximum payoff possible with every decision, and thus choose the decision that would yield the maximum payoff out of the maximums of payoffs respectively of the various alternatives. This approach is generally referred to as the *maximax rule.* An example will make the notion clearer. Suppose that there are five possible levels of demand, called the *states of nature,* θ_1, θ_2, θ_3, θ_4, and θ_5, for three different models of the same product designated a_1, a_2, and a_3, the supply of each of which represents an action, or decision, for the company. Let us assume that the total *net* profit of an action corresponding to a state of nature is shown in each of the cells of Table 16-1.

Table 16-1
Conditional payoff

States of nature	*Actions*		
	Model a_1	*Model a_2*	*Model a_3*
θ_1	1	3	10
θ_2	3	4	8
θ_3	5	6	6
θ_4	7	4	4
θ_5	9	5	2

Since the maximum profit is 9 for a_1, 6 for a_2, and 10 for a_3, and the maximum of these three maximum payoffs is 10, the decision maker will take action a_3 according to this criterion. More generally, if there are K alternative actions and max(a) stands for the maximum payoff for each action, then we have

$$\text{Maximax payoff} = \max[\max(a_1), \max(a_2), \ldots, \max(a_K)] \qquad (16\text{-}7)$$

Another decision rule is called the *maximin criterion*. This is a more pessimistic approach since the decision involves, first of all, considering all the minimum possible payoffs, among which it selects the action that will lead to the maximum among the minimums. According to this approach, the decision maker would choose action a_2 in the above example since the minimum payoff 3 is the maximum among all the three minimums, namely, 1, 3, and 2. This criterion is usually used when a decision maker is confronted with a rational and well-informed opponent in a game or in a competitive situation. Symbolically, if min (a) is the minimum payoff of each of the K alternative actions, then we have

$$\text{Maximin payoff} = \max[\min(a_1), \min(a_2), \ldots, \min(a_K)] \qquad (16\text{-}8)$$

If the payoff function is expressed in terms of losses rather than gains or profit, the same basis for selecting an alternative according to the maximin criterion is then called the *minimax* criterion. In game playing or competition, applying of these two criteria would assure the decision maker of minimizing the maximum gain that the opposition can possibly attain.

16-2-2 Minimax criterion

Application of the minimax criterion requires a payoff table, or utility function, which is a table of losses such as the costs of alternative actions to be taken or damages to be suffered corresponding to the various decisions and states of nature. Assume that for a certain business firm there are three states of nature, θ_1, θ_2, and θ_3 designating economic *prosperity*, *stability*, and *recession*, and that there are three alternative actions, a_1, a_2, and a_3 standing respectively for *increased*, *usual*, and *reduced* rate of investment open to the decision maker. Assume also that he has a loss function $l(\theta, a)$ as shown in Table 16-2.

If available information is limited to what appears in Table 16-2, what action should the decision maker take? Under the minimax criterion, it is obvious that he will take action a_2, that is, he will choose a usual rate of investment. The maximum losses for the three actions a_1, a_2, and a_3 are 7, 4, and 5, respectively, and 4 is the minimum of the three maximums. Thus we say that according to the principle of minimax loss, the decision maker should select the action such that the worst possible loss he can suffer is the least.

In general, if max(a) stands for the maximum loss by taking action a, and there are altogether K alternative actions, then the minimax loss is given by

$$\text{Minimax loss} = \min[\max(a_1), \max(a_2), \ldots, \max(a_K)] \tag{16-9}$$

The idea of loss actually consists of two different concepts, namely, *avoidable* loss and *unavoidable* loss. A loss is termed unavoidable if we have to suffer it in spite of our perfect knowledge of the states of nature. In Table 16-2, for instance, the unavoidable loss for each of the three actions is 1 even if we knew what the state of nature is. An avoidable loss is suffered because of the ignorance of the states of nature. We have an opportunity to avoid such a loss if we have a crystal ball to inform us about the state of nature. It is often referred to as *regret* to signify the "unhappiness" for having guessed θ incorrectly. It is also called *opportunity loss* since it is the difference between the loss actually suffered and the smaller loss that there is an opportunity to realize.

Incidentally, the concept of opportunity loss is quite broad, and is not only related to cost or disutility but also to profit or utility. This is to say, for example, that when the *profit having actually resulted* from a decision is smaller than the profit that *could have been realized* if the decision had been the best one possible, the *difference* between them is called the opportunity loss of the decision. It is the profit that could have been obtained but was lost due to an erroneous decision. In the discussion to follow, we shall use the term *regret* rather than *opportunity loss* for convenience.

Table 16-2
Loss $l(\theta, a)$

	Actions		
States of nature	Increased rate a_1	Usual rate a_2	Reduced rate a_3
θ_1, *prosperity*	1	3	5
θ_2, *stability*	3	1	3
θ_3, *recession*	7	4	1

Table 16-3
Regret $r(\theta, a)$

States of nature	Actions a_1	a_2	a_3
θ_1	0	2	4
θ_2	2	0	2
θ_3	6	3	0

If we subtract the lowest value in each row from each value of the same row in Table 16-2, we obtain the regret function $r(\theta, a)$ as shown in Table 16-3. Accordingly, if θ_1, θ_2, or θ_3 is the state of nature, we shall take action a_1, a_2, or a_3, respectively. The regret would be 0, since the decision is the best one possible.

We can still apply the minimax criterion as a basis of choice when a regret table is used. Instead of minimax loss, we now have *minimax regret* as the rule of decision. Table 16-3 shows that the maximum regrets for a_1, a_2, and a_3 are 6, 3, and 4, respectively. Thus the minimax regret is 3, and a_2 should be the action taken according to the minimax regret criterion. In this example the same decision is reached according to the minimax loss or the minimax regret criterion, but this is not always the case.

16-2-3 Expected loss or regret

In the above discussion it has been assumed that the decision maker knows nothing about the probability distribution over the states of nature. As we shall see shortly, if the probability for each of the states of nature is known, the course of action might be completely different.

Expected loss Suppose that the decision maker believes that the probabilities of prosperity, stability, and recession are $p_1 = 0.7, p_2 = 0.2$, and $p_3 = 0.1$, respectively; then the expected loss, or $E[l(\theta, a)]$, for each action can be computed as in Table 16-4, where the values for $l(\theta, a)$ are the same as in Table 16-2. The expected loss for action a_1, or $L(a_1)$, is the sum of the products of the losses

Table 16-4
Computation of expected loss $E(l(\theta, a))$

Probability p	States of nature	Actions a_1	a_2	a_3
$p_1 = 0.70$	θ_1	1	3	5
$p_2 = 0.20$	θ_2	3	1	3
$p_3 = 0.10$	θ_3	7	4	1
Expected loss $L(a)$		2.0	2.7	4.2

under a_1 and the corresponding probabilities for the respective states of nature; that is,

$$E[l(\theta, a_1)] = L(a_1) = 1(0.70) + 3(0.20) + 7(0.10) = 2.0$$

Similarly, the expected losses for actions a_2 and a_3 are

$$E[l(\theta, a_2)] = L(a_2) = 3(0.70) + 1(0.20) + 4(0.10) = 2.7$$

and

$$E[l(\theta, a_3)] = L(a_3) = 5(0.70) + 3(0.20) + 1(0.10) = 4.2$$

Since action a_1 yields an expected loss of only 2.0, the lowest of all, a_1 now becomes the best action for the decision maker. Symbolically, if there are K alternative actions, then we have

$$\text{Minimum expected loss} = \min[L(a_1), L(a_2), \ldots, L(a_K)] \tag{16-10}$$

The action that is associated with this loss is the best action.

Expected regret The same decision can be reached in terms of expected regret. Applying the same set of probabilities as in Table 16-4, we compute the expected regrets, or $E[r(\theta, a)]$, for the three actions as shown in Table 16-5, where the values for $r(\theta, a)$ are the same as in Table 16-3. The procedure to obtain the expected regrets is similar to the method of computing the expected losses. The expected regret is often called *risk* denoted by R; as shown in the bottom row of Table 16-5, $R(a)$ designates the risk of action a.

The best action to choose is obviously a_1 since the expected regret is the lowest for a_1.

It would hardly be necessary to point out that the *difference* between the expected loss and the expected regret for any given action is the *unavoidable loss*, which is the value subtracted from the losses for each state of nature. Thus, instead of going through the procedure of obtaining the sum of products of the regrets and the corresponding probabilities for each action, we can simply obtain the expected regret by subtracting the unavoidable loss from the expected loss for each action.

For most decision problems, the unavoidable losses are irrelevant to arriving at the best decision. For this reason, it is customary for decision makers to

Table 16-5

Computation of expected regret $E(r(\theta, a))$

Probability	*States*	*Actions*		
p	*of nature*	a_1	a_2	a_3
$p_1 = 0.70$	θ_1	0	2	4
$p_2 = 0.20$	θ_2	2	0	2
$p_3 = 0.10$	θ_3	6	3	0
Expected regret $R(a)$		1.0	1.7	3.2

compute expected regrets rather than expected losses. It is a matter of computational convenience since a regret table has at least one 0 in each row and thus calculations are simplified.

The above discussion is known as decision making *without information* since *no more* data than the loss or regret table and the probability distribution over the states of nature are provided to assist in reaching a decision. Some statisticians call this type of problem the "no-data problem." We shall now give such problems a more formal and general treatment before we embark on the discussion of the Bayesian decision rule for problems *with information*.

Suppose that there are J states of nature, $\theta_1, \theta_2, \ldots, \theta_J$, and K actions, $a_1, a_2, \ldots, a_K$, in a certain decision problem. Let r stand for regret, and $r(\theta_j, a_k)$ be the regret corresponding to the state of nature θ_j and action a_k for $j = 1, 2, \ldots, J$ and $k = 1, 2, \ldots, K$. Associated with each state of nature θ_j, there is its probability p_j, and the set of probabilities $p_1, p_2, \ldots, p_J$ is called the set of *prior*, or a priori, probabilities, which sum to unity. Thus the risks are

$$R(a_1) = \text{expected regret for } a_1 = \sum_{j=1}^{J} p_j r(\theta_j, a_1)$$

$$R(a_2) = \text{expected regret for } a_2 = \sum_{j=1}^{J} p_j r(\theta_j, a_2) \qquad (16\text{-}11)$$

$$\cdots\cdots\cdots\cdots\cdots\cdots\cdots\cdots\cdots\cdots\cdots\cdots$$

$$R(a_K) = \text{expected regret for } a_K = \sum_{j=1}^{J} p_j r(\theta_j, a_K)$$

which are shown in the bottom row of Table 16-6. The decision rule for this type of problem is therefore to select the action a that yields

$$\text{Minimum risk} = \min[R(a_1), R(a_2), \ldots, R(a_K)] \qquad (16\text{-}12)$$

The minimum risk thus obtained is called the *Bayes risk*; action associated with

Table 16-6

Generalized regret for the no-data problem, $r(\theta, a)$

A priori probability	*States of nature*	*Actions* a_1	a_2	$\cdots$	a_K
p_1	θ_1	$r(\theta_1, a_1)$	$r(\theta_1, a_2)$	$\cdots$	$r(\theta_1, a_K)$
p_2	θ_2	$r(\theta_2, a_1)$	$r(\theta_2, a_2)$	$\cdots$	$r(\theta_2, a_K)$
$\cdots$	$\cdots$	$\cdots$	$\cdots$	$\cdots$	$\cdots$
p_J	θ_J	$r(\theta_J, a_1)$	$r(\theta_J, a_2)$	$\cdots$	$r(\theta_J, a_K)$
Expected regret $R(a)$		$\sum_{J=1}^{J} p_J r(\theta_J, a_1) = R(a_1)$	$\sum_{J=1}^{J} p_J r(\theta_J, a_2) = R(a_2)$	$\cdots$	$\sum_{J=1}^{J} p_J r(\theta_J, a_K) = R(a_K)$

the Bayes risk is called the *Bayes action*. In the above example (Table 16-5) the Bayes risk is 1, and the Bayes action is a_1. Since risk, or expected regret, signifies expected *disutility*, or expected negative utility, Bayes risk means minimum expected disutility, or maximum expected utility.

Observe that among the essential elements of the Bayesian decision rule is the set of probabilities for the states of nature. Such probabilities are known as *a priori*, or prior, probabilities in contrast with *a posteriori* probabilities discussed in Sec. 3-3. They are referred to as a priori because they are the probabilities assigned to the states of nature prior to any empirical trial or experimentation to be made about the states of nature. Thus they generally reflect hunches or the "intensity of belief" on the part of the decision maker, and hence are often called *personal or subjective* probabilities. As shall be seen in the next two sections, a priori probabilities are used also in connection with the Bayesian decision rule when the problem situation requires the collection of data or information.

Exercise 16-2

1 Define the following terms as fully as you can:

a State of nature. *c* Bayes risk.
b Opportunity loss. *d* Bayes action.

2 Distinguish between the following terms:

a Avoidable and unavoidable loss.
b Loss and regret.
c Maximax and maximin criterion.

3 Explain the difference and similarity between the maximin and minimax criteria.

4 A manufacturer makes three different models of dishwashers. If the economy is in a recession, the low-priced model will have a better market. If stability is achieved, the regular model will do well, while a booming economy will enable the company to reap greater gains with the deluxe model. Let a_1, a_2, and a_3 represent the decisions of producing the low-priced, regular, and deluxe models, and θ_1, θ_2, and θ_3 represent recession, stability, and boom, respectively. Suppose that the payoff table, or $u(\theta, a)$, for this manufacturer is as follows:

Payoff table, $u(\theta, a)$

	Actions		
States of nature	*Low-priced model a_1*	*Regular model a_2*	*Deluxe model a_3*
Recession θ_1	5	4	1
Stability θ_2	4	8	5
Boom θ_3	3	6	10

a What action should the manufacturer take according to the maximax criterion? What will be his payoff?
b What will be his action and payoff if he follows the maximin criterion?

5 Suppose that in the preceding problem, the probabilities of occurrence for the states of nature, θ_1, θ_2, and θ_3, are 0.2, 0.5, and 0.3, respectively.

a Determine the expected payoff for each of the models.

b Determine the Bayes action and Bayes payoff.

6 Suppose that in Prob. 4, the payoff table is expressed in losses as follows:

Payoff table, $l(\theta, a)$

States of nature	Actions a_1	a_2	a_3
θ_1	1	4	7
θ_2	3	2	5
θ_3	5	4	3

a What action should the manufacturer take according to the minimax loss criterion?

b Change the above loss table into a regret table, or $r(\theta, a)$.

c What action should the manufacturer take according to the minimax regret criterion?

d Explain why, in this example, the minimax loss and minimax regret criteria lead to different decisions.

7 Given that the probabilities of occurrence of the three states of nature in the preceding problem remain the same as in Prob. 5, do the following:

a Compute the expected loss and regret for each of the actions available to the decision maker.

b Determine the Bayes action and Bayes risk.

8 In a town called Duopolis the weather can be classified into only two groups, namely, sunny day and rainy day. Two actions are available to Mr. Jones, a resident of the town. He may wear a sunny-day or a rainy-day outfit. Suppose that his loss table, or $l(\theta, a)$ is as follows:

Payoff table, $l(\theta, a)$

States of nature	Actions: Sunny-day outfit a_1	Rainy-day outfit a_2
Sunny day θ_1	0	5
Rainy day θ_2	6	2

a What action should Mr. Jones take according to the minimax loss criterion? What is his loss?

b What action should he take according to the minimax regret criterion? What is his regret?

c Given that he can predict the weather with perfect accuracy, what action should he take for each state of nature?

9 Suppose that in the preceding problem the town of Duopolis has kept reliable weather records for a long period of time. According to the records, 70 percent of the days were sunny and 30 percent were rainy. Determine the Bayes action and Bayes risk for Mr. Jones.

16-3
Decision with Information

We shall now examine decision rules when information beyond the specification of a priori probabilities is available. Suppose, for example, that an economic research team has been employed by a company. For the sake of simplicity, we assume that the research team issues only two types of reports: an *optimistic forecast*, designated z_1, and a *pessimistic forecast*, designated z_2. According to company experiences, it is quite consistent that if the economy turns out to be prosperous, or θ_1, the probabilities for the research reports to be optimistic and pessimistic are 0.70 and 0.30, respectively; that is, $P(z_1 | \theta_1) = 0.70$ and $P(z_2 | \theta_1) = 0.30$. If θ_2 is the state of nature, optimistic and pessimistic reports are equiprobable, or $P(z_1 | \theta_2) = 0.50$ and $P(z_2 | \theta_2) = 0.50$. On the other hand, if the state of nature is θ_3, then $P(z_1 | \theta_3) = 0.40$ and $P(z_2 | \theta_3) = 0.60$. These probabilities are therefore *conditional* probabilities, generally designated $P(z | \theta)$ or $f(z | \theta)$ since they are dependent on the state of nature. Table 16-7 shows the conditional probabilities of information.[1]

Equipped with the payoff, or regret, function as in Table 16-3 and the conditional probabilities shown in Table 16-7, we are now ready to formulate the Bayesian rules of decision making *with information*. A rule of decision here is called a *strategy*, and it advises the decision maker what action to take for

[1] Suppose that a decision situation has two states of nature θ_1 and θ_2, two alternative actions, a_1 and a_2, and two observations z_1 and z_2. If θ_1 is the state of nature, a_1 will be the correct action; if θ_2 is the state of nature, a_2 will be the correct action. The decision maker's policy is to take action a_1 if z_1 is observed, and action a_2 if z_2 is observed. Then the conditional probabilities of z, given θ, or $P(z|\theta)$, become *action* probabilities; that is, they become the probabilities of taking the respective actions, given a state of nature. It should be pointed out that α, the level of significance, and β discussed in Chap. 10 are actually action probabilities as indicated below:

	Observations	
States of nature	z_1 (a_1)	z_2 (a_2)
θ_1	$P(z_1 \| \theta_1) = 1 - \alpha$	$P(z_2 \| \theta_1) = \alpha$
θ_2	$P(z_1 \| \theta_2) = \beta$	$P(z_2 \| \theta_2) = 1 - \beta$

Table 16-7

Conditional probabilities of information, $P(z|\theta)$ or $f(z|\theta)$

States of nature	Information: Optimistic forecast z_1	Information: Pessimistic forecast z_2	Total
θ_1	0.70	0.30	1.0
θ_2	0.50	0.50	1.0
θ_3	0.40	0.60	1.0

each type of information. One possible strategy would be, for example, to take action a_1, the increased rate of investment, if the forecast is z_1, and to take action a_2, the usual rate of investment, if the forecast is z_2. Let s stand for the strategy just described; we then have $s = (a_1, a_2)$, where a_1 corresponds to z_1 and a_2 corresponds to z_2. Since three possible actions, a_1, a_2, and a_3, are available to be taken for each kind of information, and there are two different kinds of information, z_1 and z_2, the total number of possible strategies is therefore $3^2 = 9$. These nine possible strategies are enumerated in Table 16-8.

Some of these nine strategies are obviously unreasonable. For example, strategy s_7 seems to be ridiculous; in response to a forecast of prosperous economy, or z_1, the company reduces its rate of investment, or a_3, while the rate of investment is increased, or a_1, when a recession, or z_2, is predicted. Since some of the strategies are unreasonable, the primary task of the decision maker is to distinguish the "good" from the "bad" strategies, and then choose the best strategy possible to guide his decision. It is therefore imperative to provide a criterion on which the best possible strategy can be identified.

For this purpose, we proceed to evaluate all the nine strategies. This is done by first computing the expected regret $E[r(\theta, s)]$ corresponding to each state of nature, θ, and each strategy s, and then by obtaining an *average* of the expected regret for each strategy over all the states of nature.

Table 16-9 provides an illustration of the procedure of evaluating all the strategies enumerated in Table 16-8. The regret tables are reproduced from

Table 16-8

Possible strategies showing action in response to information

Information	Strategies: s_1	s_2	s_3	s_4	s_5	s_6	s_7	s_8	s_9
z_1	a_1	a_1	a_1	a_2	a_2	a_2	a_3	a_3	a_3
z_2	a_1	a_2	a_3	a_1	a_2	a_3	a_1	a_2	a_3

Table 16-9

Evaluation of strategies

θ	$r(\theta, a)$			*Conditional probabilities of information* z	*Strategy* s	*Expected regret = risk* $R(\theta, s)$
	a_1	a_2	a_3			
θ_1	0	2	4	$z_1 = 0.7$ $z_2 = 0.3$	$s_1\begin{cases} a_1 \\ a_1 \end{cases}$	$0.7(0) + 0.3(0) = 0.0$
θ_2	2	0	2	$z_1 = 0.5$ $z_2 = 0.5$	$s_1\begin{cases} a_1 \\ a_1 \end{cases}$	$0.5(2) + 0.5(2) = 2$
θ_3	6	3	0	$z_1 = 0.4$ $z_2 = 0.6$	$s_1\begin{cases} a_1 \\ a_1 \end{cases}$	$0.4(6) + 0.6(6) = 6$
θ_1	0	2	4	$z_1 = 0.7$ $z_2 = 0.3$	$s_2\begin{cases} a_1 \\ a_2 \end{cases}$	$0.7(0) + 0.3(2) = 0.6$
θ_2	2	0	2	$z_1 = 0.5$ $z_2 = 0.5$	$s_2\begin{cases} a_1 \\ a_2 \end{cases}$	$0.5(2) + 0.5(0) = 1.0$
θ_3	6	3	0	$z_1 = 0.4$ $z_2 = 0.6$	$s_2\begin{cases} a_1 \\ a_2 \end{cases}$	$0.4(6) + 0.6(3) = 4.2$
θ_1	0	2	4	$z_1 = 0.7$ $z_2 = 0.3$	$s_3\begin{cases} a_1 \\ a_3 \end{cases}$	$0.7(0) + 0.3(4) = 1.2$
θ_2	2	0	2	$z_1 = 0.5$ $z_2 = 0.5$	$s_3\begin{cases} a_1 \\ a_3 \end{cases}$	$0.5(2) + 0.5(2) = 2.0$
θ_3	6	3	0	$z_1 = 0.4$ $z_2 = 0.6$	$s_3\begin{cases} a_1 \\ a_3 \end{cases}$	$0.4(6) + 0.6(0) = 2.4$
θ_1	0	2	4	$z_1 = 0.7$ $z_2 = 0.3$	$s_4\begin{cases} a_2 \\ a_1 \end{cases}$	$0.7(2) + 0.3(0) = 1.4$
θ_2	2	0	2	$z_1 = 0.5$ $z_2 = 0.5$	$s_4\begin{cases} a_2 \\ a_1 \end{cases}$	$0.5(0) + 0.5(2) = 1.0$
θ_3	6	3	0	$z_1 = 0.4$ $z_2 = 0.6$	$s_4\begin{cases} a_2 \\ a_1 \end{cases}$	$0.4(3) + 0.6(6) = 4.8$
θ_1	0	2	4	$z_1 = 0.7$ $z_2 = 0.3$	$s_5\begin{cases} a_2 \\ a_2 \end{cases}$	$0.7(2) + 0.3(2) = 2.0$
θ_2	2	0	2	$z_1 = 0.5$ $z_2 = 0.5$	$s_5\begin{cases} a_2 \\ a_2 \end{cases}$	$0.5(0) + 0.5(0) = 0$
θ_3	6	3	0	$z_1 = 0.4$ $z_2 = 0.6$	$s_5\begin{cases} a_2 \\ a_2 \end{cases}$	$0.4(3) + 0.6(3) = 3.0$
θ_1	0	2	4	$z_1 = 0.7$ $z_2 = 0.3$	$s_6\begin{cases} a_2 \\ a_3 \end{cases}$	$0.7(2) + 0.3(4) = 2.6$
θ_2	2	0	2	$z_1 = 0.5$ $z_2 = 0.5$	$s_6\begin{cases} a_2 \\ a_3 \end{cases}$	$0.5(0) + 0.5(2) = 1.0$
θ_3	6	3	0	$z_1 = 0.4$ $z_2 = 0.6$	$s_6\begin{cases} a_2 \\ a_3 \end{cases}$	$0.4(3) + 0.6(0) = 1.2$

Table 16-9

Evaluation of strategies (Continued)

θ	$r(\theta, a)$ a_1	a_2	a_3	Conditional probabilities of information z	Strategy s	Expected regret = risk $R(\theta, s)$
θ_1	0	2	4	$z_1 = 0.7$, $z_2 = 0.3$	$s_7 \begin{cases} a_3 \\ a_1 \end{cases}$	$0.7(4) + 0.3(0) = 2.8$
θ_2	2	0	2	$z_1 = 0.5$, $z_2 = 0.5$	$s_7 \begin{cases} a_3 \\ a_1 \end{cases}$	$0.5(2) + 0.5(2) = 2.0$
θ_3	6	3	0	$z_1 = 0.4$, $z_2 = 0.6$	$s_7 \begin{cases} a_3 \\ a_1 \end{cases}$	$0.4(0) + 0.6(6) = 3.6$
θ_1	0	2	4	$z_1 = 0.7$, $z_2 = 0.3$	$s_8 \begin{cases} a_3 \\ a_2 \end{cases}$	$0.7(4) + 0.3(2) = 3.4$
θ_2	2	0	2	$z_1 = 0.5$, $z_2 = 0.5$	$s_8 \begin{cases} a_3 \\ a_2 \end{cases}$	$0.5(2) + 0.5(0) = 1.0$
θ_3	6	3	0	$z_1 = 0.4$, $z_2 = 0.6$	$s_8 \begin{cases} a_3 \\ a_2 \end{cases}$	$0.4(0) + 0.6(3) = 1.8$
θ_1	0	2	4	$z_1 = 0.7$, $z_2 = 0.3$	$s_9 \begin{cases} a_3 \\ a_3 \end{cases}$	$0.7(4) + 0.3(4) = 4.0$
θ_2	2	0	2	$z_1 = 0.5$, $z_2 = 0.5$	$s_9 \begin{cases} a_3 \\ a_3 \end{cases}$	$0.5(2) + 0.5(2) = 2.0$
θ_3	6	3	0	$z_1 = 0.4$, $z_2 = 0.6$	$s_9 \begin{cases} a_3 \\ a_3 \end{cases}$	$0.4(0) + 0.6(0) = 0$

Table 16-3, and the conditional probabilities of information are taken from Table 16-7. The recurrences of the entries of these two tables in Table 16-9 are intended for their easy identification in the process of evaluating the strategies. The evaluation of strategy s_1 appears in the top section of the table. Observe that given any state of nature, θ, there are two possible forecasts, or observations, z_1 and z_2, with their probabilities dependent on the state of nature. If, in response to each of the two kinds of information, the same action a_1 is taken, we have (a_1, a_1) which is s_1. The strategy is evaluated by computing the three expected regrets, or risks $R(\theta_1, s_1)$, $R(\theta_2, s_1)$, and $R(\theta_3, s_1)$, for the three θ's as follows:

$$R(\theta_1, s_1) = P(z_1 \mid \theta_1)r(\theta_1, a_1) + P(z_2 \mid \theta_1)r(\theta_1, a_1) = 0.7(0) + 0.3(0) = 0$$
$$R(\theta_2, s_1) = P(z_1 \mid \theta_2)r(\theta_2, a_1) + P(z_2 \mid \theta_2)r(\theta_2, a_1) = 0.5(2) + 0.5(2) = 2$$

and

$$R(\theta_3, s_1) = P(z_1 \mid \theta_3)r(\theta_3, a_1) + P(z_2 \mid \theta_3)r(\theta_3, a_1) = 0.4(6) + 0.6(6) = 6$$

In a similar manner, s_2 and s_3 are evaluated as follows:

$$R(\theta_1, s_2) = P(z_1|\theta_1)r(\theta_1, a_1) + P(z_2|\theta_1)r(\theta_1, a_2) = 0.7(0) + 0.3(2) = 0.6$$
$$R(\theta_2, s_2) = P(z_1|\theta_2)r(\theta_2, a_1) + P(z_2|\theta_2)r(\theta_2, a_2) = 0.5(2) + 0.5(0) = 1.0$$
$$R(\theta_3, s_2) = P(z_1|\theta_3)r(\theta_3, a_1) + P(z_2|\theta_3)r(\theta_3, a_2) = 0.4(6) + 0.6(3) = 4.2$$

and

$$R(\theta_1, s_3) = P(z_1|\theta_1)r(\theta_1, a_1) + P(z_2|\theta_1)r(\theta_1, a_3) = 0.7(0) + 0.3(4) = 1.2$$
$$R(\theta_2, s_3) = P(z_1|\theta_2)r(\theta_2, a_1) + P(z_2|\theta_2)r(\theta_2, a_3) = 0.5(2) + 0.5(2) = 2.0$$
$$R(\theta_3, s_3) = P(z_1|\theta_3)r(\theta_3, a_1) + P(z_2|\theta_3)r(\theta_3, a_3) = 0.4(6) + 0.6(0) = 2.4$$

In a similar way the other six strategies are evaluated as shown in Table 16-9. The expected regrets, or risks, corresponding to the three states of nature and each of the nine strategies are listed in Table 16-10. The set of probabilities in the first column is the same as in Table 16-4. Before we identify the best strategy according to Bayes, let us first use the minimax criterion again in determining what strategy to use.

Since entries in Table 16-10 are risks, or expected regrets, the minimax criterion we shall now consider is called the *criterion of minimax risk*, or *minimax expected regret*. The criterion is based on a spirit of pessimism since, according to this principle, the decision maker first identifies the *maximum risk* associated with each strategy, and then selects the strategy leading to the smallest of these maximum risks.

From Table 16-10, we find that the maximum risks for the nine strategies are as follows:

Strategy:	s_1	s_2	s_3	s_4	s_5	s_6	s_7	s_8	s_9
Maximum risk:	6.0	4.2	2.4	4.8	3.0	2.6	3.6	3.4	4.0

The minimum among these maximum risks is 2.4 associated with strategy s_3. Thus the minimax risk strategy is s_3, or (a_1, a_3).

Sometimes, more than one minimum among the maximum risks may occur. For instance, there may be two or more strategies with the same maximum risks each of which is the minimum of all. In such cases any one of these strategies

Table 16-10

Conditional expected regrets, or risks $R(\theta, s)$

A priori probabilities p	*States of nature*	Strategies s_1	s_2	s_3	s_4	s_5	s_6	s_7	s_8	s_9
$p_1 = 0.7$	θ_1	0	0.6	1.2	1.4	2.0	2.6	2.8	3.4	4.0
$p_2 = 0.2$	θ_2	2	1.0	2.0	1.0	0	1.0	2.0	1.0	2.0
$p_3 = 0.1$	θ_3	6	4.2	2.4	4.8	3.0	1.2	3.6	1.8	0

may be taken as the minimax risk strategy, or a random device may be applied to determine which one of them should be used.

In the general situation, if there are altogether K strategies, and if max $[R(\theta, s_k)]$ stands for the maximum risk for strategy s_k for $k = 1, 2, \ldots, K$, then we have

$$\text{Minimax risk} = \min\{\max[R(\theta, s_1)], \max[R(\theta, s_2)], \ldots, \max[R(\theta, s_K)]\} \qquad (16\text{-}13)$$

and the strategy associated with this risk is the minimax risk strategy.

Let us now turn our attention to the Bayes criterion for strategy selection. Before identifying the Bayes strategy, it is indispensable that we have a set of probabilities corresponding to the set of states of nature. The probabilities appearing in the first column of Table 16-10 are the same as in Table 16-4. In any case, such probabilities are a priori probabilities, which are determined on the basis of personal hunches or experiences and before any observation is made about the states of nature. It is natural that the validity of the Bayesian decision rule depends, to a great extent, on the reliability of the set of a priori probabilities.

Our task now is to determine which of these strategies is the Bayes strategy. Table 16-10 shows that some strategies have greater risks than others. For example, comparing s_7 with s_5, we find that for each of the three states of nature, the risk in s_7 is greater than that in s_5. Thus s_5 should be preferred to s_7 since it leads to greater utility. Similarly, when s_4 is compared with s_2, we find that with the exception of θ_2, where the corresponding risks are equal, the risk in s_4 is greater than that incurred by using s_2 when θ_1 or θ_3 is the state of nature. Accordingly, s_2 is preferred to s_4. We say that s_7 is *dominated* by s_5, and s_4 is *dominated* by s_2. In general, if at least one of the risks associated with the states of nature for one strategy is greater than, and the others are not less than, the corresponding risks of another strategy, the former strategy is regarded as dominated by the latter. A *dominated strategy* is commonly referred to as *inadmissible;* that is, it is deleted from the set of strategies being considered for application in a particular decision situation. Thus s_4 and s_7 are inadmissible strategies; they can be discarded or ignored. As a result, only the *admissible* strategies are considered in determining what the Bayes strategy is.

Briefly, the *Bayes risk strategy* is the one for which the *expected risk* is the *lowest*. According to this criterion, we have to apply the prior probabilities for computing the expected risk for each of the possible strategies, especially for all the admissible strategies. As usual, the expected risk for any strategy s, or $E[R(s)]$, is obtained by summing the products of the probabilities multiplied by the corresponding risks for all the states of nature. That is, if there are J states of nature, the expected risk for a strategy s is

$$E[R(s)] = \sum_{j=1}^{J} p_j R(\theta_j, s) \qquad (16\text{-}14)$$

Accordingly, the expected risk for s_1 is computed as follows:

$$E[R(s_1)] = 0.7(0) + 0.2(2) + 0.1(6) = 1.0$$

and the expected risk for s_2 is computed in a similar manner,

$$E[R(s_2)] = 0.7(0.6) + 0.2(1.0) + 0.1(4.2) = 1.04$$

Thus we obtain the expected risk for each of all the other strategies (including the inadmissible ones) as follows:

Strategy:	s_1	s_2	s_3	s_4	s_5	s_6	s_7	s_8	s_9
Expected risk:	1.00	1.04	1.48	1.66	1.70	2.14	2.72	2.76	3.20

Since the expected risk 1.00 associated with strategy s_1 is the lowest among all the expected risks for all the strategies, respectively, s_1 is the Bayes strategy. If two or more strategies all have identical lowest expected risks, they all will be considered Bayes strategies. Any one of them can be used by the decision maker as the best strategy; or a random device may be employed to determine which one to use.

Generally, if there are K strategies, $s_1, s_2, \ldots, s_K$, and if $E[R(s_k)]$ is used to designate the expected risk for the strategy s_k for $k = 1, 2, \ldots, K$, the *Bayes risk*, or $B(R)$, is given as follows:

$$B(R) = \min\{E[R(s_1)], E[R(s_2)], \ldots, E[R(s_K)]\} \tag{16-15}$$

The strategy associated with this risk is the *Bayes strategy*, or $B(s)$.

Observe that the Bayes strategy arrived at is different from the minimax strategy obtained previously from the same conditional risk table. The difference is attributable to the probabilities assigned to the states of nature. Since a knowledge of the probability distribution among the states of nature is generally helpful in assisting the decision maker to reach a more intelligent and rational decision, the Bayes strategy is usually preferred to the minimax strategy.

Exercise 16-3

1 Define the following terms as fully as you can:

a Conditional probability of information.
b Strategy.
c A priori probabilities.
d Minimax expected regret.

2 If a decision situation involves three actions and two types of information, how many possible strategies will be available to the decision maker? How many possible strategies are available in a decision situation having k actions and n types of information?

3 Distinguish between the following terms:

a Minimax risk and Bayes risk.
b Admissible and inadmissible strategy.
c Minimax risk strategy and Bayes strategy.

4 Suppose that in Probs. 4 and 5 of Exercise 16-2, the manufacturer of dishwashers engaged a consulting firm to make forecasts of housing starts in the country since housing is believed to be positively related to his own business. The firm makes two types of reports to the manufacturer, as follows: z_1, housing starts in the next year will be 1.5 million or more; z_2, housing starts in the next year will be below 1.5 million. According to his experience over a long period of time, the conditional probabilities of making these forecasts for each of the states of nature are given below.

Conditional probabilities of information, $P(z|\theta)$

States of nature	Information z_1	z_2
θ_1	0.3	0.7
θ_2	0.5	0.5
θ_3	0.8	0.2

Given that the nine strategies are designated as in Table 16-8, do the following:

a Construct a table of the conditional expected payoff similar to Table 16-10.
b Determine the maximax strategy.
c Determine the maximin strategy.
d Which strategies are dominated, or inadmissible?
e Determine the Bayes strategy and the Bayes payoff.

5 On the basis of the loss table and the a priori probabilities given in Probs. 5 and 6 of Exercise 16-2 and the conditional probabilities given in Prob. 4 above, do the following:

a Construct a table of conditional expected loss and a table of conditional expected regret.
b Determine the minimax expected loss strategy.
c Determine the minimax risk strategy.
d Determine the Bayes strategy and the Bayes risk.

6 Referring to Probs. 8 and 9 of Exercise 16-2, let us assume that Mr. Jones has a weather indicator in his home. Before leaving home each morning, he invariably takes a reading of the indicator to see whether it shows the day to be clear z_1, uncertain z_2, or foul z_3. The indicator, of course, is not a perfect weather forecaster. Past records show that for the two states of nature, the probability distributions over the three weather readings on the indicator are as follows:

Conditional probabilities of weather readings, $P(z|\theta)$

Weather of the day	Weather readings: Clear z_1	Uncertain z_2	Foul z_3
Sunny θ_1	0.7	0.2	0.1
Rainy θ_2	0.2	0.3	0.5

Let the eight strategies be designated as follows:

	Strategies							
Observations	s_1	s_2	s_3	s_4	s_5	s_6	s_7	s_8
z_1	a_1	a_1	a_1	a_1	a_2	a_2	a_2	a_2
z_2	a_1	a_1	a_2	a_2	a_1	a_1	a_2	a_2
z_3	a_1	a_2	a_1	a_2	a_1	a_2	a_1	a_2

a Construct a table of conditional expected loss and a table of conditional risk.
b Determine the minimax expected loss strategy and the minimax risk strategy.
c Find the Bayes strategy and the Bayes risk.

7 Suppose that in the preceding problem there are only two kinds of readings on the weather indicator. They are "clear" z_1 and "unclear" z_2. The conditional probabilities are given as follows:

$P(z_1 \mid \theta_1) = 0.90 \qquad P(z_2 \mid \theta_1) = 0.10$

$P(z_1 \mid \theta_2) = 0.15 \qquad P(z_2 \mid \theta_2) = 0.85$

What are the values of α and β, the type I and type II error probabilities?

16-4 Joint Probability and Bayes Strategy

The procedure of determining the Bayes strategy by evaluating all the possible strategies is a tedious and troublesome method, especially when the total number of possible strategies is large. The task is made somewhat easier if only the *admissible* strategies are considered; but this makes it necessary to compare the risks for discarding the dominated strategies. This latter procedure may prove to be even more inefficient if a considerable number of strategies must be compared. Fortunately, we can dispense with all such troubles if we make use of a table of *joint* probabilities. Then the Bayes strategy and the Bayes risk can be found by a single "stroke." We shall now first review the concept of joint probability.

16-4-1 Joint probabilities

Before proceeding, you are advised to review Sec. 3-2, particularly the general multiplication rule. This should be done because the notion of joint probability was already considered there.

It is our objective now to demonstrate how to find the joint probabilities for the example considered above. We know that the prior probabilities for

Table 16-11

Prior and conditional probabilities

Prior probabilities p	States of nature	Information	
		z_1	z_2
$p_1 = 0.7$	θ_1	$P(z_1 \mid \theta_1) = 0.7$	$P(z_2 \mid \theta_1) = 0.3$
$p_2 = 0.2$	θ_2	$P(z_1 \mid \theta_2) = 0.5$	$P(z_2 \mid \theta_2) = 0.5$
$p_3 = 0.1$	θ_3	$P(z_1 \mid \theta_3) = 0.4$	$P(z_2 \mid \theta_3) = 0.6$

the three states of nature are 0.7, 0.2, and 0.1. We also know that the conditional probabilities, or $P(z \mid \theta)$, are as shown in Table 16-7. For convenience, we present the prior and the conditional probabilities together in Table 16-11.

Table 16-11 provides the required information to construct a *joint probability* table, which is indispensable for determining the Bayes strategy without evaluating all the strategies. Recall, from Sec. 3-2, that a joint probability is the probability of the intersection of events. That is,

$$P(\theta \cap z) = P(\theta)P(z \mid \theta) \tag{16-16}$$

Accordingly, we obtain the joint probabilities as shown in Table 16-12. Observe that the sum of $P(z_1)$ and $P(z_2)$, or 0.63 + 0.37, is equal to unity since the two types of information z_1 and z_2 exhaust all possible observations.

Table 16-12

Joint probabilities of states of nature and information denoted by $P(\theta \cap z)$ or $f(\theta, z)$

States of nature	Information	
	z_1	z_2
θ_1	$P(\theta_1 \cap z_1) = p_1 P(z_1 \mid \theta_1) = 0.7(0.7) = 0.49$	$P(\theta_1 \cap z_2) = p_1 P(z_2 \mid \theta_1) = 0.7(0.3) = 0.21$
θ_2	$P(\theta_2 \cap z_1) = p_2 P(z_1 \mid \theta_2) = 0.2(0.5) = 0.10$	$P(\theta_2 \cap z_2) = p_2 P(z_2 \mid \theta_2) = 0.2(0.5) = 0.10$
θ_3	$P(\theta_3 \cap z_1) = p_3 P(z_1 \mid \theta_3) = 0.1(0.4) = 0.04$	$P(\theta_3 \cap z_2) = p_3 P(z_2 \mid \theta_3) = 0.1(0.6) = 0.06$
Total	$P(z_1) = 0.63$	$P(z_2) = 0.37$

16-4-2 Bayes strategy

For finding the Bayes strategy, we shall now use the joint probabilities $P(\theta \cap z)$, as found in Table 16-12, and the regrets as in Table 16-3, which are rewritten as follows:

$P(\theta \cap z)$

θ	z_1	z_2
θ_1	$P(\theta_1 \cap z_1) = 0.49$	$P(\theta_1 \cap z_2) = 0.21$
θ_2	$P(\theta_2 \cap z_1) = 0.10$	$P(\theta_2 \cap z_2) = 0.10$
θ_3	$P(\theta_3 \cap z_1) = 0.04$	$P(\theta_3 \cap z_2) = 0.06$
Total	0.63	0.37

$r(\theta, a)$

States of nature	*Actions*		
	a_1	a_2	a_3
θ_1	0	2	4
θ_2	2	0	2
θ_3	6	3	0

Applying these probabilities and regrets, we can compute the Bayes strategy as shown in Table 16-13.

The symbols $R(z_1, a)$ and $R(z_2, a)$ in Table 16-13 designate the risks under z_1 and z_2, respectively, for each action a. When z_1 is observed, action a_1 will be taken since the risk of taking a_1, or $R(z_1, a_1)$ is 0.44, the lowest among all the three risks under z_1. When z_2 is observed, action a_1 will again be taken since the risk $R(z_2, a_1)$ is 0.56, also the lowest among the risks under z_2. The Bayes strategy, or $B(s)$, is therefore, (a_1, a_1), the same as that arrived at in Sec. 16-3. The Bayes risk, or $B(R)$, is also equal to 1, which is now obtained by summing $R(z_1, a_1)$ and $R(z_2, a_1)$; that is,

$$B(R) = R(z_1, a_1) + R(z_2, a_1) = 0.44 + 0.56 = 1$$

You are advised to check the Bayes risk for each of the other strategies obtained by this method against the result obtained by the method in Sec. 16-3.

In the general case, let $R(z_i, a_k)$ be the risk of action a_k corresponding to the information z_i for $i = 1, 2, \ldots, n$, and $k = 1, 2, \ldots, K$, where n and K denote the total number of observations and actions, respectively. If we use min

Table 16-13

Computation of Bayes' strategy by using joint probabilities

	z_1		
θ	a_1 $P(\theta \cap z_1)r(\theta, a_1)$	a_2 $P(\theta \cap z_1)r(\theta, a_2)$	a_3 $P(\theta \cap z_1)r(\theta, a_3)$
θ_1	0.49(0) = 0.00	0.49(2) = 0.98	0.49(4) = 1.96
θ_2	0.10(2) = 0.20	0.10(0) = 0.00	0.10(2) = 0.20
θ_3	0.04(6) = 0.24	0.04(3) = 0.12	0.04(0) = 0.00
$R(z_1, a)$	0.44	1.10	2.16
	z_2		
θ	a_1 $P(\theta \cap z_2)r(\theta, a_1)$	a_2 $P(\theta \cap z_2)r(\theta, a_2)$	a_3 $P(\theta \cap z_2)r(\theta, a_3)$
θ_1	0.21(0) = 0.00	0.21(2) = 0.42	0.21(4) = 0.84
θ_2	0.10(2) = 0.20	0.10(0) = 0.00	0.10(2) = 0.20
θ_3	0.06(6) = 0.36	0.06(3) = 0.18	0.06(0) = 0.00
$R(z_2, a)$	0.56	0.60	1.04

$[(R(z_i, a_k)]$ to stand for the minimum risk associated with action a_k under information z_i for all i and k, then the Bayes risk is defined as

$$B(R) = \sum_{i=1}^{n} \min[R(z_i, a_k)] \qquad (16\text{-}17)$$

provided that the joint probabilities $P(\theta \cap z)$ are used in computing the risk for each action under each type of information.

16-4-3 Value of information

Recall that in Sec. 16-2-3 the minimum expected regret, or Bayes' risk, was 1. By the methods in Secs. 16-3 and 16-4-2, where two types of information were possibly provided, the Bayes risk remained 1. In other words, the information provided does not increase utility, and as such it has *no value* at all. The Bayes risk is 1, and action a_1 is taken regardless of whether any observation, or forecasting in this case, is made.

This, however, is not always the case. In most situations the Bayes risk computed when information is provided is smaller than the Bayes risk without information. In other words, utility may be increased by furnishing information for decision making. Among the problems following this section, cases in which information will reduce risk, or increase utility, will be provided.

Exercise 16-4

1 Explain the concept of value of information.

2 Refer to Prob. 4 of Exercise 16-3, and do the following:

a Construct a joint probability table similar to Table 16-12.
b If housing starts in the country are projected to be less than 1.5 million next year, what is the probability that the economy will be in recession? In stability? In a boom?
c Determine the Bayes strategy and the Bayes payoff.
d What is the maximum fee the manufacturer would be willing to pay for the service of the consulting firm?

3 Referring to Prob. 5 of Exercise 16-3, determine the following:

a The Bayes strategy and the Bayes risk.
b The value of the information provided by the consulting firm.

4 Referring to Prob. 6 of Exercise 16-3, do the following:

a Construct a joint probability table similar to Table 16-12.
b If the weather indicator shows that the day's weather is uncertain, what is the probability that the day will turn out to be sunny? What action should Mr. Jones take?
c How much is the weather indicator worth in terms of risks?

5 Design a decision model with three states of nature, three available actions, and two types of observations. Indicate whether the information provided is worth the cost for making the observations.

6 Design a decision model with two states of nature, two available actions, and three types of observations. Indicate whether the information provided is worth the cost for making the observations.

Chapter 17

Nonparametric methods

Most of the methods for testing hypotheses and interval estimation discussed previously were based on the assumption that the forms of the population distribution were known. In Chaps. 9 and 10, statistical inferences were developed on the basis of samples assumed to have been taken from populations having binomial, normal, or other specified types of distribution. In Chaps. 11 and 13, the t and F tests were based on assumptions such as normality of population distributions and the homogeneity of variances. Likewise, inferences concerning regression and correlation discussed in Chap. 14 were developed under the assumptions of homogeneous variances as well as normality of population distributions. Only the χ^2 tests of goodness of fit and of statistical independence considered in Chap. 12 required no specific assumptions about the distribution pattern of the population sampled, and thus may be regarded as free from assumptions about population distribution.

Moreover, nearly all the tests discussed thus far were such that some unknown values of the parameters could be estimated from quantities computed from samples randomly selected from the specified population. Hypotheses were stated in terms of the specified value or values of population parameters.

Since there are many situations in which the assumptions cannot be met, numerous statistical tests have been recently developed that do not require rigorous assumptions about the population distribution, nor do they require to have hypotheses stated in terms of specified parameter values. They are therefore called *distribution-free* or *nonparametric* tests. The two terms are not strictly synonymous. The term *distribution-free* describes a method of testing hypotheses or establishing a confidence interval that does not depend on the nature of the underlying population distribution, while the term *nonparametric* is used to imply that no hypothesis is stated in terms of specific values of parameters. Customarily, we simply use nonparametric to refer to either kind of test.

Nonparametric tests, however, should not be used if *parametric* methods, that is, those with assumptions of the underlying distribution and related characteristics, can actually apply. This is because the power of nonparametric tests is relatively low in comparison with parametric tests. For this reason, if nonparametric tests have to be used, large sample sizes are recommended. As pointed out in Chap. 10, when the sample size is enlarged, the power of the test will be increased, given the level of significance.

Nonparametric methods, nonetheless, have their advantages by being easy to apply. They are relatively simple and easy to explain and understand as compared with the parametric methods. Thus they are sometimes referred to as "short-cut" methods, and are often used, with increased sample size, even in situations where parametric assumptions are actually met.

Many nonparametric techniques have been developed in recent decades. In this chapter we shall limit our study to only the most frequently used methods. They include the median test, the sign tests, the rank-sum tests, and the rank-correlation tests.

17-1 The Median Test

Let us first consider one of the simplest of the nonparametric methods, the *median* test. This is a procedure of testing whether two or more *independent* samples are taken from populations with the same median. For the sake of simplicity, we shall limit our discussion to only two samples; but the procedure can be very easily extended to three or more samples. The median test requires no assumption about the two populations from which samples are taken except that the random variable on which the two populations are compared is continuous. This assumption is made so that the probability of a tie between any two observations on the random variable is taken as 0.

The null hypothesis to be tested is that the populations from which the two samples are taken have the same median. The alternative hypothesis is that the two populations have different medians.

To carry out the median test it is necessary to determine the median value for all the observations in the two samples. For each of the two sets of scores, one above and one below the *overall* median, the number of scores belonging to each of the two samples is counted. Let n_1 and n_2 be the numbers of observations in the two samples, respectively; we obtain a 2×2 table as follows:

Number of scores	*Group I*	*Group II*	*Total*
Above overall median	a	b	$a+b$
Below overall median	c	d	$c+d$
Total	$a+c=n_1$	$b+d=n_2$	$n_1+n_2=n$

If the null hypothesis is true, that is, if the two populations sampled have the same median, we should expect that half of each sample's scores will fall above and half will fall below the median. In other words, we should expect that $a = c = 0.5n_1$ and $b = d = 0.5n_2$. Then, when $n = n_1 + n_2$ is larger than 20 and when the *expected frequency*[1] in every cell is at least 5, we may use the χ^2 test with the test statistic defined as by Eq. (12-20). That is,

$$\chi^2 = \frac{n(|ad - bc| - n/2)^2}{(a + b)(a + c)(c + d)(b + d)}$$

which has 1 degree of freedom.

To illustrate how the above median test is used, let us suppose that we wish to determine whether the wages for male workers have the same median as wages for female workers. The null hypothesis is that the median of wages for male workers is equal to that for female workers. The alternative hypothesis is that the median of men's wages differs from that of women's wages. Suppose further that two samples of wages of 14 male workers ($n_1 = 14$) and 16 female workers ($n_2 = 16$), which are independently taken, have produced a 2×2 table as follows:

Frequency of wages	*Male Workers*	*Female Workers*	*Total*
Above overall median	9	6	15
Below overall median	5	10	15
Total	14	16	30

Substituting these quantities in Eq. (12-20), we have

$$\chi^2 = \frac{30[|9(10) - 6(5)| - \frac{30}{2}]^2}{(15)(14)(15)(16)} = \frac{30(|90 - 30| - 15)^2}{50,400} = \frac{60,750}{50,400} = 1.205$$

Since the critical χ^2 value that cuts off the upper 5 percent of the distribution with 1 degree of freedom is 3.84146, the obtained value 1.205 does not lead us to reject the null hypothesis at $\alpha = 0.05$. We conclude that the median of male workers' wages is not significantly different from the median of wages for female workers. Observe that the null hypothesis is rejected *only if* the computed value of χ^2 is sufficiently *large*. Thus, similar to the test of goodness of fit and of statistical association, the median test as discussed here is essentially a one-tailed test.

[1] See Sec. 12-4 for the method of computing expected frequencies.

A problem encountered in applying the median test is that two or more observations may have the same value as the median. Under such circumstances, we simply drop the scores identical with the median as long as such scores are not numerous or n_1 and n_2 are large.

The median test is easy and simple to apply. Because of its simplicity, it is sometimes used even in situations where parametric tests could be employed. The median test, however, is known as a low-powered test, especially when compared with the alternative parametric test. Thus the size of the samples must be sufficiently enlarged if, other things being equal, the median test is to have the same power as the comparable parametric test. The median test with three or more samples is not considered here, since it can be replaced by the Kruskal-Wallis method, a more powerful test discussed in Sec. 17-3-2.

Exercise 17-1

1 Distinguish between the following terms:

a Parametric and nonparametric tests.
b Nonparametric and distribution-free methods.

2 Discuss the advantages and disadvantages of a nonparametric test.

3 Discuss fully the procedure and rationale of a median test.

4 Suppose you wish to test whether the median of sample I with 14 observations ($n_1 = 14$) is significantly different from the median of sample II with 24-observations ($n_2 = 24$). The classification of the scores of 38 observations is given as follows:

Number of scores	*Sample I*	*Sample II*	*Total*
Above overall median	4	15	19
Below overall median	10	9	19
Total	14	24	38

Test the null hypothesis that the two samples are taken from two populations with the same median against the alternative that the two populations have different medians at $\alpha = 0.01$.

5 An identical examination was administered to two classes of students. One class contains 13 students, and the other contains 17 students. The test scores of the 30 students are given below.

Class I: 54 65 66 71 73 78 78 80 82 87 92 93 95
Class II: 51 53 54 61 64 66 67 69 71 74 76 80 81 85 89 90 94

At $\alpha = 0.05$, test the null hypothesis that the two classes of students are from populations with an identical median.

6 Suppose we wish to compare the weekly wages of electricians and meatcutters in a certain metropolitan area. The sample data on the weekly wages of 15 workers in each trade are given as follows:

Electricians	*Meatcutters*
$350	$352
351	360
351	367
355	370
361	371
365	375
366	377
369	378
370	380
375	381
375	385
382	389
387	393
390	394
392	395

At $\alpha = 0.05$, determine whether the median of the electricians' weekly wages is the same as the median of the meatcutters' weekly wages.

17-2 Tests Involving Signs of Differences

If two samples are compared to determine whether there is a significant difference between two population means, the signs of differences between paired observations from the two samples can be used for hypothesis testing. In this section we shall consider two types of tests involving signs: the sign test and the signed-rank test.

17-2-1 The sign test

As pointed out at the beginning of Chap. 11, in order to apply the t test, the population from which random samples are taken must be *normally distributed.* For testing the difference between the means of two populations discussed in Sec. 11-3-2, it was further assumed that the population variances must be *identical.* In many situations, however, either or both of these two assumptions cannot be met, and thus the t test cannot be applied. Under such circumstances, we may use a nonparametric test generally known as the sign test.

The *sign test* is based on the *signs*, positive or negative, of the differences between paired observations, not taking into consideration the magnitude of the differences. It provides a convenient alternative to the method of testing hypotheses about the difference between two means when the samples are *dependent* (see Dependent samples, Sec. 11-3-2).

Suppose, for example, that it is desired to find out whether a period of vacation would increase workers' productivity. Suppose further that for this purpose the data on weekly outputs of 22 workers from a manufacturing company for the week before and the week after the vacation are collected. Let X and Y be the weekly output *before* and *after* the vacation, respectively. The data and the signs of the differences, or $Y - X$, are shown in Table 17-1.

The null hypothesis to be tested in this case is that the period of vacation has no favorable effect on the workers' productivity. In other words, a positive

Table 17-1
Weekly outputs of 22 workers before and after vacation (units)

Workers	*Before vacation* X	*After vacation* Y	*Sign of* $Y-X$
A	83	79	−
B	85	87	+
C	75	70	−
D	91	93	+
E	80	85	+
F	75	75	0
G	90	80	−
H	65	71	+
I	78	80	+
J	85	88	+
K	83	82	−
L	75	71	−
M	78	75	−
N	80	85	+
O	82	86	+
P	88	85	−
Q	85	82	−
R	80	87	+
S	78	78	0
T	81	84	+
U	70	85	+
V	80	81	+

sign is as likely as a negative sign when the difference between the two observations of each pair is determined with the same order of subtraction maintained. This is equivalent to saying that the null hypothesis is $H_0: p = 0.5$, where p is the probability of getting a positive sign. Since the test is to determine whether there is a significant increase in the productivity of the workers, the alternative hypothesis is $H_1: p > 0.5$; therefore this is a one-tailed test. We simply count the number of positive signs and designate it as the binomial random variable X for testing the null hypothesis that $p = 0.5$. The latter is rejected if the number of positive signs, or the binomial random variable X, falls within the rejection region. Thus, at significance level α, we have

$$P(X \geq c \mid p = 0.5) = \alpha \tag{17-1}$$

where c is the critical value. (If it is desired, a two-tailed test of the probability p for the signs may also be carried out.)

The sign test is easy to apply. It is merely assumed that the differences are *independent* of one another and that the differences have 0 as their *median*, since the median is the value that divides the total signs into two equal parts when the null hypothesis is true. In our example, there are 12 plus signs, 8 minus signs, and 2 cases showing a 0 difference. The two cases with 0 difference are ignored. Consequently, we have 20 cases ($n = 20$), of which 12 are "successes." Given that $\alpha = 0.05$, the critical value c is approximately 14, since Appendix III shows that given $p = 0.5$, the probability of getting 14 or more successes in 20 trials is 0.0577. Since the observed number of successes is 12, which is less than the critical value, we do not reject the null hypothesis that a period of vacation results in no increase in workers' productivity.

The sign test can also be carried out by the method of normal approximation. We know that the mean and the standard deviation of the binomial random variable X for the above example are

$$\mu = np = 20(0.5) = 10$$

and

$$\sigma = \sqrt{npq} = \sqrt{20(0.5)(0.5)} = \sqrt{5} = 2.236$$

With the level of significance at 0.05, the critical value is

$$10 + 1.645(2.236) = 13.678$$

which is greater than the observed value 12. Thus the same conclusion is reached by the method of normal approximation to the binomial. When n is less than 10, however, it is preferable to carry out the sign test on the basis of the binomial distribution rather than normal approximation.

When the unit of measurement is the same for the paired observations, as is true in the above example, the sign test can be applied to deal with questions such as the following: *(1)* Are the values of the variable X greater than those of

Y by D units? *(2)* Are the values of X greater than those of Y by P percent? To deal with the first question, we simply add D units to each observed value of Y, and find the signs of the differences between X_i and $Y_i + D$ for $i = 1, 2, 3, \ldots, n$, where n is the number of pairs. That is, we obtain the signs of

$$X_1 - (Y_1 + D), X_2 - (Y_2 + D), \ldots, X_n - (Y_n + D)$$

The sign test is then applied as usual to determine whether the measurements of X are greater than those of Y by D units.

Similarly, the second question can be answered by first multiplying Y by $1 + P/100$ and then comparing X and $Y(1 + P/100)$ to determine the sign of the difference for each of the paired observations:

$$X_1 - \left(1 + \frac{P}{100}\right)Y_1, X_2 - \left(1 + \frac{P}{100}\right)Y_2, \ldots, X_n - \left(1 + \frac{P}{100}\right)Y_n$$

The sign test is then applied to the signs of these differences.

17-2-2 The signed-rank test

As pointed out above, the sign test ignores completely the magnitude of the difference between paired values. An improvement on the sign test was suggested by Frank Wilcoxon in 1945. The improved test, which is called *Wilcoxon's signed-rank test*, takes into consideration the magnitude of the differences. To carry out the test, the first step is to rank all the *absolute* values of the differences between the paired observations, from the smallest to the largest. That is, we assign rank 1 to the smallest difference, rank 2 to the next smallest, and so forth. Since we rank differences without regard to sign, a difference of -1 or $+1$ is assigned the same rank, which is lower than the rank assigned to a difference of either -2 or $+2$. After the differences are ranked, then the sign of the difference is affixed to each rank. The sum of the positive ranks and that of the negative ranks are then computed respectively; and the smaller sum, with the sign ignored, is the test statistic commonly designated T.

If the null hypothesis that the two populations are identical is true, we should expect the two sums to be about equal since the number of larger ranks with positive signs would be approximately equal to the number of larger ranks with negative signs. If the two sums are greatly different from each other, we should infer that the two populations are not identical. That is, we should reject the null hypothesis if either the sum of the positive ranks or the sum of the negative ranks is very small.

The test statistic T can be used for both one- and two-tailed tests. When a one-tailed test is applied, we must *anticipate* in advance the *sign* of the smaller sum of ranks in case the null hypothesis is false. If the smaller sum has a sign different from what is anticipated, we should not reject the null hypothesis. Table 17-2 shows the process of computing T for the example given for the sign

Table 17-2

Computation of test statistic T for Wilcoxon's signed-rank test

Workers	X	Y	Difference Y − X	Rank	Signed rank Negative	Signed rank Positive
A	83	79	− 4	12	−12	
B	85	87	+ 2	4		+ 4
C	75	70	− 5	15	−15	
D	91	93	+ 2	4		+ 4
E	80	85	+ 5	15		+15
F	75	75	0	—		
G	90	80	−10	19	−19	
H	65	71	+ 6	17		+17
I	78	80	+ 2	4		+ 4
J	85	88	+ 3	8		+ 8
K	83	82	− 1	1.5	− 1.5	
L	75	71	− 4	12	−12	
M	78	75	− 3	8	− 8	
N	80	85	+ 5	15		+15
O	82	86	+ 4	12		+12
P	88	85	− 3	8	− 8	
Q	85	82	− 3	8	− 8	
R	80	87	+ 7	18		+18
S	78	78	0			
T	81	84	+ 3	8		+ 8
U	70	85	+15	20		+20
V	80	81	+ 1	1.5		+ 1.5
Sum					T = −83.5	126.5

test above. Since it is a one-tailed test, and the alternative hypothesis is that a period of vacation would increase the productivity (value of Y) of the workers, the smaller sum is anticipated to have a *negative* sign.

Again, the two cases with 0 difference are ignored. When a tie in ranks occurs, we simply assign the average of the tied ranks to each of the differences with the same absolute size. For instance, the differences -1 (worker K) and $+1$ (worker V) are tied for the first two ranks; thus each is assigned the rank 1.5, which is the average of the ranks 1 and 2. Similarly, the differences $+2$ (worker B), $+2$ (worker D), and $+2$ (worker I) are tied for the next three ranks, and so their average rank, or 4, is assigned to each of these three differences. The smaller negative sum is -83.5, and therefore the test statistic T is equal to 83.5 with the sign ignored.

The value of T is referred to the T table developed by Wilcoxon for comparison with the critical value at a given level of significance. A portion of the T table is shown in Table 17-3. The table provides the critical values of T at the specified α levels 0.005, 0.01, and 0.025 for one-tailed tests, and 0.01, 0.02, and 0.05 for two-tailed tests. For a one-tailed test with $n = 20$, the critical value of T at $\alpha = 0.01$ is 43, at or *below* which lies the critical region. Since the observed value is 83.5, the null hypothesis is not rejected. That is, a period of vacation has no significant favorable effect on the productivity of the workers.

Observe that if a two-tailed test is desired, the critical value 43 is associated with a significance level 0.02 for $n = 20$.

For pairs larger than 25, the table of T values cannot be used. Fortunately, for large n the distribution of T is approximately normal, and the method of

Table 17-3
Critical values of T for Wilcoxon signed-rank test

Pairs n	Significance level α			
	One-tailed test: / Two-tailed test:	0.005 / 0.01	0.01 / 0.02	0.025 / 0.05
6		..	..	0
7		..	0	2
8		0	2	4
9		2	3	6
10		3	5	8
11		5	7	11
12		7	10	14
13		10	13	17
14		13	16	21
15		16	20	25
16		20	24	30
17		23	28	35
18		28	33	40
19		32	38	46
20		38	43	52
21		43	49	59
22		49	56	66
23		55	62	73
24		61	69	81
25		68	77	89

Source: Adapted from Frank Wilcoxon, "Some Rapid Approximate Statistical Procedures," American Cyanamid Company, New York, 1949, with the kind permission of author and publisher.

normal approximation is usually used. In fact T is approximately $N[E(T), \sigma_T^2]$ with n equal to 8 or greater. We now give the formulas for computing the mean and standard deviation of the sampling distribution of T without proof:

$$E(T) = \frac{n(n+1)}{4} \tag{17-2}$$

and

$$\sigma_T = \sqrt{\frac{n(n+1)(2n+1)}{24}} \tag{17-3}$$

The Z score is computed as follows:

$$Z = \frac{T - E(T)}{\sigma_T} \tag{17-4}$$

In the above example we have

$$E(T) = \frac{20(20+1)}{4} = 105$$

$$\sigma_T = \sqrt{\frac{20(20+1)(40+1)}{24}} = \sqrt{\frac{17{,}220}{24}} = \sqrt{717.5} = 26.78$$

and

$$Z = \frac{83.5 - 105}{26.78} = -0.83$$

which is greater than the critical value -1.645 with $\alpha = 0.05$. Accordingly, we shall not reject the null hypothesis that the period of vacation results in no increase in productivity. This result is the same as that obtained by the sign test discussed previously. However, the results can be different. In other words, with the same experiment, we may have sufficient evidence to reject the null hypothesis through the sign test, while the Wilcoxon ranked-sign test may not lead us to reject the null hypothesis, and vice versa.

Exercise 17-2

1 Explain the differences and similarities between the sign test and the test about the difference between two means.

2 Explain the rationale underlying the sign test and the signed-rank test.

3 Suppose you want to determine the effectiveness of a certain reducing diet. Seventeen persons were put on the diet; their weights before and after they tried the diet are shown below:

Persons	*Weights before*	*Weights after*	*Persons*	*Weights before*	*Weights after*
A	210	208	*J*	195	180
B	197	196	*K*	154	150
C	203	195	*L*	179	173
D	175	175	*M*	243	235
E	234	229	*N*	195	204
F	178	170	*O*	198	193
G	252	242	*P*	169	169
H	230	221	*Q*	217	210
I	190	213			

a Test the null hypothesis that the diet is not effective ($p = 0.5$) against the alternative that it is effective ($p > 0.5$) by using the sign test at $\alpha = 0.01$.

b Do the same by using the method of normal approximation.

4 It was suspected that high school graduates who plan to major in mathematics in college would score more than five points higher in a natural sciences test than in a social sciences test. The following are the test scores of 20 such students:

Students	*Natural sciences test scores*	*Social sciences test scores*	*Students*	*Natural sciences test scores*	*Social sciences test scores*
A	98	87	*K*	80	70
B	75	66	*L*	85	75
C	94	80	*M*	70	60
D	85	75	*N*	90	77
E	70	62	*O*	66	55
F	80	76	*P*	77	80
G	90	78	*Q*	76	65
H	75	65	*R*	95	88
I	65	50	*S*	68	60
J	60	60	*T*	70	60

Test the null hypothesis that the natural sciences test scores are five points higher than the social sciences test scores against the alternative that the former are higher by more than five points at $\alpha = 0.01$ by applying the sign test.

5 Referring to the preceding problem, test the null hypothesis that the natural sciences test scores are 10 percent higher than the social sciences test scores against the alternative that the former are higher by more than 10 percent.

6 Refer to Prob. 3 above. By applying Wilcoxon's T test, determine whether the diet has significantly *reduced* the weights of the persons involved, at the 1 percent level.

7 A company has set \$10,000 as the standard daily sales for its salesmen. A random sample of 30 salesmen is taken, and 20 of them have made sales over \$10,000 each and 10 have achieved a volume less than \$10,000 each during a given day. At $\alpha = 0.05$, test the null hypothesis that the salesmen's performance in terms of sales during that day just meets the standard against the alternative that:

a It is higher than the standard.
b It is different from the standard.

8 Refer to the data in Prob. 4 above. By applying the Wilcoxon signed-rank test, determine whether the natural sciences test scores are 10 percent higher than the social sciences test scores against the alternative that the former are higher by more than 10 percent at $\alpha = 0.01$.

9 Suppose that we wish to determine whether a raise in wages would increase the hourly output of workers. Let X be the hourly output in units before the raise in wages and Y be the hourly output after the raise. A sample of 20 workers yields the following data:

Workers	*X*	*Y*	*Workers*	*X*	*Y*
A	91	88	*K*	75	74
B	83	87	*L*	84	86
C	70	67	*M*	71	72
D	64	69	*N*	80	90
E	85	83	*O*	70	75
F	86	81	*P*	85	83
G	91	94	*Q*	65	75
H	66	67	*R*	75	82
I	72	76	*S*	75	65
J	60	55	*T*	65	67

Use the value of the statistic T to test the null hypothesis that the raise in wages has no effect on the hourly output of the workers against the alternative that:

a The hourly output after the raise is higher than that before the raise at $\alpha = 0.01$.
b The hourly output after the raise differs from that before the raise at $\alpha = 0.05$.

10 Do the preceding problem by the method of normal approximation.

17-3 Rank-Sum Tests

Closely related to the Wilcoxon signed-rank test are two important nonparametric tests, both of which use the sum of ranks in testing hypotheses. They are Mann-Whitney's U test and Kruskal-Wallis' H test. The two tests are similar but have different applications. First, let us examine the Mann-Whitney test.

17-3-1 The Mann-Whitney test

The Wilcoxon test has already shown how sample observations can be replaced by their ranks in significance tests. We shall in this section consider another method based on ranks. It is known as the *Mann-Whitney test*, named after the two statisticians who first described it in 1947. The test, like other nonparametric tests, does not require rigid assumptions about the populations from which samples are taken. The only assumption needed is that the values of the random variable on which two groups are to be compared are *continuously* distributed. In actual practice, however, no serious difficulty will be encountered even if this assumption is not met.

The null hypothesis to be tested is that two samples *independently* taken come from populations having the same mean. The test may be either one- or two-tailed. Thus the Mann-Whitney test is another useful alternative to the parametric t test when we wish to avoid the assumptions required under the t test. It is also referred to as the *U test* since the test statistic U is computed from sample data for testing the null hypothesis.

The U test is usually employed instead of the sign test or the Wilcoxon test when two independent samples are involved. Suppose that two samples, 1 and 2, with n_1 and n_2 observations respectively, are independently selected and the $n_1 + n_2$ scores from both samples are arranged in an array of descending or ascending order. A rank is assigned to each score according to its magnitude. That is, the lowest score is assigned rank 1, the next lowest rank 2, and so on. Then one of the two samples, say sample 1 which has n_1 observations, is chosen, and the sum of its ranks is computed. Let us call this sum R_1. The test statistic U is defined as

$$U = n_1 n_2 + \frac{n_1(n_1 + 1)}{2} - R_1 \qquad (17\text{-}5)$$

Similarly, U can be obtained by using the formula

$$U = n_1 n_2 + \frac{n_2(n_2 + 1)}{2} - R_2 \qquad (17\text{-}6)$$

where R_2 is the sum of the ranks in sample 2 with n_2 observations. The two formulas may yield two different values for U. What will be actually used is the smaller of the two U values. The larger value is designated U'. Before employing the test, we should check whether we have found U or U' by comparing it with $n_1 n_2/2$. If the resulting value is larger than $n_1 n_2/2$, it is then U', and the value U can be obtained by applying the equation

$$U = n_1 n_2 - U' \qquad (17\text{-}7)$$

For illustrating how the U test is used, let us suppose that we wish to determine whether the annual sales volume achieved by salesmen who have been graduated from college differs from the volume achieved by salesmen who have failed to obtain a college degree. Let G and F stand for the two groups of salesmen, respectively. Suppose further that a random sample of 10 college-graduated salesmen ($n_1 = 10$) and another sample of 21 salesmen without a college degree ($n_2 = 21$) are independently selected. The sales volumes and the ranks are shown in Table 17-4. Note that, for this example, $R_1 = 98$ and $R_2 = 398$. The value of U is found by substituting the observed quantities in Eq. (17-5) as follows:

$$U = 10(21) + \frac{10(10 + 1)}{2} - 98 = 167$$

Table 17-4

Annual sales volumes of college-graduated salesmen, G, and salesmen without a college degree, F, and ranks

Salesmen G	Annual sales volume, $ (in thousands)	Rank	Salesmen F	Annual sales volume, $ (in thousands)	Rank
1	82	24	1	92	31
2	75	19	2	90	29.5
3	70	15	3	90	29.5
4	65	11	4	89	28
5	60	8	5	86	27
6	58	7	6	85	26
7	50	4.5	7	83	25
8	50	4.5	8	81	22.5
9	46	3	9	81	22.5
10	42	2	10	78	21
			11	76	20
			12	73	18
			13	72	17
			14	71	16
			15	68	14
			16	67	13
			17	66	12
			18	64	10
			19	63	9
			20	52	6
			21	40	1
		$R_1 = 98$			$R_2 = 398$

which is greater than $n_1 n_2/2 = 10(21)/2 = 105$. Thus the U value that we should use is

$$U = 10(21) - 167 = 43$$

You should check that Eq. (17-6) will yield the same value for U.

For n_1 and n_2 each less than 20, the smaller value of U is referred to the U table of critical values for determining whether the null hypothesis of no difference between the two means should be rejected.[1] Since in this example n_2 is greater than 20, the normal approximation, and not the U table, will be employed.

It can be demonstrated that as n_1 and n_2 increase, the distribution of U rapidly approaches the normal distribution. In fact, when n_1 or $n_2 = 20$, U is approximately normal with a mean

$$E(U) = \frac{n_1 n_2}{2} \tag{17-8}$$

and standard deviation

$$\sigma_U = \sqrt{\frac{n_1 n_2(n_1 + n_2 + 1)}{12}} \tag{17-9}$$

Accordingly, we determine the significance of an observed U value by computing the standardized normal score

$$Z = \frac{U - E(U)}{\sigma_U}$$

When the normal approximation to the sampling distribution of U is employed, it does not matter which value of U is used since the absolute value of Z remains the same. For a one-tailed test, the critical region lies below the negative Z value when the smaller value of U is used, or lies above the positive Z value when the larger value of U is used. For a two-tailed test, as in this example, we find

$$E(U) = \frac{10(21)}{2} = 105 \qquad \text{by (17-8)}$$

and

$$\sigma_U = \sqrt{\frac{10(21)(10 + 21 + 1)}{12}} \qquad \text{by (17-9)}$$

$$= \sqrt{\frac{6{,}720}{12}} = \sqrt{560} = 23.66$$

[1] See Sidney Siegel, "Nonparametric Statistics for the Behavioral Sciences," tables *J* and *K*, appendix, McGraw-Hill Book Company, New York, 1956.

thus

$$Z = \frac{43 - 105}{23.66} = -2.62$$

If the level of significance is chosen at $\alpha = 0.01$, the critical Z values are ± 2.575. Thus we reject the null hypothesis and conclude that the annual sales volume achieved by salesmen without a college degree is not equal to the volume for salesmen with such a degree.

The U test can also be used to determine whether the dispersion of one population is the same as the dispersion of another population. That is, it can be used to test the null hypothesis that two independent samples are selected from populations with the same *variance* against the alternative hypothesis that they come from two populations with different variances. As shown previously, the scores of the combined samples are arranged in either ascending or descending order of magnitude. Then they are ranked from both the lowest and the highest scores toward the middle. The lowest score is first assigned rank 1, the highest and the second highest scores are assigned ranks 2 and 3, the second and third lowest are assigned ranks 4 and 5, and so on. The value of U is computed by applying Eqs. (17-5) or (17-6) as before. In the given example, we find that the new system of ranking gives the results shown in the accompanying table.

The rationale of the test about the difference in variability of the two populations lies in that if the population from which the sample of size n_1 (or n_2) is

Score	Source G or F	Rank	Score	Source G or F	Rank
\$40	*F*	1	\$72	*F*	30
42	*G*	4	73	*F*	27
46	*G*	5	75	*G*	26
50	*G*	8	76	*F*	23
50	*G*	9	78	*F*	22
52	*F*	12	81	*F*	19
58	*G*	13	81	*F*	18
60	*G*	16	82	*G*	15
63	*F*	17	83	*F*	14
64	*F*	20	85	*F*	11
65	*G*	21	86	*F*	10
66	*F*	24	89	*F*	7
67	*F*	25	90	*F*	6
68	*F*	28	90	*F*	3
70	*G*	29	92	*F*	2
71	*F*	31			

taken has greater (smaller) variability, R_1 (R_2) would be smaller (larger). Consequently, the value of U will be larger (smaller), which tends to result in the rejection of the null hypothesis that the two populations have the same variability.

The scores of sample G have ranks 4, 5, 8, 9, 13, 16, 21, 29, 26, and 15. The sum of these ranks is $R_1 = 146$, and the value of U is $10(21) + 10(11)/2 - 146 = 119$. Since the mean and the standard deviation of U are the same as before, we have

$$Z = \frac{119 - 105}{23.66} = 0.6$$

which is smaller than the critical Z value $+1.96$ at significance level 0.05. Thus we do not reject the null hypothesis, but conclude that the variability of the sales volume of group G does not differ significantly from that of group F.

17-3-2 The Kruskal-Wallis test

The Mann-Whitney test can be extended to situations where three or more groups are compared. It is then known as the H test, which is also called the *Kruskal-Wallis test*, named after William H. Kruskal and W. Allen Wallis, who in 1952 first suggested its use. The Kruskal-Wallis test is an extremely useful alternative to the technique of one-way analysis of variance discussed in Sec. 13-3. It is designed to determine whether K *independent* samples are taken from populations that have the same mean. Unlike the one-way analysis of variance, however, this test does not require the assumptions of normality of the population distribution and homogeneity of variance. All it assumes is that the random variable on which the various groups are to be compared is *continuously* distributed.

The null hypothesis to be tested is that the means of the K populations from which samples are taken are identical; the alternative is that they are different. To perform the test, the scores from all the K samples combined are ranked from the lowest to the highest. The lowest score is assigned rank 1, the second lowest rank 2, and the highest rank N, the total number of observations in the K samples. The sum of the ranks for the observations of each sample is then computed, and is designated R_k for $k = 1, 2, \ldots, K$. It can be shown that if n_k, the size of sample k, is larger than 5 for all k, then the test statistic H defined as

$$H = \frac{12}{N(N+1)} \sum_{k=1}^{K} \frac{R_k^2}{n_k} - 3(N+1) \tag{17-10}$$

is distributed approximately as χ^2 with $K - 1$ degrees of freedom. For testing the null hypothesis that the K samples are from populations with the same mean, we compare the value of H with the critical χ^2 value at a given level of significance.

Suppose, for example, that we wish to determine whether the wages for machinists, carpenters, and electricians in a certain community differ significantly from one another. That is, we wish to test the null hypothesis that the wages for workers in the three trades are the same against the alternative that they are different. Suppose also that three ($K = 3$) independent samples are taken respectively, and that the size of the sample, the weekly wages, and the ranks of wages in all three samples combined are as indicated in Table 17-5. Thus the total number of observations in the three samples is

$$N = 7 + 7 + 6 = 20.$$

Substituting these quantities in Eq. (17-10), we obtain

$$H = \frac{12}{20(20+1)} \sum_{k=1}^{3} \frac{R_k^{\,2}}{n_k} - 3(20+1)$$

$$= \frac{12}{420}\left(\frac{74^2}{7} + \frac{36^2}{7} + \frac{100^2}{6}\right) - 63$$

$$= 12.26$$

Referring to Appendix VIII, we find that the critical χ^2 value that cuts off the upper 1 percent of the distribution with 2 degrees of freedom is 9.21034. Since our observed H value is greater than this critical value, we reject the null hypothesis. We conclude that the wages earned by workers in the above-mentioned three trades are different at $\alpha = 0.01$.

The Kruskal-Wallis test is more powerful than a similar nonparametric test such as the median test when extended to three or more groups. Although it is somewhat less powerful than the parametric F test, it is still often used because of its simplicity and ease of application.

Table 17-5

Weekly wages and their ranks for machinists, carpenters, and electricians

Machinists		*Carpenters*		*Electricians*	
Wages, \$	*Rank*	*Wages, \$*	*Rank*	*Wages, \$*	*Rank*
310	8.5	312	10	315	11
316	12.5	307	5	325	18
322	16	304	3	330	19
316	12.5	310	8.5	332	20
318	14	300	1	323	17
305	4	308	6.5	320	15
308	6.5	301	2		
$R_1 = 74.0$		$R_2 = 36.0$		$R_3 = 100.0$	
$n_1 = 7$		$n_2 = 7$		$n_3 = 6$	

Exercise 17-3

1 To what parametric test is the Mann-Whitney test a convenient alternative? Explain fully.

2 What is the particular condition that makes it appropriate to apply the Mann-Whitney U test rather than the sign test or the Wilcoxon T test?

3 Use numerical values to verify the statement, "It does not matter which value of U, larger or smaller, is used, when the normal approximation is employed to test the hypothesis about the difference between two means."

4 Explain the rationale underlying the U test about the difference between the variabilities of two distributions.

5 Suppose that two samples, I and II, contain the following data:

Sample I	*Sample II*
3	7
8	1
2	11
5	9
4	10

Find the value of the statistic U.

6 The following data represent the annual incomes of insurance salesmen and stockbrokers, who are randomly selected from the two industries:

Insurance salesmen I	*Stockbrokers S*	
$ 8,500	$ 8,200	$10,700
9,200	9,500	11,000
9,600	9,800	11,500
7,800	9,400	16,800
8,000	9,700	12,500
9,100	9,900	9,950
10,000	8,600	10,300
13,500	12,800	14,200
12,000	13,700	13,000
	10,900	12,700
	$9,350	

a Use the U test to determine whether the annual incomes of insurance salesmen and stockbrokers are the same at $\alpha = 0.05$.

b Use the U test to determine whether the dispersions of the two distributions are the same at $\alpha = 0.05$.

7 Suppose that we wish to compare the grades received by male and female students in a certain standard test. Two randomly selected samples of 20 students each produced the following results:

Male students M points	Female students F points
51	45
90	55
68	95
83	80
65	70
75	50
71	99
85	88
79	74
84	60
87	67
72	82
76	86
92	98
69	62
91	97
63	93
71	61
78	73
59	94

a Employ the U test to check whether the mean grades of male and female students are the same at $\alpha = 0.05$.

b Employ the U test to check whether the dispersions of the two distributions are the same at $\alpha = 0.05$.

8 To what parametric test is the H test a convenient alternative? Compare these two tests as fully as you can.

9 Explain the differences and similarities between the Kruskal-Wallis H test and the Mann-Whitney U test.

10 Refer to the example in Sec. 13-3-6. Employ the H test to determine whether the means of the three populations are the same at $\alpha = 0.05$. Is your conclusion different from that in Sec. 13-3-6?

11 Refer to Prob. 4 of Exercise 13-3. Employ the H test to determine whether the mean grades of the three classes, I, II, and III, are equal.

12 Four groups, *A*, *B*, *C*, and *D*, are involved in an experiment. The items in each group are randomly selected, and the following data represent the results of the experiment:

Group A	*Group B*	*Group C*	*Group D*
57	43	87	60
81	61	69	70
67	42	58	75
64	45	82	72
96	39	90	79
80		91	76
68			
56			

Use the Kruskal-Wallis *H* test to decide whether the four groups are taken from populations with the same mean at $\alpha = 0.05$.

17-4 Rank-Correlation Methods

In discussing correlation between two variables in Chap. 14, it was assumed that the two variables had a joint normal distribution and that the conditional variance of one variable, given the other, was the same. In situations where the truth of these assumptions is doubtful, we may use other techniques generally known as the *rank-correlation methods.*

Two rank-correlation methods will be considered in this section. One is based on the *coefficient* of rank correlation, denoted by r_S, which was proposed by Carl Spearman in 1904; the other, a more recently developed test, is applied by computing the coefficient τ (a Greek letter read as "tau"), which was first described by Maurice G. Kendall in 1938. Both methods are nonparametric and are carried out by ranking the values of each variable.

Since the relation between the variables is analyzed according to the ranks of each variable, the methods can be applied to situations in which exact numerical measurements are not available. For instance, it may be very difficult to measure exactly the morale of each employee in a group, but it would be a rather simple job for the manager to rank each of them.

17-4-1 Spearman's method

We shall first consider the coefficient of rank correlation, r_S, which is more widely used. The method calls for the *sum* of the *squared differences* between each pair of ranks, after one of the two variables to be correlated is arranged in order of ranks. Then, if no tie in ranks exists, we can apply the formula for Spearman's coefficient of rank correlation to find r_S as follows:

$$r_S = 1 - \frac{6\sum_{i=1}^{n} d_i^2}{n(n^2-1)} \tag{17-11}$$

where d_i is the difference between ranks of the ith pair and n is the number of pairs included.

Let us use a simple and practical example to illustrate the application of this method. Suppose we wish to determine whether the scores received by employees in a certain test are correlated with their job performance in terms of the number of units produced for a given period. Let X be the test score, and Y be the number of units produced. Ten employees are selected; the test scores and measures of job performance are presented in Table 17-6. Ranks of the values of each variable, the differences between paired ranks, and the squares of the differences are also provided. Substituting the values of n and the sum of squared differences in Eq. (17-11), we have

$$r_S = 1 - \frac{6(24)}{10(10^2-1)} = 1 - \frac{144}{990} = 1 - 0.145 = 0.855$$

Table 17-6

Illustration of the rank-correlation method

Employees	Test score X	Rank	Units produced Y	Rank	Difference between ranks d	d^2
A	65	1	30	2	1	1
B	70	2	25	1	1	1
C	76	4	35	3	1	1
D	75	3	40	5	2	4
E	80	6	38	4	2	4
F	78	5	42	6	1	1
G	83	7	48	8	1	1
H	84	8	50	9	1	1
I	85	9	55	10	1	1
J	90	10	45	7	3	9
Total						$\sum_{i=1}^{10} d_i^2 = 24$

Spearman's rank-correlation coefficient may be tested by applying Eq. (14-31); that is,

$$t = r_S\sqrt{\frac{n-2}{1-r_S^2}}$$

which is distributed as Student's t with $n - 2$ degrees of freedom. Thus we get

$$t = 0.855\sqrt{\frac{10-2}{1-(0.855)^2}} = 0.855\sqrt{\frac{8}{0.269}}$$

$$= 0.855(5.45)$$

$$= 4.66$$

With $10 - 2 = 8$ degrees of freedom, the critical t value for a two-tailed test at $\alpha = 0.01$ is 3.355. The computed value is higher than the critical t value at $\alpha = 0.01$. We conclude that the correlation between the test scores of the employees and their job performance is significant at the 1 percent level.

The hypothesis tested is that the two variables with rank values are *independent*; that is, there is no particular association between the ranks of one variable and the ranks of the other. If the rank order of one variable agrees with that of the other, there will be a positive correlation between the rank orders, while a negative coefficient reflects a negative correlation. In the above example, some positive correlation is manifest between the two variables.

17-4-2 Kendall's method

Using the same type of data for which the Spearman rank-correlation coefficient is useful, we can compute another statistic (designated τ) called the *Kendall rank-correlation coefficient*. Like r_S, τ is a measure of the degree of correlation between two sets of ranks. Suppose that two executives, I and II, are to rank five employees, A, B, C, D, and E, in terms of their performances, and that the following rankings are obtained:

Employees	*A*	*B*	*C*	*D*	*E*
Executive I	2	3	1	5	4
Executive II	3	1	2	4	5

Let executive I's rankings be rearranged in the natural order; then we have

Employees	*C*	*A*	*B*	*E*	*D*
Executive I	1	2	3	4	5
Executive II	2	3	1	5	4

The next step in computing τ is to compare every pair of ranks assigned by II, given that I's rankings are in the natural order. Each comparison yields a score $+1$ or -1 depending on whether the pair is in natural or in reverse order. The algebraic sum of all the scores, designated S, is the numerator of the coefficient τ. We know that II's rankings of the five employees C, A, B, E, and D are, respectively, 2, 3, 1, 5, and 4. Starting with the first number, or 2, we count the number of ranks to the right that are greater; there are three such ranks, namely, 3, 5, and 4. Then we subtract from 3 the number of ranks to the right that are smaller. Since there is only one such rank, the net result is $+3 - 1 = +2$. Next we move to the second number on the left, or 3. There are two ranks to its right that are greater and one rank to its right that is smaller; the net contribution to S is therefore $+2 - 1 = +1$. The next rank, or 1, has no rank to its right that is smaller, but two ranks that are greater; thus a score $+2$ is contributed to S. Finally, the rank 5 has only one smaller rank to its right, and therefore adds -1 to S. Consequently, the total score is obtained as follows:

$$S = +2 + 1 + 2 - 1 = +4$$

Recall that the total number of combinations of n objects taken two at a time is $\binom{n}{2}$. Accordingly, the total number of possible pairs to be compared for n observations is

$$\binom{n}{2} = \frac{n!}{2!(n-2)!} = \frac{n(n-1)}{2}$$

which is also the maximum possible value for S. The sum S would reach the maximum value only if the two rankings by I and II are in complete agreement. This is the case because, when I's rankings are in natural order and every pair of II's rankings is in natural order, every pair of II's rankings will receive a score $+1$, which gives $\binom{n}{2}$ possible pairs. Observe that if the two rankings are in perfectly reverse order, the value of S would be $-\binom{n}{2}$ since each pair would receive a score -1.

The Kendall rank-correlation coefficient is the ratio of S to the total number of possible pairs; that is,

$$\tau = \frac{S}{n(n-1)\frac{1}{2}} \tag{17-12}$$

In the above illustration, there are five observations ($n = 5$). Thus we have

$$\tau = \frac{S}{n(n-1)\frac{1}{2}} = \frac{+4}{5(5-1)\frac{1}{2}} = \frac{+4}{10} = 0.40$$

The value of τ ranges from -1 to $+1$. When n is larger than 10, then τ has an approximately normal distribution with a mean

$$E(\tau) = 0 \tag{17-13}$$

and standard deviation

$$\sigma_\tau = \sqrt{\frac{2(2n+5)}{9n(n-1)}} \tag{17-14}$$

Then we may test the null hypothesis that no association or correlation exists between two sets of observations against the alternative that there is such correlation by computing the standardized normal score as follows:

$$Z = \frac{\tau - E(\tau)}{\sigma_\tau} = \frac{\tau}{\sqrt{2(2n+5)/9n(n-1)}}$$

For $n \leq 10$, the exact probability associated with the occurrence of S, and consequently τ because τ is a function of S, is found from a table provided by Kendall.[1]

To illustrate the method of normal approximation to the sampling distribution of τ, let us extend the above example to 11 employees to be ranked by I and II. Suppose that the two sets of ranks appear as follows:

Employees	*A*	*B*	*C*	*D*	*E*	*F*	*G*	*H*	*I*	*J*	*K*
I's rankings	4	2	6	1	3	7	5	10	11	8	9
II's rankings	3	2	5	4	1	11	10	9	8	6	7

After I's rankings are rearranged in their natural order, the two sets of rankings become

Employees	*D*	*B*	*E*	*A*	*G*	*C*	*F*	*J*	*K*	*H*	*I*
I's rankings	1	2	3	4	5	6	7	8	9	10	11
II's rankings	4	2	1	3	10	5	11	6	7	9	8

[1] See Maurice G. Kendall, "Rank Correlation Methods" table 1, appendix, Charles Griffin & Company, Ltd., London, 1948.

Given I's rankings in their natural order, we can compute the value of S for the corresponding rankings by II as follows:

$$S = (7-3) + (8-1) + (8-0) + (7-0) + (1-5) + (5-0) + (0-4) + (3-0) + (2-0) + (0-1)$$

$$= 4 + 7 + 8 + 7 - 4 + 5 - 4 + 3 + 2 - 1$$

$$= +27$$

Since $n = 11$ and $S = +27$, we can compute the value of τ by applying Eq. (17-12)

$$\tau = \frac{+27}{11(11-1)\frac{1}{2}} = \frac{27}{55} = 0.49$$

This gives

$$Z = \frac{0.49}{\sqrt{2[2(11)+5]/9(11)(11-1)}} = \frac{0.49}{0.2336} = 2.097$$

which is larger than 1.960, the critical z score at $\alpha = 0.05$, but smaller than 2.575, the critical z score at $\alpha = 0.01$. Thus we may reject the null hypothesis that there is no correlation between the two sets of rankings at $\alpha = 0.05$, but will not reject it at $\alpha = 0.01$.

If two or more items on either of the two sets of rankings are tied, we follow the usual procedure of assigning the average of the ranks to the tied observations. Then the same procedure is followed to compute the value of τ.

The values of r_S and τ may not be the same even if the same data are used in the computations. Nevertheless, they have the same power to detect the existence of correlation between the populations. Given a set of data, both tests will lead to the same conclusion in regard to the null hypothesis. There is, however, one advantage of τ over r_S; that is, τ can be applied to partial correlation analysis, while r_S cannot. The discussion of partial correlation is not within the scope of this book.

Exercise 17-4

1 Explain the differences and similarities between Spearman's and Kendall's rank-correlation methods.

2 Can rank-correlation methods be applied to situations in which exact numerical measurements are not available? Explain.

3 Suppose you wish to determine whether high school graduates' performance in mathematics is correlated with their performance in English. Two standard tests, one in mathematics and the other in English, are administered to a group of 12 graduates. The following data are the test scores they received:

Students	*Scores in mathematics*	*Scores in English*
A	82	79
B	75	80
C	95	90
D	99	89
E	85	91
F	70	65
G	77	67
H	60	62
I	63	61
J	66	68
K	80	81
L	89	84

a Find the value of Spearman's r_S.
b Is the correlation between the test scores in mathematics and English significant at the 1 percent level?

4 Referring to the preceding problem, find:

a The value of Kendall's τ.
b Test the null hypothesis that there is no correlation between the perfomances in mathematics and English at the 1 percent level by using the statistic τ.

5 Two judges, I and II, are to rate 11 players of the offensive team of a football club. The ratings in ranks are shown below.

Players	*Ranks given by I*	*Ranks given by II*
A	6	7
B	7	5
C	10	11
D	1	3
E	3	4
F	2	1
G	5	2
H	9	8
I	11	9
J	4	6
K	8	10

a Find the value of Spearman's r_s.
b Use r_S to decide whether there is any signicant correlation between the ratings by I and II at the 1 percent level.

6 Do the preceding problem by using Kendall's rank-correlation method.

7 Five supervisors are rated by two vice presidents, I and II, of a manufacturing company. The ratings in ranks are given below.

Supervisors	*Ranks given by I*	*Ranks given by II*
A	1	1
B	3	3
C	5	5
D	4	4
E	2	2

Find the value of Kendall's rank-correlation coefficient τ.

8 Suppose that in the preceding problem the five supervisors are rated by vice president I and the personnel manager. The latter rates *A*, *B*, *C*, *D*, and *E* as 5, 3, 1, 2, and 4, respectively. Find the value of τ.

Supplements

Supplement I
The Variance Is Minimized If Computed from the Mean

For demonstrating that the variance has the smallest value if it is computed from the mean, let us assume that a "variance" ${s_a}^2$ is calculated by subtracting an arbitrary value a from each item, squaring the difference, and then averaging. That is,

$$ {s_a}^2 = \frac{\sum_{i=1}^{n} (X_i - a)^2}{n} $$

If we subtract and then add $\bar{X}$ for each item, the value of ${s_a}^2$ will not change. Thus

$$ \begin{aligned} {s_a}^2 &= \frac{1}{n} \sum_{i=1}^{n} (X_i - \bar{X} + \bar{X} - a)^2 \\ &= \frac{1}{n} \sum_{i=1}^{n} [(X_i - \bar{X}) + (\bar{X} - a)]^2 \\ &= \frac{1}{n} \sum_{i=1}^{n} [(X_i - \bar{X})^2 + 2(\bar{X} - a)(X_i - \bar{X}) + (\bar{X} - a)^2] \\ &= \frac{1}{n} \sum_{i=1}^{n} (X_i - \bar{X})^2 + 2(\bar{X} - a) \sum_{i=1}^{n} (X_i - \bar{X}) + n(\bar{X} - a)^2 \qquad \text{by (4-7) and (4-5)} \\ &= \frac{1}{n} \sum_{i=1}^{n} (X_i - \bar{X})^2 + n(\bar{X} - a)^2 \qquad \text{by (4-9)} \\ &= s^2 + (\bar{X} - a)^2 \end{aligned} $$

The last term $(\overline{X} - a)^2$ is a squared value and must be positive or 0. When it is 0, that is, when a is identical with $\overline{X}$, s_a^2 is equal to s^2. For any other value of a, s_a^2 will be larger than s^2. In short, the variance, and hence the standard deviation, from the arithmetic mean is always smaller than from any other value.

Supplement II
$E(|X - m'|)$ Is Minimized When m' = Median of X

Let m be the median of X, and d be the difference between m and another value m'; that is, $m' = m + d$. Then, we have

$$-d \leq \begin{cases} |X - m'| - |X - m| = d & \text{for } X \leq m \\ |X - m'| - |X - m| \leq d & \text{for } m \leq X \leq m' \\ |X - m'| - |X - m| = -d & \text{for } X \geq m' \end{cases}$$

Let W be another random variable, and

$$W = \begin{cases} d & \text{for } X \leq m \\ -d & \text{for } X > m \end{cases}$$

Thus the above three relations can be summarized as follows:

$$|X - m'| - |X - m| \geq W$$

Since

$$E(W) = dP(X \leq m) + (-d)P(X > m) = d[P(X \leq m) - P(X > m)]$$

and

$$P(X \leq m) \geq 0.5$$

we have

$$E(W) \geq 0$$

Consequently,

$$E(|X - m')| - |X - m|) \geq E(W) \geq 0$$

$$E(|X - m'|) \geq E(|X - m|)$$

That is, $E(|X - m'|)$ is not less than $E(|X - m|)$ for $m' > m$. Similarly, it can be shown that $E(|X - m'|)$ is not less than $E(|X - m|)$ for $m' < m$. Thus $E(|X - m')$ is minimized only if m' is equal to m, the median of X.

Supplement III
If X and Y Are Independent, Then $\sigma_{X+Y}^2 = \sigma_X{}^2 + \sigma_Y{}^2$

The proof of Eq. (5-21) runs as follows: Since by Eq. (4-25) the mean of the sum X plus Y is the sum of the mean of X and the mean of Y, that is, $\mu_{X+Y} = \mu_X + \mu_Y$, we have

$$\begin{aligned}
\sigma_{X+Y}^2 &= E[(X + Y - \mu_{X+Y})^2] = E[(X + Y - \mu_X - \mu_Y)^2] \\
&= E\{[(X - \mu_X) + (Y - \mu_Y)]^2\} \\
&= E[(X - \mu_X)^2 + 2(X - \mu_X)(Y - \mu_Y) + (Y - \mu_Y)]^2 \\
&= E[(X - \mu_X)^2] + E[(Y - \mu_Y)^2] + 2E[(X - \mu_X)(Y - \mu_Y)] \\
&= \sigma_X{}^2 + \sigma_Y{}^2 + 2E[(X - \mu_X)(Y - \mu_Y)]
\end{aligned}$$

The last term contains the factors of 2 and the population covariance of X and Y. It can be demonstrated that if X and Y are independent, the covariance vanishes, as indicated below:

$$\begin{aligned}
E[(X - \mu_X)(Y - \mu_Y)] &= E(XY - \mu_X Y - X\mu_Y + \mu_X \mu_Y) \\
&= E(XY) - \mu_X E(Y) - E(X)\mu_Y + E(\mu_X \mu_Y)
\end{aligned}$$

The first term $E(XY)$ on the right side of this equation is equal to $E(X)E(Y)$. For proving that $E(XY) = E(X)E(Y)$ if X and Y are independent, let us suppose that X has the set of values $\{x_1, x_2, \ldots, x_j\}$ with probabilities $P(x_1), P(x_2), \ldots, P(x_j)$, and that Y has the set of values $\{y_1, y_2, \ldots, y_k\}$ with probabilities $P(y_1), P(y_2), \ldots, P(y_k)$. Then the product of $X + Y$ assumes the values $x_1y_1, x_1y_2, \ldots, x_1y_k, x_2y_1, x_2y_2, \ldots, x_2y_k, \ldots, x_jy_1, x_jy_2, \ldots, x_jy_k$ with probabilities $P(x_1)P(y_1), P(x_1)P(y_2), \ldots, P(x_1)P(y_k), P(x_2)P(y_1), P(x_2)P(y_2), \ldots, P(x_2)P(y_k), \ldots, P(x_j)P(y_1), P(x_j)P(y_2), \ldots, P(x_j)P(y_k)$. Thus

$$\begin{aligned}
E(XY) &= x_1y_1P(x_1)P(y_1) + x_1y_2P(x_1)P(y_2) + \cdots + x_1y_kP(x_1)P(y_k) \\
&\quad + x_2y_1P(x_2)P(y_1) + x_2y_2P(x_2)P(y_2) + \cdots + x_2y_2P(x_2)P(y_k) \\
&\quad + \cdots + x_jy_1P(x_j)P(y_1) + x_jy_2P(x_j)P(y_2) \\
&\quad + \cdots + x_jy_kP(x_j)P(y_k) \\
&= x_1P(x_1)[y_1P(y_1) + y_2P(y_2) + \cdots + y_kP(y_k)] + x_2P(x_2)[y_1P(y_1) \\
&\quad + y_2P(y_2) + \cdots + y_kP(y_k)] + \cdots + x_jP(x_j)[y_1P(y_1) \\
&\quad + y_2P(y_2) + \cdots + y_kP(y_k)] \\
&= [x_1P(x_1) + x_2P(x_2) + \cdots + x_jP(x_j)][y_1P(y_1) + y_2P(y_2) \\
&\quad + \cdots + y_kP(y_k)] \\
&= E(X) \times E(Y) \qquad \text{by (4-19).}
\end{aligned}$$

That is,

If X and Y are two independent random variables, then

$$E(XY) = E(X)E(Y)$$

Consequently, we have

$$\begin{aligned} E[(X - \mu_X)(Y - \mu_Y)] &= E(X)E(Y) - \mu_X \mu_Y - \mu_X \mu_Y + \mu_X \mu_Y \\ &= \mu_X \mu_Y - \mu_X \mu_Y = 0 \end{aligned}$$

Thus

$$\sigma_{X+Y}^2 = \sigma_X{}^2 + \sigma_Y{}^2$$

Supplement IV
Tchebycheff's Inequality

To prove Tchebycheff's inequality, we let the variable X have the set of values $\{x_1, x_2, \ldots, x_m\}$, the sample of n observations have a mean $\overline{X}$ and a standard deviation s, and also let

$$C = \{x: |x - \overline{X}| > ks\}$$

where k is any positive number. By the definition of sample variance, we have

$$ns^2 = \sum_{i=1}^{n} (X_i - \overline{X})^2$$

If the data are organized into a frequency distribution, then we have

$$\sum_{i=1}^{n} (X_i - \overline{X})^2 = \sum_{j=1}^{m} (x_j - \overline{X})^2 FR(x_j)$$

$$ns^2 = \sum_{j=1}^{m} (x_j - \overline{X})^2 FR(x_j) \geq \sum_{x_j \in C} (x_j - \overline{X})^2 FR(x_j)$$

Since

$$\begin{aligned} &|x_j - \overline{X}| > ks \qquad \text{for } x_j \in C \\ &(x_j - \overline{X})^2 > k^2 s^2 \end{aligned}$$

and

$$\begin{aligned} \sum_{x_j \in C} (x_j - \overline{X})^2 FR(x_j) &> \sum_{x_j \in C} k^2 s^2 FR(x_j) \\ &= k^2 s^2 \sum_{x_j \in C} FR(x_j) \qquad \text{by (4-6)} \end{aligned}$$

Thus

$$ns^2 > k^2s^2FR(C)$$

Dividing each side of this inequality by nk^2s^2 gives

$$\frac{1}{k^2} > \frac{FR(C)}{n} = RF(C)$$

Hence

$$1 - \frac{1}{k^2} \leq 1 - RF(C) = RF(|X - \overline{X}| \leq ks)$$

Supplement V
The Mean of the Binomial Random Variable

The equation $\mu = np$ can also be proved in terms of probabilities of the x's. Since $P(x) = \binom{x}{n} p^x q^{n-x}$ for $x = 0, 1, 2, \ldots, n$, we have

$$\mu = \sum_{x=0}^{n} xP(x)$$

$$= \sum_{x=0}^{n} x\binom{n}{x} p^x q^{n-x} \qquad \text{by (7-4)}$$

$$= \sum_{x=1}^{n} x\binom{n}{x} p^x q^{n-x} + 0\binom{n}{0} p^0 q^{n-0} = \sum_{x=1}^{n} x \frac{n!}{x!(n-x)!} p^x q^{n-x}$$

$$= \sum_{x=1}^{n} x \frac{n(n-1)!}{x(x-1)!(n-x)!} \times p \times p^{x-1} q^{n-x}$$

$$= np \sum_{x=1}^{n} \frac{(n-1)!}{(x-1)!(n-x)!} p^{x-1} q^{n-x}$$

Let $x = k + 1$, then $k = x - 1$, and so

$$\mu = np \sum_{k=0}^{n-1} \frac{(n-1)}{k!(n-1-k)!} p^k q^{n-1-k} = np \sum_{k=0}^{n-1} \binom{n-1}{k} p^k q^{n-1-k}$$

$$= np(1) \qquad \text{by (7-5)}$$

$$= np$$

Supplement VI
The Poisson Function

Since $P(X = x) = \binom{n}{x} p^x(1-p)^{n-x}$, we have, as n approaches infinity, p approaches 0, and np approaches μ,

$$P(X = x)$$

$$= \lim_{\substack{n\to\infty \\ p\to 0}} \binom{n}{x} p^x(1-p)^{n-x} = \lim_{\substack{n\to\infty \\ p\to 0}} \frac{n!}{x!\,(n-x)!} p^x(1-p)^{n-x}$$

$$= \lim_{\substack{n\to\infty \\ p\to 0}} \frac{n(n-1)(n-2)\cdots(n-x+1)(n-x)!}{x!\,(n-x)!} p^x(1-p)^{n-x}$$

$$= \lim_{\substack{n\to\infty \\ p\to 0}} \frac{n(n-1)(n-2)\cdots(n-x+1)}{x!\,n^x} n^x p^x(1-p)^{n-x}$$

$$= \lim_{\substack{n\to\infty \\ p\to 0}} \frac{n(n-1)(n-2)\cdots(n-x+1)}{n\times n\times n\times\cdots\times n} \frac{(np)^x}{x!}(1-p)^{n-x}$$

$$= \lim_{\substack{n\to\infty \\ p\to 0}} \frac{n}{n}\times\frac{n-1}{n}\times\frac{n-2}{n}\cdots\frac{n-x+1}{n}\frac{\mu^2}{x!}(1-p)^{n-x}$$

$$= \lim_{\substack{n\to\infty \\ p\to 0}} 1\times\left(1-\frac{1}{n}\right)\times\left(1-\frac{2}{n}\right)\times\cdots\times\left(1-\frac{x-1}{n}\right)\frac{\mu^x}{x!}(1-p)^n(1-p)^{-x}$$

$$= 1\times 1\times 1\times\cdots\times 1 \lim_{\substack{n\to\infty \\ p\to 0}} \left[(1-p)^{\frac{-1}{p}}\right]^{-np} \frac{\mu^x}{x!}(1-p)^{-x}$$

But

$$\lim_{z\to 0}(1+z)^{1/z} = e$$

where

$$z = -p$$

and

$$\lim_{p\to 0}(1-p)^{-x} = 1$$

hence

$$P(X = x) = e^{-np}\frac{\mu^x}{x!} = e^{-\mu}\frac{\mu^x}{x!}$$

Supplement VII
Biased Estimator of the Population Variance

The sample variance as defined in Eq. (5-2) is a biased estimator of the population variance, since $E(s^2) \neq \sigma^2$. This can be demonstrated as follows:

$$E(s^2) = E\left(\frac{\sum_{i=1}^{n} X_i^2}{n} - \overline{X}^2\right) = E\left(\frac{\sum_{i=1}^{n} X_i^2}{n}\right) - E(\overline{X}^2)$$

$$= \frac{1}{n} E\left(\sum_{i=1}^{n} X_i^2\right) - E(\overline{X}^2) \qquad \text{by (4-23)}$$

Let us now consider separately the two terms on the right side of this equation. By Eqs. (4-27) and (5-17), the first term becomes

$$\frac{1}{n} E\left(\sum_{i=1}^{n} X_i^2\right) = \frac{1}{n}\sum_{i=1}^{n} E(X_i^2) = \frac{1}{n}\sum_{i=1}^{n} (\sigma_{X_i}^2 + \mu^2) = \sigma_X^2 + \mu^2$$

Since the variance of the sample mean is given by

$$\sigma_{\overline{X}}^2 = E(\overline{X}^2) - [E(\overline{X})]^2 \qquad \text{by (5-17)}$$

the second term becomes

$$E(\overline{X}^2) = \sigma_{\overline{X}}^2 + \mu^2$$

Substituting these two quantities in the expression for $E(s^2)$, we have

$$E(s^2) = \sigma_X^2 + \mu^2 - (\sigma_{\overline{X}}^2 + \mu^2)$$
$$= \sigma_X^2 - \sigma_{\overline{X}}^2$$

That is, the expected value of the sample variance is equal to the difference between the population variance and the variance of the sample mean. Since the sample size cannot be infinitely large, the variance of the sample mean must not be 0; consequently, the sample variance s^2 is a *biased estimator* of the population variance σ^2.

Supplement VIII
Unbiased Estimator of the Population Variance

Since

$$\hat{s}^2 = \frac{\sum_{i=1}^{n} (X_i - \overline{X})^2}{n-1} \qquad \text{by definition}$$

we have

$$E(\hat{s}^2) = E\frac{\sum_{i=1}^{n} (X_i - \overline{X})^2}{n-1} = \frac{1}{n-1} E\left[\sum_{i=1}^{n} (X_i - \overline{X})^2\right]$$

Multiplying each side of this equation by $n - 1$, we have

$$(n-1)E(\hat{s}^2) = E\left[\sum_{i=1}^{n}(X_i - \overline{X})^2\right]$$

$$= E\left\{\sum_{i=1}^{n}[(X_i - \mu) - (\overline{X} - \mu)]^2\right\} \qquad \text{by adding and subtracting } \mu$$

$$= E\left\{\sum_{i=1}^{n}[(X_i - \mu)^2 - 2(X - \mu)(\overline{X} - \mu) + (\overline{X} - \mu)^2]\right\}$$

$$= E\left[\sum_{i=1}^{n}(X_i - \mu)^2 - 2\sum_{i=1}^{n}(X_i - \mu)(\overline{X} - \mu) + n(\overline{X} - \mu)^2\right]$$

Since $\overline{X} - \mu$ is a constant, the second term in the bracket can be written as

$$2(\overline{X} - \mu)\sum_{i=1}^{n}(X_i - \mu)$$

But

$$\sum_{i=1}^{n}(X_i - \mu) = \sum_{n=1}^{n}X_i - n\mu = n\overline{X} - n\mu = n(\overline{X} - \mu)$$

Hence

$$(n-1)E(\hat{s}^2) = E\left[\sum_{i=1}^{n}(X_i - \mu)^2 - 2n(\overline{X} - \mu)^2 + n(\overline{X} - \mu)^2\right]$$

$$= E\left[\sum_{i=1}^{n}(X_i - \mu)^2 - n(\overline{X} - \mu)^2\right]$$

$$= \sum_{i=1}^{n}E[(X_i - \mu)^2] - nE[(\overline{X} - \mu)^2]$$

$$= n\sigma^2 - n\frac{\sigma^2}{n} = (n-1)\sigma^2$$

and

$$E(\hat{s}^2) = \sigma^2$$

Let us now use a simple example to show that the sample variance $\hat{s}^2$ is the unbiased estimator of the population variance. Recall that in Sec. 6-3 the random variable X has the set of values {2, 4, 6, 8}, and that samples of two observations each are taken with replacement from this simulated population. As pointed out earlier, there are 16 *different* possible samples that could be obtained. They are listed in the table below together with their means, variances, and standard deviations. The computation of the sample means is self-explanatory. The variances are computed by using Eq. (9-4). For example, the variance of the fourth sample is

$$\frac{(2-5)^2 + (8-5)^2}{2-1} = \frac{3^2 + 3^2}{1} = 18$$

Computation of sample variances of all possible samples of size 2 for X having the set of values {2, 4, 6, 8}

Sample	Sample values X_1	X_2	Sample mean $\bar{X}$	Variances $\hat{s}^2$	Standard deviation $\hat{s}$
1	2	2	2	0	0
2	2	4	3	2	1.4142
3	2	6	4	8	2.8284
4	2	8	5	18	4.2426
5	4	2	3	2	1.4142
6	4	4	4	0	0
7	4	6	5	2	1.4142
8	4	8	6	8	2.8284
9	6	2	4	8	2.8284
10	6	4	5	2	1.4142
11	6	6	6	0	0
12	6	8	7	2	1.4142
13	8	2	5	18	4.2426
14	8	4	6	8	2.8284
15	8	6	7	2	1.4142
16	8	8	8	0	0
Totals				80	28.2840

The rest of the computations can be very easily verified.

We know that the variance of the population, or $\sigma_X{}^2$, is

$$\sigma_X{}^2 = E(X^2) - [E(X)]^2 \qquad \text{by definition}$$

$$= 2^2\left(\frac{1}{4}\right) + 4^2\left(\frac{1}{4}\right) + 6^2\left(\frac{1}{4}\right) + 8^2\left(\frac{1}{4}\right) - 5^2$$

$$= \frac{120}{4} - 25 = 30 - 25 = 5$$

The expectation of the sample variance $\hat{s}^2$ is computed by dividing the sum of the $\hat{s}^2$ values by the number of possible samples as follows:

$$E(\hat{s}^2) = \frac{80}{16} = 5$$

This average is precisely the value of the population parameter $\sigma_X{}^2$. We have thus demonstrated that $\hat{s}^2$ is the unbiased estimator of $\sigma_X{}^2$.

Observe, however, that the expected value of the standard deviation $\hat{s}$, namely

$$E(\hat{s}) = \frac{28.284}{16} = 1.78025$$

is different from σ_X, the standard deviation of the population, which is

$$\sqrt{\sigma^2} = \sqrt{5} = 2.2361$$

This is an arithmetic demonstration that the standard deviation of the sample is *not* an unbiased estimator of the population parameter σ_X.

Supplement IX
The Variances for b, Y_c, and Y_a

Since

$$Y_c = a + bX \qquad \text{and} \qquad a = \overline{Y} - b\overline{X}$$

We have

$$Y_c = \overline{Y} - b\overline{X} + bX = \overline{Y} + b(X - \overline{X})$$

The variance of the computed Y value, or $\sigma^2_{Y_c}$, can be expressed as follows:

$$\sigma^2_{Y_c} = \sigma^2_{[\overline{Y} + b(X - \overline{X})]}$$

Because of the independence between $\overline{Y}$ and b, it follows that

$$\sigma^2_{Y_c} = \sigma^2_{\overline{Y}} + \sigma^2_{b(X - \overline{X})}$$

Since, for a given X, $\sigma_Y{}^2 = \sigma^2_{Y|X}$, the first term on the right side of this equation is

$$\sigma_{\overline{Y}}{}^2 = \sigma^2_{\Sigma Y/n} = \left(\frac{1}{n}\right)^2 \sigma^2_{\Sigma Y} = \frac{1}{n^2}(\sigma^2_{Y_1} + \sigma^2_{Y_2} + \cdots + \sigma^2_{Y_n})$$

$$= \frac{1}{n^2}(\sigma^2_{Y|X} + \sigma^2_{Y|X} + \cdots + \sigma^2_{Y|X})$$

$$= \frac{1}{n^2} n\sigma^2_{Y|X} = \frac{\sigma^2_{Y|X}}{n}$$

Because $X - \overline{X}$ is a constant, the second term becomes

$$\sigma^2_{b(X - \overline{X})} = (X - \overline{X})^2 \sigma_b{}^2$$

We know, however, that

$$b = \frac{\operatorname{cov}(X, Y)}{\sum(X - \bar{X})^2/n} = \frac{\sum(X - \bar{X})(Y - \bar{Y})}{\sum(X - \bar{X})^2}$$

$$= \frac{\sum XY - \sum \bar{X}Y - \sum X\bar{Y} + n\bar{X}\bar{Y}}{\sum(X - \bar{X})^2}$$

$$= \frac{\sum(XY - \bar{X}Y) - n\bar{X}\bar{Y} + n\bar{X}\bar{Y}}{\sum(X - \bar{X})^2} = \frac{\sum Y(X - \bar{X})}{\sum(X - \bar{X})^2}$$

Thus

$$\sigma_b{}^2 = \left[\frac{1}{\sum(X - \bar{X})^2}\right]^2 \sum(X - \bar{X})^2\sigma^2_{Y|X} = \frac{\sigma^2_{Y|X}}{\sum(X - \bar{X})^2}$$

and

$$\sigma^2_{Y_c} = \frac{\sigma^2_{Y|X}}{n} + (X - \bar{X})^2 \frac{\sigma^2_{Y|X}}{\sum(X - \bar{X})^2}$$

$$= \sigma^2_{Y|X}\left[\frac{1}{n} + \frac{(X - \bar{X})^2}{\sum(X - \bar{X})^2}\right]$$

The variability of Y_c from Y means the variability of the error $Y_c - Y$. Accordingly, the variance of Y_a can be expressed as follows:

$$\sigma^2_{Y_a} = \sigma^2_{(Y_c - Y)} = \sigma_Y{}^2 + \sigma^2_{Y_c}$$

$$= \sigma^2_{Y|X} + \sigma^2_{Y|X}\left[\frac{1}{n} + \frac{(X - \bar{X})^2}{\sum(X - \bar{X})^2}\right] = \sigma^2_{Y|X}\left[1 + \frac{1}{n} + \frac{(X - \bar{X})^2}{\sum(X - \bar{X})^2}\right]$$

Since we use $\hat{s}^2_{Y|Y}$ as an estimator for $\sigma^2_{Y|X}$, the various sample variances are expressed below:

$$\hat{s}_b{}^2 = \frac{\hat{s}^2_{Y|X}}{\sum(X - \bar{X})^2} = \frac{\hat{s}^2_{Y|X}}{\sum X^2 - (\sum X)^2/n}$$

$$\hat{s}^2_{Y_c} = \hat{s}^2_{Y|X}\left[\frac{1}{n} + \frac{(X - \bar{X})^2}{\sum(X - \bar{X})^2}\right]$$

$$\hat{s}^2_{Y_a} = \hat{s}^2_{(Y - Y_c)} = \hat{s}^2_{Y|X}\left[1 + \frac{1}{n} + \frac{(X - \bar{X})^2}{\sum(X - \bar{X})^2}\right]$$

Appendixes

Appendix I
Random Digits

10 09 73 25 33	76 52 01 35 86	34 67 35 48 76	80 95 90 91 17	39 29 27 49 45
37 54 20 48 05	64 89 47 42 96	24 80 52 40 37	20 63 61 04 02	00 82 29 16 65
08 42 26 89 53	19 64 50 93 03	23 20 90 25 60	15 95 33 47 64	35 08 03 36 06
99 01 90 25 29	09 37 67 07 15	38 31 13 11 65	88 67 67 43 97	04 43 62 76 59
12 80 79 99 70	80 15 73 61 47	64 03 23 66 53	98 95 11 68 77	12 17 17 68 33
66 06 57 47 17	34 07 27 68 50	36 69 73 61 70	65 81 33 98 85	11 19 92 91 70
31 06 01 08 05	45 57 18 24 06	35 30 34 26 14	86 79 90 74 39	23 40 30 97 32
85 26 97 76 02	02 05 16 56 92	68 66 57 48 18	73 05 38 52 47	18 62 38 85 79
63 57 33 21 35	05 32 54 70 48	90 55 35 75 48	28 46 82 87 09	83 49 12 56 24
73 79 64 57 53	03 52 96 47 78	35 80 83 42 82	60 93 52 03 44	35 27 38 84 35
98 52 01 77 67	14 90 56 86 07	22 10 94 05 58	60 97 09 34 33	50 50 07 39 98
11 80 50 54 31	39 80 82 77 32	50 72 56 82 48	29 40 52 42 01	52 77 56 78 51
83 45 29 96 34	06 28 89 80 83	13 74 67 00 78	18 47 54 06 10	68 71 17 78 17
88 68 54 02 00	86 50 75 84 01	36 76 66 79 51	90 36 47 64 93	29 60 91 10 62
99 59 46 73 48	87 51 76 49 69	91 82 60 89 28	93 78 56 13 68	23 47 83 41 13
65 48 11 76 74	17 46 85 09 50	58 04 77 69 74	73 03 95 71 86	40 21 81 65 44
80 12 43 56 35	17 72 70 80 15	45 31 82 23 74	21 11 57 82 53	14 38 55 37 63
74 35 09 98 17	77 40 27 72 14	43 23 60 02 10	45 52 16 42 37	96 28 60 26 55
69 91 62 68 03	66 25 22 91 48	36 93 68 72 03	76 62 11 39 90	94 40 05 64 18
09 89 32 05 05	14 22 56 85 14	46 42 75 67 88	96 29 77 88 22	54 38 21 45 98
91 49 91 45 23	68 47 92 76 86	46 16 28 35 54	94 75 08 99 23	37 08 92 00 48
80 33 69 45 98	26 94 03 68 58	70 29 73 41 35	53 14 03 33 40	42 05 08 23 41
44 10 48 19 49	85 15 74 79 54	32 97 92 65 75	57 60 04 08 81	22 22 20 64 13
12 55 07 37 42	11 10 00 20 40	12 86 07 46 97	96 64 48 94 39	28 70 72 58 15
63 60 64 93 29	16 50 53 44 84	40 21 95 25 63	43 65 17 70 82	07 20 73 17 90

Source: From *A Million Random Digits with 100,000 Normal Deviates*, 1955, with the kind permission of the Rand Corporation.

Appendix I—*Continued*

61 19 69 04 46	26 45 74 77 74	51 92 43 37 29	65 39 45 95 93	42 58 26 05 27
15 47 44 52 66	95 27 07 99 53	59 36 78 38 48	82 39 61 01 18	33 21 15 94 66
94 55 72 85 73	67 89 75 43 87	54 62 24 44 31	91 19 04 25 92	92 92 74 59 73
42 48 11 62 13	97 34 40 87 21	16 86 84 87 67	03 07 11 20 59	25 70 14 66 70
23 52 37 83 17	73 20 88 98 37	68 93 59 14 16	26 25 22 96 63	05 52 28 25 62
04 49 35 24 94	75 24 63 38 24	45 86 25 10 25	61 96 27 93 35	65 33 71 24 72
00 54 99 76 54	64 05 18 81 59	96 11 96 38 96	54 69 28 23 91	23 28 72 95 29
35 96 31 53 07	26 89 80 93 54	33 35 13 54 62	77 97 45 00 24	90 10 33 93 33
59 80 80 83 91	45 42 72 68 42	83 60 94 97 00	13 02 12 48 92	78 56 52 01 06
46 05 88 52 36	01 39 00 22 86	77 28 14 40 77	93 91 08 36 47	70 61 74 29 41
32 17 90 05 97	87 37 92 52 41	05 56 70 70 07	86 74 31 71 57	85 39 41 18 38
69 23 46 14 06	20 11 74 52 04	15 95 66 00 00	18 74 39 24 23	97 11 89 63 38
19 56 54 14 30	01 75 87 53 79	40 41 92 15 85	66 67 43 68 06	84 96 28 52 07
45 15 51 49 38	19 47 60 72 46	43 66 79 45 43	59 04 79 00 33	20 82 66 95 41
94 86 43 19 94	36 16 81 08 51	34 88 88 15 53	01 54 03 54 56	05 01 45 11 76
98 08 62 48 26	45 24 02 84 04	44 99 90 88 96	39 09 47 34 07	35 44 13 18 80
33 18 51 62 32	41 94 15 09 49	89 43 54 85 81	88 69 54 19 94	37 54 87 30 43
80 95 10 04 06	96 38 27 07 74	20 15 12 33 87	25 01 62 52 98	94 62 46 11 71
79 75 24 91 40	71 96 12 82 96	69 86 10 25 91	74 85 22 05 39	00 38 75 95 79
18 63 33 25 37	98 14 50 65 71	31 01 02 46 74	05 45 56 14 27	77 93 89 19 36
74 02 94 39 02	77 55 73 22 70	97 79 01 71 19	52 52 75 80 21	80 81 45 17 48
54 17 84 56 11	80 99 33 71 43	05 33 51 29 69	56 12 71 92 55	36 04 09 03 24
11 66 44 98 83	52 07 98 48 27	59 38 17 15 39	09 97 33 34 40	88 46 12 33 56
48 32 47 79 28	31 24 96 47 10	02 29 53 68 70	32 30 75 75 46	15 02 00 99 94
69 07 49 41 38	87 63 79 19 76	35 58 40 44 01	10 51 82 16 15	01 84 87 69 38
09 18 82 00 97	32 82 53 95 27	04 22 08 63 04	83 38 98 73 74	64 27 85 80 44
90 04 58 54 97	51 98 15 06 54	94 93 88 19 97	91 87 07 61 50	68 47 66 46 59
73 18 95 02 07	47 67 72 62 69	62 29 06 44 64	27 12 46 70 18	41 36 18 27 60
75 76 87 64 90	20 97 18 17 49	90 42 91 22 72	95 37 50 58 71	93 82 34 31 78
54 01 64 40 56	66 28 13 10 03	00 68 22 73 98	20 71 45 32 95	07 70 61 78 13
08 35 86 99 10	78 54 24 27 85	13 66 15 88 73	04 61 89 75 53	31 22 30 84 20
28 30 60 32 64	81 33 31 05 91	40 51 00 78 93	32 60 46 04 75	94 11 90 18 40
53 84 08 62 33	81 59 41 36 28	51 21 59 02 90	28 46 66 87 95	77 76 22 07 91
91 75 75 37 41	61 61 36 22 69	50 26 39 02 12	55 78 17 65 14	83 48 34 70 55
89 41 59 26 94	00 39 75 83 91	12 60 71 76 46	48 94 97 23 06	94 54 13 74 08
77 51 30 38 20	86 83 42 99 01	68 41 48 27 74	51 90 81 39 80	72 89 35 55 07
19 50 23 71 74	69 97 92 02 88	55 21 02 97 73	74 28 77 52 51	65 34 46 74 15
21 81 85 93 13	93 27 88 17 57	05 68 67 31 56	07 08 28 50 46	31 85 33 84 52
51 47 46 64 99	68 10 72 36 21	94 04 99 13 45	42 83 60 91 91	08 00 74 54 49
99 55 96 83 31	62 53 52 41 70	69 77 71 28 30	74 81 97 81 42	43 86 07 28 34
33 71 34 80 07	93 58 47 28 69	51 92 66 47 21	58 30 32 98 22	93 17 49 39 72
85 27 48 68 93	11 30 32 92 70	28 83 43 41 37	73 51 59 04 00	71 14 84 36 43
84 13 38 96 40	44 03 55 21 66	73 85 27 00 91	61 22 26 05 61	62 32 71 84 23
56 73 21 62 34	17 39 59 61 31	10 12 39 16 22	85 49 65 75 60	81 60 41 88 80
65 13 85 68 06	87 64 88 52 61	34 31 36 58 61	45 87 52 10 69	85 64 44 72 77

38 00 10 21 76	81 71 91 17 11	71 60 29 29 37	74 21 96 40 49	65 58 44 96 98
37 40 29 63 97	01 30 47 75 86	56 27 11 00 86	47 32 46 26 05	40 03 03 74 38
97 12 54 03 48	87 08 33 14 17	21 81 53 92 50	75 23 76 20 47	15 50 12 95 78
21 82 64 11 34	47 14 33 40 72	64 63 88 59 02	49 13 90 64 41	03 85 65 45 52
73 13 54 27 42	95 71 90 90 35	85 79 47 42 96	08 78 98 81 56	64 69 11 92 02
07 63 87 79 29	03 06 11 80 72	96 20 74 41 56	23 82 19 95 38	04 71 36 69 94
60 52 88 34 41	07 95 41 98 14	59 17 52 06 95	05 53 35 21 39	61 21 20 64 55
83 59 63 56 55	06 95 89 29 83	05 12 80 97 19	77 43 35 37 83	92 30 15 04 98
10 85 06 27 46	99 59 91 05 07	13 49 90 63 19	53 07 57 18 39	06 41 01 93 62
39 82 09 89 52	43 62 26 31 47	64 42 18 08 14	43 80 00 93 51	31 02 47 31 67
59 58 00 64 78	75 56 97 88 00	88 83 55 44 86	23 76 80 61 56	04 11 10 84 08
38 50 80 73 41	23 79 34 87 63	90 82 29 70 22	17 71 90 42 07	95 95 44 99 53
30 69 27 06 68	94 68 81 61 27	56 19 68 00 91	82 06 76 34 00	05 46 26 92 00
65 44 39 56 59	18 28 82 74 37	49 63 22 40 41	08 33 76 56 76	96 29 99 08 36
27 26 75 02 64	13 19 27 22 94	07 47 74 46 06	17 98 54 89 11	97 34 13 03 58
91 30 70 69 91	19 07 22 42 10	36 69 95 37 28	28 82 53 57 93	28 97 66 62 52
68 43 49 46 88	84 47 31 36 22	62 12 69 84 08	12 84 38 25 90	09 81 59 31 46
48 90 81 58 77	54 74 52 45 91	35 70 00 47 54	83 82 45 26 92	54 13 05 51 60
06 91 34 51 97	42 67 27 86 01	11 88 30 95 28	63 01 19 89 01	14 97 44 03 44
10 45 51 60 19	14 21 03 37 12	91 34 23 78 21	88 32 58 08 51	43 66 77 08 83
12 88 39 73 43	65 02 76 11 84	04 28 50 13 92	17 97 41 50 77	90 71 22 67 69
21 77 83 09 76	38 80 73 69 61	31 64 94 20 96	63 28 10 20 23	08 81 64 74 49
19 52 35 95 15	65 12 25 96 59	86 28 36 82 58	69 57 21 37 98	16 43 59 15 29
67 24 55 26 70	35 58 31 65 63	79 24 68 66 86	76 46 33 42 22	26 65 59 08 02
60 58 44 73 77	07 50 03 79 92	45 13 42 65 29	26 76 08 36 37	41 32 64 43 44
53 85 34 13 77	36 06 69 48 50	58 83 87 38 59	49 36 47 33 31	96 24 04 36 42
24 63 73 87 36	74 38 48 93 42	52 62 30 79 92	12 36 91 86 01	03 74 28 38 73
83 08 01 24 51	38 99 22 28 15	07 75 95 17 77	97 37 72 75 85	51 97 23 78 67
16 44 42 43 34	36 15 19 90 73	27 49 37 09 39	85 13 03 25 52	54 84 65 47 59
60 79 01 81 57	57 17 86 57 62	11 16 17 85 76	45 81 95 29 79	65 13 00 48 60
03 99 11 04 61	93 71 61 68 94	66 08 32 46 53	84 60 95 82 32	88 61 81 91 61
38 55 59 55 54	32 88 65 97 80	08 35 56 08 60	29 73 54 77 62	71 29 92 38 53
17 54 67 37 04	92 05 24 62 15	55 12 12 92 81	59 07 60 79 36	27 95 45 89 09
32 64 35 28 61	95 81 90 68 31	00 91 19 89 36	76 35 59 37 79	80 86 30 05 14
69 57 26 87 77	39 51 03 59 05	14 06 04 06 19	29 54 96 96 16	33 56 46 07 80
24 12 26 65 91	27 69 90 64 94	14 84 54 66 72	61 95 87 71 00	90 89 97 57 54
61 19 63 02 31	92 96 26 17 73	41 83 95 53 82	17 26 77 09 43	78 03 87 02 67
30 53 22 17 04	10 27 41 22 02	39 68 52 33 09	10 06 16 88 29	55 98 66 64 85
03 78 89 75 99	75 86 72 07 17	74 41 65 31 66	35 20 83 33 74	87 53 90 88 23
48 22 86 33 79	85 78 34 76 19	53 15 26 74 33	35 66 35 29 72	16 81 86 03 11
60 36 59 46 53	35 07 53 39 49	42 61 42 92 97	01 91 82 83 16	98 95 37 32 31
83 79 94 24 02	56 62 33 44 42	34 99 44 13 74	70 07 11 47 36	09 95 81 80 65
32 96 00 74 05	36 40 98 32 32	99 38 54 16 00	11 13 30 75 86	15 91 70 62 53
19 32 25 38 45	57 62 05 26 06	66 49 76 86 46	78 13 86 65 59	19 64 09 94 13
11 22 09 47 47	07 39 93 74 08	48 50 92 39 29	27 48 24 54 76	85 24 43 51 59

Appendix I—*Continued*

31 75 15 72 60	68 98 00 53 39	15 47 04 83 55	88 65 12 25 96	03 15 21 91 21
88 49 29 93 82	14 45 40 45 04	20 09 49 89 77	74 84 39 34 13	22 10 97 85 08
30 93 44 77 44	07 48 18 38 28	73 78 80 65 33	28 59 72 04 05	94 20 52 03 80
22 88 84 88 93	27 49 99 87 48	60 53 04 51 28	74 02 28 46 17	82 03 71 02 68
78 21 21 69 93	35 90 29 13 86	44 37 21 54 86	65 74 11 40 14	87 48 13 72 20
41 84 98 45 47	46 85 05 23 26	34 67 75 83 00	74 91 06 43 45	19 32 58 15 49
46 35 23 30 49	69 24 89 34 60	45 30 50 75 21	61 31 83 18 55	14 41 37 09 51
11 08 79 62 94	14 01 33 17 92	59 74 76 72 77	76 50 33 45 13	39 66 37 75 44
52 70 10 83 37	56 30 38 73 15	16 52 06 96 76	11 65 49 98 93	02 18 16 81 61
57 27 53 68 98	81 30 44 85 85	68 65 22 73 76	92 85 25 58 66	88 44 80 35 84
20 85 77 31 56	70 28 42 43 26	79 37 59 52 20	01 15 96 32 67	10 62 24 83 91
15 63 38 49 24	90 41 59 36 14	33 52 12 66 65	55 82 34 76 41	86 22 53 17 04
92 69 44 82 97	39 90 40 21 15	59 58 94 90 67	66 82 14 15 75	49 76 70 40 37
77 61 31 90 19	88 15 20 00 80	20 55 49 14 09	96 27 74 82 57	50 81 69 76 16
38 68 83 24 86	45 13 46 35 45	59 40 47 20 59	43 94 75 16 80	43 85 25 96 93
25 16 30 18 89	70 01 41 50 21	41 29 06 73 12	71 85 71 59 57	68 97 11 14 03
65 25 10 76 29	37 23 93 32 95	05 87 00 11 19	92 78 42 63 40	18 47 76 56 22
36 81 54 36 25	18 63 73 75 09	82 44 49 90 05	04 92 17 37 01	14 70 79 39 97
64 39 71 16 92	05 32 78 21 62	20 24 78 17 59	45 19 72 53 32	83 74 52 25 67
04 51 52 56 24	95 09 66 79 46	48 46 08 55 58	15 19 11 87 82	16 93 03 33 61
83 76 16 08 73	43 25 38 41 45	60 83 32 59 83	01 29 14 13 49	20 36 80 71 26
14 38 70 63 45	80 85 40 92 79	43 52 90 63 18	38 38 47 47 61	41 19 63 74 80
51 32 19 22 46	80 08 87 70 74	88 72 25 67 36	66 16 44 94 31	66 91 93 16 78
72 47 20 00 08	80 89 01 80 02	94 81 33 19 00	54 15 58 34 36	35 35 25 41 31
05 46 65 53 06	93 12 81 84 64	74 45 79 05 61	72 84 81 18 34	79 98 26 84 16
39 52 87 24 84	82 47 42 55 93	48 54 53 52 47	18 61 91 36 74	18 61 11 92 41
81 61 61 87 11	53 34 24 42 76	75 12 21 17 24	74 62 77 37 07	58 31 91 59 97
07 58 61 61 20	82 64 12 28 20	92 90 41 31 41	32 39 21 97 63	61 19 96 79 40
90 76 70 42 35	13 57 41 72 00	69 90 26 37 42	78 46 42 25 01	18 62 79 08 72
40 18 82 81 93	29 59 38 86 27	94 97 21 15 98	62 09 53 67 87	00 44 15 89 97
34 41 48 21 57	86 88 75 50 87	19 15 20 00 23	12 30 28 07 83	32 62 46 86 91
63 43 97 53 63	44 98 91 68 22	36 02 40 08 67	76 37 84 16 05	65 96 17 34 88
67 04 90 90 70	93 39 94 55 47	94 45 87 42 84	05 04 14 98 07	20 28 83 40 60
79 49 50 41 46	52 16 29 02 86	54 15 83 42 43	46 97 83 54 82	59 36 29 59 38
91 70 43 05 52	04 73 72 10 31	75 05 19 30 29	47 66 56 43 82	99 78 29 34 78
94 01 54 68 74	32 44 44 82 77	59 82 09 61 63	64 65 42 58 43	41 14 54 28 20
74 10 88 82 22	88 57 07 40 15	25 70 49 10 35	01 75 51 47 50	48 96 83 86 03
62 88 08 78 73	95 16 05 92 21	22 30 49 03 14	72 87 71 73 34	39 28 30 41 49
11 74 81 21 02	80 58 04 18 67	17 71 05 96 21	06 55 40 78 50	73 95 07 95 52
17 94 40 56 00	60 47 80 33 43	25 85 25 89 05	57 21 63 96 18	49 85 69 93 26
66 06 74 27 92	95 04 35 26 80	46 78 05 64 87	09 97 15 94 81	37 00 62 21 86
54 24 49 10 30	45 54 77 08 18	59 84 99 61 69	61 45 92 16 47	87 41 71 71 98
30 94 55 75 89	31 73 25 72 60	47 67 00 76 54	46 37 62 53 66	94 74 64 95 80
69 17 03 74 03	86 99 59 03 07	94 30 47 18 03	26 82 50 55 11	12 45 99 13 14
08 34 58 89 75	35 84 18 57 71	08 10 55 99 87	87 11 22 14 76	14 71 37 11 81

27 76 74 35 84	85 30 18 89 77	29 49 06 97 14	73 03 54 12 07	74 69 90 93 10
13 02 51 43 38	54 06 61 52 43	47 72 46 67 33	47 43 14 39 05	31 04 85 66 99
80 21 73 62 92	98 52 52 43 35	24 43 22 48 96	43 27 75 88 74	11 46 61 60 82
10 87 56 20 04	90 39 16 11 05	57 41 10 63 68	53 85 63 07 43	08 67 08 47 41
54 12 75 73 26	26 62 91 90 87	24 47 28 87 79	30 54 02 78 86	61 73 27 54 54
60 31 14 28 24	37 30 14 26 78	45 99 04 32 42	17 37 45 20 03	70 70 77 02 14
49 73 97 14 84	92 00 39 80 86	76 66 87 32 09	59 20 21 19 73	02 90 23 32 50
78 62 65 15 94	16 45 39 46 14	39 01 49 70 66	83 01 20 98 32	25 57 17 76 28
66 69 21 39 86	99 83 70 05 82	81 23 24 49 87	09 50 49 64 12	90 19 37 95 68
44 07 12 80 91	07 36 29 77 03	76 44 74 25 37	98 52 49 78 31	65 70 40 95 14
41 46 88 51 49	49 55 41 79 94	14 92 43 96 50	95 29 40 05 56	70 48 10 69 05
94 55 93 75 59	49 67 85 31 19	70 31 20 56 82	66 98 63 40 99	74 47 42 07 40
41 61 57 03 60	64 11 45 86 60	90 85 06 46 18	80 62 05 17 90	11 43 63 80 72
50 27 39 31 13	41 79 48 68 61	24 78 18 96 83	55 41 18 56 67	77 53 59 98 92
41 39 68 05 04	90 67 00 82 89	40 90 20 50 69	95 08 30 67 83	28 10 25 78 16
25 80 72 42 60	71 52 97 89 20	72 68 20 73 85	90 72 65 71 66	98 88 40 85 83
06 17 09 79 65	88 30 29 80 41	21 44 34 18 08	68 98 48 36 20	89 74 79 88 82
60 80 85 44 44	74 41 28 11 05	01 17 62 88 38	36 42 11 64 89	18 05 95 10 61
80 94 04 48 93	10 40 83 62 22	80 58 27 19 44	92 63 84 03 33	67 05 41 60 67
19 51 69 01 20	46 75 97 16 43	13 17 75 52 92	21 03 68 28 08	77 50 19 74 27
49 38 65 44 80	23 60 42 35 54	21 78 54 11 01	91 17 81 01 74	29 42 09 04 38
06 31 28 89 40	15 99 56 93 21	47 45 86 48 09	98 18 98 18 51	29 65 18 42 15
60 94 20 03 07	11 89 79 26 74	40 40 56 80 32	96 71 75 42 44	10 70 14 13 93
92 32 99 89 32	78 28 44 63 47	71 20 99 20 61	39 44 89 31 36	25 72 20 85 64
77 93 66 35 74	31 38 45 19 24	85 56 12 96 71	58 13 71 78 20	22 75 13 65 18
38 10 17 77 56	11 65 71 38 97	95 88 95 70 67	47 64 81 38 85	70 66 99 34 06
39 64 16 94 57	91 33 92 25 02	92 61 38 97 19	11 94 75 62 03	19 32 42 05 04
84 05 44 04 55	99 39 66 36 80	67 66 76 06 31	69 18 19 68 45	38 52 51 16 00
47 46 80 35 77	57 64 96 32 66	24 70 07 15 94	14 00 42 31 53	69 24 90 57 47
43 32 13 13 70	28 97 72 38 96	76 47 96 85 62	62 34 20 75 89	08 89 90 59 85
64 28 16 18 26	18 55 56 49 37	13 17 33 33 65	78 85 11 64 99	87 06 41 30 75
66 84 77 04 95	32 35 00 29 85	86 71 63 87 46	26 31 37 74 63	55 38 77 26 81
72 46 13 32 30	21 52 95 34 24	92 58 10 22 62	78 43 86 62 76	18 39 67 35 38
21 03 29 10 50	13 05 81 62 18	12 47 05 65 00	15 29 27 61 39	59 52 65 21 13
95 36 26 70 11	06 65 11 61 36	01 01 60 08 57	55 01 85 63 74	35 82 47 17 08
40 71 29 73 80	10 40 45 54 52	34 03 06 07 26	75 21 11 02 71	36 63 36 84 24
58 27 56 17 64	97 58 65 47 16	50 25 94 63 45	87 19 54 60 92	26 78 76 09 39
89 51 41 17 88	68 22 42 34 17	73 95 97 61 45	30 34 24 02 77	11 04 97 20 49
15 47 25 06 69	48 13 93 67 32	46 87 43 70 88	73 46 50 98 19	58 86 93 52 20
12 12 08 61 24	51 24 74 43 02	60 88 35 21 09	21 43 73 67 86	49 22 67 78 37
19 61 27 84 30	11 66 19 47 70	77 60 36 56 69	86 86 81 26 65	30 01 27 59 89
39 14 17 74 00	28 00 06 42 38	73 25 87 17 94	31 34 02 62 56	66 45 33 70 16
64 75 68 04 57	08 74 71 28 36	03 46 95 06 78	03 27 44 34 23	66 67 78 25 56
92 90 15 18 78	56 44 12 29 98	29 71 83 84 47	06 45 32 53 11	07 56 55 37 71
03 55 19 00 70	09 48 39 40 50	45 93 81 81 35	36 90 84 33 21	11 07 35 18 03

Appendix I—*Concluded*

98 88 46 62 09	06 83 05 36 56	14 66 35 63 46	71 43 00 49 09	19 81 80 57 07
27 36 98 68 82	53 47 30 75 41	53 63 37 08 63	03 74 81 28 22	19 36 04 90 88
59 06 67 59 74	63 33 52 04 83	43 51 43 74 81	58 27 82 69 67	49 32 54 39 51
91 64 79 37 83	64 16 94 90 22	98 58 80 94 95	49 82 95 90 68	38 83 10 48 38
83 60 59 24 19	39 54 20 77 72	71 56 87 56 73	35 18 58 97 59	44 90 17 42 91
24 89 58 85 30	70 77 43 54 39	46 75 87 04 72	70 20 79 26 75	91 62 36 12 75
35 72 02 65 56	95 59 62 00 94	73 75 08 57 88	34 26 40 17 03	46 83 36 52 48
14 14 15 34 10	38 64 90 63 43	57 25 66 13 42	72 70 97 53 18	90 37 93 75 62
27 41 67 56 70	92 17 67 25 35	93 11 95 60 77	06 88 61 82 44	92 34 43 13 74
82 07 10 74 29	81 00 74 77 49	40 74 45 69 74	23 33 68 88 21	53 84 11 05 36
21 44 58 27 93	24 83 19 32 41	14 19 97 62 68	70 88 36 80 02	03 82 91 74 43
72 51 37 64 00	52 22 59 23 48	62 30 89 84 81	29 74 43 31 65	33 14 16 10 20
71 47 94 50 27	76 16 05 74 11	13 78 01 36 32	52 30 87 77 62	88 87 43 36 97
83 21 05 14 66	09 08 85 03 95	26 74 30 53 06	21 70 67 00 01	99 43 98 07 67
68 74 99 51 48	94 89 77 86 36	96 75 00 90 24	94 53 89 11 43	96 69 36 18 86
05 18 47 57 63	47 07 58 81 58	05 31 35 34 39	14 90 80 88 30	60 09 62 15 51
13 65 16 25 46	96 89 22 52 40	47 51 15 84 83	87 34 27 88 18	07 85 53 92 69
00 56 62 12 20	00 29 22 40 69	25 07 22 95 19	52 54 85 40 91	21 28 22 12 96
50 95 81 76 95	58 07 26 89 90	60 32 99 59 55	71 58 66 34 17	35 94 76 78 07
57 62 16 45 47	46 85 03 79 81	38 52 70 90 37	64 75 60 33 24	04 98 68 36 66
09 28 22 58 44	79 13 97 84 35	35 42 84 35 61	69 79 96 33 14	12 99 19 35 16
23 39 49 42 06	93 43 23 78 36	94 91 92 68 46	02 55 57 44 10	94 91 54 81 99
05 28 03 74 70	93 62 20 43 45	15 09 21 95 10	18 09 41 66 13	78 23 45 00 01
95 49 19 79 76	38 30 63 21 92	82 63 95 46 24	72 43 49 26 06	23 19 17 46 93
78 52 10 01 04	18 24 87 55 83	90 32 65 07 85	54 03 46 62 51	35 77 41 46 92
96 34 54 45 79	85 93 24 40 53	75 70 42 08 40	86 58 38 39 44	52 45 67 37 66
77 96 33 11 51	32 36 49 16 91	47 35 74 03 38	23 43 52 40 65	08 45 89 53 66
07 52 01 12 94	23 23 80 17 48	41 69 06 73 28	54 81 43 77 77	10 05 74 23 32
38 42 30 23 09	70 70 38 57 36	46 14 81 42 58	29 23 61 21 52	05 08 86 58 25
02 46 36 55 33	21 19 96 05 55	33 92 80 18 17	07 39 68 92 15	30 72 22 21 02
15 88 09 22 61	17 29 28 81 90	61 78 14 88 98	92 52 52 12 83	88 58 16 00 98
71 92 60 08 19	59 14 40 02 24	30 57 09 01 94	18 32 90 69 99	26 85 71 92 38
64 42 52 81 08	16 55 41 60 16	00 04 28 32 29	10 33 33 61 68	65 61 79 48 34
79 78 22 39 24	49 44 03 04 32	81 07 73 15 43	95 21 66 48 65	13 65 85 10 81
35 33 77 45 38	44 55 36 46 72	90 96 04 18 49	93 86 54 46 08	93 17 63 48 51
05 24 92 93 29	19 71 59 40 82	14 73 88 66 67	43 70 86 63 54	93 69 22 55 27
56 46 39 93 80	38 79 38 57 74	19 05 61 39 39	46 06 22 76 47	66 14 66 32 10
96 29 63 31 21	54 19 63 41 08	75 81 48 59 86	71 17 11 51 02	28 99 26 31 65
98 38 03 62 69	60 01 40 72 01	62 44 84 63 85	42 17 58 83 50	46 18 24 91 26
52 56 76 43 50	16 31 55 39 69	80 39 58 11 14	54 35 86 45 78	47 26 91 57 47
78 49 89 08 30	25 95 59 92 36	43 28 69 10 64	99 96 99 51 44	64 42 47 73 77
49 55 32 42 41	08 15 08 95 35	08 70 39 10 41	77 32 38 10 79	45 12 79 36 86
32 15 10 70 75	83 15 51 02 52	73 10 08 86 18	23 89 18 74 18	45 41 72 02 68
11 31 45 03 63	26 86 02 77 99	49 41 68 35 34	19 18 70 80 59	76 67 70 21 10
12 36 47 12 10	87 05 25 02 41	90 78 59 78 89	81 39 95 81 30	64 43 90 56 14

Appendix II
Binomial Probability Distributions

Table entries are values of $\binom{n}{x} p^x q^{n-x}$. Items omitted are less than 0.00005.

x	.1	.2	.3	.4	.5	.6	.7	.8	.9
					$n = 5$				
0	.5905	.3277	.1681	.0778	.0312	.0102	.0024	.0003	.0000
1	.3280	.4096	.3602	.2592	.1562	.0768	.0284	.0064	.0004
2	.0729	.2048	.3087	.3456	.3125	.2304	.1323	.0512	.0081
3	.0081	.0512	.1323	.2304	.3125	.3456	.3087	.2048	.0729
4	.0004	.0064	.0284	.0768	.1562	.2592	.3602	.4096	.3280
5	.0000	.0003	.0024	.0102	.0312	.0778	.1681	.3277	.5905
					$n = 10$				
0	.3487	.1074	.0282	.0060	.0010	.0001	.0000	.0000	.0000
1	.3874	.2684	.1211	.0403	.0098	.0016	.0001	.0000	.0000
2	.1937	.3020	.2335	.1209	.0439	.0106	.0014	.0001	.0000
3	.0574	.2013	.2668	.2150	.1172	.0425	.0090	.0008	.0000
4	.0112	.0881	.2001	.2508	.2051	.1115	.0368	.0055	.0001
5	.0015	.0264	.1029	.2007	.2461	.2007	.1029	.0264	.0015
6	.0001	.0055	.0368	.1115	.2051	.2508	.2001	.0881	.0112
7	.0000	.0008	.0090	.0425	.1172	.2150	.2668	.2013	.0574
8	.0000	.0001	.0014	.0106	.0439	.1209	.2335	.3020	.1937
9	.0000	.0000	.0001	.0016	.0098	.0403	.1211	.2684	.3874
10	.0000	.0000	.0000	.0001	.0010	.0060	.0282	.1074	.3487
					$n = 15$				
0	.2059	.0352	.0047	.0005	.0000	.0000	.0000	.0000	.0000
1	.3432	.1319	.0305	.0047	.0005	.0000	.0000	.0000	.0000
2	.2669	.2309	.0916	.0219	.0032	.0003	.0000	.0000	.0000
3	.1285	.2501	.1700	.0634	.0139	.0016	.0001	.0000	.0000
4	.0428	.1876	.2186	.1268	.0417	.0074	.0006	.0000	.0000
5	.0105	.1032	.2061	.1859	.0916	.0245	.0030	.0001	.0000
6	.0019	.0430	.1472	.2066	.1527	.0612	.0116	.0007	.0000
7	.0003	.0138	.0811	.1711	.1964	.1181	.0348	.0035	.0000
8	.0000	.0035	.0348	.1181	.1964	.1771	.0811	.0138	.0003
9	.0000	.0007	.0116	.0612	.1527	.2066	.1472	.0430	.0019
10	.0000	.0001	.0030	.0245	.0916	.1859	.2061	.1032	.0105

Source: The part of this table with $n = 5$, 10, 15, 20, and 30 is adopted from National Bureau of Standards, *Tables of the Binomial Distribution*, Applied Mathematics Series, no. 6, 1950; the part with $n = 50$ is abridged from Harry G. Romig, 50–100 *Binomial Tables*, New York, John Wiley & Sons, Inc., 1953, with the kind permission of the publishers.

Appendix II—*Continued*

x	.1	.2	.3	.4	.5	.6	.7	.8	.9
					$n = 15$ (*continued*)				
11	.0000	.0000	.0006	.0074	.0417	.1268	.2186	.1876	.0428
12	.0000	.0000	.0001	.0016	.0139	.0634	.1700	.2501	.1285
13	.0000	.0000	.0000	.0003	.0032	.0219	.0916	.2309	.2669
14	.0000	.0000	.0000	.0000	.0005	.0047	.0305	.1319	.3432
15	.0000	.0000	.0000	.0000	.0000	.0005	.0047	.0352	.2059
					$n = 20$				
0	.1216	.0115	.0008	.0000	.0000	.0000	.0000	.0000	.0000
1	.2702	.0576	.0068	.0005	.0000	.0000	.0000	.0000	.0000
2	.2852	.1369	.0278	.0031	.0002	.0000	.0000	.0000	.0000
3	.1901	.2054	.0716	.0123	.0011	.0000	.0000	.0000	.0000
4	.0898	.2182	.1304	.0350	.0046	.0003	.0000	.0000	.0000
5	.0319	.1746	.1789	.0746	.0148	.0013	.0000	.0000	.0000
6	.0089	.1091	.1916	.1244	.0370	.0049	.0002	.0000	.0000
7	.0020	.0545	.1643	.1659	.0739	.0146	.0010	.0000	.0000
8	.0004	.0222	.1144	.1797	.1201	.0355	.0039	.0001	.0000
9	.0001	.0074	.0654	.1597	.1602	.0710	.0120	.0005	.0000
10	.0000	.0020	.0308	.1171	.1762	.1171	.0308	.0020	.0000
11	.0000	.0005	.0120	.0710	.1602	.1597	.0654	.0074	.0001
12	.0000	.0001	.0039	.0355	.1201	.1797	.1144	.0222	.0004
13	.0000	.0000	.0010	.0146	.0739	.1659	.1643	.0545	.0020
14	.0000	.0000	.0002	.0049	.0370	.1244	.1916	.1091	.0089
15	.0000	.0000	.0000	.0013	.0148	.0746	.1789	.1746	.1319
16	.0000	.0000	.0000	.0003	.0046	.0350	.1304	.2182	.0898
17	.0000	.0000	.0000	.0000	.0011	.0123	.0716	.2054	.1901
18	.0000	.0000	.0000	.0000	.0002	.0031	.0278	.1369	.2852
19	.0000	.0000	.0000	.0000	.0000	.0005	.0068	.0576	.2702
20	.0000	.0000	.0000	.0000	.0000	.0000	.0008	.0115	.1216
					$n = 30$				
0	.0424	.0012	.0000	.0000	.0000	.0000	.0000	.0000	.0000
1	.1413	.0093	.0003	.0000	.0000	.0000	.0000	.0000	.0000
2	.2277	.0337	.0018	.0000	.0000	.0000	.0000	.0000	.0000
3	.2361	.0785	.0072	.0003	.0000	.0000	.0000	.0000	.0000
4	.1771	.1325	.0208	.0012	.0000	.0000	.0000	.0000	.0000
5	.1023	.1723	.0464	.0041	.0001	.0000	.0000	.0000	.0000
6	.0474	.1795	.0829	.0115	.0006	.0000	.0000	.0000	.0000
7	.0180	.1538	.1219	.0263	.0019	.0000	.0000	.0000	.0000
8	.0058	.1106	.1501	.0505	.0055	.0001	.0000	.0000	.0000
9	.0016	.0676	.1573	.0823	.0133	.0006	.0000	.0000	.0000
10	.0004	.0355	.1416	.1152	.0280	.0020	.0000	.0000	.0000

x	.1	.2	.3	.4	.5	.6	.7	.8	.9
				n = 30 (*continued*)					
11	.0001	.0161	.1103	.1396	.0509	.0054	.0001	.0000	.0000
12	.0000	.0064	.0749	.1474	.0805	.0129	.0005	.0000	.0000
13	.0000	.0022	.0444	.1360	.1115	.0269	.0015	.0000	.0000
14	.0000	.0007	.0231	.1101	.1354	.0489	.0042	.0000	.0000
15	.0000	.0002	.0106	.0783	.1445	.0783	.0106	.0002	.0000
16	.0000	.0000	.0042	.0489	.1354	.1101	.0231	.0007	.0000
17	.0000	.0000	.0015	.0269	.1115	.1360	.0444	.0022	.0000
18	.0000	.0000	.0005	.0129	.0805	.1474	.0749	.0064	.0000
19	.0000	.0000	.0001	.0054	.0509	.1396	.1103	.0161	.0001
20	.0000	.0000	.0000	.0020	.0280	.1152	.1416	.0355	.0004
21	.0000	.0000	.0000	.0006	.0133	.0823	.1573	.0676	.0016
22	.0000	.0000	.0000	.0001	.0055	.0505	.1501	.1106	.0058
23	.0000	.0000	.0000	.0000	.0019	.0263	.1219	.1538	.0180
24	.0000	.0000	.0000	.0000	.0006	.0115	.0829	.1795	.0474
25	.0000	.0000	.0000	.0000	.0001	.0041	.0464	.1723	.1023
26	.0000	.0000	.0000	.0000	.0000	.0012	.0208	.1325	.1771
27	.0000	.0000	.0000	.0000	.0000	.0003	.0072	.0785	.2361
28	.0000	.0000	.0000	.0000	.0000	.0000	.0018	.0337	.2277
29	.0000	.0000	.0000	.0000	.0000	.0000	.0003	.0093	.1413
30	.0000	.0000	.0000	.0000	.0000	.0000	.0000	.0012	.0424
				n = 50					
0	.0052	.0000	.0000	.0000	.0000	.0000	.0000	.0000	.0000
1	.0286	.0002	.0000	.0000	.0000	.0000	.0000	.0000	.0000
2	.0779	.0011	.0000	.0000	.0000	.0000	.0000	.0000	.0000
3	.1386	.0044	.0000	.0000	.0000	.0000	.0000	.0000	.0000
4	.1809	.0128	.0001	.0000	.0000	.0000	.0000	.0000	.0000
5	.1849	.0295	.0006	.0000	.0000	.0000	.0000	.0000	.0000
6	.1541	.0554	.0018	.0000	.0000	.0000	.0000	.0000	.0000
7	.1076	.0870	.0048	.0000	.0000	.0000	.0000	.0000	.0000
8	.0643	.1169	.0111	.0002	.0000	.0000	.0000	.0000	.0000
9	.0333	.1364	.0220	.0005	.0000	.0000	.0000	.0000	.0000
10	.0152	.1398	.0386	.0014	.0000	.0000	.0000	.0000	.0000
11	.0061	.1271	.0602	.0035	.0000	.0000	.0000	.0000	.0000
12	.0022	.1033	.0838	.0076	.0001	.0000	.0000	.0000	.0000
13	.0007	.0755	.1050	.0147	.0003	.0000	.0000	.0000	.0000
14	.0002	.0499	.1189	.0260	.0008	.0000	.0000	.0000	.0000
15	.0001	.0299	.1223	.0415	.0020	.0000	.0000	.0000	.0000

Appendix II—*Concluded*

x	.1	.2	.3	.4	.5	.6	.7	.8	.9
				n = 50 (*continued*)					
16	.0000	.0164	.1147	.0606	.0044	.0000	.0000	.0000	.0000
17	.0000	.0082	.0983	.0808	.0087	.0001	.0000	.0000	.0000
18	.0000	.0038	.0772	.0987	.0160	.0003	.0000	.0000	.0000
19	.0000	.0016	.0558	.1109	.0270	.0009	.0000	.0000	.0000
20	.0000	.0006	.0370	.1146	.0419	.0020	.0000	.0000	.0000
21	.0000	.0002	.0227	.1091	.0598	.0043	.0000	.0000	.0000
22	.0000	.0001	.0128	.0959	.0788	.0084	.0001	.0000	.0000
23	.0000	.0000	.0067	.0778	.0960	.0154	.0002	.0000	.0000
24	.0000	.0000	.0032	.0584	.1080	.0259	.0006	.0000	.0000
25	.0000	.0000	.0014	.0405	.1123	.0405	.0014	.0000	.0000
26	.0000	.0000	.0006	.0259	.1080	.0584	.0032	.0000	.0000
27	.0000	.0000	.0002	.0154	.0960	.0778	.0067	.0000	.0000
28	.0000	.0000	.0001	.0084	.0788	.0959	.0128	.0001	.0000
29	.0000	.0000	.0000	.0043	.0598	.1091	.0227	.0002	.0000
30	.0000	.0000	.0000	.0020	.0419	.1146	.0370	.0006	.0000
31	.0000	.0000	.0000	.0009	.0270	.1109	.0558	.0016	.0000
32	.0000	.0000	.0000	.0003	.0160	.0987	.0772	.0038	.0000
33	.0000	.0000	.0000	.0001	.0087	.0808	.0983	.0082	.0000
34	.0000	.0000	.0000	.0000	.0044	.0606	.1147	.0164	.0000
35	.0000	.0000	.0000	.0000	.0020	.0415	.1223	.0299	.0001
36	.0000	.0000	.0000	.0000	.0008	.0260	.1189	.0499	.0002
37	.0000	.0000	.0000	.0000	.0003	.0147	.1050	.0755	.0007
38	.0000	.0000	.0000	.0000	.0001	.0076	.0838	.1033	.0022
39	.0000	.0000	.0000	.0000	.0000	.0035	.0602	.1271	.0061
40	.0000	.0000	.0000	.0000	.0000	.0014	.0386	.1398	.0152
41	.0000	.0000	.0000	.0000	.0000	.0005	.0220	.1364	.0333
42	.0000	.0000	.0000	.0000	.0000	.0002	.0111	.1169	.0643
43	.0000	.0000	.0000	.0000	.0000	.0000	.0048	.0870	.1076
44	.0000	.0000	.0000	.0000	.0000	.0000	.0018	.0554	.1541
45	.0000	.0000	.0000	.0000	.0000	.0000	.0006	.0295	.1849
46	.0000	.0000	.0000	.0000	.0000	.0000	.0001	.0128	.1809
47	.0000	.0000	.0000	.0000	.0000	.0000	.0000	.0044	.1386
48	.0000	.0000	.0000	.0000	.0000	.0000	.0000	.0011	.0779
49	.0000	.0000	.0000	.0000	.0000	.0000	.0000	.0002	.0286
50	.0000	.0000	.0000	.0000	.0000	.0000	.0000	.0000	.0052

Appendix III
Cumulative Probabilities for Binomial Distributions

$$CP(x) = \sum_{k=0}^{x} \binom{n}{k} p^k q^{n-k}$$

x	.1	.2	.3	.4	.5	.6	.7	.8	.9
					n = 5				
0	.5905	.3277	.1681	.0778	.0312	.0102	.0024	.0003	.0000
1	.9185	.7373	.5283	.3370	.1874	.0870	.0308	.0067	.0004
2	.9914	.9421	.8370	.6826	.5000	.3174	.1631	.0579	.0085
3	.9995	.9933	.9693	.9130	.8124	.6630	.4718	.2627	.0814
4	.9999	.9997	.9977	.9898	.9686	.9222	.8320	.6723	.4094
5	1.0000	1.0000	1.0000	1.0000	1.0000	1.0000	1.0000	1.0000	1.0000
					n = 10				
0	.3487	.1074	.0282	.0060	.0010	.0001	.0000	.0000	.0000
1	.7361	.3758	.1493	.0463	.0108	.0017	.0001	.0000	.0000
2	.9298	.6778	.3828	.1672	.0547	.0123	.0015	.0001	.0000
3	.9872	.8791	.6496	.3822	.1719	.0548	.0105	.0009	.0000
4	.9984	.9672	.8497	.6330	.3770	.1664	.0473	.0064	.0001
5	.9999	.9936	.9526	.8337	.6231	.3670	.1502	.0328	.0016
6	1.0000	.9991	.9894	.9452	.8282	.6178	.3503	.1209	.0128
7	1.0000	.9999	.9984	.9877	.9454	.8328	.6171	.3222	.0702
8	1.0000	1.0000	.9998	.9983	.9893	.9537	.8506	.6242	.2639
9	1.0000	1.0000	.9999	.9999	.9991	.9940	.9717	.8926	.6513
10	1.0000	1.0000	1.0000	1.0000	1.0000	1.0000	1.0000	1.0000	1.0000
					n = 15				
0	.2059	.0352	.0047	.0005	.0000	.0000	.0000	.0000	.0000
1	.5491	.1671	.0352	.0052	.0005	.0000	.0000	.0000	.0000
2	.8160	.3980	.1268	.0271	.0037	.0003	.0000	.0000	.0000
3	.9445	.6481	.2968	.0905	.0176	.0019	.0001	.0000	.0000
4	.9873	.8357	.5154	.2173	.0593	.0093	.0007	.0000	.0000
5	.9978	.9389	.7215	.4032	.1509	.0338	.0037	.0001	.0000
6	.9997	.9819	.8687	.6098	.3036	.0950	.0153	.0008	.0000
7	1.0000	.9957	.9498	.7869	.5000	.2131	.0501	.0043	.0000
8	1.0000	.9992	.9846	.9050	.6964	.3902	.1321	.0181	.0003
9	1.0000	.9999	.9962	.9662	.8491	.5968	.2784	.0611	.0022
10	1.0000	1.0000	.9992	.9907	.9407	.7827	.4845	.1643	.0127

Source: The part of this table with $n = 5$, 10, 15, 20, and 30 is adopted from National Bureau of Standards, *Tables of Binomial Distribution*, Applied Mathematics Series, no. 6, 1950; the part with $n = 50$ is abridged from Harry G. Romig, 50–100 *Binomial Tables*, New York, John Wiley & Sons, Inc., 1953, with the kind permission of the publishers.

Appendix III—*Continued*

x	.1	.2	.3	.4	.5	.6	.7	.8	.9
					n = 15 (*continued*)				
11	1.0000	1.0000	.9999	.9981	.9824	.9095	.7031	.3519	.0555
12	1.0000	1.0000	.9999	.9997	.9963	.9729	.8731	.6020	.1840
13	1.0000	1.0000	1.0000	1.0000	.9995	.9948	.9647	.8329	.4509
14	1.0000	1.0000	1.0000	1.0000	1.0000	.9995	.9952	.9648	.7941
15	1.0000	1.0000	1.0000	1.0000	1.0000	1.0000	1.0000	1.0000	1.0000
					n = 20				
0	.1216	.0115	.0008	.0000	.0000	.0000	.0000	.0000	.0000
1	.3918	.0691	.0076	.0005	.0000	.0000	.0000	.0000	.0000
2	.6770	.2060	.0354	.0036	.0002	.0000	.0000	.0000	.0000
3	.8671	.4114	.1070	.0159	.0013	.0000	.0000	.0000	.0000
4	.9569	.6296	.2374	.0509	.0059	.0003	.0000	.0000	.0000
5	.9888	.8042	.4163	.1255	.0207	.0016	.0000	.0000	.0000
6	.9977	.9133	.6079	.2499	.0577	.0065	.0002	.0000	.0000
7	.9997	.9678	.7722	.4158	.1316	.0211	.0012	.0000	.0000
8	.9999	.9900	.8866	.5955	.2517	.0566	.0051	.0001	.0000
9	1.0000	.9974	.9520	.7552	.4119	.1276	.0171	.0006	.0000
10	1.0000	.9994	.9828	.8723	.5881	.2447	.0479	.0026	.0000
11	1.0000	.9999	.9948	.9433	.7483	.4044	.1133	.0100	.0001
12	1.0000	1.0000	.9987	.9788	.8684	.5841	.2277	.0322	.0005
13	1.0000	1.0000	.9997	.9934	.9423	.7500	.3920	.0867	.0025
14	1.0000	1.0000	.9999	.9983	.9793	.8744	.5836	.1958	.0114
15	1.0000	1.0000	1.0000	.9996	.9941	.9490	.7625	.3704	.0433
16	1.0000	1.0000	1.0000	.9999	.9987	.9840	.8920	.5886	.1331
17	1.0000	1.0000	1.0000	1.0000	.9998	.9963	.9645	.7940	.3232
18	1.0000	1.0000	1.0000	1.0000	1.0000	.9994	.9923	.9309	.6084
19	1.0000	1.0000	1.0000	1.0000	1.0000	.9999	.9992	.9885	.8784
20	1.0000	1.0000	1.0000	1.0000	1.0000	1.0000	1.0000	1.0000	1.0000
					n = 30				
0	.0424	.0012	.0000	.0000	.0000	.0000	.0000	.0000	.0000
1	.1837	.0105	.0003	.0000	.0000	.0000	.0000	.0000	.0000
2	.4114	.0442	.0021	.0000	.0000	.0000	.0000	.0000	.0000
3	.6475	.1227	.0093	.0003	.0000	.0000	.0000	.0000	.0000
4	.8246	.2552	.0301	.0015	.0000	.0000	.0000	.0000	.0000
5	.9269	.4275	.0765	.0056	.0001	.0000	.0000	.0000	.0000
6	.9743	.6070	.1594	.0171	.0007	.0000	.0000	.0000	.0000
7	.9923	.7608	.2813	.0434	.0026	.0000	.0000	.0000	.0000
8	.9981	.8714	.4314	.0939	.0081	.0001	.0000	.0000	.0000
9	.9996	.9390	.5887	.1762	.0214	.0007	.0000	.0000	.0000
10	.9999	.9745	.7303	.2914	.0494	.0027	.0000	.0000	.0000

x	.1	.2	.3	.4	.5	.6	.7	.8	.9
				n = 30 (continued)					
11	1.0000	.9906	.8406	.4310	.1003	.0081	.0001	.0000	.0000
12	1.0000	.9970	.9155	.5784	.1808	.0210	.0006	.0000	.0000
13	1.0000	.9992	.9599	.7144	.2923	.0479	.0021	.0000	.0000
14	1.0000	.9999	.9830	.8245	.4277	.0968	.0063	.0000	.0000
15	1.0000	1.0000	.9936	.9028	.5722	.1751	.0169	.0002	.0000
16	1.0000	1.0000	.9978	.9518	.7076	.2852	.0400	.0009	.0000
17	1.0000	1.0000	.9993	.9788	.8191	.4212	.0844	.0031	.0000
18	1.0000	1.0000	.9998	.9917	.8998	.5686	.1593	.0095	.0000
19	1.0000	1.0000	.9999	.9970	.9505	.7082	.2696	.0256	.0001
20	1.0000	1.0000	1.0000	.9990	.9785	.8234	.4112	.0611	.0005
21	1.0000	1.0000	1.0000	.9998	.9918	.9057	.5685	.1287	.0021
22	1.0000	1.0000	1.0000	.9999	.9973	.9562	.7186	.2393	.0079
23	1.0000	1.0000	1.0000	1.0000	.9992	.9825	.8405	.3931	.0259
24	1.0000	1.0000	1.0000	1.0000	.9998	.9940	.9234	.5726	.0733
25	1.0000	1.0000	1.0000	1.0000	.9999	.9981	.9698	.7449	1.756
26	1.0000	1.0000	1.0000	1.0000	1.0000	.9985	.9906	.8774	.3527
27	1.0000	1.0000	1.0000	1.0000	1.0000	.9998	.9978	.9559	.5888
28	1.0000	1.0000	1.0000	1.0000	1.0000	1.0000	.9996	.9896	.8165
29	1.0000	1.0000	1.0000	1.0000	1.0000	1.0000	.9999	.9989	.9578
30	1.0000	1.0000	1.0000	1.0000	1.0000	1.0000	1.0000	1.0000	1.0000
				n = 50					
0	.0052	.0000	.0000	.0000	.0000	.0000	.0000	.0000	.0000
1	.0338	.0002	.0000	.0000	.0000	.0000	.0000	.0000	.0000
2	.1117	.0013	.0000	.0000	.0000	.0000	.0000	.0000	.0000
3	.2503	.0057	.0000	.0000	.0000	.0000	.0000	.0000	.0000
4	.4312	.0185	.0002	.0000	.0000	.0000	.0000	.0000	.0000
5	.6161	.0480	.0007	.0000	.0000	.0000	.0000	.0000	.0000
6	.7702	.1034	.0025	.0000	.0000	.0000	.0000	.0000	.0000
7	.8779	.1904	.0073	.0001	.0000	.0000	.0000	.0000	.0000
8	.9421	.3073	.0183	.0002	.0000	.0000	.0000	.0000	.0000
9	.9755	.4437	.0402	.0008	.0000	.0000	.0000	.0000	.0000
10	.9906	.5836	.0789	.0022	.0000	.0000	.0000	.0000	.0000
11	.9968	.7107	.1390	.0057	.0000	.0000	.0000	.0000	.0000
12	.9990	.8139	.2229	.0133	.0002	.0000	.0000	.0000	.0000
13	.9997	.8894	.3279	.0280	.0005	.0000	.0000	.0000	.0000
14	.9999	.9393	.4468	.0540	.0013	.0000	.0000	.0000	.0000
15	1.0000	.9692	.5692	.0955	.0033	.0000	.0000	.0000	.0000
16	1.0000	.9856	.6839	.1561	.0077	.0001	.0000	.0000	.0000
17	1.0000	.9937	.7822	.2369	.0164	.0002	.0000	.0000	.0000

Appendix III—*Concluded*

x	.1	.2	.3	.4	.5	.6	.7	.8	.9
					n = 50 (*continued*)				
18	1.0000	.9975	.8594	.3356	.0325	.0005	.0000	.0000	.0000
19	1.0000	.9991	.9152	.4465	.0595	.0014	.0000	.0000	.0000
20	1.0000	.9997	.9522	.5610	.1013	.0034	.0000	.0000	.0000
21	1.0000	.9990	.9749	.6701	.1611	.0076	.0000	.0000	.0000
22	1.0000	1.0000	.9877	.7660	.2399	.0160	.0001	.0000	.0000
23	1.0000	1.0000	.9944	.8438	.3360	.0314	.0003	.0000	.0000
24	1.0000	1.0000	.9976	.9022	.4439	.0573	.0009	.0000	.0000
25	1.0000	1.0000	.9991	.9427	.5561	.0987	.0024	.0000	.0000
26	1.0000	1.0000	.9997	.9686	.6641	.1562	.0056	.0000	.0000
27	1.0000	1.0000	.9999	.9840	.7601	.2340	.0123	.0000	.0000
28	1.0000	1.0000	1.0000	.9924	.8389	.3299	.0251	.0001	.0000
29	1.0000	1.0000	1.0000	.9966	.8987	.4390	.0478	.0003	.0000
30	1.0000	1.0000	1.0000	.9986	.9405	.5535	.0848	.0009	.0000
31	1.0000	1.0000	1.0000	.9995	.9675	.6644	.1406	.0025	.0000
32	1.0000	1.0000	1.0000	.9998	.9836	.7631	.2178	.0063	.0000
33	1.0000	1.0000	1.0000	.9999	.9923	.8439	.3161	.0143	.0000
34	1.0000	1.0000	1.0000	1.0000	.9967	.9045	.4308	.0308	.0000
35	1.0000	1.0000	1.0000	1.0000	.9987	.9460	.5532	.0607	.0001
36	1.0000	1.0000	1.0000	1.0000	.9995	.9720	.6721	.1106	.0003
37	1.0000	1.0000	1.0000	1.0000	.9998	.9867	.7771	.1861	.0010
38	1.0000	1.0000	1.0000	1.0000	1.0000	.9943	.8610	.2893	.0032
39	1.0000	1.0000	1.0000	1.0000	1.0000	.9978	.9211	.4164	.0094
40	1.0000	1.0000	1.0000	1.0000	1.0000	.9992	.9598	.5563	.0245
41	1.0000	1.0000	1.0000	1.0000	1.0000	.9998	.9817	.7927	.0579
42	1.0000	1.0000	1.0000	1.0000	1.0000	.9999	.9927	.8095	.1221
43	1.0000	1.0000	1.0000	1.0000	1.0000	1.0000	.9975	.8966	.2298
44	1.0000	1.0000	1.0000	1.0000	1.0000	1.0000	.9993	.9520	.3839
45	1.0000	1.0000	1.0000	1.0000	1.0000	1.0000	.9998	.9815	.5688
46	1.0000	1.0000	1.0000	1.0000	1.0000	1.0000	1.0000	.9943	.7497
47	1.0000	1.0000	1.0000	1.0000	1.0000	1.0000	1.0000	.9987	.8883
48	1.0000	1.0000	1.0000	1.0000	1.0000	1.0000	1.0000	.9998	.9662
49	1.0000	1.0000	1.0000	1.0000	1.0000	1.0000	1.0000	1.0000	.9948
50	1.0000	1.0000	1.0000	1.0000	1.0000	1.0000	1.0000	1.0000	1.0000

Appendix IV
Table of Square Roots

	0	1	2	3	4	5	6	7	8	9
1.0	1.0000	1.0050	1.0100	1.0149	1.0198	1.0247	1.0296	1.0344	1.0392	1.0440
10.	3.1623	3.1780	3.1937	3.2094	3.2249	3.2404	3.2558	3.2711	3.2863	3.3015
1.1	1.0488	1.0536	1.0583	1.0630	1.0677	1.0721	1.0770	1.0817	1.0863	1.0909
11.	3.3166	3.3317	3.3466	3.3615	3.3764	3.3912	3.4059	3.4205	3.4351	3.4496
1.2	1.0954	1.1000	1.1045	1.1091	1.1136	1.1180	1.1225	1.169	1.1314	1.1358
12.	3.4641	3.4785	3.4928	3.5071	3.5214	3.5355	3.5496	3.5637	3.5777	3.5917
1.3	1.1402	1.1446	1.1489	1.1533	1.1576	1.1619	1.1662	1.105	1.1747	1.1790
13.	3.6056	3.6194	3.6332	3.6469	3.6606	3.6742	3.6878	3.7014	3.7148	3.7283
1.4	1.1832	1.1874	1.1916	1.1958	1.2000	1.2042	1.2083	1.2124	1.2166	1.2207
14.	3.7417	3.7550	3.7683	3.7815	3.7947	3.8079	3.8210	3.8341	3.8471	3.8601
1.5	1.2247	1.2288	1.2329	1.2369	1.2410	1.2450	1.2490	1.2530	1.2570	1.2610
15.	3.8730	3.8859	3.8987	3.9115	3.9243	3.9370	3.9497	3.9623	3.9749	3.9875
1.6	1.2649	1.2689	1.2728	1.2767	1.2806	1.2845	1.2884	1.2923	1.2961	1.3000
16.	4.0000	4.0125	4.0249	4.0373	4.0497	4.0620	4.0743	4.0866	4.0988	4.1110
1.7	1.3038	1.3077	1.3115	1.3153	1.3191	1.3229	1.3266	1.3304	1.3342	1.3379
17.	4.1231	4.1352	4.1473	4.1593	4.1713	4.1833	4.1952	4.2071	4.2190	4.2308
1.8	1.3416	1.3454	1.3491	1.3528	1.3565	1.3061	1.3638	1.3675	1.3711	1.3748
18.	4.2426	4.2544	4.2661	4.2778	4.2895	4.3012	4.3128	4.3243	4.3359	4.3474
1.9	1.3784	1.3820	1.3856	1.3892	1.3928	1.3964	1.4000	1.4036	1.4071	1.4107
19.	4.3589	4.3704	4.3818	4.3932	4.4045	4.4159	4.4272	4.4385	4.4497	4.4609
2.0	1.4142	1.4177	1.4213	1.4248	1.4283	1.4318	1.4353	1.4387	1.4422	1.4457
20.	4.4721	4.4833	4.4944	4.5056	4.5166	4.5277	4.5387	4.5497	4.5607	4.5717
2.1	1.4491	1.4526	1.4560	1.4595	1.4629	1.4663	1.4697	1.4731	1.4765	1.4799
21.	4.5826	4.5935	4.6043	4.6152	4.6260	4.6368	4.6476	4.6583	4.6690	4.6797

Source: From W. J. Dixon and F. J. Massey, Jr., "Introduction to Statistical Analysis," 3d ed., McGraw-Hill Book Company, New York, 1969. By permission of the authors and publishers.

Appendix IV—*Continued*

	0	1	2	3	4	5	6	7	8	9
2.2	1.4832	1.4866	1.4900	1.4933	1.4967	1.5000	1.5033	1.5067	1.5100	1.5133
22.	4.6904	4.7011	4.7117	4.7223	4.7329	4.7434	4.7539	4.7645	4.7749	4.7854
2.3	1.5166	1.5199	1.5232	1.5264	1.5297	1.5330	1.5362	1.5395	1.5427	1.5460
23.	4.7958	4.8062	4.8166	4.8270	4.8374	4.8477	4.8580	4.8683	4.8785	4.8888
2.4	1.5492	1.5524	1.5556	1.5588	1.5620	1.5652	1.5684	1.5716	1.5748	1.5780
24.	4.8990	4.9092	4.9193	4.9295	4.9396	4.9497	4.9598	4.9699	4.9800	4.9900
2.5	1.5811	1.5843	1.5875	1.5906	1.5937	1.5960	1.6000	1.6031	1.6062	1.6093
25.	5.0000	5.0100	5.0200	5.0299	5.0398	5.0498	5.0596	5.0695	5.0794	5.0892
2.6	1.6125	1.6155	1.6186	1.6217	1.6248	1.6279	1.6310	1.6340	1.6371	1.6401
26.	5.0990	5.1088	5.1186	5.1284	5.1381	5.1478	5.1575	5.1672	5.1769	5.1865
2.7	1.6432	1.6462	1.6492	1.6523	1.6553	1.6583	1.6613	1.6643	1.6673	1.6703
27.	5.1962	5.2058	5.2154	5.2249	5.2345	5.2440	5.2536	5.2631	5.2726	5.2820
2.8	1.6733	1.6763	1.6793	1.6823	1.6852	1.6882	1.6912	1.6941	1.6971	1.7000
28.	5.2915	5.3009	5.3104	5.3198	5.3292	5.3385	5.3479	5.3572	5.3666	5.3759
2.9	1.7029	1.7059	1.7088	1.7117	1.7146	1.7176	1,7205	1.7234	1.7263	1.7292
29.	5.3852	5.3944	5.4037	5.4129	5.4222	5.4314	5.4406	5.4498	5.4589	5.4681
3.0	1.7321	1.7349	1.7378	1.7407	1.7436	1.7464	1.7493	1.7521	1.7550	1.7578
30.	5.4772	5.4863	5.4955	5.5045	5.5136	5.5227	5.5317	5.5408	5.5498	5.5588
3.1	1.7607	1.7635	1.7664	1.7692	1.7720	1.7748	1.7776	1.7804	1.7833	1.7861
31.	5.5678	5.5767	5.5857	5.5946	5.6036	5.6125	5.6214	5.6303	5.6391	5.6480
3.2	1.7889	1.7961	1.7944	1.7972	1.8000	1.8028	1.8055	1.8083	1.8111	1.8138
32.	5.6569	5.6657	5.6745	5.6833	5.6921	5.7009	5.7096	5.7184	5.7271	5.7359
3.3	1.8166	1.8193	1.9221	1.8248	1.8276	1.8303	1.8330	1.8358	1.8385	1.8412
33.	5.7446	5.7533	5.7619	5.7706	5.7793	5.7879	5.7966	5.8052	5.8138	5.8224
3.4	1.8439	1.8466	1.8493	1.8520	1.8547	1.8574	1.8601	1.8628	1.8655	1.8682
34.	5.8310	5.8395	5.8481	5.8566	5.8652	5.8737	5.8822	5.8907	5.8992	5.9076
3.5	1.8708	1.8735	1.8762	1.8788	1.8815	1.8841	1.8868	1.8894	1.8921	1.8947
35.	5.9161	5.9245	5.9330	5.9414	5.9498	5.9582	5.9666	5.9749	5.9833	5.9917

	0	1	2	3	4	5	6	7	8	9
3.6	1.8974	1.9000	1.9026	1.9053	1.9079	1.9105	1.9131	1.9157	1.9183	1.9209
36.	6.0000	6.0083	6.0166	6.0249	6.0332	6.0415	6.0498	6.0581	6.0663	6.0745
3.7	1.9235	1.9261	1.9287	1.9313	1.9339	1.9365	1.9391	1.9416	1.9442	1.9468
37.	6.0828	6.0910	6.0992	6.1074	6.1156	6.1237	6.1319	6.1400	6.1482	6.1563
3.8	1.9494	1.9519	1.9545	1.9570	1.9596	1.9621	1.9647	1.9672	1.9698	1.9723
38.	6.1644	6.1725	6.1806	6.1887	6.1968	6.2048	6.2129	6.2209	6.2290	6.2370
3.9	1.9748	1.9774	1.9799	1.9824	1.9849	1.9875	1.9900	1.9925	1.9950	1.9975
39.	6.2450	6.2530	6.2610	6.2690	6.2769	6.2849	6.2929	6.3008	6.3087	6.3166
4.0	2.0000	2.0025	2.0050	2.0075	2.0100	2.0125	2.0149	2.0174	2.0199	2.0224
40.	6.3246	6.3325	6.3403	6.3482	6.3561	6.3640	6.3718	6.3797	6.3875	6.3953
4.1	2.0248	2.0273	2.0298	2.0322	2.0347	2.0372	2.0396	2.0421	2.0445	2.0469
41.	6.4031	6.4109	6.4187	6.4265	6.4343	6.4420	6.4498	6.4576	6.4653	6.4730
4.2	2.0494	2.0518	2.0543	2.0567	2.0591	2.0616	2.0640	2.0664	2.0688	2.0712
42.	6.4807	6.4885	6.4962	6.5038	6.5115	6.5192	6.5269	6.5345	6.5422	6.5498
4.3	2.0736	2.0761	2.0785	2.0809	2.0833	2.0857	2.0881	2.0905	2.0928	2.0952
43.	6.5574	6.5651	6.5727	6.5803	6.5879	6.5955	6.6030	6.6106	6.6182	6.6257
4.4	2.0976	2.1000	2.1024	2.1048	2.1071	2.1095	2.1119	2.1142	2.1166	2.1190
44.	6.6332	6.6408	6.6483	6.6558	6.6633	6.6708	6.6783	6.6858	6.6933	6.7007
4.5	2.1213	2.1237	2.1260	2.1284	2.1307	2.1331	2.1354	2.1378	2.1401	2.1424
45.	6.7082	6.7157	6.7231	6.7305	6.7380	6.7454	6.7528	6.7602	6.7676	6.7750
4.6	2.1448	2.1471	2.1494	2.1517	2.1541	2.1564	2.1587	2.1610	2.1633	2.1656
46.	6.7823	6.7897	6.7971	6.8044	6.8118	6.8191	6.8264	6.8337	6.8411	6.8484
4.7	2.1679	2.1703	2.1726	2.1749	2.1772	2.1794	2.1817	2.1840	2.1863	2.1886
47.	6.8557	6.8629	6.8702	6.8775	6.8848	6.8920	6.8993	6.9065	6.9138	6.9210
4.8	2.1909	2.1932	2.1954	2.1977	2.2000	2.2023	2.2045	2.2068	2.2091	2.2113
48.	6.9282	6.9354	6.9426	6.9498	6.9570	6.9642	6.9714	6.9785	6.9857	6.9929
4.9	2.2136	2.2159	2.2181	2.2204	2.2226	2.2249	2.2271	2.2293	2.2316	2.2338
49.	7.0000	7.0071	7.0143	7.0214	7.0285	7.0356	7.0427	7.0498	7.0569	7.0640

Appendix IV—*Continued*

	0	1	2	3	4	5	6	7	8	9
5.0	2.2361	2.2383	2.2405	2.2428	2.2450	2.2472	2.2494	2.2517	2.2539	2.2561
50.	7.0711	7.0781	7.0852	7.0922	7.0993	7.1063	7.1134	7.1204	7.1274	7.1344
5.1	2.2583	2.2605	2.2627	2.2650	2.2672	2.2694	2.2716	2.2738	2.2760	2.2782
51.	7.1414	7.1484	7.1554	7.1624	7.1694	7.1764	7.1833	7.1903	7.1972	7.2042
5.2	2.2804	2.2825	2.2847	2.2869	2.2891	2.2913	2.2935	2.2956	2.2978	2.3000
52.	7.2111	7.2180	7.2250	7.2319	7.2388	7.2457	7.2526	7.2595	7.2664	7.2732
5.3	2.3022	2.3043	2.3065	2.3087	2.3108	2.3130	2.3152	2.3173	2.3195	2.3216
53.	7.2801	7.2870	7.2938	7.3007	7.3075	7.3141	7.3212	7.3280	7.3348	7.3417
5.4	2.3238	2.3259	2.3281	2.3302	2.3324	2.3345	2.3367	2.3388	2.3409	2.3431
54.	7.3485	7.3553	7.3621	7.3689	7.3756	7.3824	7.2892	7.3959	7.4027	7.4095
5.5	2.3452	2.3473	2.3495	2.3516	2.3537	2.3558	2.3580	2.3601	2.3622	2.3643
55.	7.4162	7.4229	7.4297	7.4364	7.4431	7.4498	7.4565	7.4632	7.4699	7.4766
5.6	2.3664	2.3685	2.3707	2.3728	2.3749	2.3770	2.3791	2.3812	2.3833	2.3854
56.	7.4833	7.4900	7.4967	7.5033	7.5100	7.5166	7.5233	7.5299	7.5366	7.5432
5.7	2.3875	2.3896	2.3917	2.3937	2.3958	2.3979	2.4000	2.4021	2.4042	2.4062
57.	7.5498	7.5565	7.5631	7.5697	7.5763	7.5829	7.5895	7.5961	7.6026	7.6092
5.8	2.4083	2.4104	2.4125	2.4145	2.4166	2.4187	2.4207	2.4228	2.4249	2.4269
58.	7.6158	7.6223	7.6289	7.6354	7.6420	7.6485	7.6551	7.6616	7.6681	7.6746
5.9	2.4290	2.4310	2.4331	2.4352	2.4372	2.4393	2.4413	2.4434	2.4454	2.4474
59.	7.6811	7.6877	7.6942	7.7006	7.7071	7.7136	7.7201	7.7266	7.7330	7.7395
6.0	2.4495	2.4515	2.4536	2.4556	2.4576	2.4597	2.4617	2.4637	2.4658	2.4678
60.	7.7460	7.7524	7.7589	7.7653	7.7717	7.7782	7.7846	7.7910	7.7974	7.8038
6.1	2.4698	2.4718	2.4739	2.4759	2.4779	2.4799	2.4819	2.4839	2.4860	2.4880
61.	7.8102	7.8166	7.8230	7.8294	7.8358	7.8422	7.8486	7.8549	7.8613	7.8677
6.2	2.4900	2.4920	2.4940	2.4960	2.4980	2.5000	2.5020	2.5040	2.5060	2.5080
62.	7.8740	7.8804	7.8867	7.8930	7.8994	7.9057	7.9120	7.9183	7.9246	7.9310
6.3	2.5100	2.5120	2.5140	2.5159	2.5179	2.5199	2.5219	2.5239	2.5259	2.5278
63.	7.9373	7.9436	7.9498	7.9561	7.9624	7.9687	7.9750	7.9812	7.9875	7.9937

	0	1	2	3	4	5	6	7	8	9
6.4	2.5298	2.5318	2.5338	2.5357	2.5377	2.5397	2.5417	2.5436	2.5456	2.5475
64.	8.0000	8.0062	8.0125	8.0187	8.0250	8.0312	8.0374	8.0436	8.0498	8.0561
6.5	2.5495	2.5515	2.5534	2.5554	2.5573	2.5593	2.5612	2.5632	2.5652	2.5671
65.	8.0623	8.0685	8.0747	8.0808	8.0870	8.0932	8.0994	8.1056	8.1117	8.1179
6.6	2.5690	2.5710	2.5729	2.5749	2.5768	2.5788	2.5807	2.5826	2.5846	2.5865
66.	8.1240	8.1302	8.1363	8.1425	8.1486	8.1548	8.1609	8.1670	8.1731	8.1792
6.7	2.5884	2.5904	2.5923	2.5942	2.5962	2.5981	2.6000	2.6019	2.6038	2.6058
67.	8.1854	8.1915	8.1976	8.2037	8.2098	8.2158	8.2219	8.2280	8.2341	8.2401
6.8	2.6077	2.6096	2.6115	2.6134	2.6153	2.6173	2.6192	2.6211	2.6230	2.6249
68.	8.2462	8.2523	8.2583	8.2644	8.2704	8.2765	8.2825	8.2885	8.2946	8.3006
6.9	2.6268	2.6287	2.6306	2.6325	2.6344	2.6363	2.6382	2.6401	2.6420	2.6439
69.	8.3066	8.3126	8.3187	8.3247	8.3307	8.3367	8.3427	8.3487	8.3546	8.3606
7.0	2.6458	2.6472	2.6495	2.6514	2.6533	2.6552	2.6571	2.6589	2.6608	2.6627
70.	8.3666	8.3726	8.3785	8.3845	8.3905	8.3964	8.4024	8.4083	8.4143	8.4202
7.1	2.6646	2.6665	2.6683	2.6702	2.6721	2.6739	2.6758	2.6777	2.6796	2.6811
71.	8.4261	8.4321	8.4380	8.4439	8.4499	8.4558	8.4617	8.4676	8.4735	8.4791
7.2	2.6833	2.6851	2.6870	2.6889	2.6907	2.6926	2.6944	2.6963	2.6981	2.7000
72.	8.4853	8.4912	8.4971	8.5029	8.5088	8.5147	8.5264	8.5264	8.5328	8.5381
7.3	2.7019	2.7037	2.7055	2.7074	2.7092	2.7111	2.7129	2.7148	2.7166	2.7185
73.	8.5440	8.5499	8.5557	8.5615	8.5674	8.5732	8.5790	8.5849	8.5907	8.5965
7.4	2.7203	2.7221	2.2740	2.7258	2.7276	2.7295	2.7313	2.7331	2.7350	2.7368
74.	8.6023	8.6081	8.6139	8.6197	8.6255	8.6313	8.6371	8.6429	8.6487	8.6545
7.5	2.7386	2.7404	2.7423	2.7441	2.7459	2.7477	2.7495	2.7514	2.7532	2.7550
75.	8.6603	8.6660	8.6718	8.6776	8.6833	8.6891	8.6948	8.7006	8.7063	8.7121
7.6	2.7568	2.7586	2.7604	2.7622	2.7641	2.7659	2.7677	2.7695	2.7713	2.7731
76.	8.7178	8.7235	8.7293	8.7350	8.7407	8.7464	8.7521	8.7579	8.7636	8.7693
7.7	2.7749	2.7767	2.7785	2.7803	2.7821	2.7839	2.7857	2.7875	2.7893	2.7911
77.	8.7750	8.7807	8.7864	8.7920	8.7977	8.8034	8.8091	8.8148	8.8204	8.8261

Appendix IV—*Continued*

	0	1	2	3	4	5	6	7	8	9
7.8	2.7928	2.7946	2.7964	2.7982	2.8000	2.8018	2.8036	2.8054	2.8071	2.8089
78.	8.8318	8.8374	8.8431	8.8487	8.8544	8.8600	8.8657	8.8713	8.8769	8.8826
7.9	2.8107	2.8125	2.8142	2.8160	2.8178	2.8196	2.8213	2.8231	2.8249	2.8267
79.	8.8882	8.8938	8.8994	8.9051	8.9107	8.9163	8.9219	8.9275	8.9331	8.9387
8.0	2.8284	2.8302	2.8320	2.8337	2.8355	2.8373	2.8390	2.8408	2.8425	2.8113
80.	8.9443	8.9499	8.9554	8.9610	8.9666	8.9722	8.9778	8.9833	8.9889	8.9911
8.1	2.8460	2.8478	2.8496	2.8513	2.8531	2.8548	2.8566	8.8583	2.8601	2.8618
81.	9.0000	9.0056	9.0111	9.0167	9.0222	9.0277	9.0333	9.0388	9.0443	9.0199
8.2	2.8636	2.8653	2.8671	2.8688	2.8705	2.8723	2.8740	2.8758	2.8775	2.8792
82.	9.0554	9.0609	9.0664	9.0719	9.0774	9.0830	9.0885	9.0940	9.0995	9.10 9
8.3	2.8810	2.8827	2.8844	2.8862	2.8879	2.8896	2.8914	2.8931	2.8948	2.8965
83.	9.1104	9.1159	9.1214	9.1269	9.1324	9.1378	9.1433	9.1488	9.1542	9.1597
8.4	2.8983	2.9000	2.9017	2.9034	2.9052	2.9069	2.9086	2.9103	2.9120	2.9138
84.	9.1652	9.1706	9.1761	9.1815	9.1869	9.1924	9.1978	9.2033	9.2087	9.2141
8.5	2.9155	2.9172	2.9180	2.9206	2.9223	2.9240	2.9257	2.9275	2.9292	2.9309
85.	9.2195	9.2250	9.2304	9.2358	9.2412	9.2466	9.2520	9.2574	9.2628	9.2682
8.6	2.9326	2.9343	2.9360	2.9377	2.9394	2.9411	2.9428	2.9445	2.9462	2.9479
86.	9.2736	9.2790	9.2844	9.2898	9.2952	9.3005	9.3059	9.3113	9.3167	9.3220
8.7	2.9496	2.9513	2.9530	2.9547	2.9563	2.9580	2.9597	2.9614	2.9631	2.9648
87.	9.3274	9.3327	9.3381	9.3434	9.3488	9.3541	9.3595	9.3648	9.3702	9.3755
8.8	2.9665	2.9682	2.9698	2.9715	2.9732	2.9749	2.9766	2.9783	2.9799	2.9816
88.	9.3808	9.3862	9.3915	9.3968	9.4021	9.4074	9.4128	9.4181	9.4234	9.4287
8.9	2.9833	2.9850	2.9866	2.9883	2.9900	2.9917	2.9933	2.9950	2.9967	2.9983
89.	9.4340	9.4393	9.4446	9.4499	9.4552	9.4604	9.4657	9.4710	9.4763	9.4816
9.0	3.0000	3.0017	3.0033	3.0050	3.0067	3.0083	3.0100	3.0116	3.0133	3.0150
90.	9.4868	9.4921	9.4974	9.5026	9.5079	9.5131	9.5184	9.5237	9.5289	9.5341
9.1	3.0166	3.0183	3.0199	3.0216	3.0232	3.0249	3.0265	3.0282	3.0299	3.0315
91.	9.5394	9.5446	9.5499	9.5551	9.6503	9.5656	9.5708	9.5760	9.5812	9.5864

	0	1	2	3	4	5	6	7	8	9
9.2	3.0332	3.0348	3.0364	3.0381	3.0397	3.0414	3.0430	3.0447	3.0463	3.0480
92.	9.5917	9.5969	9.6021	9.6073	9.6125	9.6177	9.6229	9.6281	9.6333	9.6385
9.3	3.0496	3.0512	3.0529	3.0545	3.0561	3.0578	3.0594	3.0610	3.0627	3.0643
93.	9.6437	9.6488	9.6540	9.6592	9.6644	9.6695	9.6747	9.6799	9.6850	9.6902
9.4	3.0659	3.0676	3.0692	3.0708	3.0725	3.0741	3.0757	3.0773	3.0790	3.0806
94.	9.6954	9.7005	9.7057	9.7108	9.7160	9.7211	9.7263	9.7314	9.7365	9.7417
9.5	3.0822	3.0838	3.0854	3.0871	3.0887	3.0903	3.0919	3.0935	3.0952	3.0968
95.	9.7468	9.7519	9.7570	9.7622	9.7673	9.7724	9.7775	9.7826	9.7877	9.7929
9.6	3.0984	3.1000	3.1016	3.1032	3.1048	3.1064	3.1081	3.1097	3.1113	3.1129
96.	9.7980	9.8031	9.8082	9.8133	9.8184	9.8234	9.8334	9.8285	9.8387	9.8438
9.7	3.1145	3.1161	3.1177	3.1193	3.1209	3.1225	3.1241	3.1257	3.1273	3.1289
97.	9.8489	9.8539	9.8590	9.8641	9.8691	9.8742	9.8793	9.8843	9.8894	9.8944
9.8	3.1305	3.1321	3.1337	3.1353	3.1369	3.1385	3.1401	3.1417	3.1432	3.1448
98.	9.8995	9.9045	9.9096	9.9146	9.9197	9.9247	9.9298	9.9348	9.9398	9.9448
9.9	3.1464	3.1480	3.1496	3.1512	3.1528	3.1544	3.1559	3.1575	3.1591	3.1607
99.	9.9499	9.9549	9.9599	9.9649	9.9700	9.9750	9.9800	9.9850	9.9900	9.9950

Appendix V
The Poisson Distribution

Values of $e^{-\mu} \dfrac{\mu^x}{x!}$

x \ μ	.1	.2	.3	.4	.5	.6	.7	.8	.9	1.0
0	.90484	.81873	.74082	.67032	.60653	.54881	.49659	.44933	.40657	.36788
1	.09048	.16375	.22225	.26813	.30327	.32929	.34761	.35946	.36591	.36788
2	.00452	.01637	.03334	.05363	.07582	.09879	.12166	.14379	.16466	.18394
3	.00015	.00109	.00333	.00715	.01264	.01976	.02839	.03834	.04940	.06131
4	.00000	.00005	.00025	.00072	.00158	.00296	.00497	.00767	.01111	.01533
5	.00000	.00000	.00002	.00006	.00016	.00036	.00070	.00123	.00200	.00307
6		.00000	.00000	.00000	.00001	.00004	.00008	.00016	.00030	.00051
7					.00000	.00000	.00001	.00002	.00004	.00007
8						.00000	.00000	.00000	.00000	.00001
9								.00000	.00000	.00000

x \ μ	1.1	1.2	1.3	1.4	1.5	1.6	1.7	1.8	1.9	2.0
0	.33287	.30119	.27253	.24660	.22313	.20190	.18268	.16530	.14957	.13534
1	.36616	.36143	.35429	.34524	.33470	.32303	.31056	.29754	.28418	.27067
2	.20139	.21686	.23029	.24166	.25102	.25843	.26398	.26778	.26997	.27067
3	.07384	.08674	.09979	.11278	.12551	.13783	.14959	.16067	.17098	.18045
4	.02031	.02602	.03243	.03947	.04707	.05513	.06357	.07230	.08122	.09022
5	.00447	.00625	.00843	.01105	.01412	.01764	.02162	.02603	.03086	.03609
6	.00082	.00125	.00183	.00258	.00353	.00470	.00612	.00781	.00977	.01203
7	.00013	.00021	.00034	.00052	.00076	.00108	.00149	.00201	.00265	.00344
8	.00002	.00003	.00006	.00009	.00014	.00022	.00032	.00045	.00063	.00086
9	.00000	.00000	.00001	.00001	.00002	.00004	.00006	.00009	.00013	.00019
10	.00000	.00000	.00000	.00000	.00000	.00001	.00001	.00002	.00003	.00004
11			.00000	.00000	.00000	.00000	.00000	.00000	.00000	.00001
12							.00000	.00000	.00000	.00000
13										.00000

x \ μ	2.1	2.2	2.3	2.4	2.5	2.6	2.7	2.8	2.9	3.0
0	.12246	.11080	.10026	.09072	.08209	.07427	.06721	.06081	.05502	.04979
1	.25716	.24377	.23060	.21772	.20521	.19311	.18146	.17027	.15957	.14936
2	.27002	.26814	.26518	.26127	.25652	.25105	.24496	.23838	.23137	.22404
3	.18901	.19664	.20331	.20901	.21376	.21757	.22047	.22248	.22366	.22404
4	.09923	.10815	.11690	.12541	.13360	.14142	.14881	.15574	.16215	.16803
5	.04168	.04759	.05377	.06020	.06680	.07354	.08036	.08721	.09405	.10082
6	.01459	.01745	.02061	.02408	.02783	.03187	.03616	.04070	.04546	.05041
7	.00438	.00548	.00677	.00826	.00994	.01184	.01395	.01628	.01883	.02160
8	.00115	.00151	.00195	.00248	.00311	.00385	.00471	.00570	.00683	.00810
9	.00027	.00037	.00050	.00066	.00086	.00111	.00141	.00177	.00220	.00270
10	.00006	.00008	.00011	.00016	.00022	.00029	.00038	.00050	.00064	.00081
11	.00001	.00002	.00002	.00003	.00005	.00007	.00009	.00013	.00017	.00022
12	.00000	.00000	.00000	.00001	.00001	.00001	.00002	.00003	.00004	.00006
13	.00000	.00000	.00000	.00000	.00000	.00000	.00000	.00001	.00001	.00001
14				.00000	.00000	.00000	.00000	.00000	.00000	.00000

x \ μ	3.1	3.2	3.3	3.4	3.5	3.6	3.7	3.8	3.9	4.0
0	.04505	.04076	.03688	.03337	.03020	.02732	.02472	.02237	.02024	.01832
1	.13965	.13044	.12172	.11347	.10569	.09837	.09148	.08501	.07894	.07326
2	.21646	.20870	.20083	.19290	.18496	.17706	.16923	.16152	.15394	.14653
3	.22368	.22262	.22091	.21862	.21579	.21247	.20872	.20459	.20012	.19537
4	.17335	.17809	.18225	.18582	.18881	.19122	.19307	.19436	.19512	.19537
5	.10748	.11398	.12029	.12636	.13217	.13768	.14287	.14771	.15219	.15629
6	.05553	.06079	.06616	.07160	.07710	.08261	.08810	.09355	.09892	.10419
7	.02459	.02779	.03119	.03478	.03855	.04248	.04657	.05078	.05511	.05954
8	.00953	.01112	.01287	.01478	.01686	.01912	.02154	.02412	.02687	.02977
9	.00328	.00395	.00472	.00558	.00656	.00765	.00885	.01018	.01164	.01323
10	.00102	.00126	.00156	.00190	.00230	.00275	.00328	.00387	.00454	.00529
11	.00029	.00037	.00047	.00059	.00073	.00090	.00110	.00134	.00161	.00192
12	.00007	.00010	.00013	.00017	.00021	.00027	.00034	.00042	.00052	.00064
13	.00002	.00002	.00003	.00004	.00006	.00007	.00010	.00012	.00016	.00020
14	.00000	.00001	.00001	.00001	.00001	.00002	.00003	.00003	.00004	.00006
15	.00000	.00000	.00000	.00000	.00000	.00000	.00001	.00001	.00001	.00002
16		.00000	.00000	.00000	.00000	.00000	.00000	.00000	.00000	.00000

Appendix V—*Continued*

x \ μ	4.1	4.2	4.3	4.4	4.5	4.6	4.7	4.8	4.9	5.0
0	.01657	.01500	.01357	.01228	.01111	.01005	.00910	.00823	.00745	.00674
1	.06785	.06298	.05835	.05402	.04999	.04624	.04275	.03950	.03649	.03369
2	.13929	.13226	.12544	.11885	.11248	.10635	.10046	.09481	.08940	.08422
3	.19037	.18517	.17980	.17431	.16872	.16307	.15738	.15169	.14602	.14037
4	.19513	.19442	.19328	.19174	.18981	.18753	.18493	.18203	.17887	.17747
5	.16000	.16331	.16622	.16873	.17083	.17252	.17383	.17475	.17529	.17547
6	.10933	.11432	.11913	.12373	.12812	.13227	.13617	.13980	.14315	.14622
7	.06404	.06859	.07318	.07777	.08236	.08692	.09143	.09586	.10021	.10444
8	.03282	.03601	.03933	.04278	.04633	.04998	.05371	.05752	.06138	.06528
9	.01495	.01680	.01879	.02091	.02316	.02554	.02805	.03068	.03342	.03627
10	.00613	.00706	.00808	.00920	.01042	.01175	.01318	.01472	.01637	.01813
11	.00228	.00269	.00316	.00368	.00426	.00491	.00563	.00642	.00729	.00824
12	.00078	.00094	.00113	.00135	.00160	.00188	.00221	.00257	.00298	.00343
13	.00025	.00030	.00037	.00046	.00055	.00067	.00080	.00095	.00112	.00132
14	.00007	.00009	.00011	.00014	.00018	.00022	.00027	.00033	.00039	.00047
15	.00002	.00003	.00003	.00004	.00005	.00007	.00008	.00010	.00013	.00016
16	.00001	.00001	.00001	.00001	.00002	.00002	.00002	.00003	.00004	.00005
17	.00000	.00000	.00000	.00000	.00000	.00001	.00001	.00001	.00001	.00001
18	.00000	.00000	.00000	.00000	.00000	.00000	.00000	.00000	.00000	.00000
19						.00000	.00000	.00000	.00000	.00000

x \ μ	6.0	7.0	8.0	9.0	10.0	11.0	12.0	13.0	14.0	15.0
0	.00248	.00091	.00034	.00012	.00005	.00002	.00001	.00000	.00000	.00000
1	.01487	.00638	.00268	.00111	.00045	.00018	.00007	.00003	.00001	.00000
2	.04462	.02234	.01073	.00500	.00227	.00101	.00044	.00019	.00008	.00003
3	.08924	.05213	.02873	.01499	.00757	.00370	.00177	.00083	.00038	.00017
4	.13385	.09123	.05725	.03374	.01892	.01019	.00531	.00269	.00133	.00065
5	.16062	.12772	.09160	.06073	.03783	.02242	.01274	.00699	.00373	.00194
6	.16062	.14900	.12214	.09109	.06306	.04109	.02548	.01515	.00870	.00484
7	.13768	.14900	.13959	.11712	.09008	.06458	.04368	.02814	.01739	.01037
8	.10326	.13038	.13959	.13175	.11260	.08879	.06552	.04573	.03044	.01944
9	.06884	.10140	.12408	.13175	.12511	.10852	.08736	.06605	.04734	.03241
10	.04130	.07098	.09926	.11858	.12511	.11938	.10484	.08587	.06628	.04861
11	.02253	.04517	.07219	.09702	.11374	.11938	.11437	.10148	.08436	.06629
12	.01126	.02635	.04813	.07276	.09478	.10943	.11437	.10994	.09842	.08286
13	.00520	.01419	.02962	.05038	.07291	.09259	.10557	.10994	.10599	.09561
14	.00223	.00709	.01692	.03238	.05208	.07275	.09049	.10209	.10599	.10243
15	.00089	.00331	.00903	.01943	.03472	.05335	.07239	.08847	.09892	.10243
16	.00033	.00145	.00451	.01093	.02170	.03668	.05429	.07189	.08656	.09603
17	.00012	.00060	.00212	.00579	.01276	.02373	.03832	.05497	.07128	.08473
18	.00004	.00023	.00094	.00289	.00709	.01450	.02555	.03970	.05544	.07061
19	.00001	.00009	.00040	.00137	.00373	.00840	.01614	.02716	.04085	.05575
20	.00000	.00003	.00016	.00062	.00187	.00462	.00968	.01766	.02860	.04181
21	.00000	.00001	.00006	.00026	.00089	.00242	.00553	.01093	.01906	.02986
22	.00000	.00000	.00002	.00011	.00040	.00121	.00302	.00646	.01213	.02036
23		.00000	.00001	.00004	.00018	.00058	.00157	.00365	.00738	.01328
24			.00000	.00002	.00007	.00027	.00079	.00198	.00431	.00830
25			.00000	.00001	.00003	.00012	.00038	.00103	.00241	.00498
26				.00000	.00001	.00005	.00017	.00051	.00130	.00287
27				.00000	.00000	.00002	.00008	.00025	.00067	.00160
28					.00000	.00001	.00003	.00011	.00034	.00086
29						.00000	.00001	.00005	.00016	.00044
30						.00000	.00001	.00002	.00008	.00022
31						.00000	.00000	.00001	.00003	.00011
32							.00000	.00000	.00001	.00005
33								.00000	.00001	.00002
34								.00000	.00000	.00001
35									.00000	.00000

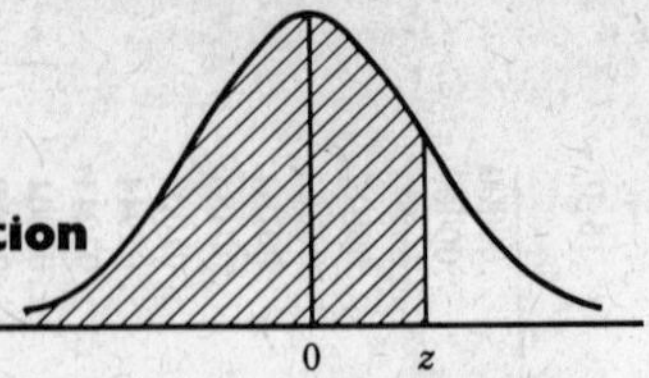

Appendix VI
Cumulative Standard Normal Distribution

Values of $F(z)$

z	.00	.01	.02	.03	.04	.05	.06	.07	.08	.09
.0	.5000	.5040	.5080	.5120	.5160	.5199	.5239	.5279	.5319	.5359
.1	.5398	.5438	.5478	.5517	.5557	.5596	.5636	.5675	.5714	.5753
.2	.5793	.5832	.5871	.5910	.5948	.5987	.6026	.6064	.6103	.6141
.3	.6179	.6217	.6255	.6293	.6331	.6368	.6406	.6443	.6480	.6517
.4	.6554	.6591	.6628	.6664	.6700	.6736	.6772	.6808	.6844	.6879
.5	.6915	.6950	.6985	.7019	.7054	.7088	.7123	.7157	.7190	.7224
.6	.7257	.7291	.7324	.7357	.7389	.7422	.7454	.7486	.7517	.7549
.7	.7580	.7611	.7642	.7673	.7704	.7734	.7764	.7794	.7823	.7852
.8	.7881	.7910	.7939	.7967	.7995	.8023	.8051	.8078	.8106	.8133
.9	.8159	.8186	.8212	.8238	.8264	.8289	.8315	.8340	.8365	.8389
1.0	.8413	.8438	.8461	.8485	.8508	.8531	.8554	.8577	.8599	.8621
1.1	.8643	.8665	.8686	.8708	.8729	.8749	.8770	.8790	.8810	.8830
1.2	.8849	.8869	.8888	.8907	.8925	.8944	.8962	.8980	.8997	.9015
1.3	.9032	.9049	.9066	.9082	.9099	.9115	.9131	.9147	.9162	.9177
1.4	.9192	.9207	.9222	.9236	.9251	.9265	.9279	.9292	.9306	.9319
1.5	.9332	.9345	.9357	.9370	.9382	.9394	.9406	.9418	.9429	.9441
1.6	.9452	.9463	.9474	.9484	.9495	.9505	.9515	.9525	.9535	.9545
1.7	.9554	.9564	.9573	.9582	.9591	.9599	.9608	.9616	.9625	.9633
1.8	.9641	.9649	.9656	.9664	.9671	.9678	.9686	.9693	.9699	.9706
1.9	.9713	.9719	.9726	.9732	.9738	.9744	.9750	.9756	.9761	.9767
2.0	.9772	.9778	.9783	.9788	.9793	.9798	.9803	.9808	.9812	.9817
2.1	.9821	.9826	.9830	.9834	.9838	.9842	.9846	.9850	.9854	.9857
2.2	.9861	.9864	.9868	.9871	.9875	.9878	.9881	.9884	.9887	.9890
2.3	.9893	.9896	.9898	.9901	.9904	.9906	.9909	.9911	.9913	.9916
2.4	.9918	.9920	.9922	.9925	.9927	.9929	.9931	.9932	.9934	.9936
2.5	.9938	.9940	.9941	.9943	.9945	.9946	.9948	.9949	.9951	.9952
2.6	.9953	.9955	.9956	.9957	.9959	.9960	.9961	.9962	.9963	.9964
2.7	.9965	.9966	.9967	.9968	.9969	.9970	.9971	.9972	.9973	.9974
2.8	.9974	.9975	.9976	.9977	.9977	.9978	.9979	.9979	.9980	.9981
2.9	.9981	.9982	.9982	.9983	.9984	.9984	.9985	.9985	.9986	.9986
3.0	.9987	.9987	.9987	.9988	.9988	.9989	.9989	.9989	.9990	.9990
3.1	.9990	.9991	.9991	.9991	.9992	.9992	.9992	.9992	.9993	.9993
3.2	.9993	.9993	.9994	.9994	.9994	.9994	.9994	.9995	.9995	.9995
3.3	.9995	.9995	.9995	.9996	.9996	.9996	.9996	.9996	.9996	.9997
3.4	.9997	.9997	.9997	.9997	.9997	.9997	.9997	.9997	.9997	.9998

z	1.282	1.645	1.960	2.326	2.575	3.090	3.291	3.891	4.417
$F(z)$	.90	.95	.975	.99	.995	.999	.9995	.99995	.999995
$2[1 - F(z)]$	.20	.10	.05	.02	.01	.002	.001	.0001	.00001

Source: Adopted by permission from A. M. Mood, *Introduction to the Theory of Statistics*, New York, McGraw-Hill Book Company, 1950.

Appendix VII
Percentage Points of the *t* Distribution

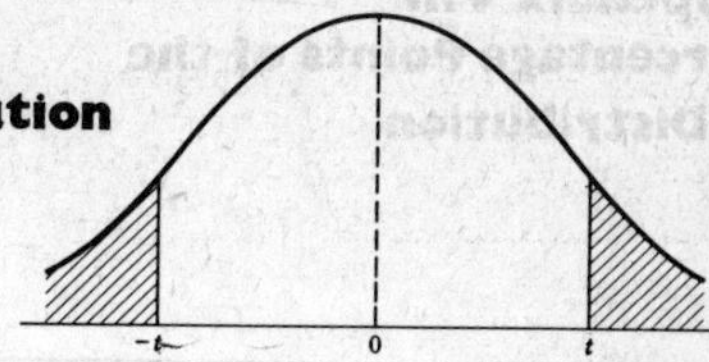

ν	$Q = .4$ $2Q = .8$	.25 .5	.1 .2	.05 .1	.025 .05	.01 .02	.005 .01	.0025 .005	.001 .002	.0005 .001
1	.325	1.000	3.078	6.314	12.706	31.821	63.657	127.32	318.31	636.62
2	.289	.816	1.886	2.920	4.303	6.965	9.925	14.089	22.327	31.598
3	.277	.765	1.638	2.353	3.182	4.541	5.841	7.453	10.214	12.924
4	.271	.741	1.533	2.132	2.776	3.747	4.604	5.598	7.173	8.610
5	.267	.727	1.476	2.015	2.571	3.365	4.032	4.773	5.893	6.869
6	.265	.718	1.440	1.943	2.447	3.143	3.707	4.317	5.208	5.959
7	.263	.711	1.415	1.895	2.365	2.998	3.499	4.029	4.785	5.408
8	.262	.706	1.397	1.860	2.306	2.896	3.355	3.833	4.501	5.041
9	.261	.703	1.383	1.833	2.262	2.821	3.250	3.690	4.297	4.781
10	.260	.700	1.372	1.812	2.228	2.764	3.169	3.581	4.144	4.587
11	.260	.697	1.363	1.796	2.201	2.718	3.106	3.497	4.025	4.437
12	.259	.695	1.356	1.782	2.179	2.681	3.055	3.428	3.930	4.318
13	.259	.694	1.350	1.771	2.160	2.650	3.012	3.372	3.852	4.221
14	.258	.692	1.345	1.761	2.145	2.624	2.977	3.326	3.787	4.140
15	.258	.691	1.341	1.753	2.131	2.602	2.947	3.286	3.733	4.073
16	.258	.690	1.337	1.746	2.120	2.583	2.921	3.252	3.686	4.015
17	.257	.689	1.333	1.740	2.110	2.567	2.898	3.222	3.646	3.965
18	.257	.688	1.330	1.734	2.101	2.552	2.878	3.197	3.610	3.922
19	.257	.688	1.328	1.729	2.093	2.539	2.861	3.174	3.579	3.883
20	.257	.687	1.325	1.725	2.086	2.528	2.845	3.153	3.552	3.850
21	.257	.686	1.323	1.721	2.080	2.518	2.831	3.135	3.527	3.819
22	.256	.686	1.321	1.717	2.074	2.508	2.819	3.119	3.505	3.792
23	.256	.685	1.319	1.714	2.069	2.500	2.807	3.104	3.485	3.767
24	.256	.685	1.318	1.711	2.064	2.492	2.797	3.091	3.467	3.745
25	.256	.684	1.316	1.708	2.060	2.485	2.787	3.078	3.450	3.725
26	.256	.684	1.315	1.706	2.056	2.479	2.779	3.067	3.435	4.707
27	.256	.684	1.314	1.703	2.052	2.473	2.771	3.057	3.421	3.690
28	.256	.683	1.313	1.701	2.048	2.467	2.763	3.047	3.408	3.674
29	.256	.683	1.311	1.699	2.045	2.462	2.756	3.038	3.396	3.659
30	.256	.683	1.310	1.697	2.042	2.457	2.750	3.030	3.385	3.646
40	.255	.681	1.303	1.684	2.021	2.423	2.704	2.971	3.307	3.551
60	.254	.679	1.296	1.671	2.000	2.390	2.660	2.915	3.232	3.460
120	.254	.677	1.289	1.658	1.980	2.358	2.617	2.860	3.160	3.373
∞	.253	.674	1.282	1.645	1.960	2.326	2.576	2.807	3.090	3.291

$Q = 1 - P(t \mid \nu)$ is the upper-tail area of the distribution for ν degrees of freedom, appropriate for use in a single-tail test. For a two-tail test, $2Q$ must be used.

Source: This table is reproduced with the kind permission of Prof. E. S. Pearson and the trustees of Biometrika from E. S. Pearson and H. O. Hartley (eds.), *The Biometrika Tables for Statisticians*, vol. 1, *Biometrika*, 1966.

Appendix VIII
Percentage Points of the χ^2 Distribution

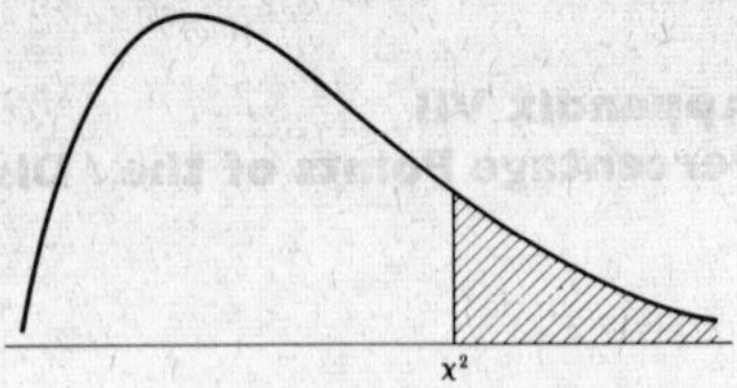

ν \ Q	0.995	0.990	0.975	0.950	0.900	0.750	0.500
1	$392704 . 10^{-10}$	$157088 . 10^{-9}$	$982069 . 10^{-9}$	$393214 . 10^{-8}$	0.0157908	0.1015308	0.454936
2	0.0100251	0.0201007	0.0506356	0.102587	0.210721	0.575364	1.38629
3	0.0717218	0.114832	0.215795	0.351846	0.584374	1.212534	2.36597
4	0.206989	0.297109	0.484419	0.710723	1.063623	1.92256	3.35669
5	0.411742	0.554298	0.831212	1.145476	1.61031	2.67460	4.35146
6	0.675727	0.872090	1.23734	1.63538	2.20413	3.45460	5.34812
7	0.989256	1.239043	1.68987	2.16735	2.83311	4.25485	6.34581
8	1.34441	1.64650	2.17973	2.73264	3.48954	5.07064	7.34412
9	1.73493	2.08790	2.70039	3.32511	4.16816	5.89883	8.34283
10	2.15586	2.55821	3.24697	3.94030	4.86518	6.73720	9.34182
11	2.60322	3.05348	3.81575	4.57481	5.57778	7.58414	10.3410
12	3.07382	3.57057	4.40379	5.22603	6.30380	8.43842	11.3403
13	3.56503	4.10692	5.00875	5.89186	7.04150	9.29907	12.3398
14	4.07467	4.66043	5.62873	6.57063	7.78953	10.1653	13.3393
15	4.60092	5.22935	6.26214	7.26094	8.54676	11.0365	14.3389
16	5.14221	5.81221	6.90766	7.96165	9.31224	11.9122	15.3385
17	5.69722	6.40776	7.56419	8.67176	10.0852	12.7919	16.3382
18	6.26480	7.01491	8.23075	9.39046	10.8649	13.6753	17.3379
19	6.84397	7.63273	8.90652	10.1170	11.6509	14.5620	18.3377
20	7.43384	8.26040	9.59078	10.8508	12.4426	15.4518	19.3374
21	8.03365	8.89720	10.28293	11.5913	13.2396	16.3444	20.3372
22	8.64272	9.54249	10.9823	12.3380	14.0415	17.2396	21.3370
23	9.26043	10.19567	11.6886	13.0905	14.8480	18.1373	22.3369
24	9.88623	10.8564	12.4012	13.8484	15.6587	19.0373	23.3367
25	10.5197	11.5240	13.1197	14.6114	16.4734	19.9393	24.3366
26	11.1602	12.1981	13.8439	15.3792	17.2919	20.8434	25.3365
27	11.8076	12.8785	14.5734	16.1514	18.1139	21.7494	26.3363
28	12.4613	13.5647	15.3079	16.9279	18.9392	22.6572	27.3362
29	13.1211	14.2565	16.0471	17.7084	19.7677	23.5666	28.3361
30	13.7867	14.9535	16.7908	18.4927	20.5992	24.4776	29.3360
40	20.7065	22.1643	24.4330	26.5093	29.0505	33.6603	39.3353
50	27.9907	29.7067	32.3574	34.7643	37.6886	42.9421	49.3349
60	35.5345	37.4849	40.4817	43.1880	46.4589	52.2938	59.3347
70	43.2752	45.4417	48.7576	51.7393	55.3289	61.6983	69.3345
80	51.1719	53.5401	57.1532	60.3915	64.2778	71.1445	79.3343
90	59.1963	61.7541	65.6466	69.1260	73.2911	80.6247	89.3342
100	67.3276	70.0649	74.2219	77.9295	82.3581	90.1332	99.3341
X	−2.5758	−2.3263	−1.9600	−1.6449	−1.2816	−0.6745	0.0000

Source: This table is reproduced with the kind permission of Prof. E. S. Pearson and the trustees of Biometrika from E. S. Pearson and H. O. Hartley (eds.), *The Biometrika Tables for Statisticians*, vol. 1. *Biometrika*, 1966.

ν \ Q	0.250	0.100	0.050	0.025	0.010	0.005	0.001
1	1.32330	2.70554	3.84146	5.02389	6.63490	7.87944	10.828
2	2.77259	4.60517	5.99146	7.37776	9.21034	10.5966	13.816
3	4.10834	6.25139	7.81473	9.34840	11.3449	12.8382	16.266
4	5.38527	7.77944	9.48773	11.1433	13.2767	14.8603	18.467
5	6.62568	9.23636	11.0705	12.8325	15.0863	16.7496	20.515
6	7.84080	10.6446	12.5916	14.4494	16.8119	18.5476	22.458
7	9.03715	12.0170	14.0671	16.0128	18.4753	20.2777	24.322
8	10.2189	13.3616	15.5073	17.5345	20.0902	21.9550	26.125
9	11.3888	14.6837	16.9190	19.0228	21.6660	23.5894	27.877
10	12.5489	15.9872	18.3070	20.4832	23.2093	25.1882	29.588
11	13.7007	17.2750	19.6751	21.9200	24.7250	26.7568	31.264
12	14.8454	18.5493	21.0261	23.3367	26.2170	28.2995	32.909
13	15.9839	19.8119	22.3620	24.7356	27.6882	29.8195	34.528
14	17.1169	21.0641	23.6848	26.1189	29.1412	31.3194	36.123
15	18.2451	22.3071	24.9958	27.4884	30.5779	32.8013	37.697
16	19.3689	23.5418	26.2962	28.8454	31.9999	34.2672	39.252
17	20.4887	24.7690	27.5871	30.1910	33.4087	35.7185	40.790
18	21.6049	25.9894	28.8693	31.5264	34.8053	37.1565	42.312
19	22.7178	27.2036	30.1435	32.8523	36.1909	38.5823	43.820
20	23.8277	28.4120	31.4104	34.1696	37.5662	39.9968	45.315
21	24.9348	29.6151	32.6706	35.4789	38.9322	41.4011	46.797
22	26.0393	30.8133	33.9244	36.7807	40.2894	42.7957	48.268
23	27.1413	32.0069	35.1725	38.0756	41.6384	44.1813	49.728
24	28.2412	33.1962	36.4150	39.3641	42.9798	45.5585	51.179
25	29.3389	34.3816	37.6525	40.6465	44.3141	46.9279	52.618
26	30.4346	35.5632	38.8851	41.9232	45.6417	48.2899	54:052
27	31.5284	36.7412	40.1133	43.1945	46.9629	49.6449	55.476
28	32.6205	37.9159	41.3371	44.4608	48.2782	50.9934	56.892
29	33.7109	39.0875	42.5570	45.7223	49.5879	52.3356	58.301
30	34.7997	40.2560	43.7730	46.9792	50.8922	53.6720	59.703
40	45.6160	51.8051	55.7585	59.3417	63.6907	66.7660	73.402
50	56.3336	63.1671	67.5048	71.4202	76.1539	79.4900	86.661
60	66.9815	74.3970	79.0819	83.2977	88.3794	91.9517	99.607
70	77.5767	85.5270	90.5312	95.0232	100.425	104.215	112.317
80	88.1303	96.5782	101.879	106.629	112.329	116.321	124.839
90	98.6499	107.565	113.145	118.136	124.116	128.299	137.208
100	109.141	118.498	124.342	129.561	135.807	140.169	149.449
X	+0.6745	+1.2816	+1.6449	+1.9600	+2.3263	+2.5758	+3.0902

Appendix IX
Percentage Points of the F Distribution

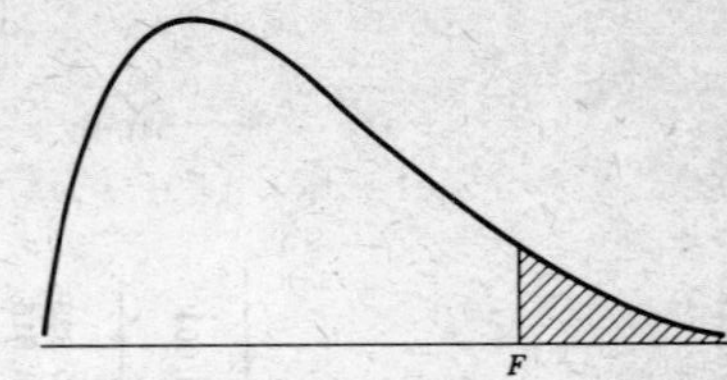

Upper 10 percent points

ν_2 \ ν_1	1	2	3	4	5	6	7	8	9	10	12	15	20	24	30	40	60	120	∞
1	39.86	49.50	53.59	55.83	57.24	58.20	58.91	59.44	59.86	60.19	60.71	61.22	61.74	62.00	62.26	62.53	62.79	63.06	63.33
2	8.53	9.00	9.16	9.24	9.29	9.33	9.35	9.37	9.38	9.39	9.41	9.42	9.44	9.45	9.46	9.47	9.47	9.48	9.49
3	5.54	5.46	5.39	5.34	5.31	5.28	5.27	5.25	5.24	5.23	5.22	5.20	5.18	5.18	5.17	5.16	5.15	5.14	5.13
4	4.54	4.32	4.19	4.11	4.05	4.01	3.98	3.95	3.94	3.92	3.90	3.87	3.84	3.83	3.82	3.80	3.79	3.78	3.76
5	4.06	3.78	3.62	3.52	3.45	3.40	3.37	3.34	3.32	3.30	3.27	3.24	3.21	3.19	3.17	3.16	3.14	3.12	3.10
6	3.78	3.46	3.29	3.18	3.11	3.05	3.01	2.98	2.96	2.94	2.90	2.87	2.84	2.82	2.80	2.78	2.76	2.74	2.72
7	3.59	3.26	3.07	2.96	2.88	2.83	2.78	2.75	2.72	2.70	2.67	2.63	2.59	2.58	2.56	2.54	2.51	2.49	2.47
8	3.46	3.11	2.92	2.81	2.73	2.67	2.62	2.59	2.56	2.54	2.50	2.46	2.42	2.40	2.38	2.36	2.34	2.32	2.29
9	3.36	3.01	2.81	2.69	2.61	2.55	2.51	2.47	2.44	2.42	2.38	2.34	2.30	2.28	2.25	2.23	2.21	2.18	2.16
10	3.29	2.92	2.73	2.61	2.52	2.46	2.41	2.38	2.35	2.32	2.28	2.24	2.20	2.18	2.16	2.13	2.11	2.08	2.06
11	3.23	2.86	2.66	2.54	2.45	2.39	2.34	2.30	2.27	2.25	2.21	2.17	2.12	2.10	2.08	2.05	2.03	2.00	1.97
12	3.18	2.81	2.61	2.48	2.39	2.33	2.28	2.24	2.21	2.19	2.15	2.10	2.06	2.04	2.01	1.99	1.96	1.93	1.90
13	3.14	2.76	2.56	2.43	2.35	2.28	2.23	2.20	2.16	2.14	2.10	2.05	2.01	1.98	1.96	1.93	1.90	1.88	1.85
14	3.10	2.73	2.52	2.39	2.31	2.24	2.19	2.15	2.12	2.10	2.05	2.01	1.96	1.94	1.91	1.89	1.86	1.83	1.80
15	3.07	2.70	2.49	2.36	2.27	2.21	2.16	2.12	2.09	2.06	2.02	1.97	1.92	1.90	1.87	1.85	1.82	1.79	1.76
16	3.05	2.67	2.46	2.33	2.24	2.18	2.13	2.09	2.06	2.03	1.99	1.94	1.89	1.87	1.84	1.81	1.78	1.75	1.72
17	3.03	2.64	2.44	2.31	2.22	2.15	2.10	2.06	2.03	2.00	1.96	1.91	1.86	1.84	1.81	1.78	1.75	1.72	1.69
18	3.01	2.62	2.42	2.29	2.20	2.13	2.08	2.04	2.00	1.98	1.93	1.89	1.84	1.81	1.78	1.75	1.72	1.69	1.66
19	2.99	2.61	2.40	2.27	2.18	2.11	2.06	2.02	1.98	1.96	1.91	1.86	1.81	1.79	1.76	1.73	1.70	1.67	1.63
20	2.97	2.59	2.38	2.25	2.16	2.09	2.04	2.00	1.96	1.94	1.89	1.84	1.79	1.77	1.74	1.71	1.68	1.64	1.61
21	2.96	2.57	2.36	2.23	2.14	2.08	2.02	1.98	1.95	1.92	1.87	1.83	1.78	1.75	1.72	1.69	1.66	1.62	1.59
22	2.95	2.56	2.35	2.22	2.13	2.06	2.01	1.97	1.93	1.90	1.86	1.81	1.76	1.73	1.70	1.67	1.64	1.60	1.57
23	2.94	2.55	2.34	2.21	2.11	2.05	1.99	1.95	1.92	1.89	1.84	1.80	1.74	1.72	1.69	1.66	1.62	1.59	1.55
24	2.93	2.54	2.33	2.19	2.10	2.04	1.98	1.94	1.91	1.88	1.83	1.78	1.73	1.70	1.67	1.64	1.61	1.57	1.53
25	2.92	2.53	2.32	2.18	2.09	2.02	1.97	1.93	1.89	1.87	1.82	1.77	1.72	1.69	1.66	1.63	1.59	1.56	1.52
26	2.91	2.52	2.31	2.17	2.08	2.01	1.96	1.92	1.88	1.86	1.81	1.76	1.71	1.68	1.65	1.61	1.58	1.54	1.50
27	2.90	2.51	2.30	2.17	2.07	2.00	1.95	1.91	1.87	1.85	1.80	1.75	1.70	1.67	1.64	1.60	1.57	1.53	1.49
28	2.89	2.50	2.29	2.16	2.06	2.00	1.94	1.90	1.87	1.84	1.79	1.74	1.69	1.66	1.63	1.59	1.56	1.52	1.48
29	2.89	2.50	2.28	2.15	2.06	1.99	1.93	1.89	1.86	1.83	1.78	1.73	1.68	1.65	1.62	1.58	1.55	1.51	1.47
30	2.88	2.49	2.28	2.14	2.05	1.98	1.93	1.88	1.85	1.82	1.77	1.72	1.67	1.64	1.61	1.57	1.54	1.50	1.46
40	2.84	2.44	2.23	2.09	2.00	1.93	1.87	1.83	1.79	1.76	1.71	1.66	1.61	1.57	1.54	1.51	1.47	1.42	1.38
60	2.79	2.39	2.18	2.04	1.95	1.87	1.82	1.77	1.74	1.71	1.66	1.60	1.54	1.51	1.48	1.44	1.40	1.35	1.29
120	2.75	2.35	2.13	1.99	1.90	1.82	1.77	1.72	1.68	1.65	1.60	1.55	1.48	1.45	1.41	1.37	1.32	1.26	1.19
∞	2.71	2.30	2.08	1.94	1.85	1.77	1.72	1.67	1.63	1.60	1.55	1.49	1.42	1.38	1.34	1.30	1.24	1.17	1.00

Source: This table is reproduced with the kind permission of Prof. E. S. Pearson and the trustees of Biometrika from E. S. Pearson and H.O.

Upper 5 percent points

ν_2 \ ν_1	1	2	3	4	5	6	7	8	9	10	12	15	20	24	30	40	60	120	∞
1	161.4	199.5	215.7	224.6	230.2	234.0	236.8	238.9	240.5	241.9	243.9	245.9	248.0	249.1	250.1	251.1	252.2	253.3	254.3
2	18.51	19.00	19.16	19.25	19.30	19.33	19.35	19.37	19.38	19.40	19.41	19.43	19.45	19.45	19.46	19.47	19.48	19.49	19.50
3	10.13	9.55	9.28	9.12	9.01	8.94	8.89	8.85	8.81	8.79	8.74	8.70	8.66	8.64	8.62	8.59	8.57	8.55	8.53
4	7.71	6.94	6.59	6.39	6.26	6.16	6.09	6.04	6.00	5.96	5.91	5.86	5.80	5.77	5.75	5.72	5.69	5.66	5.63
5	6.61	5.79	5.41	5.19	5.05	4.95	4.88	4.82	4.77	4.74	4.68	4.62	4.56	4.53	4.50	4.46	4.43	4.40	4.36
6	5.99	5.14	4.76	4.53	4.39	4.28	4.21	4.15	4.10	4.06	4.00	3.94	3.87	3.84	3.81	3.77	3.74	3.70	3.67
7	5.59	4.74	4.35	4.12	3.97	3.87	3.79	3.73	3.68	3.64	3.57	3.51	3.44	3.41	3.38	3.34	3.30	3.27	3.23
8	5.32	4.46	4.07	3.84	3.69	3.58	3.50	3.44	3.39	3.35	3.28	3.22	3.15	3.12	3.08	3.04	3.01	2.97	2.93
9	5.12	4.26	3.86	3.63	3.48	3.37	3.29	3.23	3.18	3.14	3.07	3.01	2.94	2.90	2.86	2.83	2.79	2.75	2.71
10	4.96	4.10	3.71	3.48	3.33	3.22	3.14	3.07	3.02	2.98	2.91	2.85	2.77	2.74	2.70	2.66	2.62	2.58	2.54
11	4.84	3.98	3.59	3.36	3.20	3.09	3.01	2.95	2.90	2.85	2.79	2.72	2.65	2.61	2.57	2.53	2.49	2.45	2.40
12	4.75	3.89	3.49	3.26	3.11	3.00	2.91	2.85	2.80	2.75	2.69	2.62	2.54	2.51	2.47	2.43	2.38	2.34	2.30
13	4.67	3.81	3.41	3.18	3.03	2.92	2.83	2.77	2.71	2.67	2.60	2.53	2.46	2.42	2.38	2.34	2.30	2.25	2.21
14	4.60	3.74	3.34	3.11	2.96	2.85	2.76	2.70	2.65	2.60	2.53	2.46	2.39	2.35	2.31	2.27	2.22	2.18	2.13
15	4.54	3.68	3.29	3.06	2.90	2.79	2.71	2.64	2.59	2.54	2.48	2.40	2.33	2.29	2.25	2.20	2.16	2.11	2.07
16	4.49	3.63	3.24	3.01	2.85	2.74	2.66	2.59	2.54	2.49	2.42	2.35	2.28	2.24	2.19	2.15	2.11	2.06	2.01
17	4.45	3.59	3.20	2.96	2.81	2.70	2.61	2.55	2.49	2.45	2.38	2.31	2.23	2.19	2.15	2.10	2.06	2.01	1.96
18	4.41	3.55	3.16	2.93	2.77	2.66	2.58	2.51	2.46	2.41	2.34	2.27	2.19	2.15	2.11	2.06	2.02	1.97	1.92
19	4.38	3.52	3.13	2.90	2.74	2.63	2.54	2.48	2.42	2.38	2.31	2.23	2.16	2.11	2.07	2.03	1.98	1.93	1.88
20	4.35	3.49	3.10	2.87	2.71	2.60	2.51	2.45	2.39	2.35	2.28	2.20	2.12	2.08	2.04	1.99	1.95	1.90	1.84
21	4.32	3.47	3.07	2.84	2.68	2.57	2.49	2.42	2.37	2.32	2.25	2.18	2.10	2.05	2.01	1.96	1.92	1.87	1.81
22	4.30	3.44	3.05	2.82	2.66	2.55	2.46	2.40	2.34	2.30	2.23	2.15	2.07	2.03	1.98	1.94	1.89	1.84	1.78
23	4.28	3.42	3.03	2.80	2.64	2.53	2.44	2.37	2.32	2.27	2.20	2.13	2.05	2.01	1.96	1.91	1.86	1.81	1.76
24	4.26	3.40	3.01	2.78	2.62	2.51	2.42	2.36	2.30	2.25	2.18	2.11	2.03	1.98	1.94	1.89	1.84	1.79	1.73
25	4.24	3.39	2.99	2.76	2.60	2.49	2.40	2.34	2.28	2.24	2.16	2.09	2.01	1.96	1.92	1.87	1.82	1.77	1.71
26	4.23	3.37	2.98	2.74	2.59	2.47	2.39	2.32	2.27	2.22	2.15	2.07	1.99	1.95	1.90	1.85	1.80	1.75	1.69
27	4.21	3.35	2.96	2.73	2.57	2.46	2.37	2.31	2.25	2.20	2.13	2.06	1.97	1.93	1.88	1.84	1.79	1.73	1.67
28	4.20	3.34	2.95	2.71	2.56	2.45	2.36	2.29	2.24	2.19	2.12	2.04	1.96	1.91	1.87	1.82	1.77	1.71	1.65
29	4.18	3.33	2.93	2.70	2.55	2.43	2.35	2.28	2.22	2.18	2.10	2.03	1.94	1.90	1.85	1.81	1.75	1.70	1.64
30	4.17	3.32	2.92	2.69	2.53	2.42	2.33	2.27	2.21	2.16	2.09	2.01	1.93	1.89	1.84	1.79	1.74	1.68	1.62
40	4.08	3.23	2.84	2.61	2.45	2.34	2.25	2.18	2.12	2.08	2.00	1.92	1.84	1.79	1.74	1.69	1.64	1.58	1.51
60	4.00	3.15	2.76	2.53	2.37	2.25	2.17	2.10	2.04	1.99	1.92	1.84	1.75	1.70	1.65	1.59	1.53	1.47	1.39
120	3.92	3.07	2.68	2.45	2.29	2.17	2.09	2.02	1.96	1.91	1.83	1.75	1.66	1.61	1.55	1.50	1.43	1.35	1.25
∞	3.84	3.00	2.60	2.37	2.21	2.10	2.01	1.94	1.88	1.83	1.75	1.67	1.57	1.52	1.46	1.39	1.32	1.22	1.00

Appendix IX—*Continued*

Upper 2.5 percent points

ν_2 \ ν_1	1	2	3	4	5	6	7	8	9	10	12	15	20	24	30	40	60	120	∞
1	647.8	799.5	864.2	899.6	921.8	937.1	948.2	956.7	963.3	968.6	976.7	984.9	993.1	997.2	1001	1006	1010	1014	1018
2	38.51	39.00	39.17	39.25	39.30	39.33	39.36	39.37	39.39	39.40	39.41	39.43	39.45	39.46	39.46	39.47	30.48	39.49	39.50
3	17.44	16.04	15.44	15.10	14.88	14.73	14.62	14.54	14.47	14.42	14.34	14.25	14.17	14.12	14.08	14.04	13.99	13.95	13.90
4	12.22	10.65	9.98	9.60	9.36	9.20	9.07	8.98	8.90	8.84	8.75	8.66	8.56	8.51	8.46	8.41	8.36	8.31	8.26
5	10.01	8.43	7.76	7.39	7.15	6.98	6.85	6.76	6.68	6.62	6.52	6.43	6.33	6.28	6.23	6.18	6.12	6.07	6.02
6	8.81	7.26	6.60	6.23	5.99	5.82	5.70	5.60	5.52	5.46	5.37	5.27	5.17	5.12	5.07	5.01	4.96	4.90	4.85
7	8.07	6.54	5.89	5.52	5.29	5.12	4.99	4.90	4.82	4.76	4.67	4.57	4.47	4.42	4.36	4.31	4.25	4.20	4.14
8	7.57	6.06	5.42	5.05	4.82	4.65	4.53	4.43	4.36	4.30	4.20	4.10	4.00	3.95	3.89	3.84	3.78	3.73	3.67
9	7.21	5.71	5.08	4.72	4.48	4.32	4.20	4.10	4.03	3.96	3.87	3.77	3.67	3.61	3.56	3.51	3.45	3.39	3.33
10	6.94	5.46	4.83	4.47	4.24	4.07	3.95	3.85	3.78	3.72	3.62	3.52	3.42	3.37	3.31	3.26	3.20	3.14	3.08
11	6.72	5.26	4.63	4.28	4.04	3.88	3.76	3.66	3.59	3.53	3.43	3.33	3.23	3.17	3.12	3.06	3.00	2.94	2.88
12	6.55	5.10	4.47	4.12	3.89	3.73	3.61	3.51	3.44	3.37	3.28	3.18	3.07	3.02	2.96	2.91	2.85	2.79	2.72
13	6.41	4.97	4.35	4.00	3.77	3.60	3.48	3.39	3.31	3.25	3.15	3.05	2.95	2.89	2.84	2.78	2.72	2.66	2.60
14	6.30	4.86	4.24	3.89	3.66	3.50	3.38	3.29	3.21	3.15	3.05	2.95	2.84	2.79	2.73	2.67	2.61	2.55	2.49
15	6.20	4.77	4.15	3.80	3.58	3.41	3.29	3.20	3.12	3.06	2.96	2.86	2.76	2.70	2.64	2.59	2.52	2.46	2.40
16	6.12	4.69	4.08	3.73	3.50	3.34	3.22	3.12	3.05	2.99	2.89	2.79	2.68	2.63	2.57	2.51	2.45	2.38	2.32
17	6.04	4.62	4.01	3.66	3.44	3.28	3.16	3.06	2.98	2.92	2.82	2.72	2.62	2.56	2.50	2.44	2.38	2.32	2.25
18	5.98	4.56	3.95	3.61	3.38	4.22	3.10	3.01	2.93	2.87	2.77	2.67	2.56	2.50	2.44	2.38	2.32	2.26	2.19
19	5.92	4.51	3.90	3.56	3.33	3.17	3.05	2.96	2.88	2.82	2.72	2.62	2.51	2.45	2.39	2.33	2.27	2.20	2.13
20	5.87	4.46	3.86	3.51	3.29	3.13	3.01	2.91	2.84	2.77	2.68	2.57	2.46	2.41	2.35	2.29	2.22	2.16	2.09
21	5.83	4.42	3.82	3.48	3.25	3.09	2.97	2.87	2.80	2.73	2.64	2.53	2.42	2.37	2.31	2.25	2.18	2.11	2.04
22	5.79	4.38	3.78	3.44	3.22	3.05	2.93	2.84	2.76	2.70	2.60	2.50	2.39	2.33	2.27	2.21	2.14	2.08	2.00
23	5.75	4.35	3.75	3.41	3.18	3.02	2.90	2.81	2.73	2.67	2.57	2.47	2.36	2.30	2.24	2.18	2.11	2.04	1.97
24	5.72	4.32	3.72	3.38	3.15	2.99	2.87	2.78	2.70	2.64	2.54	2.44	2.33	2.27	2.21	2.15	2.08	2.01	1.94
25	5.69	4.29	3.69	3.35	3.13	2.97	2.85	2.75	2.68	2.61	2.51	2.41	2.30	2.24	2.18	2.12	2.05	1.98	1.91
26	5.66	4.27	3.67	3.33	3.10	2.94	2.82	2.73	2.65	2.59	2.49	2.39	2.28	2.22	2.16	2.09	2.03	1.95	1.88
27	5.63	4.24	3.65	3.31	3.08	2.92	2.80	2.71	2.63	2.57	2.47	2.36	2.25	2.19	2.13	2.07	2.00	1.93	1.85
28	5.61	4.22	3.63	3.29	3.06	2.90	2.78	2.69	2.61	2.55	2.45	2.34	2.23	2.17	2.11	2.05	1.98	1.91	1.83
29	5.59	4.20	3.61	3.27	3.04	2.88	2.76	2.67	2.59	2.53	2.43	2.32	2.21	2.15	2.09	2.03	1.96	1.89	1.81
30	5.57	4.18	3.59	3.25	3.03	2.87	2.75	2.65	2.57	2.51	2.41	2.31	2.20	2.14	2.07	2.01	1.94	1.87	1.79
40	5.42	4.05	3.46	3.13	2.90	2.74	2.62	2.53	2.45	2.39	2.29	2.18	2.07	2.01	1.94	1.88	1.80	1.72	1.64
60	5.29	3.93	3.34	3.01	2.79	2.63	2.51	2.41	2.33	2.27	2.17	2.06	1.94	1.88	1.82	1.74	1.67	1.58	1.48
120	5.15	3.80	3.23	2.89	2.67	2.52	2.39	2.30	2.22	2.16	2.05	1.94	1.82	1.76	1.69	1.61	1.53	1.43	1.31
∞	5.02	3.69	3.12	2.79	2.57	2.41	2.29	2.19	2.11	2.05	1.94	1.83	1.71	1.64	1.57	1.48	1.39	1.27	1.00

Upper 1 percent points

ν_2 \ ν_1	1	2	3	4	5	6	7	8	9	10	12	15	20	24	30	40	60	120	∞
1	4052	4999.5	5403	5625	5764	5859	5928	5981	6022	6056	6106	6157	6209	6235	6261	6287	6313	6339	6366
2	98.50	99.00	99.17	99.25	99.30	99.33	99.36	99.37	99.39	99.40	99.42	99.43	99.45	99.46	99.47	99.47	99.48	99.49	99.50
3	34.12	30.82	29.46	28.71	28.24	27.91	27.67	27.49	27.35	27.23	27.05	26.87	26.69	26.60	26.50	26.41	26.32	26.22	26.13
4	21.20	18.00	16.69	15.98	15.52	15.21	14.98	14.80	14.66	14.55	14.37	14.20	14.02	13.93	13.84	13.75	13.65	13.56	13.46
5	16.26	13.27	12.06	11.39	10.97	10.67	10.46	10.29	10.16	10.05	9.89	9.72	9.55	9.47	9.38	9.29	9.20	9.11	9.02
6	13.75	10.92	9.78	9.15	8.75	8.47	8.26	8.10	7.98	7.87	7.72	7.56	7.40	7.31	7.23	7.14	7.06	6.97	6.88
7	12.25	9.55	8.45	7.85	7.46	7.19	6.99	6.84	6.72	6.62	6.47	6.31	6.16	6.07	5.99	5.91	5.82	5.74	5.65
8	11.26	8.65	7.59	7.01	6.63	6.37	6.18	6.03	5.91	5.81	5.67	5.52	5.36	5.28	5.20	5.12	5.03	4.95	4.86
9	10.56	8.02	6.99	6.42	6.06	5.80	5.61	5.47	5.35	5.26	5.11	4.96	4.81	4.73	4.65	4.57	4.48	4.40	4.31
10	10.04	7.56	6.55	5.99	5.64	5.39	5.20	5.06	4.94	4.85	4.71	4.56	4.41	4.33	4.25	4.17	4.08	4.00	3.91
11	9.65	7.21	6.22	5.67	5.32	5.07	4.89	4.74	4.63	4.54	4.40	4.25	4.10	4.02	3.94	3.86	3.78	3.69	3.60
12	9.33	6.93	5.95	5.41	5.06	4.82	4.64	4.50	4.39	4.30	4.16	4.01	3.86	3.78	3.70	3.62	3.54	3.45	3.36
13	9.07	6.70	5.74	5.21	4.86	4.62	4.44	4.30	4.19	4.10	3.96	3.82	3.66	3.59	3.51	3.43	3.34	3.25	3.17
14	8.86	6.51	5.56	5.04	4.69	4.46	4.28	4.14	4.03	3.94	3.80	3.66	3.51	3.43	3.35	3.27	3.18	3.09	3.00
15	8.68	6.36	5.42	4.89	4.56	4.32	4.14	4.00	3.89	3.80	3.67	3.52	3.37	3.29	3.21	3.13	3.05	2.96	2.87
16	8.53	6.23	5.29	4.77	4.44	4.20	4.03	3.89	3.78	3.69	3.55	3.41	3.26	3.18	3.10	3.02	2.93	2.84	2.75
17	8.40	6.11	5.18	4.67	4.34	4.10	3.93	3.79	3.68	3.59	3.46	3.31	3.16	3.08	3.00	2.92	2.83	2.75	2.65
18	8.29	6.01	5.09	4.58	4.25	4.01	3.84	3.71	3.60	3.51	3.37	3.23	3.08	3.00	2.92	2.84	2.75	2.66	2.57
19	8.18	5.93	5.01	4.50	4.17	3.94	3.77	3.63	3.52	3.43	3.30	3.15	3.00	2.92	2.84	2.76	2.67	2.58	2.49
20	8.10	5.85	4.94	4.43	4.10	3.87	3.70	3.56	3.46	3.37	3.23	3.09	2.94	2.86	2.78	2.69	2.61	2.52	2.42
21	8.02	5.78	4.87	4.37	4.04	3.81	3.64	3.51	3.40	3.31	3.17	3.03	2.88	2.80	2.72	2.64	2.55	2.46	2.36
22	7.95	5.72	4.82	4.31	3.99	3.76	3.59	3.45	3.35	3.26	3.12	2.98	2.83	2.75	2.67	2.58	2.50	2.40	2.31
23	7.88	5.66	4.76	4.26	3.94	3.71	3.54	3.41	3.30	3.21	3.07	2.93	2.78	2.70	2.62	2.54	2.45	2.35	2.26
24	7.82	5.61	4.72	4.22	3.90	3.67	3.50	3.36	3.26	3.17	3.03	2.89	2.74	2.66	2.58	2.49	2.40	2.31	2.21
25	7.77	5.57	4.68	4.18	3.85	3.63	3.46	3.32	3.22	3.13	2.99	2.85	2.70	2.62	2.54	2.45	2.36	2.27	2.17
26	7.72	5.53	4.64	4.14	3.82	3.59	3.42	3.29	3.18	3.09	2.96	2.81	2.66	2.58	2.50	2.42	2.33	2.23	2.13
27	7.68	5.49	4.60	4.11	3.78	3.56	3.39	3.26	3.15	3.06	2.93	2.78	2.63	2.55	2.47	2.38	2.29	2.20	2.10
28	7.64	5.45	4.57	4.07	3.75	3.53	3.36	3.23	3.12	3.03	2.90	2.75	2.60	2.52	2.44	2.35	2.26	2.17	2.06
29	7.60	5.42	4.54	4.04	3.73	3.50	3.33	3.20	3.09	3.00	2.87	2.73	2.57	2.49	2.41	2.33	2.23	2.14	2.03
30	7.56	5.39	4.51	4.02	3.70	3.47	3.30	3.17	3.07	2.98	2.84	2.70	2.55	2.47	2.39	2.30	2.21	2.11	2.01
40	7.31	5.18	4.31	3.83	3.51	3.29	3.12	2.99	2.89	2.80	2.66	2.52	2.37	2.29	2.20	2.11	2.02	1.92	1.80
60	7.08	4.98	4.13	3.65	3.34	3.12	2.95	2.82	2.72	2.63	2.50	2.35	2.20	2.12	2.03	1.94	1.84	1.73	1.60
120	6.85	4.79	3.95	3.48	3.17	2.96	2.79	2.66	2.56	2.47	2.34	2.19	2.03	1.95	1.86	1.76	1.66	1.53	1.38
∞	6.63	4.61	3.78	3.32	3.02	2.80	2.64	2.51	2.41	2.32	2.18	2.04	1.88	1.79	1.70	1.59	1.47	1.32	1.00

Answers to odd-numbered problems

Exercise O-1

1 Numerical data assembled and classified to present meaningful information.

5 A population is the totality of all potential observations in a given problem situation. A sample usually contains only a small portion of all the possible observations in the population. In the rating of TV programs, for example, the population consists of the potential responses of all the TV viewers, while the sample consists of only the responses of those TV viewers who are actually interviewed about their preferences.

7 *a* The responses of the 50 housewives who have been interviewed.

b The potential responses of all housewives in the United States.

c The sample proportion is the sample statistic; its value is $\frac{3}{5} = 0.6$. The population parameter is the true proportion of all American housewives who prefer that brand of detergent.

d (*1*) Method of descriptive statistics.

(*2*) Method of statistical inference.

9 Scientific investigation involves generalizations on the basis of empirical evidence. This is exactly what inferential statistics is designed to accomplish. See Sec. O-2-1 for suggestions about the contributions of inferential statistics to modern scientific explorations.

Exercise O-2

5 *a* By establishing a relation between the test scores received by job applicants and their job performances after they are hired, it is possible to estimate an applicant's future job performance from the score he receives in the test.

b An automatic machine is designed to produce a given product with certain specifications. A sample is periodically taken to test whether the products turned out by the machine meet the specifications.

c A survey of a sample of potential customers is conducted to determine their attitude toward a new product. On the basis of the results of the survey, inference may be made about the market potential of the product.

d Let a real estate appraiser appraise a set of properties, and then the same properties be appraised by another appraiser. Determine whether the average appraised value by one appraiser is significantly different from that by the other.

e A number of income tax returns are randomly selected and audited to determine the proportion of returns that contain errors. Inference may be made about the proportion of all returns that contain errors.

7 The original report of the study fails to show whether other ethnic groups are also predominately city dwellers. Since city dwellers are known to have higher suicide rates, it is erroneous to compare one ethnic group who are mainly city dwellers with other groups who are not.

Exercise 1-1

1 $L = \{s, t, a, i, c\}$

b $D = \{7, 8, 0\}$.

c $S = \{s, t, a, i, c, l, n, f, e, r\}$.

d $P =$ {Nixon, Johnson, Kennedy, Eisenhower, Truman}.

3 *a* All are different. $X = \{-2\}$; $Y = \{0, -2\}$; $Z = \{+1, -2\}$.

b A and C are the same, and B is different.

c G and H are the same, and F is different.

5 *a* Finite. *d* Infinite.

b Infinite. *e* Finite.

c Infinite.

7 (Alternative)

$A = \{p, q, r, 7\}$; $A \subseteq U$. $C = \{p, q\}$; $C \subseteq U$.

$B = \{p, q, r\}$; $B \subseteq U$. $D = \{p, r\}$; $D \subseteq U$.

9 For $I = \{a, b, c\}$, the subsets of I are:

a $\{a, b, c\}$. *d* $\{b, c\}$. *g* $\{c\}$.

b $\{a, b\}$. *e* $\{a\}$. *h* $\emptyset$.

c $\{a, c\}$. *f* $\{b\}$.

11 If S contains n elements, there are 2^n subsets of S, including the null set $\emptyset$ and the universal set S.

15 *a* A and B are overlapping subsets; so are B and C.
b A and C are disjoint subsets.
c A and C partition I.

17 *a* $\{0, 1, 2, 3, 4, 5, 6, 7, 8, 9\}$.
b $\{3, 5, 7\}$.
c $\{0, 2, 3, 4, 6, 7, 8\}$.
d $\{3, 5, 7\}$.
e $\{0, 2, 4, 6, 8\}$.
f $\{3, 7\}$.
g $\{0, 1, 2, 3, 4, 5, 6, 7, 8, 9\}$.
h $\{3, 5, 7\}$.
i $\{0, 2, 4, 6, 8\}$.
j $\{0, 1, 2, 3, 4, 5, 6, 7, 8, 9\}$.
k $\{3, 7\}$.
l $\emptyset$.

21 *a* $\{d\}$.
b $\{a\}$.
c $\{a, b, c, d\}$.
d $\{b, c\}$.
e $\{a, b, c\}$.
f $\{b, c, d\}$.
g $\emptyset$.
h $\{a, d\}$.
i $\{d\}$.
j $\{a, d\}$.
k $\emptyset$.
l $\{a, c\}$.

Exercise 1-2

3 *a* Yes.
b No.
c No.
d No.
e No.
f Yes.

5 $P = \{4, 5, 6\}$.

7 *a* $x = 1$; $y = 3$.
b $x = 3$; $y = 2$.
c $x = 2$; $y = 1$.
d $x = 2$; $y = 4$.

9 1,000.

11 *a* A function.
b A function.
c Not a function.
d A function.
e Not a function.
f Not a function.
g A function.
h Not a function.

13 No, because each student is paired with more than one grandparent.

15 Yes. Range of age might be from 16 to 60.

17 $-6 \le y \le 14$.

19 *a* $3x + 5$.
b $y + 4$.
c $12y + 11$.
d $9z - 4$.

21 For *a*, *b*, and *c*: x is the independent variable, and y is the dependent variable.
For *d*: x and y are the independent variables, and z is the dependent variable.

Exercise 2-1

3 The event of obtaining two heads is a compound event since it consists of three elementary events, namely, (HHT), (HTH), and (THH).

5 $S_1 = \{(TTT), (TTH), (THT), (HTT)\}$ and $S_2 = \{(HHT), (HTH), (THH), (HHH)\}$.

11 *a* Region 2. *c* Region 7.
b Region 4. *d* Regions 5, 6, 7, and 8.

Exercise 2-2

3 $2^4 = 16$ ways. **9** 360,360. **15** 792.

5 18 ways. **11** 720. **17** $\frac{52!}{13!\,39!}$.

7 20. **13** 2,730.

19 *a* 142,506. *b* 51,300. *c* 15,756.

Exercise 2-3

1 0.75. **3** 0.15.

5 *a* $\frac{1}{2}$. *b* $\frac{1}{8}$. *c* $\frac{3}{4}$.

7 $\frac{4}{15}$. **9** $\frac{4}{635{,}013{,}559{,}600}$. **11** $\frac{4\binom{13}{5}}{\binom{52}{5}}$.

Exercise 3-1

3 *a* $\frac{2}{52}$. *c* $\frac{12}{52}$. *e* $\frac{4}{52}$.
b $\frac{14}{52}$. *d* $\frac{16}{52}$. *f* $\frac{16}{52}$.

5 0.15. **7** $\frac{1}{3}$.

Exercise 3-2

3 *a* $\frac{1}{24}$. *c* $\frac{1}{6}$. *e* $\frac{1}{6}$.
b $\frac{1}{24}$. *d* $\frac{1}{2}$.

5 *a* $\frac{1}{52}$. *c* $\frac{9}{16}$. *e* $\frac{3}{169}$.
b $\frac{1}{169}$. *d* $\frac{1}{169}$.

7 *a* $\frac{15}{16}$. *d* $\frac{25}{64}$. *g* $\frac{49}{64}$.
b 1. *e* $\frac{25}{64}$. *h* $\frac{15}{64}$.
c $\frac{39}{64}$. *f* $\frac{39}{64}$.

9 *a* No.
b $P(R_1 | C) = 0.20, \qquad P(R_2 | C') = \dfrac{13}{30}, \qquad P(R_3 | C) = 0.20.$

11 *a* $\frac{1}{5}$. *c* $\frac{3}{10}$.
b $\frac{1}{15}$. *d* $\frac{1}{5}$.

13 *a* $\frac{2}{9}$. *c* $\frac{1}{15}$. *e* $\frac{14}{90}$.
b $\frac{5}{18}$. *d* $\frac{1}{90}$.

Exercise 3-3

3 The probability that the person actually has the disease is $\frac{0.008}{0.0575} = \frac{80}{575} = \frac{16}{115}$.

5 The probability that this man will buy an insurance policy is $\frac{0.12}{0.26} = \frac{6}{13}$. The probability that a person who has an income of \$50,000 will buy an insurance policy is $\frac{0.18}{0.74} = \frac{9}{37}$.

7 The probability that the drug was A is $P(A | D) = \frac{0.225}{0.300} = \frac{225}{300} = \frac{3}{4}$. The probability that the drug was B is $P(B | D) = \frac{0.075}{0.300} = \frac{1}{4}$.

9 $P(L | T) = \frac{0.12}{0.37} = \frac{12}{37}, \qquad P(S | T) = \frac{0.15}{0.37} = \frac{15}{37}, \qquad P(N | T) = \frac{0.10}{0.37} = \frac{10}{37}.$

Exercise 3-4

5 *a* $f(x) = \dfrac{1}{b - a} = \dfrac{1}{4 - 2} = \dfrac{1}{2}.$

$$F(x) = \begin{cases} 0 & \text{for } x < 2 \\ \dfrac{x-2}{4-2} = \dfrac{x-2}{2} & \text{for } 2 \le x \le 4 \\ 1 & \text{for } x > 4. \end{cases}$$

Exercise 4-1

5

x	$FR(x)$
53	1
56	4
59	7
62	13
65	15
68	7
71	3
74	0
Total	50

7 Graphs have a visual appeal and are easy to comprehend.

9

x	$FR(x)$
80	5
85	12
90	13
95	22
100	30
105	35
110	32
115	20
120	15
125	10
130	6
Total	200

11

x	$FR(x)$
0	2
1	11
2	18
3	12
4	6
5	1
Total	50

Exercise 4-2

3 *a* $-15.$ *b* 45. *c* 120.

5 *a* $\sum_{i=1}^{4} X_i^2.$ *b* $a\sum_{i=1}^{5} iX_i.$ *c* $c\sum_{i=1}^{n} (X_i + i).$
d $\sum_{i=1}^{5} (X_i - i)i^2.$

7 *a* 6. *b* 22. *c* 30. *d* 210.

9 $\bar{X} = 105.025.$

11 *a* 2.24. *b* 2.24.

15 *a* 3. *b* 21.

Exercise 4-3

1 *a* 6. *b* Median = 8, mode = 7.

3 Median = 3.83, mode = 3.84.

9 *a* Med = 105.07. *b* $Q_3 = 112.75.$ *c* $P_{90} = 121.17.$
d Mo = 105.625.

11 Med = 2. The median is smaller than the mean.

13 Med = 9,269.2, Mo = 9823.5.

15 *a* $\frac{27}{13}.$ *c* 3.2. *e* $\frac{50,000}{11,111}.$
b $\frac{24}{7}.$ *d* 4.

Exercise 4-4

1 The expected value of a random variable is the long-run average of the random variable.

3 1.5. **7** \$0.5. **11** \$6 million.

5 3 milligrams. **9** $E(W) = 5.5.$ **13** -2 dollars.

Exercise 5-1

3 *a* AD = 2.4. *b* AD = 2.2.
c The average deviation from the median is smaller than the average deviation from the mean.

5 Mean = 65, AD = 17.3.

Exercise 5-2

3 Salesman I seems to be more consistent with respect to the volume of sales.

5 $\overline{X} = 65$, $s^2 = 384.5$, $s = 19.6$.

7 When the variance is 1, it is equal to the standard deviation.

9 $s^2 = 138.625$, $s = 11.8$.

11 $s^2 = 1.2224$, $s = 1.106$.

Exercise 5-3

1 $s^2 = \frac{38}{11}$, $s = 1.8574$.

3 *a* $s_Z{}^2 = 31.0909$, $s_Z = 5.57$.
b If $Z = bX$, then $s_Z = bs_X$.

5 *a* $\overline{X} = 4$, Med $= 2$. *b* $s^2 = 10$. *c* $s^2 = 14$.
d The variance from the mean is smaller than the variance from the median.

7 The condition is that the variables X and Y are independent. Under such a condition the covariance of X and Y will be equal to zero.

9 *a* (*1*) $\overline{Y} = 244$. (*2*) $s_Y{}^2 = 6{,}400$. (*3*) $s_Y = 80$.
b (*1*) $\overline{Y} = -44$. (*2*) $s_Y{}^2 = 400$. (*3*) $s_Y = 20$.
c (*1*) $\overline{Y} = 4$. (*2*) $s_Y{}^2 = \frac{25}{16}$. (*3*) $s_Y = \frac{5}{4}$.

11 $s_Y = 16$.

Exercise 5-4

3 $\sigma_W{}^2 = 4\frac{11}{12}$.

5 *a* $\sigma_X{}^2 = 2.92$. *b* $\sigma_Y{}^2 = 242.47$.

7 $\sigma_X{}^2 = 0.75$, $\sigma_Y{}^2 = 3.75$.

9 9.

Exercise 5-5

1 No.

3 *a* $k = 10$. *c* $k = 2.58$. *e* 0.
b $k = 2.24$. *d* $k = 1.41$. *f* $\frac{1}{9}$.

5 4.56 and 100.68.

7 *a* At least 75 percent. *b* At most $\frac{1}{9}$.

Exercise 6-1

1 All the potential responses of the viewers of the TV program.

3 *a* The elementary event in such an experiment is a reading of the spot on which the pointer of the dial rests after being spun. The population is the universal set of all the possible distinct elementary events.

b The elementary event is the response from an individual voter in an interview. The population is the set of all registered voters.

c An elementary event is the expressed preference of any housewife. The population is the set of all housewives in the country.

5 (Alternative)

a A survey of teen-age girls in the United States to determine their preference concerning a particular kind of cosmetic product.

b To determine the extent of air contamination after an atomic bomb test in the atmosphere. In each case, we select a sample, on the basis of which a decision is made.

7 *a* Infinite. The population is the totality of observations of all screws that have been produced, or will be produced, by the machine. Theoretically, there is no limit to the number of screws that can be produced.

b Finite. The population is made up of all the employees of the company.

c Infinite. The population is made up of all the potential weights, which can be obtained indefinitely.

9 *a* $\frac{1}{65,780}$. *b* $\frac{1}{65,780}$. *c* $\frac{1}{65,780}$.

11 *a* $\frac{1}{252}$. *b* Same as in *a*. *c* Same as in *a*.

Exercise 6-2

3 *a* The population is all the potential observations of this product that might be produced. A systematic sampling seems to be appropriate in this situation.

b The population is all the potential responses of the housewives in this metropolitan area. A comparatively efficient sampling plan would be a systematic sampling design by using the area's telephone directories.

5 Assign a number to each voter in the roster of registered voters of the city. Then use the random digit table, the goldfish bowl, or the systematic sampling method to select the sample.

9 Assign the numbers 1, 2, 3, . . ., 500,000 to the 500,000 subscribers. Use any random device to select a number from among the numbers 1, 2, 3, . . ., 10. If the number 5 is selected, then the subscribers with the numbers 5, 15, 25, 35, . . ., 499,995 are included in the sample. They amount to 10 percent of the total 500,000 subscribers.

11 *a* First stage: Draw a simple random sample from among all the school districts in the United States.
b Second stage: Draw a simple random sample from among the schools in each of the districts included in the sample obtained in *a* above.
c Third stage: Draw a simple random sample from the faculty of each of the schools included in the sample obtained in *b* above.
All those included in all the samples in *c* above constitute the required sample.

Exercise 6-3

3 $n = 2{,}500$. **5** $n = 40{,}000$. **7** $n = 144$.

9 *b* $\sigma_X{}^2 = 0.625$, $\sigma_X = 0.79$.

Exercise 7-1

3 The mean of Y for these 30 tosses is $\frac{18}{30} = 0.6$. It is not a population mean. It is the mean of the sample of 30 observations. The population mean is $E(Y) = p$, the probability that heads will occur.

5 $E(Y) = p = 0.8$, $\sigma_Y{}^2 = 0.16$.

7 $E(W) = p = 0.7$, $\sigma_W{}^2 = 0.21$.

Exercise 7-2

3 *a* 0.1445. *c* 0.0494. *e* 0.8995.
b 0.0280. *d* 0.9786.

5 *a* 0.6475. *b* 0.5886. *c* 0.6409.

9 *a* 0.0015. *b* 0.0015.

11 *a* $q^3 + 3pq^2 + 3p^2q + p^3$.
b $q^5 + 5pq^4 + 10p^2q^3 + 10p^3q^2 + 5p^4q + p^5$.
c $q^{10} + 10pq^9 + 45p^2q^8 + 120p^3q^7 + 210p^4q^6 + 252p^5q^5 + 210p^6q^4 + 120p^7q^3 + 45p^8q^2 + 10p^9q + p^{10}$.

Exercise 7-3

1 $E(X) = 25$, $\sigma_X{}^2 = 12.5$.

3 $E(X) = 10, \quad \sigma_X^2 = 5.$

5 $n = 3 \quad E(X) = 1.5, \quad \sigma_X^2 = 0.75.$
$n = 5 \quad E(X) = 2.5, \quad \sigma_X^2 = 1.25.$
$n = 10 \quad E(X) = 5, \quad \sigma_X^2 = 2.5.$

7 $E(X) = 3, \quad \sigma_X^2 = 1.5.$

9 *a* $E(X - 6) = 0.$ *c* $E(X^2) = 38.4.$ *e* $E[(X - 6)^2] = 2.4.$

b $E\left(\frac{X}{10}\right) = \frac{6}{10}.$ *d* $E\left(\frac{X-6}{10}\right) = 0.$

11 $\sigma_Z^2 = 1\frac{1}{6}.$

Exercise 7-4

1 Given the parameters n and p, the probability distribution of the sample proportion is identical with that of the binomial random variable, except that the possible values of the sample proportion are 0, $1/n$, $2/n$, $3/n$, . . . , n/n, and the possible values of the binomial random variable are 0, 1, 2, 3, . . . , n.

3 $E(\hat{p}) = 0.5, \quad \sigma_{\hat{p}}^2 = 0.025.$

5 0.0480.

7 $E(\hat{p}) = 0.4, \quad \sigma_{\hat{p}}^2 = 0.0048.$

Exercise 7-5

1 The hypergeometric distribution should be used when the population is finite and sampling is taken without replacement.

3 $P(X \geq 3) = \frac{2{,}502}{3{,}003}.$

5 $$\frac{\binom{50}{5}\binom{30}{3}\binom{20}{2}}{\binom{100}{10}}.$$

7 *a* $\frac{19}{42}.$ *b* $\frac{\binom{4}{x}\binom{6}{4-x}}{\binom{10}{4}}.$ *c* $E(X) = 1.6.$

Exercise 7-6

3 $P(X = 0) = 0.13534$, $P(X = 3) = 0.18045$,
$P(X = 1) = 0.27067$, $P(X = 4) = 0.09022$.
$P(X = 2) = 0.27067$,

5 *a* 0.00674. *c* 0.12465.
b 0.03369. *d* 0.73498.

7 As n increases, the Poisson distribution changes from positive skewness to a more symmetrical pattern. As n approaches infinity, the Poisson distribution approaches the continuous normal distribution.

9 *a* 0.36788. *b* 0.08030. *c* 0.98101.

Exercise 8-1

1 The parameters of the normal density function are the mean μ and the variance σ^2. The mean determines the location of the center of the distribution, while the variance determines the particular shape of the distribution.

9 *a* 0.9760. *b* 0.9987.

Exercise 8-2

1 *a* 0.3085. *c* 0.6915. *e* 0.4332. *g* 0.6898.
b 0.9452. *d* 0.0548. *f* 0.1832. *h* 0.6976.

3 *a* −0.655. *c* 1.175. *e* 0.842.
b −0.265. *d* −2.94. *f* −1.15.

5 *a* 0.3085. *c* 0.0948.
b 0.9332. *d* 0.5468.

7 82.692. **9** 34.71 months.

Exercise 8-3

3 W is $N(0, 31)$.

7 *a* 0.0062. *c* 0.9394. *e* 0.9974.
b 0.0668. *d* 0.7672. *f* 0.9544.

9 $E(\overline{X}) = 3\frac{1}{2}$, $\sigma_{\overline{X}}^2 = 0.0292$.

11 *a* Approximately normal. *b* $\sigma_{\hat{p}}^2 = 0.0025$.

13 0.9759.

15 *a* Approximately normal.
b $E(D) = 100$, $\sigma_D^2 = 4{,}900$, $\sigma_D = 70$.

Exercise 8-4

1 It is necessary because binomial tables may not be available. For large *n*, such tables are simply nonexistent.

3 *a* 0.6982. *b* 0.6978.

5 *a* 0.3158. *b* 0.3436.

7 *a* 0.8413. *c* 0.8185.
b 0.7475. *d* 0.4522.

Exercise 9-1

3 *a* $E(\overline{X}) = 3$. *b* $E(\hat{s}^2) = \frac{8}{3}$.

5 Yes, because its value depends on chance factors for its occurrence. Even if the sample size remains unchanged, a different sample is very likely to result in a different value for the estimator, which cannot be predetermined.

7 $\hat{s}^2 = \dfrac{n}{n-1} s^2$. **9** $\hat{s}^2 = \frac{172}{9}$.

Exercise 9-2

1

p	$P(\hat{p} = \frac{30}{50} \mid p)$
0.1	0
0.2	0
0.3	0
0.4	0.0020
0.5	0.0419
0.6	0.1146
0.7	0.0370
0.8	0.0006
0.9	0

3 *a*

p	$P(X \leq 10)$
0.1	0.9906
0.2	0.5836
0.3	0.0789
0.4	0.0022
0.5	0
0.6	0
0.7	0
0.8	0
0.9	0

b

p	$P(X \geq 40)$
0.1	0
0.2	0
0.3	0
0.4	0
0.5	0
0.6	0.0022
0.7	0.0789
0.8	0.5836
0.9	0.9906
1.0	1.0000

c

p	$P(X \geq 20)$
0.1	0
0.2	0.0009
0.3	0.0848
0.4	0.5535
0.5	0.9405
0.6	0.9986
0.7	1.0000
0.8	1.0000
0.9	1.0000
1.0	1.0000

d *p*	$P(20 \leq X \leq 30)$
0.1	0
0.2	0.0009
0.3	0.0848
0.4	0.5521
0.5	0.8810
0.6	0.5521
0.7	0.0848
0.8	0.0009
0.9	0

e *p*	$P(30 \leq X \leq 35)$
0.1	0
0.2	0
0.3	0
0.4	0.0034
0.5	0.1000
0.6	0.5070
0.7	0.5054
0.8	0.0604
0.9	0.0001

5

p	$P(X = 25)$
0.1	0
0.2	0
0.3	0.0014
0.4	0.0405
0.5	0.1123
0.6	0.0405
0.7	0.0014
0.8	0
0.9	0

Exercise 9-3

1 A point estimate is a single value, while an interval estimate is a set of values which the population parameter to be estimated may possibly assume. The latter is more useful because it provides a measure of precision of the estimate.

5 *a* (*1*) $-2.79 < \mu < 3.79$. (*2*) $-2.064 < \mu < 3.064$.
b (*1*) $-0.79 < \mu < 5.79$. (*2*) $-0.064 < \mu < 5.064$.
c (*1*) $0.589 < \mu < 1.411$. (*2*) $0.6795 < \mu < 1.3205$.
d (*1*) $-0.072 < \mu < 0.092$. (*2*) $-0.054 < \mu < 0.074$.

7 *a* $142.16 < \mu < 157.84$. *b* 143.42.

9 243.42 pounds.

Exercise 9-4

1 The difference $\bar{D}$ is a random variable because $\bar{X}_1$ and $\bar{X}_2$ are random variables and their values depend on chance factors for their occurrences. Accordingly, the values of $\bar{D}$ depend on chance factors for their occurrences.

3 $4.45 < \delta < 15.55$

5 *a* $1.53 < \delta < 4.47$. *b* $1.07 < \delta < 4.93$.

7 *a* $1.08 < \delta < 8.92$. *b* $-0.15 < \delta < 10.15$.

Exercise 9-5

1 *a* $\frac{30.4}{100} < p < \frac{49.6}{100}$. *b* $\frac{27.4}{100} < p < \frac{52.6}{100}$.

3 $0.46 < p < 0.74$.

5 *a* $0.24 < p < 0.56$. *d* $0.34 < p < 0.66$.
b $0.11 < p < 0.29$. *e* $0.49 < p < 0.71$.
c $0.19 < p < 0.41$. *f* $0.71 < p < 0.89$.

7 $p = \hat{p} \pm z_c\sqrt{\hat{p}(1 - \hat{p})/n}$.

9 $0.46 < p < 0.72$.

Exercise 10-1

3 A one-tailed test is conducted when the null hypothesis is stated as $\theta \geq \theta_0$ (or $\theta \leq \theta_0$) and the alternative as $\theta < \theta_0$ (or $\theta > \theta_0$); and under such a test, we should reject H_0 if the observed value is very low (very high). A one-tailed test is implied when the principal question has terms such as *increased, decreased, more than,* or *less than.*

5 A one-tailed test is called *directional* because the rejection region lies on a particular side, or in a direction, from the specified value under H_0. A two-tailed test is called *nondirectional* because the null hypothesis is rejected as long as the observed value is sufficiently large or small, regardless of on what side, or in what direction, the observed value may lie from the specified value under H_0.

7 They are inversely related. That is, as α increases, β will decrease, and vice versa, under the assumption that other things such as sample size and variance remain unchanged.

9 H_0: $\mu = 55$, H_1: $\mu > 55$.

Exercise 10-2

3 *a* Critical value = 14.1125; do not reject H_0.
b Critical value = 15.815; do not reject H_0.

5 *a* Critical value = 106.978; do not reject H_0.
b $1 - \beta = 0.8430$.

7 *a* Critical value = 11; do not reject H_0.
b Critical value = 9; reject H_0.

9 Critical value = 30.368; reject the claim of the supplier. Power of the test = 0.591.

11 At $\alpha = 0.0082$, $1 - \beta = 0.8713$. At $\alpha = 0.0495$, $1 - \beta = 0.9744$.

13 Critical value = 1,541.08; the evidence does not support the manufacturer's claim.

Exercise 10-3

3 *a* Critical value = 5.8875; do not reject H_0.
b Critical value = 14.1125; reject H_0.
c Critical values are 14.9 and 5.1; reject H_0.

5 Critical value = 106.978; do not reject H_0. The 95 percent confidence interval for μ is $100.12 < \mu < 111.88$.

7 *a* $n = 71$. *c* $n = 28{,}561$.
b $n = 1{,}159$. *d* $n = 80{,}000$.

9 Critical values are 17.94 and 12.06; reject H_0. The 95 percent confidence limits are 12 ± 2.94, or 9.06 and 14.94.

11 *a* Critical values are 21 and 9; do not reject H_0.
b Critical value = 20; reject H_0.

13 *a* At $\alpha = 0.05$, critical value = 0.8753; reject H_0.
b At $\alpha = 0.01$, critical value = 0.86511; reject H_0.

Exercise 10-4

3

μ	$1-\beta$	β
97	0	1.0
100	0.01 (α)	0.99 ($1-\alpha$)
103	0.0924	0.9076
106	0.3722	0.6278
109	0.7498	0.2502
112	0.9529	0.0471
115	0.9962	0.0038

5 The critical values for X, the number of heads obtained, are 9 and 21.

p	$1-\beta$	β
0.1	0.9996	0.0004
0.2	0.9390	0.0610
0.3	0.5887	0.4113
0.4	0.1772	0.8228
0.5	0.0429 (α)	0.9571 ($1-\alpha$)
0.6	0.1773	0.8227
0.7	0.5888	0.4112
0.8	0.9389	0.0611
0.9	0.9995	0.0005

9 Critical value = 0.8753.

p	$1-\beta$	β
0.9	0.05 (α)	0.95 ($1-\alpha$)
0.8903	0.1587	0.8413
0.8753	0.500	0.500
0.8603	0.8413	0.1587
0.8453	0.9772	0.0228
0.8303	0.9987	0.0013

Exercise 11-1

3 *a* $t = 2.602.$ *d* $t = -1.753.$
b $t = 1.753.$ *e* $t = -1.341.$
c $t = -2.131.$

7 *a* 1.725. *c* −1.753.
b 2.457. *d* −1.812.

Exercise 11-2

1 *a* Critical value = 10.8225.
b Critical value = −0.8225.
c Critical value = −5.8225.
d Critical values are 5.980 and 4.020.

3 *a* Critical value = 6.05325; reject H_0.
b Critical value = 5.9511; do not reject H_0.
c Critical values are 6.594 and 6.006; reject H_0.
d Critical values are 6.68625 and 5.91375; do not reject H_0.

5 *a* $\beta = 0.6718.$ *b* $\beta = 0.8699.$

7 At $\alpha = 0.01$, critical value = 181.978; do not reject H_0.
At $\alpha = 0.05$, critical value = 179.935; reject H_0.

9 *a* Critical values are 2.368 and 1.632; do not reject H_0.
b Critical values are 2.266 and 1.734; do not reject H_0.

Exercise 11-3

1 The basic reason for testing about the significance of difference between two means is to see if a group (sample) under some kind of treatment is significantly different from another group (sample) not under the treatment or under a different treatment. In other words, it is designed to answer the question whether any treatment really makes the difference.

3 Critical values are 3.92 and −3.92; reject H_0.

5 At $\alpha = 0.01$, $t = (95 - 97)/3.08 = -0.649$. Since the critical t ratios are −2.921 and 2.921, we do not reject the null hypothesis.
At $\alpha = 0.05$, the critical t ratios are ± 2.120. We do not reject the null hypothesis.

7 Critical value = 32.178; do not reject the null hypothesis.

9 $t = 1.089$. Since the critical values of $t = \pm 2.131$, do not reject H_0. The 95 percent confidence interval for δ is $-0.96 < \delta < 2.96$. The 99 percent confidence interval for δ is $-1.7 < \delta < 3.7$.

Exercise 12-1

3 Yes, it is possible that the computed χ^2 value is too small, because if our alternative hypothesis is $H_1: \theta < \theta_0$, the critical value would be a value that cuts off the lower α of the χ^2 distribution.

5 *a* $\chi^2 = 24.9958$.
b $\chi^2 = 30.5779$.
c $\chi^2 = 8.54676$.
d $\chi^2 = 7.26094$.
e $\chi^2 = 5.22935$.

7 $P(\chi^2_{(128)} \geq 152.8) = 0.0606$.

Exercise 12-2

1 $49.5 > \sigma^2 > 9.34$.

3 $45.1 > \sigma^2 > 8.87$.

5 $\chi^2_{(29,0.025)} = 45.7223$. $\chi^2_{(29,0.975)} = 16.0471$.
Since $ns^2/\sigma^2 = 33$; do not reject H_0. The 95 percent confidence interval for σ^2 is $514.1 > \sigma^2 > 180.44$.

7 The power of a test is $1 - \beta$, the probability of rejecting a null hypothesis when it is false (a correct decision). As the sample size increases, it increases the power of the test, or the probability of making a correct decision, when the null hypothesis is false.

Exercise 12-3

1 Critical value $= 21.666$. Since the observed value $= 4.60$, we conclude that the observed distribution is not significantly different from the expected distribution.

3 Critical value $= 7.81473$. Since the observed value is 4.143, we do not reject H_0.

5 Critical value $= 16.8119$. Since the observed χ^2 value is 4.73, we do not reject H_0.

7 Critical value = 11.0705. Observed value = 7.2234; do not reject the null hypothesis.

Exercise 12-4

3 Observed value = 10.398. Since the critical value = 9.21034, we conclude that age is related to absenteeism.

5 Observed value = 1.714. Critical χ^2 value = 3.84146. Since the observed value is smaller than the critical value, we conclude that the vaccine had no significant effect on the recovery.

Exercise 13-1

5 *a* 6.37. *c* 2.77. *e* 0.227.
b 4.30. *d* 2.35. *f* 0.426.

9 X_1 and X_2 must be normally distributed. X_1 and X_2 must be independent. $\sigma_1{}^2 = \sigma_2{}^2$.
ν_1 is associated with the larger variance, or the numerator of the F ratio.
ν_2 is associated with the smaller variance, or the denominator of the F ratio.

Exercise 13-2

1 Observed value of $F = 2.25$. Since the critical value of $F = 1.98$, we reject the null hypothesis.

3 Critical values are 2.27 and 0.44. Since $F = \frac{0.03}{0.04} = 0.75$, we do not reject H_0.

5 Observed value of $F = 1.15$. Since the critical value = 0.467, we do not reject H_0.

7 *a* Observed value of $F = 1.20$. Since the critical value of $F = 2.29$, we do not reject H_0.
b Critical values are 2.70 and 0.41. Since $F = 1.2$, we do not reject H_0.

Exercise 13-3

5 Critical F value = 5.29; do not reject the null hypothesis. F ratio = 3.03.

9 Critical F value = 3.89; do not reject the null hypothesis. F ratio = 3.84.

11 No. There may be some treatment effects. The effects, however, are not *significant* at the level specified.

Exercise 13-4

3 No. No significant difference in the row, column, or interaction effects does not really mean that they are all zero (i.e., the variability is accounted for merely by the random error). There may be some effect, but at the specified significance level, it is not significant. If the ratios are less than, but close to, the critical values, another experiment might be in order.

5 *a* Observed value of $F = 3.15$. Critical value of $F = 4.26$; do not reject the null hypothesis.
b Observed value of $F = 66.67$. Critical value of $F = 4.26$; reject the null hypothesis.
c Observed value of $F = 1.12$. Critical value of $F = 3.63$; we conclude that there is no interaction effect.

9 *a* Column effect: MSC/MSE = 1.03. $F_{(0.05;3,20)} = 3.10$; do not reject the null hypothesis.
b Row effect: MSR/MSE = 3.33. $F_{(0.05;4,20)} = 2.87$; reject the null hypothesis.
c Interaction effect: MSI/MSE = 0.28. $F_{(0.05;12,20)} = 2.28$; do not reject the null hypothesis.

Exercise 14-1

5 $\sigma_{Y|X} = 2.5$. The 95.45 confidence limits for $\mu_{Y|X}$ are $\mu_{Y|X} \pm 5$.

Exercise 14-2

3 $Y = -174.612 + 4.94X$. When $X = 67$, then $Y = 156.4$ pounds.

7 They both are based on sample information. In point estimation of a population mean, however, the sample statistic, which is a single value, is used as an estimate of the population parameter. In point estimation of a population's regression line, a line computed from a sample is used as the estimate.

Exercise 14-3

3 *a* $Y_c = 10.37 + 0.188X$.
b When $X = 400$, then $Y_c = 85.57$. The 99 percent confidence interval for $\mu_{Y|X}$ is $70.84 < \mu_{Y|X} < 100.3$.

c $s_{Y|X} = 25.35.$ $s_{Y_a} = 25.984.$
The 99 percent confidence interval for Y_a at $X = 400$: $18.79 < Y_a < 152.35$.

5 When a line in the form $Y_c = a + bX$ is used to estimate the conditional mean $\mu_{Y|X}$, it is called point estimation. When a confidence band such as the symmetrical confidence limits shown in Fig. 14-5 is established to estimate the true conditional mean $\mu_{Y|X}$, it is called interval estimation.

Exercise 14-4

3 $\hat{s}_{Y|X} = 10.42,$ $\hat{s}_b = 0.69.$
The 95 percent confidence limits for β are $3.49 < \beta < 6.39$.

5 *a* $b = -0.95,$ $a = 9.75.$
b $\hat{s}_{Y|X} = 0.91,$ $\hat{s}_b = 0.12,$ $-t_{(7, 0.01)}\hat{s}_b = -0.36.$
Reject H_0 because $b = -0.95$, which is less than -0.36.
c The 99 percent confidence limits for β are $-1.37 < \beta < -0.53$.

7 The 95 percent confidence interval for β therefore is $0.746 < \beta < 1.314$.

Exercise 14-5

3 *a* $r = 0.86,$ *b* $t_{(18, 0.025)} = \pm 2.101,$ $t = 7.16.$
Since the computed t value is larger than the critical t value, we reject H_0.

7 $s_X = 162.6,$ $s_Y = 39.5,$ $r = 0.77,$ $b = 0.188.$
In Prob. 3 of Exercise 14-3, $b = 0.188$.

9 *a* $s_X = 2.87,$ $s_Y = 3.10,$ $r = 0.96.$
b $t = r\sqrt{(n-2)/(1-r^2)} = 9.6.$
At $\alpha = 0.05$, $t_{(8, 0.025)} = 2.306$; at $\alpha = 0.01$, $t_{(8, 0.005)} = 3.355$.
Reject H_0 at both $\alpha = 0.05$ and $\alpha = 0.01$ since 9.6 is larger than both 2.306 and 3.355.

Exercise 15-1

3 (Alternative)
a Number of customers who come to a certain store.
b Number of passengers airlifted by an airline company.
c Number of deposits received by a commercial bank.
d Sales by a gift shop.

5 *a* The cyclical component and secular trend are used, because the demand for computers follows the cyclical variation of the general economy along with the long-term growth.

b The seasonal component is used, because the demand for airline seating space depends on seasonal events such as holidays and weekends.

c The cyclical and seasonal components are used since both influence sales during a period of 1 year. For example, a mild recession may be anticipated in a few months, and changes in climatic conditions may affect its business.

d The secular trend is used, since only the long-term growth of the company in question is of interest.

Exercise 15-2

3 *b* $Y = 8.733 + 0.624X$.

c (*1*)

Quarters	*Seasonal index*
I	93.44
II	114.22
III	104.74
IV	87.60

(*2*)

Quarter	*Deseasonalized sales*
I	39.60
II	35.02
III	36.28
IV	42.24

(*3*)

Year	*Quarter*	*Cyclical C*	*Irregular I*
1962	1		
	2	142.09	104.95
	3	122.60	95.46
	4	99.98	101.68
1963	1	84.36	96.33
	2	72.34	97.00
	3	64.66	101.44
	4	66.13	88.04

Year	*Quarter*	*Cyclical C*	*Irregular I*
1964	1	67.67	110.21
	2	70.69	99.26
	3	59.91	112.39
	4	53.46	78.97
1965	1	56.08	90.64
	2	70.04	107.34
	3	91.83	91.94
	4	105.50	109.84
1966	1	121.27	95.82
	2	125.88	104.51
	3	137.36	94.53
	4	137.61	109.50
1967	1	137.21	96.44
	2	131.10	98.11
	3	133.53	99.11
	4	122.91	113.60
1968	1	113.88	84.97
	2	100.92	104.28
	3	99.16	101.62
	4	96.00	95.29
1969	1	95.36	100.40
	2	98.87	99.99
	3	96.12	106.12
	4	96.22	90.94
1970	1	94.17	104.64
	2	95.19	101.34
	3	95.28	95.05
	4	98.99	99.81
1971	1	101.49	106.04
	2	99.05	98.07
	3	95.93	96.31
	4	95.78	102.59
1972	1	95.88	100.83
	2	93.55	99.10
	3	96.02	95.05
	4	100.97	103.09
1973	1	101.73	105.73
	2	98.81	94.67
	3	99.35	95.94
	4		

(4) Year	Quarter	Predicted quarterly sales
1974	1	36.73
	2	45.61
	3	42.48
	4	36.08

Exercise 15-3

3 *a* Simple aggregate price index = 85.8. Simple average of price relatives = 80.56.

b Weighted aggregative price index = 86.1. Weighted average of price relatives = 86.1.

c Weighted average price index = 86.2. Weighted average of price relatives = 86.2.

7

Year	*Consumer price index (1971 = 100)*	*Average monthly salary, 1971 pesos*	*Actual percentage change*	*Real percentage change*
1967	71.5	1,050	0	0
1968	78.5	1,000	+4.7	−4.8
1969	82	1,000	+4.5	0
1970	93	914	+3.7	−8.6
1971	100	980	+15.3	+7.2
1972	114	965	+12.3	−1.5
1973	132	940	+12.7	−2.6
1974	146	958	+12.9	+1.9

Exercise 16-1

3 Since the expected value of the game is negative, I should *not* play the game. The maximum amount of money I am willing to pay to avoid the game will be 50¢.

7 *a* The expected gain per game is 0.
Yes, I should play the game; this is a fair game. I am a risk lover, and a risk lover would take a fair bet.

b The expected gain per game is −3,000. The weighted average of utilities for this game is 0.3 which is less than 0.5, the utility of keeping the money with certainty. Accordingly, I should not play this game. This game is not a fair one.

c The expected gain per game is −1,000. The weighted average of utilities for this game is 0.575 which is greater than 0.5, the utility of keeping the money with certainty. I should, therefore, play the game even though the game is an unfair one to me.

9 *a* The expected payoff from a_1 is 3. The expected payoff from a_2 is 3. Therefore, the decision maker is indifferent to which action to take.

b His expected payoff is 3.

11 The $2,000 that Mr. Smith possesses gives him a utility of 3. If he invests, his expected utility will be 3.6. Accordingly, he should invest.

Exercise 16-2

5 *a*

		Payoff		
$P(\theta)$	θ	a_1	a_2	a_3
0.2	θ_1	5	4	1
0.5	θ_2	4	8	5
0.3	θ_3	3	6	10

Computation of expected payoff

θ	a_1	a_2	a_3
θ_1	1.0	0.8	0.2
θ_2	2.0	4.0	2.5
θ_3	0.9	1.8	3.0
Expected payoff	3.9	6.6	5.7

b The Bayes action is a_2, which has the highest expected payoff, or the Bayes payoff, 6.6.

7 *a* *Computation of expected loss*

θ	a_1	a_2	a_3
θ_1	0.2	0.8	1.4
θ_2	1.5	1.0	2.5
θ_3	1.5	1.2	0.9
Expected loss	3.2	3.0	4.8

Computation of expected regret

θ	a_1	a_2	a_3
θ_1	0.0	0.6	1.2
θ_2	0.5	0.0	1.5
θ_3	0.6	0.3	0.0
Expected regret	1.1	0.9	2.7

b Bayes' action is a_2, the Bayes' risk is 0.9.

9 Bayes' action is a_1, and Bayes' risk is 1.2.

Exercise 16-3

5 *a* *Conditional expected loss* $E[sl(\theta, s)]$

θ	s_1	s_2	s_3	s_4	s_5	s_6	s_7	s_8	s_9
θ_1	1	3.1	5.2	1.9	4	6.1	2.8	4.9	7
θ_2	3	2.5	4.0	2.5	2	3.5	4.0	3.5	5
θ_3	5	4.8	4.6	4.2	4	3.8	3.4	3.2	3

Conditional expected regret $E[r(\theta, s)] = R(\theta, s)$

θ	s_1	s_2	s_3	s_4	s_5	s_6	s_7	s_8	s_9
θ_1	0	2.1	4.2	0.9	3.0	5.1	1.8	3.9	6.0
θ_2	1.0	0.5	2.0	0.5	0	1.5	2.0	1.5	3.0
θ_3	2.0	1.8	1.6	1.2	1.0	0.8	0.4	0.2	0

b The minimax expected loss strategy is s_5 or s_7.
c The minimax risk strategy is s_4.
d The Bayes strategy is s_4, and the Bayes risk 0.79.

7 $\alpha = 0.10$. $\beta = 0.15$.

Exercise 16-4

3 *a* Bayes' strategy: (a_2, a_1). Bayes' risk = 0.79 (this is the same as obtained in Prob. 5 of Exercise 16-3).
b The consulting firm's service reduced the risk from 0.9 to 0.79. Thus, the value of information is $0.9 - 0.79 = 0.11$.

Exercise 17-1

5 Observed value of $\chi^2 = 0.543$. Critical value $\chi^2_{(1, 0.05)} = 3.84146$. Accordingly, we do not reject H_0.

Exercise 17-2

3 *a* Critical value = 12. Since $X = 13$, we reject H_0.
b Computed Z value is $(13 - 7.5)/1.936 = 2.84$. Decision rule: Reject H_0 if $Z \geq 2.326$. Therefore, we reject H_0.

5 For $\alpha = 0.0207$, critical value = 15; reject H_0: $p = 0.5$.
For $\alpha = 0.059$, critical value = 16; reject H_0, also.

7 *a* Decision rule: Reject H_0 if $X \geq 20$, since $P(X \geq 20 | p = 0.5, n = 30) = 0.0495 \approx \alpha$. The observed X is 20; therefore we reject H_0 at the 5 percent level.

b Decision rule: Reject H_0 if $X \geq 21$ or $X \leq 9$. Since $X = 20$, we do not reject H_0 at the 5 percent level.

9 *a* Since $T = 79$ is larger than $T_{(20, 0.01)} = 43$, we do not reject the null hypothesis.

b Since $T = 79$ is larger than $T_{(20, 0.025)} = 52$, we do not reject H_0.

Exercise 17-3

3 Verification by using the example in Sec. 17-3-1.
$U' = 167$, $U = 43$, $E(U) = 105$, $\sigma_U = 23.66$,
$Z = (43 - 105)/23.66 = -2.62$, or $Z = (167 - 105)/23.66 = +2.62$.
Thus, when U' is used, the upper tail constitutes the critical region; when U is used, the lower tail is the critical region.

5 $U = 6$.

7 *a* $Z = (186 - 200)/36.968 = -0.392$. Decision rule: Reject H_0 if $Z \geq 1.96$ or $Z \leq -1.96$. Since the observed Z value is -0.392, we do not reject H_0.

b Using normal approximation, we have $Z = (112 - 200)/36.968 = -2.2$. Decision rule: Reject H_0 if $Z \geq \pm 1.960$ or $Z \leq -1.960$. Therefore, we reject H_0.

11 $H = 9.91$.
$\alpha = 0.01$. Decision rule: Reject H_0 if $H \geq \chi^2_{(2, 0.01)} = 9.21$. Accordingly, we reject H_0.

Exercise 17-4

3 *a* $r_S = 0.881$.

b $t = 5.889$. Critical values $t_{(10, 0.005)} = \pm 3.169$. Decision: Reject H_0.

5 *a* $r_S = 0.845$.

b $t = 4.74$. Decision rule: Reject H_0 if $t \geq 3.250$ or $t \leq -3.250$. Decision: Reject H_0.

7 $\tau = 1$.

INDEX

INDEX